GROUND FREEZING 88
VOLUME 1

5th GROUND FREEZING
5th INTERNATIONAL SYMPOSIUM
ISGF 88
NOTTINGHAM
ENGLAND

PROCEEDINGS OF THE FIFTH INTERNATIONAL SYMPOSIUM ON GROUND FREEZING / NOTTINGHAM / 26-28 JULY 1988

Ground Freezing 88

Edited by
R.H.JONES & J.T.HOLDEN
University of Nottingham, UK

VOLUME ONE

A.A.BALKEMA / ROTTERDAM / BROOKFIELD / 1988

INTERNATIONAL ORGANISING COMMITTEE

H.L.Jessberger (Chairman), Department of Civil Engineering, Ruhr University, Bochum, FR Germany
R.H.Jones (Secretary), Department of Civil Engineering, University of Nottingham, Nottingham, UK
E.J. Chamberlain, US Army Cold Regions Research and Engineering Laboratory, Hanover, New Hampshire, USA
M.Fremond, Laboratoire Central des Ponts & Chaussees (LCPC), Paris, France
S.Kinosita, Institute of Low Temperature Science, Hokkaido University, Sapporo, Japan
B.Ladanyi, Northern Engineering Centre, Ecole Polytechnique, Montreal, Canada
A.V.Sadovsky, Gersevanov Research Institute of Bases and Underground Structures, Gosstroy, Moscow, USSR
J.A.Shuster, Geofreeze Corp., Lorton, Virginia, USA
Yu Xiang, Central Coal Mining Research Institute, Beijing, People's Republic of China

NATIONAL ORGANISING COMMITTEE

R.H.Jones (Chairman), Department of Civil Engineering, University of Nottingham
J.S.Harris (Secretary), Geotechnical Consultant, Nottingham
F.A.Auld (Treasurer), Cementation Mining Ltd, Doncaster
M.J.Bell, British Coal HQ Technical Branch, Burton-on-Trent, Staffordshire
L.Crystal, Consulting Civil and Structural Engineer, Pinner, Middlesex
G.P.Daw, T.H. Engineering Services Ltd, Rickmansworth, Hertfordshire
H.C.English, Husband & Company, Sheffield
J.T.Holden, Department of Theoretical Mechanics, University of Nottingham

The texts of the various papers in this volume were set individually by typists under the supervision of each of the authors concerned.

Published by
A.A.Balkema, P.O.Box 1675, Rotterdam, Netherlands
A.A.Balkema Publishers, Old Post Road, Brookfield, VT 05036, USA

For the complete set of two volumes: ISBN 90 6191 824 3
For volume 1: ISBN 90 6191 825 1
For volume 2: ISBN 90 6191 826 X

Printed in the Netherlands

5th International Symposium on Ground Freezing, Jones & Holden (eds)
© 1988 Balkema, Rotterdam. ISBN 90 6191 824 3

Contents

2. *Mechanical properties*

3. *Engineering design*

4. *Case histories*

5th International Symposium on Ground Freezing, Jones & Holden (eds)
© 1988 Balkema, Rotterdam. ISBN 90 6191 824 3

Preface

This volume contains the text of most of the papers accepted for presentation at the Fifth International Symposium on Ground Freezing held on 26th to 28th July 1988 at the Royal Moat House International Hotel, Nottingham, England. The papers are primarily concerned with the theory, design and practice of artificial ground freezing (AGF) which was first used over one hundred years ago as a construction expedient in shaft sinking. Since then, AGF has been used in many parts of the world to provide temporary support and/or to control the flow of ground water during the construction of shafts, tunnels and other excavations. It has also been applied to other situations such as underpinning. In the last two decades technological advances have resulted in improved design methods and closer control and monitoring during construction. These developments have led to both greater economy and safety and also to an extension of the range of applicability of AGF methods.

The series of international symposia on ground freezing have provided a useful forum in which developments in ground freezing techniques could be discussed. Throughout, efforts have been made to relate theory more closely with practice and the symposia have attracted participants from a wide range of backgrounds and activities. The publication of the proceedings (see p. X) has contributed significantly to the available literature on ground freezing. The present symposium is no exception and Volume 1 contains a total of 48 papers (including four invited states of the art), on the following topics: Heat and mass transfer (17 papers), mechanical properties (11 papers), engineering design (10 papers) and case histories (10 papers).

Innovations are the inclusions of an author index and a keyword index which it is hoped will enhance the usefulness of the volume.

Volume 2 of the proceedings, to be published after the symposium, will contain some further papers, general reports and summaries of the poster presentations.

Our thanks are due to the authors (the majority of whom met the copy dates), the papers sub-committee (J.K.G.Boucher, R.J.Kettle, G.W.Tuffs, E.Waller, A.J.Wills and until his untimely death J.F.C.Thoelen), the members of the National and International Committees and all others who helped by giving of their time to referee the draft papers. Many of the 42 referees (from 11 countries) made very detailed contributions which led to the clarification of both the content and language of the papers. Finally we wish to thank our colleagues in the Department of Civil Engineering and the Department of Theoretical Mechanics of the University of Nottingham for their co-operation and support.

Ron Jones & John Holden
Editors

AVAILABILITY OF ISGF PAPERS

1st ISGF, 1978, Bochum, FR Germany
Original proceedings in 2 volumes (now out of print). However, all papers reprinted (but with different page numbers) by Elsevier as both *Engineering Geology, Vol. 13* and *Developments in Geotechnical Engineering, No 26* (Editor: H.L.Jessberger).

2nd ISGF, 1980, Trondheim, Norway
Originally published in two volumes (preprints and proceedings) by the Norwegian Institute of Technology, Trondheim. Papers marked # were reprinted (but with different page numbers) by Elsevier as both *Engineering Geology, Vol. 18* and *Developments in Geotechnical Engineering, No 28* (Eds: P.E.Frivik, N.Janbu, R.Saetersdal & L.I.Finborud).

3rd ISGF, 1982, Hanover, N.H., USA
Proceedings published by USA CRREL as *CRREL Special Report 82-16.*

4th ISGF, 1985, Sapporo, Japan
Most papers appear in *Ground Freezing* published by A.A.Balkema, Rotterdam. Five papers and all the 'posters' were published as *Ground Freezing, 1985, Volume 2* by Hokkaido University Press, Sapporo, Japan. Editors of both volumes were Seiito Kinosita and Masami Fukuda. (NB. The Balkema volume is often referred to as Volume 1 although this is not part of the official title).

5th ISGF, 1988, Nottingham, England
Where papers in this symposium are referenced, mainly by the state of the art authors, page numbers may be found from the author index.

Volume 2 containing late papers and poster summaries will be published by Balkema after the symposium.

1. Heat and mass transfer

5th International Symposium on Ground Freezing, Jones & Holden (eds)
 ISBN 90 6191 824 3

State of the art: Heat and mass transfer in freezing soils

B.D.Kay & E.Perfect
Department of Land Resource Science, University of Guelph, Guelph, Ontario, Canada

ABSTRACT: Significant advances have been made in the past decade in our understanding of processes which contribute to the movement of heat and mass in soils when freezing occurs. These advances have been facilitated by the introduction of new technology for measuring critical parameters. There has also been considerable progress in integrating the description of these processes in models which are capable of accurately predicting heat and mass transport under an increasing range of conditions. This paper attempts to define our current understanding of heat and mass transport and to identify future research needs. Areas requiring further research include: defining the distribution of stresses among the various components of freezing soils either under a load or under confined conditions, adapting current concepts of heat and mass transport in "disturbed" soils to soils with well defined structure, characterizing the redistribution of solutes and the separation of charge when soils freeze, clarifying the roles of solute distribution and electrical potentials on water flow and developing the capability to describe and predict variability in heat and mass transport under field conditions.

1 INTRODUCTION

The purpose of this paper is to define our current understanding of the processes associated with heat and mass transfer in freezing soils and to document some of the research which has contributed to this understanding. The paper primarily focusses on progress in the past decade. It builds on a comparable review prepared for the Second International Symposium on Ground Freezing (Loch, 1980). In assembling material for this paper the authors acknowledge that lags in communication between researchers and simple oversights may have resulted in some works not being included. The authors apologize in advance for any such omissions.

Research is reviewed in three areas. The response of soil to subzero temperatures is examined under static conditions. This provides a base for the subsequent discussion on the dynamics of heat and mass transfer in freezing soils. The development of simulation models to describe ground freezing is then reviewed. The paper concludes with a brief discussion on future research needs.

2 STATIC CONDITIONS

Soil components are considered under static or equilibrium conditions as a preface to the discussion of dynamic conditions. Many of the insights gained in studying soils under equilibrium conditions at subzero temperatures have been fundamental to studies on the dynamics of heat and mass transport in freezing soils. Of paramount importance is the assumption that under dynamic conditions, "local" equilibrium exists. Relations between variables which can be experimentally determined under static conditions (e.g. between intensive variables such as temperature and extensive variables such as the amount of unfrozen water) are assumed to be equally applicable, on a local basis, under dynamic conditions when hysteresis is taken into account. The validity of this assumption depends on the magnitude of the departure of the system from an overall equilibrium condition. There have been few studies on frozen soils directed towards a rigorous test of this assumption.

The magnitude of the gradients under field conditions, and the size of the fluxes involved lead to an intuitive conclusion that relations between variables should apply under both static and dynamic conditions in frozen soils. However, Nakano and Horiguchi (1985) have questioned the validity of local phase equilibrium at an advancing freezing front. These authors suggested that supercooling between the frost front and the unfrozen soil could explain their data, although no direct evidence of supercooling was presented. Although this work accentuates the need for rigorous testing of the assumption of local equilibrium, the authors of this paper have proceeded on the belief that characterization of soil components under equilibrium conditions provides a fundamental basis for understanding soil components under dynamic conditions.

The soil system at subzero temperatures will be considered to be made up of heat, water in the vapor, liquid and solid states, solutes (including charged and uncharged species) and the soil matrix. Although biological materials are an integral component of soil at the earth's surface, their involvement in ground freezing will not be considered in this paper.

2.1 Heat

The heat content of a soil is related to its temperature, apparent heat capacity, and the latent heat contribution associated with transformation between liquid water and ice phases. The latent heat contribution arising from changes in the water content of the vapor phase and the heat capacity of the vapor phase can be shown to be negligible in size relative to values of other components. The heat capacity of the components in soil are additive. The change in heat content, dH, of a soil can therefore be described by:

$$dH = c_w \rho_w \theta_w dT + c_i \rho_i \theta_i dT + c_m \rho_m \theta_m dT + H_f \rho_w d\theta_w \qquad (1)$$

where c is the specific heat capacity, ρ is the density, θ is the volume fraction; subscripts w, i and m refer to the liquid water, ice and mineral material respectively, and H_f is the partial specific latent heat of fusion. The specific heat capacity of soil components changes linearly with temperature (Kay and Goit, 1975) and $\overline{H}_f$ changes with θ_w at low water contents (Groenevelt and Kay, 1974). However, most of the heat and water transport in non-saline freezing soils occurs within one or at most two degrees below 0°C. Under these circumstances errors introduced by assuming constant values of c and H_f are insignificant. The errors become larger when changes in the heat content of frozen soil are considered at much lower temperatures. Increasing attention paid to saline soils will generate the need for accurate thermal characterization of brine solutions (e.g. Thurmond and Brass, 1987). Such studies will utilize the theoretical framework described above and will obviously have to be tied to studies on unfrozen water contents in saline materials.

2.2 Water

2.2.1 Liquid phase

The energy status and the quantity of water remaining in the liquid phase at subzero temperatures is crucial to the physical, chemical, and mechanical properties of frozen soil materials. The energy status of unfrozen water is described by an equation relating the liquid pressure of this water to the osmotic potential, the ice pressure, the temperature and the latent heat of fusion based on the Clapeyron equation:

$$\overline{V}_i p_i - \overline{V}_w (p_w - \pi) = -\overline{H}_f \frac{T-T_o}{T_o}, \qquad (2)$$

where $\overline{V}_i$, $\overline{V}_w$ are the partial specific volumes of ice and water respectively
p_i, p_w are the ice pressure and hydrostatic (tensiometer) pressures respectively
π is the osmotic pressure associated with leachable solutes
$\overline{H}_f$ is the partial specific latent heat of fusion and
T, To are the freezing temperatures (°K) of water in soil and in the bulk state respectively

Loch (1978) emphasized that eqn.(2) is based on equilibrium conditions and thus strictly applies only under static or steady state dynamic conditions, i.e. conditions under which local pressure and temperature are not changing with time. The range of transient conditions (i.e. local pressures and temperature changing with time) under which eqn.(2) can be employed without introducing an unacceptable level of error is uncertain.

Eqn.(2) is most easily applied when the soil is maintained in an unconstrained condition; at equilibrium p_i will be zero. However, when the soil is constrained or an overburden imposed, the distribution of this stress between soil solids, ice and liquid water must be determined. Miller (1978) applied the Terzaghi concepts of effective stress (σ_e) and neutral stress (σ_n) to saturated frozen soil, thus requiring the use of a stress partition function, χ:

$$\sigma_n = \sigma_e + [\chi p_w + (1-\chi p_i)] \quad (3)$$

Snyder and Miller (1985a) then combined Aitchison's extension of the classic Haines-Fisher model of capillary cohesion with Griffith's theory of the fracture of cracked elastic solids in order to describe tensile failure. These authors illustrate how the parameter χ can be calculated from measurements of tensile strength as a function of pore water pressure. The theory presented by Snyder and Miller is strictly applicable to unstructured soils, i.e. soils without aggregates or peds. Snyder and Miller (1985b) have outlined a pneumatic technique to measure the tensile strength of unsaturated unfrozen soils, thereby allowing calculation of χ by employing the above theory.

Groenevelt and Kay (1981) developed mechanical and thermodynamical models, using an alternative approach, which resulted in equations describing the distribution of effective stress in unsaturated frozen soils and a framework for the calculation of χ. Experimentally determined relations between void ratio (m^3 pore space/m^3 solid) and moisture ratio (m^3 water/m^3 solid) under different load pressures are required for this calculation. These developments significantly enhance our ability to predict the pressure potential of liquid water in frozen soil experiencing a confining pressure. Further work will be required to extend these concepts to structured soils which already possess a three dimensional network of planes of weakness, created mechanically by previous freezing and/or drying events.

The liquid water content of frozen soil is a function of the total liquid pressure potential ($p_w - \pi$). The nature of the relationship between these variables for saturated soils with total liquid water potentials between 0 and 1.5 MPa is related to the characteristics of the pores (Homshaw, 1980) and the concentration of solutes. At lower potentials the amount of unfrozen water is less influenced by porosity and more strongly influenced by the characteristics of the mineral particles such as specific surface area (Anderson and Tice, 1972). The geometry of the soil matrix and the chemical characteristics of the soil particle surfaces influence unfrozen water through menisci, the electrical double layer and bonding by different mechanisms at a molecular level. These influences cumulatively determine p_w in an unloadedsituation. Although studies to evaluate the role of different mechanisms is of scientific interest, such studies have not provided insights on how water flow in freezing soil might be described more effectively than by simply using p_w.

Techniques to measure the unfrozen water content are critical to the characterization of frozen soil materials in many studies. A comprehensive review of different methods was presented by Anderson and Morgenstern (1973). Significant developments which have occurred since that time include refinement of the use of nuclear magnetic resonance (NMR) and the introduction of dielectric measurements using time domain reflectometry (TDR). Oliphant and Tice (1982) obtained a very good correlation between unfrozen water contents across a range of subzero temperatures using pulsed NMR and differential scanning calorimetry methods. The presence of ferromagnetic minerals in soils can cause difficulties in the measurement of unfrozen water content using the NMR technique. Tice and Oliphant (1984) have introduced a demagnetizing procedure to permit accurate measurement of unfrozen water contents in soils containing magnetic particles. Xu et al. (1985) describe a method of speeding up the measurement of unfrozen water content as a function of temperature in unconfined samples using NMR.

Patterson and Smith (1981) introduced the use of TDR to measure unfrozen water contents. This technique utilizes differences in the dielectric constant between liquid water and all solid material in frozen soil to measure unfrozen water contents. It appears to be rapid and reliable. Oliphant (1985) has compared unfrozen water contents measured by NMR and TDR and found that the best fit between the data was obtained when an average dielectric constant for water which was somewhat less than that for bulk water was used. Both TDR and NMR measurements are rapid, but TDR measurements have the added advantage that they can be readily made under field conditions (Smith, 1985).

Values of liquid water pressure in frozen soils are required to establish the characteristic relation between liquid water pressure and liquid water content. In addition, values of liquid water pressure are crucial to understanding the dynamics of water flow in freezing soils. The value of liquid water pressure is normally calculated using eqn.(2) rather than measured directly. This is due to the lack of suitable instrumentation. However, uncertainties related to the use of eqn.(2) under transient conditions or in the presence of confining pressure provide incentive to develop methods to measure p_w directly.

McKim et al. (1976) outlined a method using a tensiometer containing ethylene glycol-water solution which could operate below 0°C. An obvious concern with this method relates to the tensiometer membrane: if this was not perfectly semipermeable then ethylene glycol would diffuse into the frozen soil, and the value of π in the unfrozen water adjacent to the tensiometer would increase. This would immediately result in a readjustment of p_w since $(p_w - \pi)$ must remain constant as long as p_i and T remain constant [eqn.(2)]. A subsequent report (McGaw et al., 1983) indicated that ethylene glycol does not diffuse into the soil water. These authors then measured p_w in a sandy silt under transient conditions and related the potentials measured in the soil at subzero temperatures to unfrozen water contents measured in the same soil under static conditions using NMR. An analysis of the data suggests that the observed values of p_w are much smaller than would be predicted from eqn.(2) assuming π and p_i are zero. Lack of coincidence between observed and calculated values of p_w may be due to errors in the above assumptions, the application of eqn.(2) to transient conditions, or a reduction in the observed values of p_w due to diffusion of ethylene glycol out of the tensiometer. The importance of accurate measurements of p_w justify more work in this area.

2.2.2 Vapor

The relative vapor presure can be used to calculate the energy status of water in the vapor phase. Under static conditions the water vapor is normally only examined as a means of learning more about the ice and liquid water phases.

2.2.3 Ice

The energy status of ice is also described by eqn.(2). Measurements of p_i are readily made providing the ice exists as large segregated bodies (lenses). However, values of p_i when ice exists as pore ice are not readily measured and it is ice in this condition that is of greatest interest in studying water flow in freezing soils. Recent advances in the measurement of relative humidity and vapor pressure offer the potential for such measurements being employed to measure p_i.

Ice contents are normally measured as the difference between total water content and unfrozen water content. Measurements of total water content may be either destructive (i.e. sectioning and gravimetric determination of water content) or nondestructive, depending on the method employed. New non-destructive laboratory techniques which have been introduced include, dual gamma-ray attenuation (Goit et al., 1978) and thermal neutron radiography (Clark and Kettle, 1985). Such measurements will give values on a "macroscopic" scale averaged over a volume of the sample, the magnitude of which, is dependent on the technique employed.

Miller (1973) suggested that the formation of ice in unsaturated pores can lead to instabilities, with the ice crystals either evanescing or the pore spontaneously filling with ice. Colbeck (1982, 1985) examined the configuration and fabrics of ice in frozen glass beads and sands with a low initial water content and limited water supply. He concluded that pores did seem to fill in an unstable manner. Different ice shapes and crystal distributions were observed, indicating a mixture of kinetic crystal growth processes and equilibrium constraints.

2.3 Solutes

Interest in the behaviour of solutes in freezing soils has grown as a consequence of current interest in subsea permafrost and saline soils. Interest in solutes in non-saline environments has also grown as the concept of unfrozen water content in frozen soils becomes more widely accepted by researchers whose traditional focus has been on unfrozen soils.

Solutes tend to be excluded from ice and concentrated in the unfrozen water. The behaviour of solutes is of particular importance to the energy status of unfrozen water [i.e. eqn.(2)] and consequently to the unfrozen water content (Patterson and Smith, 1985; Tice et al., 1984). The influence of solutes on liquid water pressure and on unfrozen water content can have major effects on the transport of water in freezing soils. This will be discussed in greater detail later. High concentrations of solutes in the unfrozen water created by the exclusion of solutes from ice may result in adsorption of solutes by the mineral material (Nelson and Romkens, 1972) and/or in chemical reactions (e.g. Lahav and Anderson, 1973; Christianson and Cho, 1983). A common feature of concentrating solutes in unfrozen water is the formation of precipitates or crystals as a consequence of exceeding the solubility limit (Wada and Nagasato, 1983). The phase relations of brines become very important under such conditions. Adsorption, reaction and crystallization must be considered when calculations are made of the increase in concentration of solutes in unfrozen water and values of osmotic pressure (eqn.2). There is considerable opportunity to expand our understanding of the behaviour of a wide range of solutes under static conditions.

3 DYNAMICS

The dynamics of freezing soils involve the transport of heat, water, solutes and charge. A discussion of the transport mechanisms is facilitated by a distinction between direct and coupled transport. Direct transport of a component arises from a gradient in some property of that component. For instance, heat flow occurs in response to a temperature gradient, liquid water flow occurs in response to a liquid water pressure gradient, solute flow arises as a consequence of a solute concentration gradient and charge flow or current arises as a consequence of a gradient in electrical potential. Coupled flow occurs when flow of one component arises from the gradient in a property of another component; coupled flow of one component can be potentially caused by a gradient in a property of each of the other components. For instance, coupled water flow can arise from a gradient in temperature, solute concentration and/or electrical potential. Transport by these mechanisms will supplement transport caused by a direct gradient in liquid pressure. Similar examples of coupled flow exist for each of the other components of frozen soil. Processes responsible for the transport of heat and mass by a combination of direct and coupled flow have been examined in less detail in frozen soil than has been the case for unfrozen soils, although the basic framework has been established for many years (e.g. Groenevelt and Bolt, 1969). This section will review transport of each component of the soil system; processes responsible for both direct and coupled transport will be considered.

A full description of the transport processes should be independent of time and space, i.e. description of a process should remain the same whether it occurs over a short term or a long term, in the laboratory or in the field, in a short column or a long column. However, the relative significance of different processes may vary with scale. For instance, Akagawa et al. (1985) acknowledge the importance of time scale in defining 3 stages of frost heaving: transient, steady state and long term. The external constraints on the system being investigated in the laboratory compared to field conditions, and the thermal and hydrologic conditions imposed in the laboratory relative to field conditions may also bear on the relative significance of the different processes. An additional complication is the temporal and spatial variability in boundary conditions as well as the temporal and spatial variability in the properties of the system itself (Perfect et al., 1988). Consequently, scales of time and space will need to be acknowledged as research on processes responsible for heat and mass transport proceeds.

3.1 Transport of heat

Research on the thermal aspects of ground freezing up to 1980 has been summarized by Frivik (1981) in his state-of-the-art report for the 2nd International Symposium on Ground Freezing. Farouki (1982) has reviewed the methods currently available for estimating the thermal properties of frozen soil. In addition, Lunardini (1982) provides an excellent general treatment of heat transfer in cold climates. Discussion in this section will focus upon the boundary conditions associated with heat flow, the processes responsible for heat flow and methods to characterize heat flow.

The magnitude of heat flow which occurs in freezing soils depends on boundary conditions. When freezing occurs under artificial conditions, either in the laboratory or the field, boundary conditions are normally well defined. Under natural conditions, however, measurements such as ground surface temperature, or the flow of heat across the ground surface boundary are often lacking. Studies to assess the importance of ground cover (i.e. snow, vegetation) have been carried out and provide excellent qualitative information on the role of boundary conditions on heat flow (Goodrich, 1982; Benoit and Mostaghimi, 1985). In the absence of such information, the upper boundary can be shifted from the soil surface to the lower atmosphere and the radiative and turbulent transport of sensible and latent heat characterized (Fahey and Thompson, 1982). Although the processes in these energy balance calculations are well understood, there remains considerable spatial and temporal variability in the transport coefficients.

The direct process of heat transport in freezing soils is by conduction as a consequence of a temperature gradient. Coupled flow arises primarily from gradients in properties of different components which cause water flow and result in the transport of sensible and latent heat. The proportionality constant (transport characteristic) relating heat flow to a temperature gradient is the thermal conductivity. Considerable effort has been directed to both theoretical studies and empirical measurements of this parameter. A summary of methods used to characterize thermal conductivity was presented by Johansen and Frivik (1980). Research which has been carried out since that time (e.g. Riseborough et al., 1983) has been directed to refining these techniques.

The variation in thermal conductivity with subzero temperature clearly indicates the importance of the variation in unfrozen water content with temperature (thermal conductivity of frozen soil decreasing as unfrozen water content increases). Work has continued in an attempt to improve predictions of soil thermal conductivity from knowledge of the thermal conductivities of individual soil components (e.g. Gori, 1983) and to provide additional experimental evidence of the variation in thermal conductivity with temperature (e.g. Inaba, 1983; Sawada and Ohno, 1985; McCabe and Kettle, 1985). However, most of these measurements have been made at temperatures sufficiently below the onset of freezing of soil water that the coupled transport of sensible and latent heat by the convective flow of water would not be large. Few measurements have been made at temperatures where water flow may be important.

The significance of water flow, due to a temperature gradient on measurements of thermal conductivity was assessed by Kay et al. (1981). These authors incorporated the transport of heat due to water migration into an "apparent thermal conductivity" term and suggested that at temperatures close to zero the transport of latent heat contributed by water flow in frost susceptible soils will be much greater than the transport of heat by conductive flow alone, causing the "apparent thermal conductivity" to rise dramatically with increasing temperature. Under such conditions the analytical solution to equations normally used to calculate thermal conductivity from measurements using a line heat source would have to be reformulated. Currently, only limited data are available to evaluate this theory. The influence of temperature gradients on the coupled transport of heat due to water flow caused by gradients in solute concentration and/or electrical potential remain unexplored.

Studies on heat flow have been carried out primarily in non-saline soils. However, Osterkamp (1987) has recently developed theory to describe the advance of a freezing front into saline materials. He assumes that no coupled flow occurs (i.e. unfrozen water and brine do not move) and that there is no salt rejection by the advancing freezing front. Cary (1987) has developed a theory to describe the coupled flow of heat, water and solutes but does not consider exclusion of solutes by the advancing freezing front or the possibility of water flow (and therefore heat flow) due to osmotic pressure gradients.

Field measurements of the amount of heat which has been removed from soil are norm ally based on temperature profiles or measurement of the total depth of freezing. Viereck and Lev (1983) report reliable, long-term use of the CRREL frost tube to monitor annual freeze-thaw cycles in the active layer. Recently, two new experimental methods have been described for locating the frozen-unfrozen interface in soils. Baker et al. (1982) used TDR measurements made with vertical and horizontal transmission lines, to estimate frost penetration in the field; the errors were equal to those associated with temperature measurements. Hayhoe et al. (1986) describe the development of an electrical frost probe, based upon an apparent differential capacitance technique.

3.2 Transport of water

Williams (1984) has reviewed the experimental and theoretical evidence for transport of water in partially frozen soil. This section will emphasize the processes responsible for the transport of soil water, in the liquid, vapor and solid phases at subzero temperatures. Considerable research has also been conducted since 1980 on various boundary conditions. The physics of snowmelt infiltration into seasonally frozen soil has been examined by Kane and Stein (1983). Outward flux of water vapor from soil by evaporation and sublimation has been investigated (Aguirre-Puente et al., 1985; Smith and Burn, 1986). However, these subjects, are beyond the scope of this paper and will not be considered further.

3.2.1 Liquid water

Water flow in frozen soils containing exclusively pore ice can be described by Darcy's Law. This is the direct process with the driving force being the gradient in liquid pressure and the transport characteristic being the hydraulic conductivity. According to Eqn.(2), a temperature gradient within partially frozen soil can induce a gradient of liquid pressure. Further coupled processes leading to water transport arise from the gradients in electrical potential and solute concentration. Iwata (1980) has formulated the relation between flow rate and gradients of water pressure, solute concentration and electrical potential during freezing. Of major importance is the recognition that each gradient term is related to the water flux by a different transport characteristic.

Laboratory measurements of direct and coupled transport in soil which is entirely below 0°C have been made (Oliphant et al., 1983; Aguirre-Puente and Gruson, 1983; Xu et al., 1985). Values of apparent hydraulic conductivity for the frozen soil have been derived in various ways by several authors and are reasonably consistent. Typically, the apparent hydraulic conductivity function decreases rapidly over the temperature range 0 to -1°C, from unfrozen values in the order of 10^{-8} m.s^{-1} to values of between 10^{-12} and 10^{-14} m.s^{-1} (Horiguchi and Miller, 1983). Below -1°C conductivity values appear to remain relatively constant (Ratkje et al., 1982; Yoneyama et al., 1983, Ishizaki et al., 1985). Nakano et al. (1982) report the development of a laboratory procedure for measuring the soil-water diffusivity of frozen soil under isothermal conditions. In a subsequent series of papers, these authors use this method to investigate the effects of temperature and ice content on water movement through the soil matrix (Nakano et al., 1983, 1984a,b; Nakano and Tice, 1987). Recently, Smith (1985) has made use of time domain reflectometry to estimate the apparent hydraulic conductivity of frozen soil under field conditions.

The relatively large number of recent field studies on freezing-induced redistribution is due in part, to the development of improved methods for the non-destructive determination of soil moisture changes in situ. The neutron moderation technique in particular, has gained wide acceptance among soil physicists and is becoming more common in soil freezing studies (Sheppard et al., 1981; Gray et al., 1985; Gray and Granger, 1986). However, major drawbacks with this method are the large sampling volume (which effectively homogenizes any discontinuities due to ice segregation) and the assumption of a constant bulk density (i.e. no major structural changes occur as a result of freezing). Solutions to the later problem might be found in the literature on shrinking and swelling clay soils (Vertisols), to which the neutron method has been extensively applied (Greacen, 1981). Hayhoe and Bailey (1985) report monitoring changes in the total and unfrozen water content of seasonally frozen soil using a combination of both TDR and neutron moderation techniques.

3.2.2 Transport of Vapor

Transport of water in frozen soil may also take place in the vapor phase. However, evidence is lacking on the relative importance of the vapor mode of mass transfer (Gray and Granger, 1986) and the extent of coupling which occurs. Nakano et al. (1984c) have introduced a method for measuring vapour diffusivity when ice is absent, using small amounts of a soluble organic tracer (Diethylphthalate). Data are presented on the relative magnitudes of liquid and vapour diffusion in soil at subzero temperatures.

3.2.3 Transport of Ice

A significant component of the total mass transport in frozen soils appears to be due to ice movement by thermally-induced regelation. This is a process by which isolated soil particles embedded in ice migrate up a temperature gradient, or vice versa in the case of a fixed soil matrix. The latter concept forms the basis of the "rigid-ice" model of frost heaving proposed by R.D. Miller and co-workers. Miller (1984) gives an excellent qualitative exposition of the regelation phenomenon as it applies to ice motion in porous media. A theoretical analysis of such movement is provided by Walder (1986). Experimental data are scanty. Wood and Williams (1985a) report measurements of the apparent hydraulic conductivity of ice as determined in a modified "ice sandwich permeameter". Their coefficients of proportionality decrease exponentially by two orders of magnitude over the temperature range 0 to -0.3°C. Recently Ohrai and Yamamoto (1985) have published long-term observations of ice lens growth behind a stationary freezing front, suggesting redistribution of soil ice by thermally-induced regelation.

3.2.4 Accumulation of Water

The transport of water in the liquid, solid or vapor phases leads to changes in water content. Zones of water accumulation may result in soil with a very high water (ice) content but without visible ice lenses; this condition is commonly encountered near the ground surface under seasonal freezing (Sheppard et al., 1981). Alternatively, water may accumulate as soil-free ice in the form of needle ice or ice lenses. Needle ice is a transient surficial phenomenon, occurring when thermal conditions are such that ice crystals cannot penetrate beyond the first few millimetres (Meentemeyer and Zippin, 1981). In contrast, ice lensing within the soil profile produces a rhythmic or stratified pattern of ice bodies. These may be laminar in shape, oriented at right angles to the direction of heat flow or may be more polygonal in shape with some lenses oriented almost parallel to the direction of heat flow. In layered soils, ice lensing occurs on the face of the finest textured layer encountered by the freezing front (Penner, 1986). Yershov et al. (1983) have reviewed and classified cryogenic structures arising from ice accumulation.

The accumulation of water (as ice) in homogeneous soil was initially described in terms of a divergence of water flux (Harlan, 1973). Although this formulation could account for ice enriched zones it could not account for the formation of ice lenses which were free of soil grains. In addition, it provided no mechanism of distributing an applied load among soil components thereby making it difficult to calculate p_w (a requirement for the prediction of the water flux). Both of these limitations were removed in a rather elegant manner by Miller (1978) through the introduction of a stress partition function and the requirement that each ice lens form at a specific location where the effective stress is zero or the neutral stress equals the applied load (see Section 2.2.1). Miller's formulation is valid for saturated non-colloidal (unstructured) soils which do not contain leachable solutes.

Detailed analysis of cores of frozen silt loam, obtained from the field at different times during the winter, suggests that very little air exists behind the freezing front (Kay et al., 1985), indicating that the initiation of ice lenses normally occurs under saturated conditions. If this is the case for all soils exhibiting ice accumulation, the challenge facing researchers will be to extend Miller's theories to soils which contain colloidal material, to soils which are structured (i.e. having cementing material between grains) and to soils which contain solutes. Extension to structured soils will presumably lead to a more complete description of processes leading to the formation of polygonal ice structures and ice lenses which, in the extreme, are oriented parallel to the direction of heat flow. A modification of existing theories is also required to explain ice segregation in saline soils. Chamberlain (1983) has shown that the effect of solutes is to create a thick active freezing zone with many small ice lenses. Layers of brine-rich unfrozen soil exist between lenses.

The existence of a transition zone containing pore ice and unfrozen water (the 'frozen fringe') separating segregating ice from the zero (°C) isotherm was originally proposed by Miller (1972) and has been documented (e.g. Booth, 1981; Ishizaki et al., 1985). The growth of the lens adjoining this zone depends on the properties of the zone (Nakano, 1986; Ishizaki and Nishio, 1985; Takeda et al., 1985) and the cooling rate (Horiguchi, 1987a). Observations by Smith (1985) and Ohrai and Yamamota (1985) suggest ice lenses may grow behind the lens adjacent to the frozen fringe. If this occurs without an adjustment in temperature profiles, i.e. giving rise to the simultaneous growth of ice lenses in two or more locations, an additional component of heave would be introduced. Such a situation is not easily explained with existing theory. Ice segregation within already frozen soil has been investigated by Wood and Williams (1985b) and Williams and Wood (1985). The stress distribution in confined 'frost heaving' samples was measured and internal pressures assumed to be generated by the growth of ice lenses taking place against the resistance offered by the surrounding semi-rigid frozen soil. However, it is difficult to ascribe pressures measured macroscopically to microscopic processes. Clearly, the mechanism of ice accumulation behind the warmest ice lens needs further examination.

Ice segregation is of immense engineering significance. Since the rate of formation of ice lenses is related, in part, to soil properties, researchers continue to search for a reliable method with which to evaluate the frost susceptibility of soils. The abundance of index tests developed to date is evidence of a general lack of success in defining satisfactory criteria. Chamberlain (1981) provides a comprehensive review of over one hundred methods for determining the frost susceptibility of soil. A survey of similar breadth can be found in 'Frost i Jord', Nr. 22 (Nov. 1981), which contains details of the criteria used to classify the frost susceptibility of soils in seven different countries. The methods can be divided into three categories: those based on particle-size characteristics, those based on soil-water relationships and those which involve direct frost heave testing. From a theoretical study of the ice segregation mechanism, Rieke et al. (1983) identified the specific surface area as being fundamentally related to a soil's frost susceptibility.

Freezing results in a rearrangement of the soil matrix. Upon thawing the residual effects of frost action on various soil physical properties become apparent. These changes have important agricultural and engineering implications. Furthermore, they are responsible for an alteration in the heat and mass transport characteristics of soil during subsequent freeze-thaw cycles (Efimov et al., 1981). A recent review article by Vliet-Lanoe (1984) provides a much needed overview of the diversity of residual changes attributable to seasonal frost penetration, and their significance from a soil genesis perspective.

3.3 Solutes

Solutes can be redistributed in soils by several processes. Solutes are excluded by ice formation (Leung and Carmichael, 1984) resulting in accumulation of solutes in front of an advancing freezing front (Mahar et al., 1982; Kadlec, 1984). Solutes are also transported upwards to the frost line as water is drawn into the freezing zone (Campbell and Biederbeck, 1982; Gray and Granger, 1986). The accumulation of solutes at the freezing front and in the unfrozen water behind the freezing front will give rise to concentration gradients resulting in diffusion. The concentration gradient also creates a gradient in freezing point in soil water on the warm side of the freezing front. When the decrease in freezing point, due to the influx of solutes by diffusion, at some distance ahead of the freezing front is slower than the drop in temperature, conditions favoring ice nucleation in this zone are created. The freezing front "jumps over" these solute enriched bands, entrapping them in a frozen matrix. The extent to which such zones exist will depend on the initial solute concentration and the rate of freezing. This phenomenon has been studied extensively in the formation of brine layers in sea ice.

Kay and Groenevelt (1983) developed theory to describe the freezing induced redistribution of a non-reactive solute in sand (i.e. a matrix with very little unfrozen water behind the freezing front and therefore no water flow into the frozen zone). Comparison of theory with laboratory data provided good evidence for the formation of pockets of solute. Cary (1987) has described diffusive and convective transport of solutes associated with water flow due to a gradient in liquid pressure but does not consider exclusion of solutes by ice. Further work is required to develop a comprehensive theory which includes all the processes for solute redistribution and which accounts for precipitation, adsorption and transformation of solutes as soils freeze.

3.4 Charge

Charge separation gives rise to electrical potentials. Charge separation may be expected to arise in freezing soils from several processes. Charge separation at an advancing freezing front is caused by preferential incorporation of ion constituents of one sign into the solid phase (the Workman-Reynolds effect). The magnitude and sign of the potential difference depends on the concentration and kinds of solute species involved. Electrical neutrality is restored by a flux of hydronium and hydroxyl ions. Charge separation can also occur within the freezing zone as water moves in thin films and displaces ions in the electrical double layer thereby creating a "streaming potential". These processes may account for the electrical potentials which have been recorded as soils freeze or thaw (Hanley and Rao, 1982; Parameswaran and Mackay, 1983; Parameswaran et al., 1985). The effects of such potentials on the coupled transport of ions and water are not well documented.

4 MODELS

The purpose of modeling is two fold: to improve our understanding of the physics of a particular process and/or to predict the outcome of that process given certain boundary conditions. Models referred to as deterministic models are developed from theoretical considerations by choosing those processes which are deemed to be most relevant to a perceived problem and then formulating the mathematical relationships governing them. In contrast, models referred to as stochastic models, employ a 'black box' approach, whereby statistical considerations are substituted for a detailed knowledge of the physics involved. They are essentially a management tool for estimating the outcome of a variable process within specified confidence limits. A combination of both approaches is known as stochastic-deterministic modelling.

4.1 Deterministic Models

Deterministic models of heat and mass transfer in freezing soil vary in complexity. Mechanistic models, at one end of the spectrum, aim at a detailed physics-based description of all known processes and interactions. Non-mechanistic models, at the other end, make use of certain simplifying assumptions to omit complex or speculative relationships. This is usually done as a practical or numerical convenience.

A practical model of frost heave is required to predict the magnitude and rate of displacement as a function of boundary conditions and characteristics of the freezing system. In this context, the 'segregation potential' theory expounded by Konrad and Morgenstern (1981) can be viewed as a non-mechanistic, deterministic model of frost heave, developed for engineering needs. Microphysical relationships within the frozen fringe are represented by two macroscopic parameters (the overall hydraulic conductivity and the temperature gradient associated with the final ice lens) which can be deduced from controlled tests. In order to extend the model to conditions during transient freezing, further simplications were necessary, including the assumption of no convective heat transfer (Konrad and Morgenstern, 1982). Several authors have employed the segregation potential theory to analyze frost heave data from field and laboratory tests (e.g. Nixon, 1982; Knutsson et al., 1985). Their results suggest that it is possible to compute the segregation potential of a soil from easily performed index tests. However, this concept may be applicable only during the transient stage of freezing. Ishizaki and Nishio (1985) showed that heave rate could either increase or decrease with an increase in temperature gradient depending on how the gradient was established and at what stage in the heaving process the analysis was done. This raises questions concerning the generality of the segregation potential concept.

Takashi (1982) developed a theory to describe frost heave which has provided the base for the majority of subsequent Japanese models used for engineering purposes. The model incorporated Darcy's Law to describe water flow, Terzaghi's consolidation theory, and an empirically measured function describing ice segregation rate. On the basis of more than 200 laboratory frost heave experiments, the ice segregation rate was found to be proportional to the temperature gradient in the frozen zone, the difference between the upper limit of heaving pressure and the effective stress, and inversely proportional to the confining stress.

Most current deterministic models of coupled heat and mass transfer in freezing soil take a more mechanistic approach (O'Neill, 1983). All address, in one way or another, three phenomena critical to the frost heaving process. These are: (1) the microscopic relationship between ice and unfrozen water within soil pores, (2) transport of mass (ice, water and vapor) from the unfrozen soil to a growing ice lens, and (3) transport of sensible and latent heat, and release of latent heat during freezing. In general, microscopic conditions are represented by macroscopic variables and microscopic processes by macroscopic equations. Thus, most models are "built-up" from a microphysical analysis of the relevant subprocesses. Mathematical solution of the governing equations is complicated by interactions (coupling) between components. This requires the introduction of certain simplifying boundary conditions for an analytical solution. Otherwise, numerical methods which preserve the phase discontinuity by insuring that a coordinate boundary always lies on it must be employed (Berg, 1984).

At the very least, all mechanistic frost heave models must include the large latent heat effects that accompany an ice-water phase change. Among the simplest of recent models of simultaneous heat and moisture fluxes in freezing soil, are those of Takagi (1982), Dewen (1983) and Yanagisawa and Yao (1985). These authors make use of a variety of assumptions and approximations to reduce the partial differential equations for coupled transport to a problem of heat conduction with phase change, which can then be solved analytically. Outcalt (1980) has presented a simple energy balance model of ice segregation. However, this approach is somewhat limited due to mathematical oversimplification of the microphysical processes involved.

Several models have focussed on the microscopic interaction between a growing ice lens and unfrozen water films surrounding soil mineral particles. The adsorption force theory of frost heaving proposed by Takagi (1980) concerns itself with the disequilibrium between these two phases. During freezing, the system acts strongly to enforce approximate local equilibrium. Specifically, water is drawn into the frozen fringe to replenish the unfrozen water films and pre-existing ice is lifted away causing heave. Takagi (1980) also theorizes that the adsorbed films exhibit extraordinary flow properties and can build up internal "solid-like" stresses. This conceptualization has caused some controversy regarding the exact nature of the water in adsorbed films (Miller, 1980). Kuroda (1985) has presented a model in which the characteristics of menisci and Van der Waals forces are used to define the energy status (i.e. p_w) of unfrozen water. Recently, Horguchi (1987b) has proposed an "osmotic" model for soil freezing which effectively treats p_w in eqn.(2) as an osmotic pressure.

Another type of mechanistic model is based upon the assumption that the functional relationship between liquid water content and water potential in unfrozen unsaturated soil can be used for partially frozen air-free soil. This analog is then carried over to the functional relationship between hydraulic conductivity and water potential. Early mathematical models by Harlan (1973) and Guymon and Luthin (1974) made use of this approach, and in the 1980's a number of researchers have adopted essentially the same strategy. Jame and Norum (1980), for example, use a modified version of Harlan's model to simulate coupled heat and moisture flow in greater detail. They employ a diffusivity formulation, which is solved numerically by the finite difference method. Unfortunately, when segregated ice is present (instead of air) the diffusivity form of Darcy's law does not hold, and must be continually readjusted through a highly non-linear, empirical "impedance factor".

Although different algorithms are used for sub-processes, the one-dimensional model developed by Guymon et al. (1980) follows similar lines. This model solves the governing equations for coupled heat and moisture transport numerically, by either the subdomain method or the finite element method (Hromadka et al., 1982). The hydraulic conductivity function employed must be corrected for the presence of ice lenses by forcing model output to fit laboratory test results at certain times. Simulations have been verified against field data of frost heaving (Guymon et al., 1983). A multi-year model of coupled flow and ice segregation near the base of the active layer in arctic soils has been developed by Outcalt (1982). The model generates massive ground ice in response to temporal and spatial variations in the surface moisture regime. Recently, Cary (1987) has presented a numerical model that simulates coupled flow of heat, water and solutes in unsaturated freezing soil. The model includes simplifying assumptions regarding solutes (referred to earlier) and also assumes that overburden pressure and liquid pressure are additive. In all of the above models, ice segregation is essentially dictated by the assumptions made in their liquid flow equations.

The most physically complex models are those by Hopke (1980), Gilpin (1980) and O'Neill and Miller (1985). These are based on equations proposed by Miller (1978) to represent coupled processes in the frozen fringe. The Clapeyron equation, a capillary-type relationship and a freezing characteristic curve are used to relate temperature, liquid and ice pressures, and phase composition, in two highly non-linear partial differential equations. Of central importance is the potential jump due to curvature of the liquid water-ice-air interfaces. Mass transfer above the warmest lens is considered negligible. All assume a structureless soil with no hysteresis. Each model uses a slightly different numerical method to solve the coupled flow equations and an extremely fine nodal spacing is required to represent the mechanisms adequately. Heave is computed in terms of the initiation of individual ice lenses, their growth history and the redistribution of water between lenses. All three models can handle the expulsion of water during rapid initial freezing. The effect of overburden is also included. In general, predictions of lens distribution in time and space, water expulsion and its reversal, and overburden effect are qualitatively reasonable.

Although no single model has gained universal acceptance, the 'rigidice' model of O'Neill and Miller (1985) is being used by an increasing number of researchers. This is perhaps the most comprehensive mechanistic frost heave model currently available. Holden (1983) has simplified the governing equations using 'quasi-static' approximations for the temperature and pore water potential profiles. His reformulation gives rise to a pair of nonlinear differential equations which can be readily solved using standard numerical techniques. The model appears to describe cumulative heave, time lapse between ice lens initiation and the thickness of individual lenses successfully. Holden et al. (1985) report further development of this approach. Energy and mass conservation equations are introduced into the approximate profiles. This leads to a straight forward temperature calculation and the solution of two ordinary differential equations.

Efforts to model microphysical phenomena during a typical freezing episode, have been limited by the need to compute individual life histories for all possible ice lenses. Thus, simulation requires a large amount of computer time and adjustment of numerical tactics to reduce iterations while avoiding instabilities. As a result, a number of modellers have employed the methods of continuum mechanics to bypass explicit characterization of microscopic variables and processes. Coupled heat and mass transfer are described in terms of a continuum of responses to a continuum of boundary conditions for a zone of lensing. Black and Miller (1985) report the development of a continuum approach for predicting the heaving behaviour of freezing soils based upon the 'rigidice' model. Similarly, Nakano and Horiguchi (1984) and Blanchard and Fremond (1985) have used continuum mechanics to simulate coupled heat and water transport associated with frost heaving.

A final mechanistic approach to modelling coupled heat and moisture movement in freezing soil, is based on the theory of irreversible thermodynamics. An irreversible thermodynamic description of transport automatically defines the overall macroscopic conditions of the system. Such a description is very general and includes simultaneous vapor, liquid water, solute and heat fluxes in response to temperature, vapor, liquid water and ice potential gradients. Equations describing the thermodynamics of irreversible processes were first adapted to transport in partially frozen soil by Kay and Groenevelt (1974) and Groenevelt and Kay (1974). Other irreversible thermodynamic models of frost heave have been presented by Forland and Ratkje (1980) and Derjaguin and Churaev (1986). Until recently, experimental verification of the developed theory was lacking. However, Kung and Steenhuis (1986) report good agreement between laboratory data and computer simulations of heat and moisture transfer in a partially frozen non-heaving soil, based upon Kay and Groenevelt's original theory. Further experimental collaboration is provided by Ratkje et al. (1982) and Derjaguin and Churaev (1986). In general, the theory of thermodynamics of irreversible processes appears to deserve wider application by soil frost modellers.

4.2 Stochastic Models

Deterministic models require extensive adaptation in order to deal with field conditions which exhibit considerable spatial and temporal variability. In this context, some form of stochastic approach is required. In other words, a deterministic simulation is performed within a stochastic model of input parameter variations. Uncertainty in model output can then be expressed in terms of confidence intervals, based upon the variation in predictors and boundary conditions.

Guymon et al. (1981) advocate a stochastic-determinisitic approach to accommodate both spatial and temporal variability in mathematical models of soil freezing phenomena. These authors have incorporated a deterministic analysis of coupled heat and mass transfer into a probabilistic model based upon Rosenblueth's method. Input parameters are varied within reported levels of sampling error and spatial variation for two frost susceptible soils. Comparison of the resultant simulations with field measurements suggested a beta probability distribution for maximum soil displacements caused by ice segregation. Coefficients of variation for predicted frost heave ranged from 9 to 107% depending on the soil simulated, the magnitude of parameter variations and the surcharge condition. The most important input parameter was moisture transport, the coefficient of variation for simulated frost heave being approximately equal to that for the unfrozen hydraulic conductivity.

A similar combination of probabilistic and deterministic approaches has been used by Zuzel et al. (1986) to model the temporal variability of frozen soil in north-central Oregon. Soil frost was simulated using a physically-based, deterministic heat flow model described by Cary (1982). The temporal characteristics of frozen soil were simulated using 30 years of historical weather records as model inputs. Output was analyzed stochastically, using standard hydrological methods. The number of freeze-thaw cycles varied from 1 to 7 per year and the number of days the soil was frozen ranged from 6 to 116 per year. Model derived frequencies were approximated with a gamma distribution and a normal distribution, respectively. These frequency distributions were then used to calculate exceedence probabilities and recurrence intervals. This type of approach is useful in erosion prediction and as a planning tool for construction in areas where frost heaving is a problem.

5 RESEARCH NEEDS

Research in the next decade will be directed to closing some of the remaining gaps in our understanding of the freezing process and to enhancing our predictive capability under field conditions. Specific areas requiring further research include:

- characterizing the behavior of solutes dominating saline soils and marine permafrost at subzero temperatures;
- developing a theoretical base to describe freezing induced redistribution of solutes and the concomitant impact on the redistribution of water;
- assessing the extent and impact of charge separation and the development of electrical potentials in freezing soil on water, solute and heat transport;
- determining the limiting circumstances under which relations between variables under static conditions fail to apply under dynamic conditions;
- adapting existing theories to describe ice segregation in non-colloidal, uncemented soil materials to soils which contain colloidal material and are strongly structured;
- introducing a stochastic dimension into deterministic models in order to account for the variability in heat and mass transfer under field conditions;
- clarifying the mechanism and extent of ice segregation behind the warmest ice lens and determining the relationships governing long-term, as distinct from seasonal frost heaving;
- continuing development of methods to measure the liquid and ice pressure in frozen soils.

These research needs were identified on the basis of a comprehensive review of our current understanding of heat and mass transport in freezing soil.

REFERENCES *

Aguirre-Puente, J. and J. Gruson. 1983. Measurement of permeabilities of frozen soils. Permafrost: Fourth International Conference Proceedings. National Academy Press, Washington, D.C., U.S.A. p.5-9.

*See page X concerning ISGF papers

Aguirre-Puente, J., M. Sakly, L.E. Goodrich and G. Lambrinos. 1985. Experimental measurements and a numerical method for ice sublimation. Fourth International Symposium on Ground Freezing. Sapporo, Japan. Vol. 2, p.1-8.

Akagawa, S., Y. Yamamoto and S. Hashimoto. 1985. Frost heave characteristics and scale effect of stationary frost heave. Fourth Int. Symp. on Ground Freezing. Sapporo, Japan. Vol. 1, p.137-146.

Anderson, D.M. and N.R. Morgenstern. 1973. Physics, chemistry and mechanics of frozen ground. In Permafrost: The North American Contribution to the Second International Conference. National Academy of Sciences, Washington, D.C., U.S.A. p.257-288..

Anderson, D.M. and A.R. Tice. 1972. Predicting unfrozen water contents in frozen soils from surface area measurements. Highway Res. Rec. 373: 12-18.

Baker, T.H.W., J.L. Davis, H.N. Hayhoe and G.C. Topp. 1982. Locating the frozen-unfrozen interface in soils using time-domain reflectometry. Canadian Geotechnical Journal 19: 511-517.

Benoit, G.R. and S. Mostaghimi. 1985. Modeling soil frost under three tillage systems. Transactions of the A.S.A.E. 28: 1499-1505.

Berg, R.L. 1984. Status of numerical models for heat and mass transfer in frost-susceptible soils. Permafrost: Fourth International Conference. Final Proceedings. National Academy Press, Washington, D.C., U.S.A. p.67-71.

Black, P.B. and R.D. Miller. 1985. A continuum approach to modelling of frost heaving. In: Freezing and thawing of soil-water systems. D.M. Anderson and P.J. Williams (eds.). Technical Council on Cold Regions Engineering Monograph. A.S.C.E. p.36-43.

Blanchard, D. and M. Fremond. 1985. Soils frost heaving and thaw settlement. Fourth International Symposium on Ground Freezing. Sapporo, Japan. Vol. 1, p.209-216.

Booth, D.B. 1981. Macroscopic behaviour of freezing saturated silty soils. Cold Regions Science and Technology 4: 163-174.

Campbell, C.A. and V.O. Biederbeck. 1982. Changes in mineral N and numbers of bacteria and actinomycetes during two years under wheat-fallow in southwestern Saskatchewan. Can. J. Soil Sci. 59: 271-276.

Cary, J.W. 1982. Amount of soil ice predicted from weather observations. Agricultural Meteorology 27: 35-43.

Cary, J.W. 1987. A new method for calculating frost heave including solute effects. Water Resources Res. 23: 1620-1624.

Chamberlain, E.J. 1981. Frost susceptibility, review of index tests. Monograph 81-2. U.S. Army Cold Regions Research and Engineering Laboratory, Hanover, NH, 110 pp.

Chamberlain, E.J. 1983. Frost heave of saline soils. Permafrost: Fourth International Conference, Proceedings. National Academy Press, Washington, D.C., U.S.A. p.121-126.

Christianson, C.B. and C.M. Cho. 1983. Chemical denitrification of nitrate in frozen soils. Soil Sci. Soc. Amer. J. 47: 38-42.

Clark, C. and R. Kettle. 1985. Thermal neutron radiography for studying mass transfer in partially frozen soil. Fourth International Symposium on Ground Freezing. Sapporo, Japan. Vol. 2, p.109-114.

Colbeck, S.C. 1982. Configuration of ice in frozen media. Soil Science 133: 116-123.

Colbeck, S.C. 1985. A technique for observing freezing fronts. Soil Science 139: 13-20.

Derjaguin, B.V. and N.V. Churaev. 1986. Flow of nonfreezing water interlayers and frost heaving. Cold Regions Science and Technology 12: 57-66.

Dewen, D. 1983. Physical nature of frost processes and a research method. Permafrost: Fourth International Conference, Proceedings. National Academy Press, Washington, D.C., U.S.A. p.221-225.

Efimov, S.S., N.N. Kozhevnikov, A.S. Kurilko, M. Nikitina and A.V. Stepanov. 1981. Influence of cyclic freezing-thawing on heat and mass transfer characteristics of clay soil. Engineering Geology 18: 147-152.

Fahey, B.D. and R.D. Thompson. 1982. Energy balance and ground thermal characteristics during periods of soil freezing at Guelph, Ontario, Canada. Archives for Meteorology, Geophysics and Bioclimatology B 31: 113-126.

Farouki, O.T. 1982. Thermal properties of soils relevant to ground freezing-design techniques for their estimation. Proceedings of the Third Intl. Symposium on Ground Freezing, Hanover, New Hampshire, U.S.A. CRREL Special Rept. 82-16. p.139-146.

Forland, T. and S.R. Ratkje. 1980. Irreversible thermodynamic treatment of frost heave. Second Intl. Symp. on

Ground Freezing. Trondheim, Norway. Norwegian Institute of Technology. p.611-617. #

Frivik, P.E. 1981. State-of-the-art report. Ground freezing: thermal properties, modelling of processes and thermal design. Engineering Geology 18: 115-133.

Frost, I. Jord. 1981. Frost-susceptibility of soils, criteria from several countries. Nr. 22. 63 pp.

Gilpin, R.R. 1980. A model for the prediction of ice lensing and frost heave in soils. Water Resources Research 16: 918-930.

Goit, J.B., P.H. Groenevelt, B.D. Kay and J.P.G. Loch. 1978. The applicability of dual gamma scanning to freezing soils and the problem of stratification. Soil Sci. Soc. Amer. J. 42: 858-863.

Goodrich, L.E. 1982. The influence of snow cover on the ground thermal regime. Canadian Geotechnical Journal 19: 421-432.

Gori, F. 1983. A theoretical model for predicting the effective thermal conductivity of unsaturated frozen soils. Permafrost: Fourth International Conference, Proceedings. National Academy Press, Washington, D.C., U.S.A. p.363-368.

Gray, D.M. and R.J. Granger. 1986. In situ measurements of moisture and salt movement in freezing soils. Can. J. Earth Sci. 23: 696-704.

Gray, D.M., R.J. Granger and G.E. Dyck. 1985. Overwinter soil moisture changes. Transactions of the American Society of Agricultural Engineers 28: 442-447.

Grecean, E.L. (Ed.) 1981. Soil water assessment by the neutron method. CSIRO. Adelaide, Australia, 140 pp.

Groenevelt, P.H. and G.H. Bolt. 1969. Non-equilibrium thermodynamics of the soil-water system. J. of Hydrology 7: 358-388.

Groenevelt, P.H. and B.D. Kay. 1974. On the interaction of water and heat transport in frozen and unfrozen soils: II The liquid phase. Soil Science Society of America Proceedings 38: 400-404.

Groenevelt, P.H. and B.D. Kay. 1981. On pressure distribution and effective stress in unsaturated soils. Can. J. Soil Sci. 61: 431-443.

Guymon, G.L., R.L. Berg and T.V. Hromadka. 1983. Field tests of a frost heave model. Permafrost: Fourth International Conference, Proceedings. National Academy Press, Washington, D.C., U.S.A. p.409-414.

Guymon, G.L., M.E. Harr, R.L. Berg and T.V. Hromadka. 1981. A probabilistic-deterministic analysis of one-dimensional ice segregation in a freezing soil column. Cold Regions Science and Technology 5: 127-140.

Guymon, G.L., T.V. Hromadka and R.L. Berg. 1980. A one-dimensional frost heave model based upon simulation of simultaneous heat and water flux. Cold Regions Science and Technology 3: 253-262.

Guymon, G.L. and J.N. Luthin. 1974. A coupled heat and moisture transport model for arctic soils. Journal of Water Resources Research 10: 995-1003.

Hanley, T. and S.R. Rao. 1982. Electrical freezing potentials and the migration of moisture and ions in freezing soil. Proceedings of the Fourth Canadian Permafrost Conference, Ottawa, Ontario, Canada. National Research Council of Canada p.453-458.

Harlan, R.L. 1973. Analysis of coupled heat-fluid transport in partially frozen soil. Journal of Water Resources Res. 9: 1314-1323.

Hayhoe, H.N. and W.G. Bailey. 1985. Monitoring changes in total and unfrozen water content in seasonally frozen soil using time domain reflectometry and neutron moderation techniques. Water Resources Research 21: 1077-1084.

Hayhoe, H.N., A.R. Mack, E.J. Brach and D. Dalchin. 1986. Evaluation of the electrical frost probe. Journal of Agricultural Engineering Research 33: 281-287.

Holden, J.T. 1983. Approximate solutions for Miller's theory of secondary heave. Permafrost: Fourth International Conference, Proceedings. National Academy Press, Washington, D.C., U.S.A. p.498-503.

Holden, J.T., D. Piper and R.H. Jones. 1985. Some developments of a rigid-ice model of frost heave. Fourth International Symposium on Ground Freezing. Sapporo, Japan. Vol. 1, p.93-99.

Homshaw, L.G. 1980. Freezing and melting temperature hysteresis of water in porous materials: application to the study of pore form. J. Soil Sci. 31: 399-414.

Hopke, S. 1980. A model for frost heave including overburden. Cold Regions Science and Technology 3: 111-127.

Horiguchi, K. 1987a. Effect of cooling rate on freezing of a saturated soil. Cold Regions Science and Technology 14: 147-153.

Horiguchi, K. 1987b. An osmotic model for soil freezing. Cold Regions Science and Technology 14: 13-22.

Horiguchi, K. and R.D. Miller. 1983. Hydraulic conductivity functions of frozen materials. Permafrost: Fourth International Conference, Proceedings. National Academy Press, Washington, D.C. U.S.A. p.504-508.

Hromadka, T.V., G.L. Guymon and R.L. Berg. 1982. Sensitivity of a frost heave model to the method of numerical simulation. Cold Regions Science and Technology 6: 1-10.

Inaba, H. 1983. Experimental study on thermal properties of frozen soils. Cold Regions Science and Technology 8: 181-187.

Ishizaki, T. and N. Nishio. 1985. Experimental study of final ice lens growth in partially frozen saturated soil. Fourth International Symposium on Ground Freezing. Sapporo, Japan. Vol. 1, p.71-78.

Ishizaki, T., K. Yoneyama and N. Nishio. 1985. X-ray technique for observation of ice lens growth in partially frozen, saturated soil. Cold Regions Science and Technology 11: 213-221.

Iwata, S. 1980. Driving force for water migration in frozen clayey soil. Soil Science and Plant Nutrition 26: 215-227.

Jame, Y. and D.I. Norum. 1980. Heat and mass transfer in a freezing unsaturated porous medium. Water Resources Research 16: 811-819.

Johansen, O. and P.E. Frivik. 1980. Thermal properties of soils and rock materials. Second Intl. Symp. on Ground Freezing. Tronheim, Norway. Norwegian Institute of Technology. p.427-453.

Kadlec, R.H. 1984. Freezing-induced vertical solute movement in peat. Proceedings 7th International Peat Congress, Volume IV, Irish National Peat Committee p.248-256.

Kane, D.L. and J. Stein. 1983. Water movement into seasonally frozen soils. Water Resources Research 19: 1547-1557.

Kay, B.D., M. Fukuda, H. Izuta and M.I. Sheppard. 1981. The importance of water migration in the measurement of the thermal conductivity of unsaturated frozen soils. Cold Regions Sci. and Tech. 5: 95-106.

Kay, B.D. and J.B. Goit. 1975. Temperature-dependent specific heats of dry soil materials. Can. Geotech. J. 12: 209-212.

Kay, B.D. and P.H. Groenevelt. 1974. On the interaction of water and heat transport in frozen and unfrozen soils: I Basic theory; the vapor phase. Soil Science Society of America Proceedings 38: 395-400.

Kay, B.D. and P.H. Groenevelt. 1983. The redistribution of solutes in freezing soil: exclusion of solutes. Permafrost: Fourth International Conference Proceedings, Washington, D.C., U.S.A. National Academy Press p.584-588.

Kay, B.D., C.D. Grant and P.H. Groenevelt. 1985. Significance of ground freezing on soil bulk density under zero tillage. Soil Science Society of America Journal 49: 973-978.

Knutsson, S., L. Domaschuk and N. Chandler. 1985. Analysis of large scale laboratory and in situ frost heave tests. Fourth International Symposium on Ground Freezing. Sapporo, Japan. Vol. 1, p.65-70.

Konrad, J.M. and N.R. Morgenstern. 1981. The segregation potential of a freezing soil. Canadian Geotechnical Journal 18: 482-491.

Konrad, J.M. and N.R. Morgenstern. 1982. Prediction of frost heave in the laboratory during transient freezing. Canadian Geotechnical Journal 19: 250-259.

Kung, S.K. and T.S. Steenhuis. 1986. Heat and moisture transfer in a partly frozen non-heaving soil. Soil Science Society of America Journal 50: 1114-1122.

Kuroda, T. 1985. Theoretical study of frost heaving-kinetic process at water layer between ice lens and soil particles. Fourth International Symposium on Ground Freezing. Sapporo, Japan. Vol. 1, p.39-46.

Lahav, N. and D.M. Anderson. 1973. Montmorillonite-benzidine reactions in the frozen and dry states. Clays and Clay Minerals 21: 137-139.

Leung, W.K.S. and G.R. Carmichael. 1984. Solute redistribution during normal freezing. Water, Air and Soil Pollution 21: 141-150.

Loch, J.P.G. 1978. Thermodynamic equilibrium between ice and water in porous media. Soil Sci. 126: 77-80.

Loch, J.P.G. 1980. Frost action in soils: state of the art. Proceedings of the Second International Symposium on Ground Freezing, Norwegian Institute of Technology, Trondheim, Norway. p. 581-596.#

Lunardini, V.J. 1982. Heat transfer in cold climates. Van Nostrand Reinhold, N.Y.

Mahar, L., T.S. Vinson and R. Wilson. 1982. Effects of salinity on freezing of granular soils. (Abstract) Proceedings of the Third Intl. Symp. on Ground Freezing. Hanover, NH, U.S.A. CRREL Special Rept. 82-16. p.149.

McCabe, E.Y. and R.J. Kettle. 1985. Thermal aspects of frost action. Fourth Intl. Symp. on Ground Freezing. Sapporo, Japan. Vol. 1. p.47-54.

McKim, H., R.L. Berg, R. McGaw, R. Atkins and J.W. Ingersoll. 1976. Development of a remote reading tensiometer transducer system for use in subfreezing temperature. Proc., Second Conf. on Soil Water Problems in Cold Regions. Edmonton, Canada.

McGaw, R., R.L. Berg and J.W. Ingersoll. 1983. An investigation of transient processes in an advancing zone of freezing. Proc. Fourth Intl. Conf. on Permafrost. Fairbanks, U.S.A. Nat. Acad. Press. p.821-825.

Meentemeyer, V. and J. Zippin. 1981. Soil moisture and texture controls of selected parameters of needle ice growth. Earth Surface Processes and Landforms 6: 113-125.

Miller, R.D. 1972. Freezing and heaving of saturated and unsaturated soils. Highway Research Record 393: 1-11.

Miller, R.D. 1973. Soil freezing in relation to pore water pressure and temperature. Permafrost. Second Intl. Conf. Nat. Acad. Sci., Washington, D.C., U.S.A. p.344-352.

Miller, R.D. 1978. Frost heaving in non-colloidal soils. Proc. Third Int'l. Conf. Permafrost, Edmonton, Canada. National Research Council of Canada, Ottawa, Canada. p.707-713.

Miller, R.D. 1980. The adsorbed film controversy. Cold Regions Science and Technology. 3: 83-86.

Miller, R.D. 1984. Thermally induced regelation: A qualitative discussion. Permafrost: Fourth International Conference. Final Proceedings. National Academy Press, Washington, D.C., U.S.A. p.61-63.

Nakano, Y., A. Tice, J. Oliphant and T. Jenkins. 1982. Transport of water in frozen soil. I. Experimental determination of soil-water diffusivity under isothermal conditions. Advances in Water Resources 5: 221-226.

Nakano, Y., A. Tice, J. Oliphant and T. Jenkins. 1983. Transport of water in frozen soil. II. Effects of ice on the transport of water under isothermal conditions. Advances in Water Resources 6: 15-26.

Nakano, Y., A. Tice and J. Oliphant. 1984a. Transport of water in frozen soil. III. Experiments on the effects of ice content. Advances in Water Resources 7: 28-34.

Nakano, Y., A. Tice and J. Oliphant. 1984b. Transport of water in frozen soil. IV. Analysis of experimental results on the effects of ice content. Advances in Water Resources 7: 58-66.

Nakano, Y., A.R. Tice and T.F. Jenkins. 1984c. Transport of water in frozen soil: V. Method for measuring the vapor diffusivity when ice is absent. Advances in Water Resources 7: 172-179.

Nakano, Y. and A.R. Tice. 1987. Transport of water in frozen soil. VI. Effects of temperature. Advances in Water Resources (in press)

Nakano, Y. and K. Horiguchi. 1984. Role of heat and water transport in frost heaving of fine-grained porous media under negligible overburden pressure. Advances in Water Resources 7: 93-102.

Nakano, Y. and K. Horiguchi. 1985. Role of phase equilibrium in frost heave of fine-grained soil under negligible overburden pressure. Advances in Water Resources 8: 50-68.

Nakano, Y. 1986. On the stable growth of segregated ice in freezing soil under negligible overburden pressure. Advances in Water Resources 9: 223-235.

Nelson, D.W. and M.J.M. Romkens. 1972. Suitability of freezing as a method of preserving runoff samples for analysis for soluble phosphate. J. Environmental Qual. 1: 323-324.

Nixon, J.F. 1982. Field frost heave predictions using the segregation potential concept. Canadian Geotechnical Journal 19: 526-529.

Ohrai, T. and H. Yamamoto. 1985. Growth and migration of ice lenses in partially frozen soil. Fourth International Symposium on Ground Freezing. Sapporo, Japan. Vol. 1, p.79-84.

Oliphant, J.L. and A.R. Tice. 1982. Comparison of unfrozen water contents measured by DSC and NMR. Proc. Third Intl. Symp. on Ground Freezing. Hanover, New Hampshire. CRREL Special Rept. 82-16. p.115-123.

Oliphant, J.L. 1985. A model for dielectric constants of frozen soils. Freezing and Thawing of Soil-Water Systems. American Soc. Civil Engineers. New York, New York. p.46-57.

Oliphant, J.L., A.R. Tice and Y. Nakano. 1983. Water migration due to a temperature gradient in frozen soil. Permafrost: Fourth International Conference, Proceedings. National Academy Press, Washington, D.C., U.S.A. p.951-956.

O'Neill, K. 1983. The physics of mathematical frost heave models: A review. Cold Regions Science and Technology 6: 275-291.

O'Neill, K. and R.D. Miller. 1985. Exploration of a rigid ice model of frost heave. Water Resources Research 21: 281-296.

Osterkamp, T.E. 1987. Freezing and thawing of soils and permafrost containing unfrozen water or brine. Water Resources Res. 23: 2279-2285.

Outcalt, S.I. 1980. A simple energy budget model of ice segregation. Cold Regions Science and Technology 3: 145-151.

Outcalt, S.I. 1982. Modeling near-surface ground ice in Alaska. Physical Geography 3: 123-147.

Patterson, D.E. and M.W. Smith. 1981. The measurement of unfrozen water content by time domain reflectometry: Laboratory test on frozen samples. Canadian Geotech. J. 18: 131-144.

Patterson, D.E. and M.W. Smith. 1985. Unfrozen water content in saline soils: results using time domain reflectometry. Can. Geotech. J. 22: 93-101.

Parameswaran, V.R. and J.R. MacKay. 1983. Field measurements of electrical freezing potentials in permafrost areas. Permafrost: Fourth International Conference, Proceedings. National Academy Press, Washington, D.C., U.S.A. p.962-967.

Parameswaran, V.R., G.H. Johnston and J.R. MacKay. 1985. Electrical potentials developed during thawing of frozen ground. Fourth International Symposium on Ground Freezing. Sapporo, Japan. Vol. 1, p.9-16.

Penner, E. 1986. Ice lensing in layered soils. Canadian Geotechnical Journal 23: 334-340.

Perfect, E., R.D. Miller and B. Burton. 1988. Spatial variation in seasonal frost heave cycles. Proceedings V International Permafrost Conference. Trondheim, Norway. (in press)

Ratkje, S.K., H. Yamamoto, T. Takashi, T. Ohrai and J. Okamoto. 1982. The hydraulic conductivity of soils during frost heave. Proceedings of the Third Intl. Symposium on Ground Freezing. Hanover, NH, U.S.A. CRREL Special Report 82-16, p. 131-138.

Rieke, R.D., T.S. Vinson and D.W. Mageau. 1983. The role of specific surface area and related index properties in the frost heave susceptibility of soils. Permafrost: Fourth International Conference Proceedings. National Academy Press, Washington, D.C., U.S.A. p.1066--1071.

Riseborough, D.W., M.W. Smith and D.H. Halliwell. 1983. Determination of the thermal properties of frozen soils. Permafrost: Fourth International Conference Proceedings. National Academy Press, Washington, D.C., U.S.A. p.1072-1077.

Sawada, S. and T. Ohno. 1985. Laboratory studies on thermal conductivity of clay, silt and sand in frozen and unfrozen states. Fourth Intl. Symposium on Ground Freezing. Sapporo, Japan. Vol. 2, p.53-58.

Sheppard, M.I., B.D. Kay and J.P.G. Loch. 1981. The coupled transport of water and heat in freezing soils: A field study. Canadian Journal of Soil Science 61: 417-429.

Smith, M.W. 1985. Observations of soil freezing and frost heave at Inuvik, Northwest Territories, Cañada. Canadian Journal of Earth Sciences 22: 283-290.

Smith, M.W. and C.R. Burn. 1986. Outward flux of vapour from frozen soils at Mayo, Yukon, Canada: Results and interpretation. Cold Regions Science and Technology 13: 143-152.

Snyder, V.A. and R.D. Miller. 1985a. Tensile strength of unsaturated soil. Soil Sci. Soc. Amer. J. 49: 58-65.

Snyder, V.A. and R.D. Miller. 1985b. A pneumatic fracture method for measuring the tensile strength of unsaturated soils. Soil Sci. Soc. Amer. J. 49: 1369-1374.

Takagi, S. 1980. The adsorption force theory of frost heaving. Cold Regions Science and Technology 3: 57-81.

Takagi, S. 1982. Initial stage of the formation of soil-laden ice lenses. Proc. Third Intl. Symposium on Ground Freezing. Hanover, NH. CRREL Special Report 82-16, p.223-232.

Takeda, K., J. Nakazawa and S. Kinosita. 1985. Thermal condition for ice lens formation in soil freezing. Fourth International Symposium on Ground Freezing. Sapporo, Japan. Vol. 2, p.89-94.

Takashi, T. 1982. Analysis of frost heave mechanisms. Seiken Co. Ltd., Tokyo, Japan. 123 pp.

Thurmond, V.L. and G.W. Brass. 1987. Geochemistry of freezing brines. Low temperature properties of sodium chloride. CRREL Rept. 87-13. U.S. Army Cold Regions Research and Engineering Lab., Hanover, New Hampshire, U.S.A.

Tice, A.R., Zhu Yuanlin and J.L. Oliphant. 1984. The effects of solute salts on the unfrozen water contents of the Lanzhou, P.R.C. silt. CRREL Rept.

84-16. U.S. Army Cold Regions Research and Engineering Lab., Hanover, New Hampshire, U.S.A.
Tice, A.R. and J.L. Oliphant. 1984. The effects of magnetic particles on the unfrozen water content of frozen soils determined by nuclear magnetic resonance. Soil Sci. 138: 63-73.
Viereck, L.A. and D.J. Lev. 1983. Long-term use of frost tubes to monitor the annual freeze-thaw cycle in the active layer. Permafrost: Fourth International Conference Proceedings. National Academy Press, Washington, D.C., U.S.A. p.1309-1314.
Vliet-Lanoe. B. Van. 1984. Frost effects in soils. Soils and Quaternary Landscape Evolution (edited by J. Boardman), Chichester, U.K. John Wiley & Sons p.117-158.
Wada, S.A. and Nagasato, 1983. Formation of silica microplates by freezing dilute silica acid solutions. Soil Sci. Plant Nutr. 29: 93-95.
Walder, J.S. 1986. Motion of sub-freezing ice past particles, with applications to wire regelation and frozen soils. Journal of Glaciology 32: 404-414.
Williams, P.J. 1984. Moisture migration in frozen soils. Permafrost: Fourth International Conference. Final Proceedings. National Academy Press, Washington, D.C., U.S.A. p.64-66.
Williams, P.J. and J.A. Wood. 1985. Internal stresses in frozen ground. Canadian Geotechnical Journal 22: 413-416.
Wood, J.A. and P.J. Williams. 1985a. Further experimental investigation of regelation flow with an ice sandwich permeameter. In: Freezing and Thawing of Soil-Water Systems. D.M. Anderson and P.J. Williams (eds.). Technical Council on Cold Regions Engineering Monograph. A.S.C.E. 97 pp.
Wood, J.A. and P.J. Williams. 1985b. Stress distribution in frost heaving soils. Fourth International Symposium on Ground Freezing. Sapporo, Japan. Vol. 1, p.165-172.
Xu Xiaozu, J.L. Oliphant and A.R. Tice. 1985. Experimental study on factors affecting water migration in frozen morin clay. Fourth International Symposium on Ground Freezing. Sapporo, Japan. Vol. 1, p.123-128.
Xu Xiaozu, J.L. Oliphant and A.R. Tice. 1985. Prediction of unfrozen water contents in frozen soils by a two-point or one-point method. Fourth Intl. Symp. Ground Freezing, Sapporo, Japan. Vol.2. p.83-87.
Yanagisawa, E. and Y.J. Yao. 1985. Moisture movement in freezing soils under constant temperature condition. Fourth International Symposium on Ground Freezing. Sapporo, Japan. Vol. 1, p.85-92.
Yershov, E.D., Y.P. Lebedenko, O.M. Yazynin, Y.M. Chuvilin, V.N. Sokolov, V.V. Rogev and V.V. Kondakov. 1983. A review of cryogenic structure and texture in fine-grained rocks and soils. Permafrost: Fourth International Conference, Proceedings. National Academy Press, Washington, D.C., U.S.A. p.1440-1444.
Yoneyama, K., T. Ishizaki and N. Nishio. 1983. Water redistribution measurements in partially frozen soil by x-ray technique. Permafrost: Fourth International Conference Proceedings. National Academy Press, Washington, D.C., U.S.A. p.1445-1450.
Zuzel, J.F., J.L. Pikul and R.N. Greenwalt. 1986. Point probability distributions of frozen soil. Journal of Climate and Applied Meteorology 25: 1681-1686.

5th International Symposium on Ground Freezing, Jones & Holden (eds)
© 1988 Balkema, Rotterdam. ISBN 90 6191 824 3

Evaluation of the X-ray radiography efficiency for heaving and consolidation observation

Satoshi Akagawa
Shimizu Corporation, Tokyo, Japan
(Presently: USA CRREL)

ABSTRACT: Two step-freeze tests were conducted during which 136 radiographs were taken. These were used to test the feasibility of utilizing X-rays as a nondestructive method for observing changes in a soil's physical and mechanical properties due to frost heave. The radiographs were analyzed using computer image processing techniques to measure the position of lead spheres embedded in the soil column and to examine the spatial distribution of intensity changes of the transmitted X-rays. An analysis of the attenuation properties of frozen soil is presented. A linear correlation between the soil's attenuation of X-rays and the amount of heave and consolidation is determined. This relationship is utilized to compute strain distribution profiles.

KEYWORDS: clays, frost heave, image processing, laboratory tests, X-ray

1 INTRODUCTION

Physical and mechanical changes in a frost-susceptible soil are mainly studied using step-freeze heave tests. Interpretation of the results is not straightforward, though, because of the possible consolidation of the unfrozen soil resulting from the strong suction created at the freezing front (Akagawa, 1988a). It is necessary to use additional techniques that can measure the local consolidation in order to interpret the behavior of the soil in these step-freeze tests. This paper discusses the nondestructive technique of radiography and demonstrates how it can be used to infer the strain distribution within the freezing soil necessary in the interpretation of the step-freeze tests.

Generally speaking, an incident X-ray is attenuated by passing through a medium as:

$$I_{out} = e^{-\mu T} I_{in} \qquad (1)$$

where I_{out} and I_{in} are the measured incident and transmitted X-ray strength, respectively, μ is linear absorption coefficient of the medium, and T is its thickness; μ is related to density ρ by:

$$\mu/\rho = \text{Const.} \qquad (2)$$

Thus the attenuation of X-rays passing through soil will depend upon its density.

During soil freezing, porosity and therefore density will change due to frost heave and consolidation, as shown in Fig.1. The measured attenuation of X-rays passing through the soil should then vary according to local density fluctuations resulting from frost heave. For the case of normal incidence of X-rays to the soil, a simple set of equations based upon eq (1) describe the attenuation of X-rays in the three cases arising in step-freeze tests: initial unfrozen soil, heaving soil, and consolidating soil.

Initial unfrozen soil:

$$I_{out} = \frac{Lv \cdot e^{-\mu_w T} + Ls \cdot e^{-\mu_s T}}{L} I_{in} \qquad (3)$$

where subscripts w and s of μ refer to water and soil, respectively and L = Lv + Ls.

Heaving soil:

$$I_{out} = \frac{Lw \cdot e^{-\mu_w T} + Li \cdot e^{-\mu_i T} + Ls \cdot e^{-\mu_s T} + \Delta h \cdot e^{-\mu_i T}}{L + \Delta h} I_{in} \qquad (4)$$

where subscript i of μ refers to ice, and $L+\Delta h=Lw+Li+Ls+\Delta h$.

Consolidating soil:

$$I_{out}=\frac{Ls\cdot e^{-\mu_s T}+(Lv-\Delta Lv)\cdot e^{-\mu_w T}}{L-\Delta Lv}I_{in}. \quad (5)$$

Then the difference in X-ray attenuation attributable to heaving is shown as eq (6) by subtracting eq (3) from eq (4):

$$\frac{\Delta I_{out}}{I_{in}}=A\cdot\frac{Li}{L}+\frac{Lw\cdot A+Ls\cdot B}{L}\cdot\frac{\epsilon}{1+\epsilon} \quad (6)$$

where $A=(e^{-\mu_i T}-e^{-\mu_w T})$, $B=(e^{-\mu_i T}-e^{-\mu_s T})$ and $\epsilon=\Delta h/L$.

The X-ray strength difference ascribable to consolidation is shown as eq (7) by subtracting eq (3) from eq (5):

$$\frac{\Delta I_{out}}{I_{in}}=C\cdot\frac{Ls}{L}\cdot\frac{\epsilon}{1+\epsilon} \quad (7)$$

where $C=(e^{-\mu_s T}-e^{-\mu_w T})$ and $\epsilon=-\Delta Lv/L$.

Since $\mu i=\mu w$, eq (6) becomes:

$$\frac{\Delta I_{out}}{I_{in}}=B\cdot\frac{Ls}{L}\cdot\frac{\epsilon}{1+\epsilon}. \quad (6a)$$

Inspection of eq (6a) for heaving sections and eq (7) for consolidating sections shows a simple linear relationship between X-ray attenuation and the amount of heave or consolidation. Thus, by knowing the initial X-ray attenuation intensity distribution of the soil system, subsequent attenuation intensity is directly related to the heave or consolidation state of the soil system.

The following sections describe the experimental setup and resulting attenuation changes due to step-freeze tests. From these data, and displacements of the embedded lead spheres, the strain of the soil system is evaluated.

2 EXPERIMENT

Two step-freeze frost heave tests were conducted utilizing the test cell shown in Fig. 2. Specimen temperature was controlled by brine circulated in the pedestals. The boundary temperature conditions were fixed during each test. Test conditions are shown in Table 1. The physical, mechanical and thermal properties of the soil utilized in the tests are summarized in Table 2. The burette supplied the required pore water. Lead spheres were installed in the soil sample at the initial positions shown in Fig. 3a. Each sphere contained a thermocouple (Fig. 3b) for precise temperature measurement.

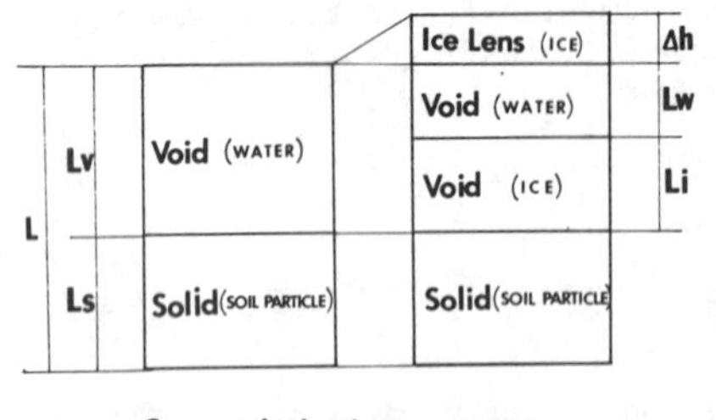

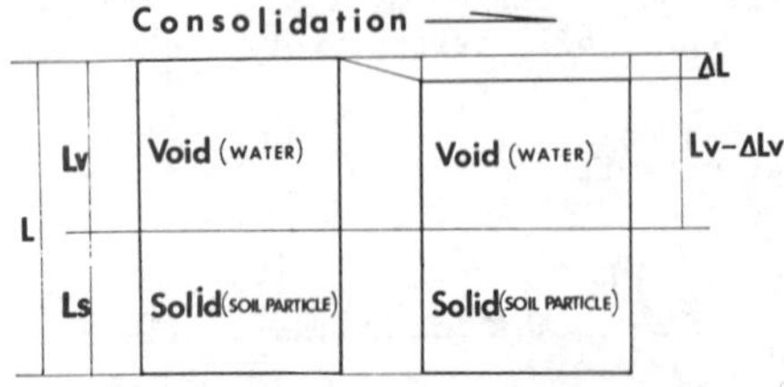

Fig. 1 Schematic structures of heaving soil and consolidating soil.

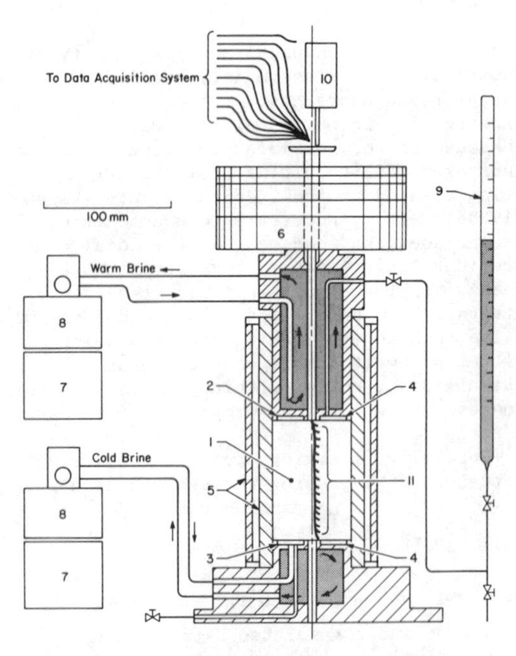

Fig. 2 Frost heave test apparatus.

Table 1. Test conditions

TEST NAME	PRESSURE		TEMPERATURE	
	Overburden (gauge pressure in KPa)	Pore water (gauge pressure in KPa)	Cold side (°C)	Warm side (°C)
Test A	60	0	-5.9	2.2
Test B	110	0	-5.5	3.0

Table 2. Summary of soil properties

% by weight		
Clay (0.002 mm)		45.0 (%)
Silt (0.002-0.074 mm)		41.0
Sand (0.074-2 mm)		14.0
Gravel(2 mm)		0
Atterberg Limits		
Wp		32.4 (%)
WL		101.0
Ip		68.7
Specific Gravity		2.48
Wet Density		1.45 (cm^3/g)
Void Ratio		
Consolidation Yield Stress	Pc	2.48 (MPa)
Consolidation Coefficent	Pv	5.78×10^{-2} (cm/min)
Compression Index	Cc	1.233
Permeability		
by Consolidation Test	K	1.77×10^{-6} (cm/min)
by Permeability Test	K	4.61×10^{-6} (cm/min)
Volumetric Unfrozen Water Content by TDR at -0.2°C		60.7 (%)
-0.3		53.2
-0.4		38.0
-0.6		29.8
-0.75		27.7
-1.0		25.1
-1.5		22.1
-2.0		20.2
-3.0		18.3
-5.0		14.6
-7.0		13.1
-10.0		12.3
Thermal Conductivity by Thermal Probe Method at 24°C		6.0 (cal/cm·hr·°C)
-5		12.0
-10		16.1

Relatively weak X-rays (100 kV, 5 mA) were used in the tests. Radiographs were taken every hour for the first 24 hours and then the frequency was decreased as shown in Table 3. Two radiographs were taken each time at two angles (Fig. 4) to allow three-dimensional analysis of the lead spheres.

Each radiograph was digitized with a resolution of 640 x 400 8-bit pixels. Intensity distortion due to a point X-ray source caused by different thicknesses over the image was compensated for by subtracting the initial (time 0) intensity of a radiograph image from the image at the time in question. Intensity distortion due to film development and printing (photochemical conversion) and digitization (photoelectric conversion) were also ignored. This procedure for radiograph image analysis was described further by Akagawa (1988b).

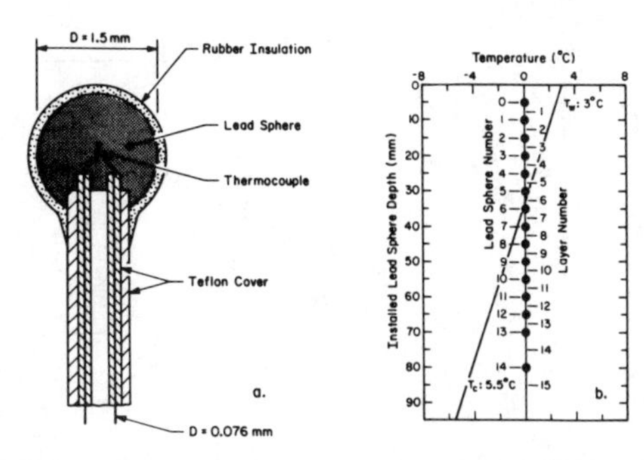

Fig. 3 Schematic drawings of installed lead spheres' location and structure of the lead sphere.

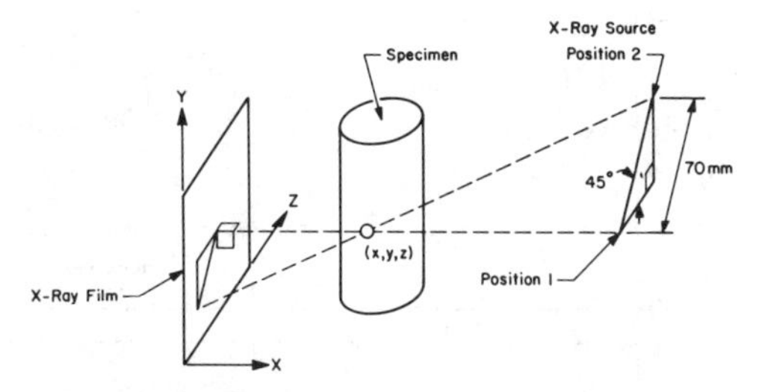

Fig. 4. Schematic drawing of radiography configuration.

Table 3. Time listing of X-ray radiography

TEST NAME	TIME (hour)
Test A	every hour from -0.5 to 23.5 and *(44.5, 141.5, 304.5, 524.5, 664.5, 832.5 and 976.5)
Test B	every hour from -1 to 24 and 28, 46, (70, 118, 166, 172, 214, 286, 382, 388, 502 and 646)

* Radiographs taken at the times shown in ( ) are not used in this paper.

3 TEST RESULTS

Measured amounts of heave, heave rate and

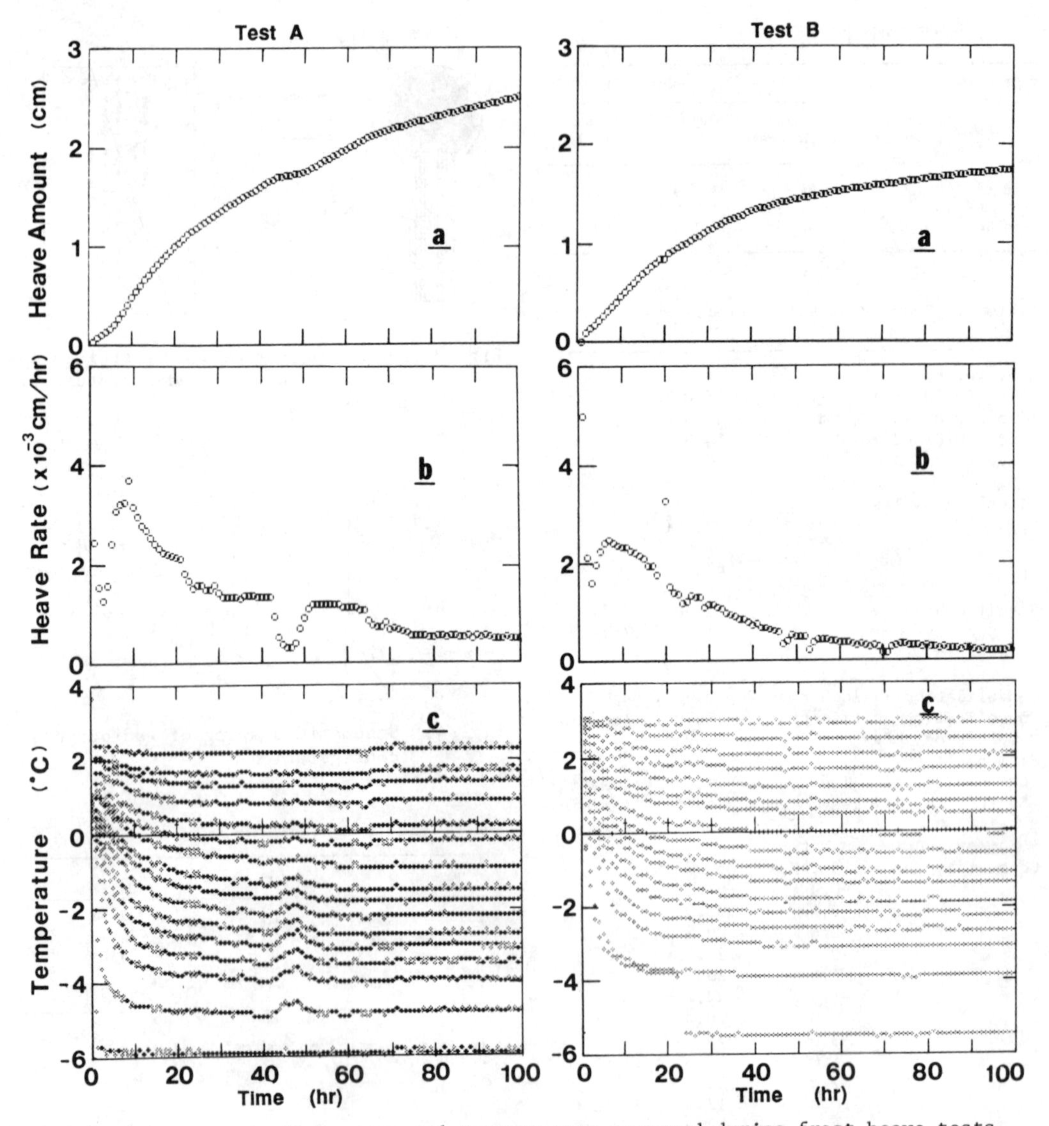

Fig. 5. Heave amount, heave rate and temperatures measured during frost heave tests.

temperature for each test are shown in Fig. 5. The difference in heave amount and rate between the two tests is a consequence of the different overburden pressures. The larger overburden used in Test B reduced the amount and rate of heave, as was to be expected.

The coordinates of each lead sphere were determined by the image processing techniques of Akagawa (1988a and 1988b) with a confidence interval of ±0.053 mm and a probability content of 68.3 %. The state of strain in the soil sample between two adjacent lead spheres was determined by:

$$\epsilon(i,n) = \frac{\Delta h(i,n)}{D(n)}$$

$$= \frac{\{Z(i,n)-Z(i,n+1)\}-\{Z(0,n)-Z(0,n+1)\}}{Z(0,n) - Z(0,n+1)} \quad (8)$$

where $\Delta h(i,n)$ is accumulated deformation in the (n)th soil layer at a time t = i, D(n) is the initial distance between the (n)th and (n+1)th lead spheres at time t = 0, Z(i, n) is the Z coordinate (direction parallel to the heat flow) of the (n)th lead sphere at time t=i. Fig. 6 is a representation of

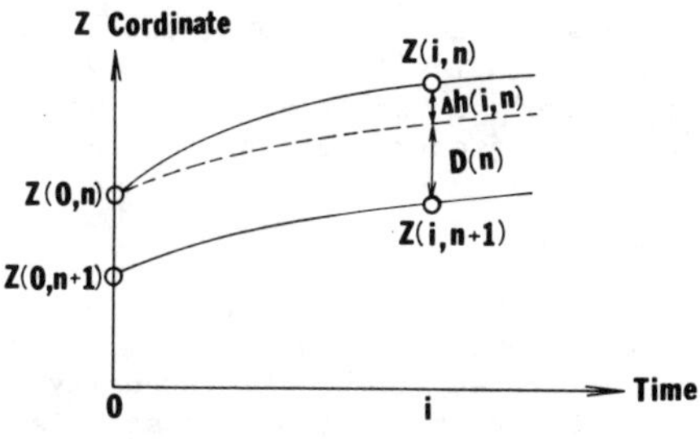

Fig. 6 Definition of strain in a layer.

the displacement of the lead spheres with time.

Typical calculated strains for each subsection of soil are shown as the heights of columns in Fig. 7, with measured temperature profiles given as broken lines. Each figure exhibits both expansive and shrinkage strain. The expansive strain is in a region where the temperature is below the segregation temperature of the warmest ice lens (Ts) and may be interpreted to be a result of heave. Similarly, shrinkage strain lies

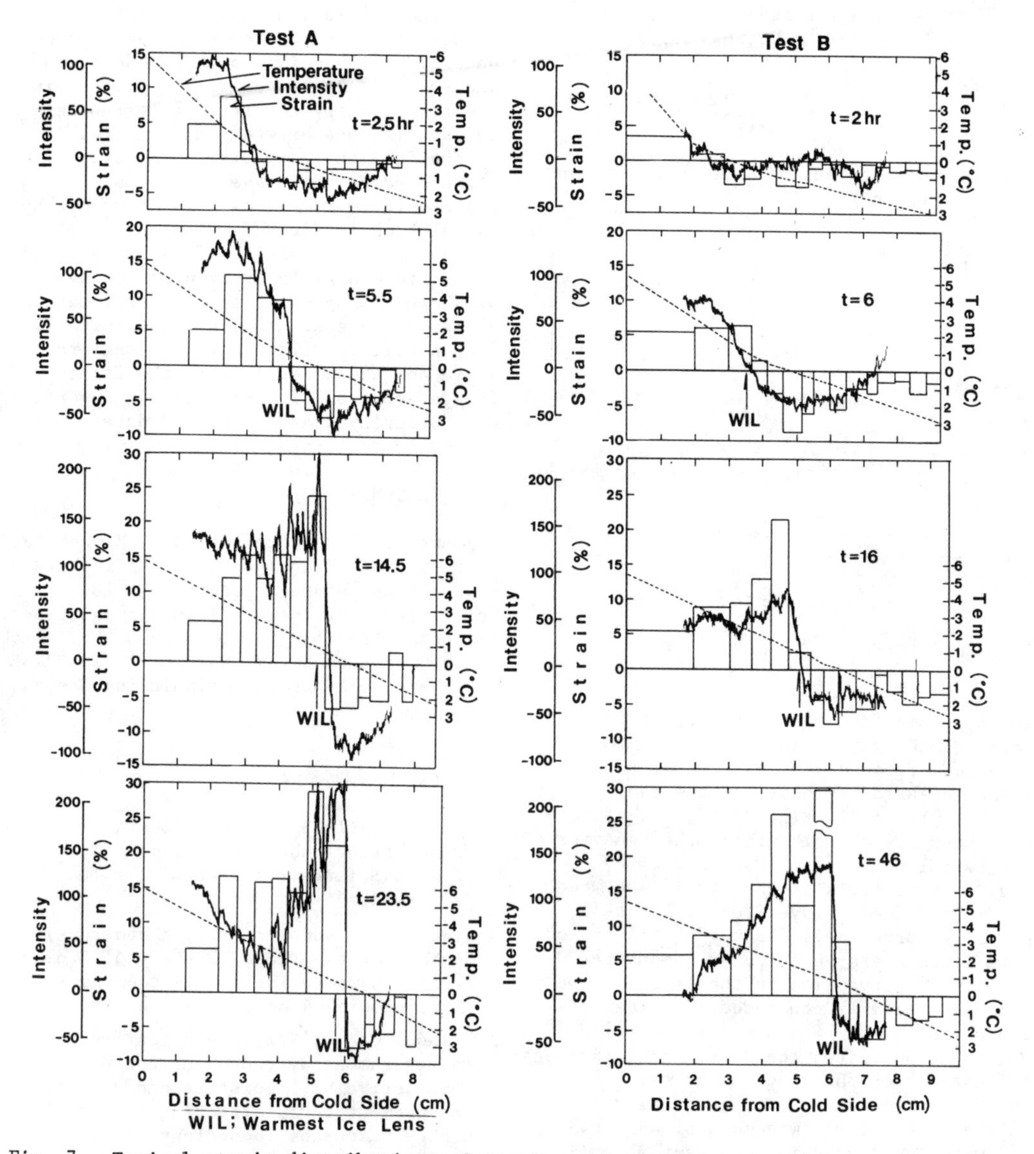

Fig. 7. Typical strain distributions, intensity and temperature profiles and location of the warmest ice lens. (Positive strain is compression.)

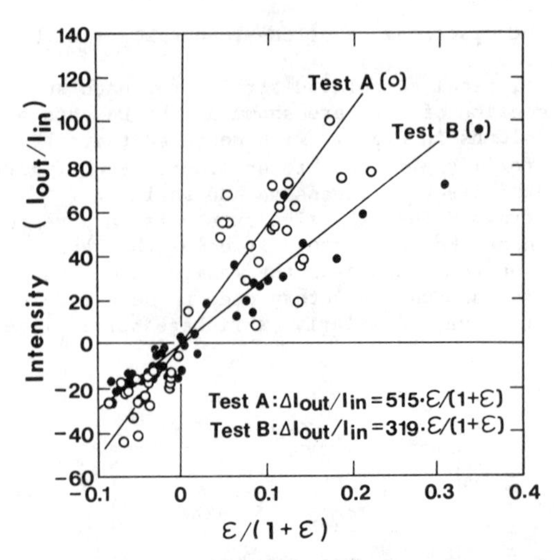

Fig. 8 Correlations between intensity and strain.

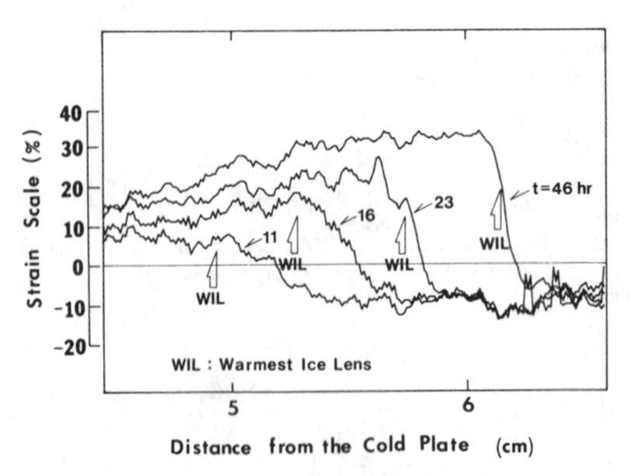

Fig. 9 Typical strain profiles.

in the warm, frozen fringe and unfrozen regions, and may be interpreted to be a consequence of consolidation.

The measured intensity of the radiographs is also plotted in Fig. 7. The magnitude of the intensity closely mimics the behavior of the calculated strain, as was desired. Fig. 8 shows plots of the strain and mean intensity of each layer for both tests. The data from both tests clearly display a positive linear relationship, as demonstrated by the linear curve fits in the figures. It can be reasonably concluded that the relationships in eqs (6a) and (7) are valid. Fig. 9 demonstrates the strain distributions of Test B over the specimen at times t = 11, 16, 23 and 46 hours. The conversion from intensity to strain was made by eqs (6a) and (7) with the regression constant shown in Fig. 8. Ice lenses are recognized as positive peaks in the strain profiles. The growth of thin ice lenses can be evaluated quantitatively by subtracting the strain values of each corresponding ice lens.

4 CONCLUSION

A relation between X-ray strength, which has been substituted for by the intensity of the X-ray film, and density of the soil is studied in relation to the frost heaving and consolidation observed in frozen and unfrozen soil.

A linear relationship was determined between the X-ray attenuation and apparent density of the soil, as was shown in eq (6a) and eq (7). By comparison of the strain distribution in the freezing soil determined by the distance between adjacent lead spheres and the intensity of X-rays in corresponding soil layers, eq (6a) and eq (7) are found to be valid as shown by the correlation in Fig. 8.

Then, the intensity profiles were interpreted to a density profile of the sample, as shown in Fig. 9. This interpretation allows us to evaluate ice lens growth in a frozen soil in relation to the temperature field and the water flow in the frozen fringe and the unfrozen soil by converting the intensity data to density data.

ACKNOWLEDGMENT

The author thanks Professors S. Kinoshita, F. Fukuda and K.Horiguchi of the Institute of Low Temperature Science, Hokkaido University, for their valuable suggestions. The author is also indebted to Dr. P. Black, Dr. R. Berg and Dr. Y. Nakano of USA CRREL for their comments and stimulating discussions.

REFERENCES

Akagawa, S. (1988a). Experimental study of frozen fringe characteristics. Cold Regions Science and Technology (in preparation).

Akagawa, S. (1988b). X-ray photography method for experimental study of frozen fringe characteristics of freezing soil. CRREL Special Report (in preparation).

Ayorinde, O.A. (1983). Application of dual-energy gamma-ray technique for non-destructive soil moisture and density measurement during freezing. Journal of Energy Resources Technology, Vol.105, pp38-42.

5th International Symposium on Ground Freezing, Jones & Holden (eds)
© 1988 Balkema, Rotterdam. ISBN 90 6191 824 3

Salt redistribution during laboratory freezing of saline sand columns

G.C.Baker & T.E.Osterkamp
Geophysical Institute, University of Alaska, Fairbanks, Alaska, USA

ABSTRACT: Columns of saline sands were frozen at constant rates in a specially designed apparatus to investigate the redistribution of salt during freezing. A silica sand with a grain size range from about 250 to 600 μm, saturated with 35 parts per thousand (ppt) NaCl solution, was used for the experiments. Measurements of the electrical conductivity of the soil solution were made on the columns, *in situ*, and on samples obtained by sectioning the columns. Moisture content profiles were determined for the columns. Temperature profiles were measured along the longitudinal axes of the columns during freezing. The experiments included freezing in both the upward and downward directions. Significant salt rejection and brine drainage occurred with downward freezing while there was none with upward freezing. The amount of salt rejected from the freezing region increased with decreasing freezing rate. No evidence of banding was observed on the scale of the sectional sampling used (1 or 2 cm). Salt movement in the partially frozen regions of the columns during downward freezing appeared to be occurring, primarily, by gravity drainage. In the thawed regions, convection, probably by salt fingering, appeared to be the primary mechanism of salt movement.

INTRODUCTION

Salt redistribution during the freezing of saline soils is of importance for both engineering and scientific reasons. For example, development of the offshore petroleum resources of the Beaufort Sea requires information on seabed freezing for the construction of subsea pipelines and engineering design efforts in northern coastal areas require information on ground freezing where the soils are likely to be saline (Heuer et al., 1983). The evaluation of procedures for treating soil freezing problems may also require an understanding of salt redistribution during freezing (Sheeran et al., 1976). From a scientific viewpoint, the physical mechanisms of salt rejection, salt movement, and prediction of salt concentration profiles during freezing are of immediate interest. An understanding of the evolution of subsea permafrost will require an understanding of salt redistribution during seabed freezing (Baker and Osterkamp, in press).

Salt redistribution during soil freezing has not been investigated extensively. Banin and Anderson (1974) have provided some of the thermodynamic relationships. Sheeran and Yong (1975) and Sheeran et al. (1976) investigated long-term effects associated with the application of salts to railroad grades in northern regions. Hallet (1978) argued, on theoretical grounds, that there should be a characteristic ice lens spacing when certain fine-grained soils freeze. In laboratory experiments on the freezing of sands containing fines, Chamberlain (1983) found that a type of banding developed consisting of alternating layers of ice-bonded and unbonded soil. Kay and Groenvelt (1983) conducted a field study and laboratory freezing experiments on sand columns. The field study showed that salt redistribution occurred during natural downward freezing and it was thought that the salt transport was convective. The laboratory study showed that there was no effective salt exclusion from the freezing soil for upward freezing. Wilson (1983) and Mahar et al. (1983) investigated salt redistribution during freezing of gravelly sand and concluded that the degree of salt redistribution depended on the soil solution salinity and the thermal gradient. They also suggested that salt transport should occur in the partially frozen region if a sufficient thermal gradient, and hence, a density gradient (density decreasing with depth) is maintained.

Salt redistribution during soil freezing is a very complex phenomenon which is governed by soil type, soil moisture content, concentration and composition of the soil solution, and the conditions under which the freezing occurs (e.g., temperature, thermal gradient, rate of freezing, direction of freezing, etc.). This paper summarizes the results of unidirectional laboratory freezing experiments with saline sand columns in the downward and upward directions at constant freezing rates.

EXPERIMENTAL DETAILS

The columns used in these experiments were constructed from plastic (tenite butyrate) tubing 6.35 cm O.D., 42 cm long with a 0.32 cm wall thickness. A latex rubber tube was attached to the bottom to saturate the sand and to act as an expansion chamber during freezing. The column was capped and suspended in a refrigerated box as shown in Figure 1. A synchronous motor

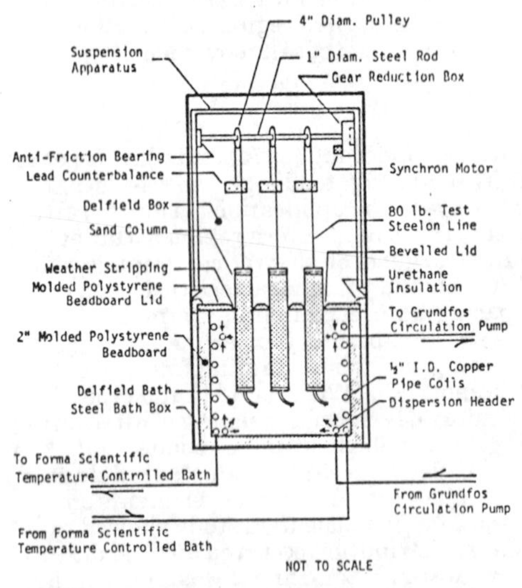

Figure 1. Diagram of apparatus used in laboratory freezing experiments.

was used to simultaneously raise or lower six columns in a controlled temperature bath at the bottom of the box. Freezing in the downward or upward direction could be achieved by adjusting the box and bath temperatures and the direction of motion. One of the six columns was instrumented to measure the *in situ* electrical conductivity of the soil solution and the temperature along the longitudinal axis of the column. The electrodes of the *in situ* electrical conductivity cells were made of sintered bronze filters glued to the wall of the column. Ten cells were placed along the column. Temperatures were measured using seventeen thermistors placed in a 9.5 mm diameter tube which was installed along the longitudinal axis of the column.

A 30 grit silica sand was sieved so that it would pass the 30, but not the 60 sieve size which yields a grain size range from 250 to 600 µm. The sand columns were saturated from the bottom with a 35 ppt (parts per thousand by weight) NaCl solution. Five of them were suspended in the freezing apparatus along with the instrumented column. Downward freezing was achieved by immersing the columns in the bath at -0.6°C and pulling the columns upward into the refrigerated box at about -14°C. Upward freezing was achieved by placing the columns in the box, maintained at about 1.0°C, and lowering them into the bath, maintained at about -15°C. Constant downward freezing rates of 0.10, 0.20, 0.50, 0.97, and 1.93 cm/day and an upward rate of 0.97 cm/day were used. The columns were initially suspended in the freezing apparatus overnight so that several centimeters at the top of the columns were frozen at the start of the experiment.

At various times during the freezing experiment, one of the five columns was removed from the apparatus and cut into sections, 1 or 2 cm in length, which were analyzed for the electrical conductivity of their soil solution using a dilution method (Baker, 1987a). Moisture contents were also measured on these samples. The temperature of the instrumented column and the electrical conductivity of the *in situ* soil solution were monitored daily. Measured values of the electrical conductivity of the soil solution were converted to salinity using an algorithm developed by Baker (1987b). The position of the ice-bonded interface (IBI) was visually determined during sectioning of the columns.

RESULTS AND DISCUSSION

Profiles of the bulk soil solution salinity, S_B, determined from the sectioned columns for downward freezing at a rate of 0.97 cm/day are shown in Figure 2. The position of the IBI, which was observed during sampling, is shown by the dash-dot line and symbol. Differences in values for S_B between the samples obtained by sectioning the columns and *in situ* electrical conductivity

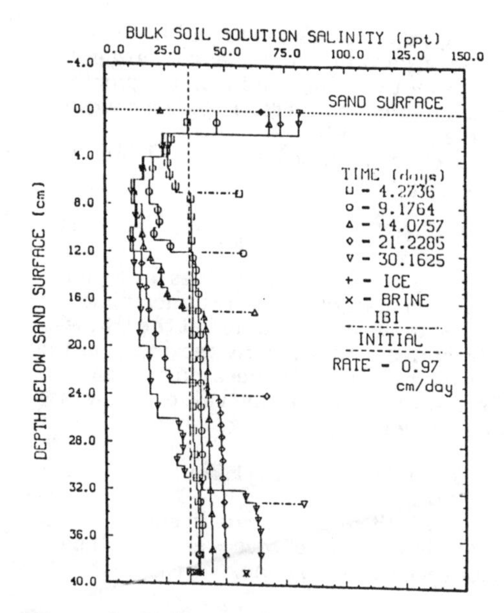

Figure 2. Bulk soil solution salinity, S_B, versus depth for an experiment with a downward freezing rate of 0.97 cm/day.

measurements were usually on the order of a few percent. These profiles are typical of those for other downward freezing rates and show that:

1. Values for S_B in the partially frozen region were significantly less than the initial soil solution salinity indicating that there was a substantial amount of salt redistribution during freezing.

2. At the top of the columns, values for S_B increased with time which may have been due to an upward expulsion of soil solution and/or drying of the top layer of soil by sublimation or evaporation.

3. Values for S_B in the partially frozen region decreased with distance immediately behind the IBI (for a few centimeters to more than ten centimeters) and then became nearly constant.

4. After the IBI passed a fixed point in the column, values for S_B at that point decreased with time. It appears that brine was draining from the partially frozen region, probably by gravity drainage, although some upward expulsion may have been occurring.

5. Values for S_B in the thawed region were always greater than the initial soil solution salinity and nearly uniform, but generally increasing slightly with depth. In addition, values for S_B in the thawed region increased with time. These results indicate that the salt rejected from the ice phase and draining as brine from the partially frozen region was mixing vertically throughout the thawed region. Since diffusion cannot account for such rapid salt transport and mixing, then the profile indicates that convection of the soil solution must have been occurring in the thawed region.

6. A layer with increased salinity (greater than the underlying thawed region) just ahead of the IBI does not exist which indicates that the convective velocity of salt movement must have been greater than the maximum freezing rate (1.93 cm/day). Subsequent experiments showed that salt fingering processes were capable of producing soil solution velocities on the order of centimeters per hour and indicated that salt fingering may be responsible for this convection (Baker and Osterkamp, in press).

7. According to modeling results, brine drainage must have been the primary mode of salt transport in the partially frozen region and that brine expulsion could not account for the observed changes in the salinity profiles with time.

8. Salt banding was not observed visually nor was it indicated by the salinity measurements on the scale of sectional sampling which was 1 or 2 cm. If banding was present, it must have occurred on a much smaller scale.

One experiment was performed with freezing in the upward direction at a rate of 0.97 cm/day. The resulting salinity profiles were uniform with depth and did not show any evidence of salt redistribution during freezing. This indicates that gravity driven brine drainage has a significant effect on salt redistribution during downward freezing for the freezing rates and salt concentrations used in these experiments.

The temperature profiles (for the experiment in Figure 2) are shown in Figure 3. Results shown in Figures 2 and 3 cannot be directly compared since the presence of the thermally conductive temperature probe in the instrumented column may have displaced the position of the IBI as much as several centimeters downwards. The temperature distribution would have also been displaced. These temperature profiles in the instrumented column show considerable curvature with depth. A variety of factors such as radial heat flow, variable thermal properties in the partially frozen region, and distributed latent heat effects (Osterkamp, 1987) may give rise to this type of curvature. The effect of these factors on salt redistribution is unknown.

The temperature profiles for the instrumented column shown in Figure 3 were used to calculate the unfrozen soil solution

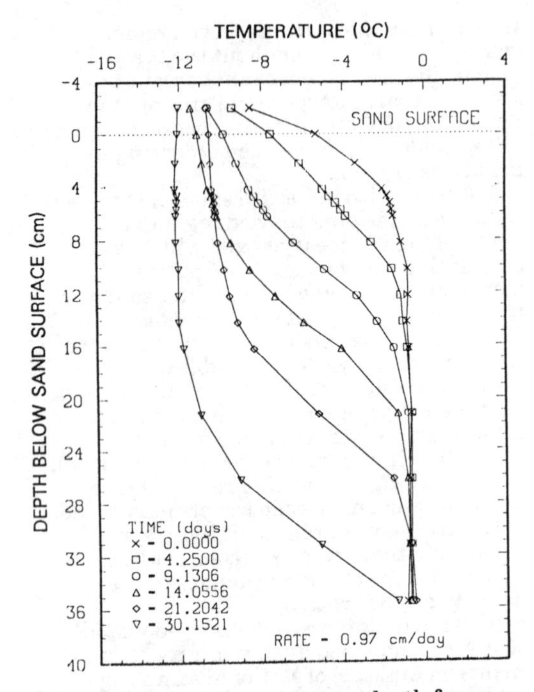

Figure 3. Temperature versus depth for an experiment with a downward freezing rate of 0.97 cm/day.

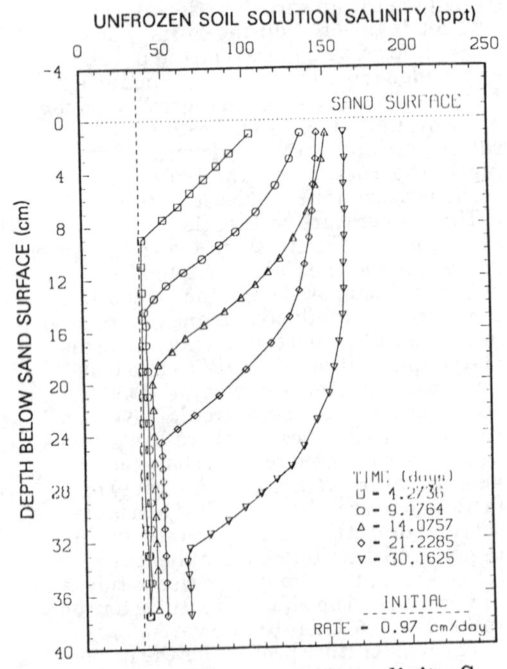

Figure 4. Unfrozen soil solution salinity, S_u, versus depth for an experiment with a downward freezing rate of 0.97 cm/day.

salinity, S_u, profiles shown in Figure 4. These large values for S_u lead to large values for the brine volume in the partially frozen region and therefore to large values for the permeability.

The effective distribution coefficient, k, which is given by

$$k = S_F/S_T \tag{1}$$

where S_F and S_T are the values for S_B for the frozen material and the underlying thawed material, evaluated at the IBI, was determined for downward freezing. For the freezing rates used in these experiments, the best fit to the average values of k was obtained with

$$k = 0.5593 + 0.1341 \ln(v) \tag{2}$$

where v is the freezing rate in cm/day. Additional details of these results will be published at a later time.

CONCLUSIONS

Significant salt redistribution occurred during the downward freezing of saline sand columns while none occurred during upward freezing. Salt redistribution increased with decreasing freezing rates. In the partially frozen regions, salt movement was primarily by gravity drainage although some brine expulsion may have occurred. In the thawed region, a convective process, probably salt fingering, was responsible for movement of the salt. Salt banding was not observed on the scale of the sectional sampling of the columns (1 or 2 cm).

ACKNOWLEDGMENTS

This research was supported by the National Science Foundation, Division of Polar Programs, Polar Earth Science under Grant DPP83-12026, by the U.S. Geological Survey, Dept. of the Interior under Award No. 14-08-0001-G1305, and by State of Alaska funds. We wish to thank J. P. Gosink for comments on an earlier version of this manuscript.

REFERENCES

Baker, G. C., Salt redistribution during freezing of saline sand columns with applications to subsea permafrost, Ph.D. Thesis, 232 p., University of Alaska-Fairbanks, Fairbanks, Alaska, September 1987a.

Baker, G. C., Electrical conductivity, freezing temperature, and salinity relationships for seawater and sodium chloride solutions for the salinity range from 0 to over 200 ppt, Geophysical Institute Report UAG R-309, 87 p., University of Alaska-Fairbanks, Fairbanks, Alaska, 1987b.

Baker, G. C. and T. E. Osterkamp, Implications of salt fingering processes for salt movement in coarse-grained subsea permafrost, Cold Regions Sci. and Tech., in press.

Banin, A. and D. M. Anderson, Effects of salt concentration changes during freezing on the unfrozen water content of porous materials, Water Resour. Res.,10(1), 124-128, 1974.

Chamberlain, E. J., Frost heave of saline soils, Proc. of the Fourth Int. Conf. on Permafrost, Fairbanks, Alaska, pp. 121-126, Nat. Acad. Sci., Washington, D. C., 1983.

Hallet, B., Solute redistribution in freezing ground, Proc. of the Third Int. Conf. on Permafrost, Edmonton, Canada, pp. 86-91, Nat. Res. Council, Ottawa, Canada, 1978.

Heuer, C. E., J. B. Caldwell and B. Zamsky, Design of seafloor pipelines for permafrost thaw settlement, Proc. of the Fourth Inter. Conf. on Permafrost, Fairbanks, Alaska, pp. 486-491, Nat. Acad. Sci., Washington, D.C., 1983.

Kay, B. D. and P. H Groenevelt, The redistribution of solutes in freezing soil: exclusion of solutes, Proc. of the Fourth Int. Conf. on Permafrost, Fairbanks, Alaska, pp. 583-588, Nat. Acad. Sci., Washington, D. C., 1983.

Mahar, L. J., R. M. Wilson and T. S. Vinson, Physical and numerical modeling of uniaxial freezing in a saline gravel, Proc. of the Fourth Int. Conf. on Permafrost, Fairbanks, Alaska, pp. 773-778, Nat. Acad. Sci., Washington, D. C., 1983.

Osterkamp, T. E., Freezing and thawing of soils and permafrost containing unfrozen water or brine, Water Resour. Res., 23(12), 2279-2285, 1987.

Sheeran, D. J. and N. R. Yong, Water and salt redistribution in freezing soils, Proc. of the First Conf. on Soil-Water Problems in Cold Regions, Calgary, Canada, pp. 58-69, Dept. of Water Sci. and Eng., Univ. California-Davis, Davis, CA., 1975.

Sheeran, D. J., C. J. Dalton and N. R. Yong, Field experimentation with chemical alleviation of frost damage, Proc. of the Second Conf. on Soil-Water Problems in Cold Regions, Edmonton, Canada, pp. 173-185, Dept. of Water Sci. and Eng., Univ. of California-Davis, Davis, CA., 1976.

Wilson, R. C., Solute redistribution and freezing rates in a coarse-grained soil with saline pore water, M. S. Thesis, 212 p., Oregon State University, Corvallis, OR, August, 1983.

5th International Symposium on Ground Freezing, Jones & Holden (eds)
© 1988 Balkema, Rotterdam. ISBN 90 6191 824 3

On salt heave of saline soil

X.B.Chen, G.Q.Qiu, Y.Q.Wang & W.K.Shen
Lanzhou Institute of Glaciology & Geocryology, Academia Sinica, People's Republic of China

ABSTRACT: Heavy saline soil is widely distributed in the North-west, Inner Mongolia and elsewhere in China, it is harmful to transportation and hydraulic and civil engineering construction in these areas. A series of experiments were conducted in the laboratory for understanding the physico-mechanical properties of two kinds of heavy saline soil, including clay and loam. Experimental results show that the heavy saline soils, rich in sodium sulphate, will increase in volume due to salt migration and the recrystallization of solutes during cooling process, the so called salt heave. The salt heave is not reversible and the salt heave ratio depends on the temperature interval, initial density, water content, cooling rate, overburden surcharge etc.

Keywords: salt heave

1 INTRODUCTION

Salt in heavy saline soil will migrate from the warm side to the cold side and from the wet end to the dry end under temperature and moisture gradients. A repeated crystallization and dissolution occurs during seasonal temperature fluctuations. All the above phenomena cause the soil expansion and are called salt heave. Salt heave has a harmful effect on transportation, hydraulic and civil engineering works in North-west, Inner Mongolia and North-east China where heavy saline soil is widely distributed (Lo, 1980). The engineers in highway and railway departments have done much work in situ and obtained much experience in dealing with the problems. Recently, saline ground problems have been taken into acount with the development of coastal regions. Internationally, similar problems heave been studied in Arctic off-shore areas because of petroleum exploitation (Chamberlain,1983; Mahar,et al., 1983).

In order to explore the mechanism of salt heave, basic experiments have been conducted in the laboratory of the effect of the effect of temperature and moisture changes in heavy saline soils, such as the phenomenon of salt redistribution, the nonreversibility of salt heave process,and the main factors affecting on salt heave, including temperature interval, density, moisture content, cooling rate and overburden surcharge. Some of the experimental results have been applied to highway design to prevent salt heave damage and are discussed briefly in this paper.

2 SAMPLES AND THE TEST METHOD

Samples were obtained from the Dunhuang Basin at the lower leaches of the Shule River, Gansu Province. Their physical and chemical properties are shown in Table 1. Number I is clay, Number II is loam. They are saline soils of Cl-SO_4-Na-Ca type,7.57% and 15.78% in salinity respectively. It was observed that the minimum temperature in the natural soil profile was -10 °C, which was much higher than the freezing point of soil (T_f is -16.74 °C for clay with a moisture of 22.2% and is -25.26 °C for loam with moisture of 16.0%), so that these soils would not be generally frozen in field condition during the winter.

Samples were compacted in layers in plexiglass cylinders, 13.9 cm in height and with an internal diameter of 11.8 cm at the top and 11.0 cm at the bottom. The internal wall of container and the sample were separated by vasiline and a membrane

Table 1 Physical and Chemical Properties of Soil

No.	% of particle size (in mm)					Soil type	Wp %	Wl %
	.05	.05-.01	.01-.005	.005-.002	.002			
I	1.7	18.2	20.0	23.7	36.4	clay	18.6	27.8
II	48.5	17.8	8.3	9.0	16.4	loam	16.0	21.4

Chemical composition in average (m.e./100g)

No.	Cl^-	$HCo_3^-+CO_3^{--}$	SO_4^{--}	Ca^{++}	Mg^{++}	Na^+	K^+
I	79.41	1.19	42.42	27.99	8.04	85.10	1.89
II	139.30	1.83	110.93	65.89	17.35	166.12	2.70

to reduce the frictional forces.

Specimens are put in a temperature-controllable box to conduct the cooling-warming tests. Some of them are insulated laterally, so that the cooling process is mainly unidirectional downward. Temperatures are monitored by means of thermocouples, deformation was detected by a displacement gauge. Redistribution of salt and moisture after the cooling testa was examined by layers.

3 EXPERIMENTAL RESULTS AND ANALYSES

3.1 Salt redistribution during cool-warming processes

Fig.1 and Fig.2 are the profiles of ion migration in cooling saline clay and loam respectively under the action of temperature gradient. The results show that salt migrates towards the colder side during cooling.

From Fig.1 and 2, we know that an intensive salt heave zone produced by ion migration and recrystallization is located in the up layer of samples with a thickness of 2 cm. It is obvious that the increment of Na^+, K^+ is the greatest in cation migration, the next is Mg^{++}, and in anion movement, the increment of Cl^- is the most intensive, the next one is SO_4^{--}.

3.2 Main factors affecting on salt heave

1. Temperature intervals: During the cooling process, salt heave behavior varies at different temperature intervals. If the salt heave in a unit height of soil column is defined as the salt heave ratio (%) and is used to express the intensity of heave, then it can be seen that the clay or loam have characteristic temperature intervals at which the peak value of salt heave ratio occurs. Statistics for 17 groups of clay tests and 33 groups of loam test results are shown in Table 2.

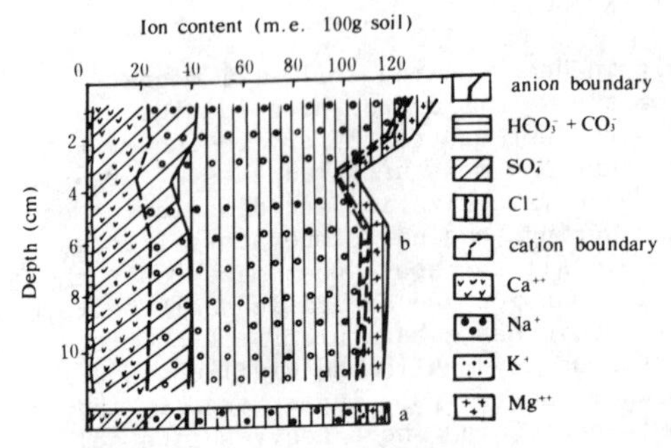

Fig.1 Ion migration in cooling saline clay under the action of temperature gradient (a--ion distribution before cooling; b--ion distribution after cooling)

In the clay case, the largest salt heave ratio occurs in the interval of +1 to -1 °C, although some samples begin to heave at 4 to 6 °C, and at intervals of -4 to -6 °C and -9 to -11 °C. For the loam case, as the temperature cools down to about 9 °C, it is possible for salt heave to occur, but the most intensive salt heave occurs in an area of +4 to +7 °C. Furthermore, the specimens have a stronger salt heave

Table 2 Frequency of peak value of salt heave in different temperature intervals

The first peak area

Temp.°C / No.	0-1	1-2	2-3	3-4	4-5	5-6	6-7	7-8	8-9	9-10
I	35			7	30	23				
II	9				31	39	9			12

The second peak area

Temp.°C / No.	0 – -1	-1 – -2	-2 – -3	-3 – -4	-4 – -5	-5 – -6	-6 – -7	-7 – -8	-8 – -9	-9 – -10	-10 – -11
I	13				37					37	13
II					33	23			3	25	16

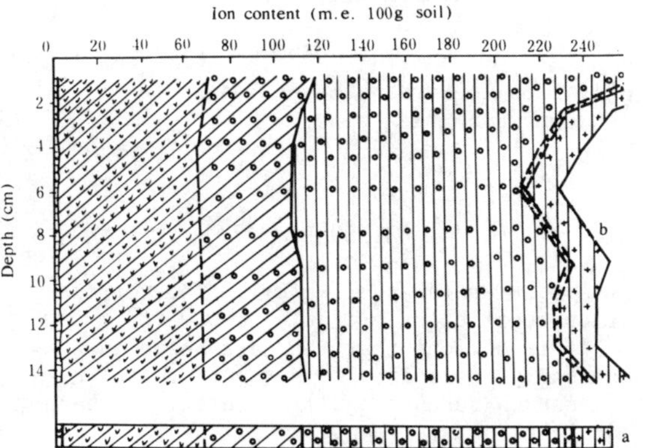

Fig.2 Ion migration in cooling saline loam under the action of temperature gradient (meaning of symbols as the same as Fig.1)

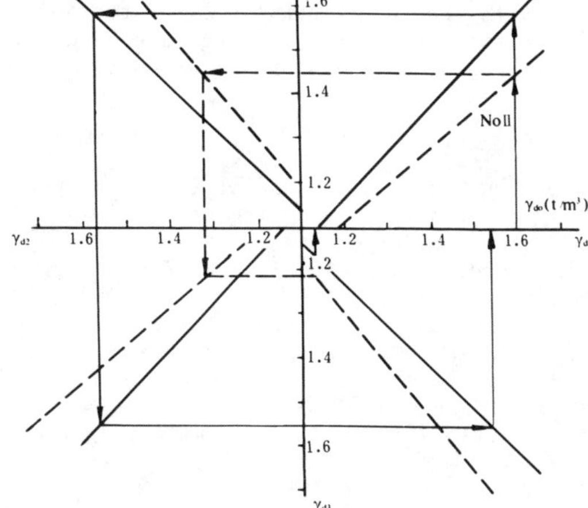

Fig.3 Density change of clay and loam after cool-warming cycles (1--one cycle; 2--two cycles; 3--three cycles; 4--four cycles)

at in intervals of -4 to -6 °C and -9 to -11 °C.

2. Density: Salt heave during the cooling process will lower the soil density. This is closely related to the initial density γ_{do}. Results of cooling tests from +15 °C down to -15 °C with a cooling rate of V_T= 0.05 °C/hr are selected for comparison.

In the clay case, as the initial water content is about 20 %, it is known by statistics based on the least square method that there is a linear relationship between dry density after cooling (γ_d) and the initial density before cooling (γ_{do}) (see Fig.3 and Equation 1).

$$\gamma_d = -0.0101 + 1.056\,\gamma_{do} \qquad (1)$$

The correlation coefficient of Equ.1 is R=0.993.

By Equ.1 it is known that if $\gamma_{do} > 1.7 t/m^3$, the difference between γ_d and γ_{do} is very small, so the clay behaves as a nonsalt-heaved one.

Density change of loam during cooling test is shown in Fig.3 and Equ.2. The initial water content is 16 to 18 %.

$$\gamma_d = 0.107 + 0.836\ \gamma_{do} \qquad (2)$$

The correlation coefficient R of Equ.2 is 0.927.

Equation 1 indicates that in the clay case, γ_d will decrease with an increase of γ_{do}, in other words, an increase in γ_d can restrain the salt heave; while in the loam case, based on Equation 2, since $\Delta\gamma_d$ increases with γ_{do}, the salt heave can not be controlled by means of increasing γ_{do}. However, experiments show that the density change tends to be smaller up when the cooling rate is as fast as 0.5 °C/hr and the γ_{do} 1.7 t/m³. Under such a condition, the salt heave behavior of loam could be improved by controlling the initial density.

From Fig.4 and 5, we know that the density of both saline clay and loam will decrease after each cool-warming cycle.The decrement of γ_d for loam is very large while for clay it is relative small.

3. Cooling Rate: Saline loam is taken as an example. When it is 1.62 to 1.67 t/m³ in initial dry density, 16 to 18 % in water content, the relationship between the salt heave ratio η and the cooling rate V_T can be expressed as Fig.4 and Equation 3.

$$\eta = 1.431\ V_T^{-0.613} \qquad (3)$$

where, η --salt heave ratio, %;
V_T = dT/dt----cooling rate, °C/hr;
t--time, hour.

The correlation coefficient R of Equ.3 is 0.805.

In form, the Equation 3 is similar to the relationship between frost heave ratio and frost penetration rate (Chen, et al., 1983). Fig.4 and Equ.3 show that the salt heave ratio will acutely increase with the decrease in cooling rate. Therefore,when we study the physical properties of saline soil, it is necessary to control the corresponding cooling rate.

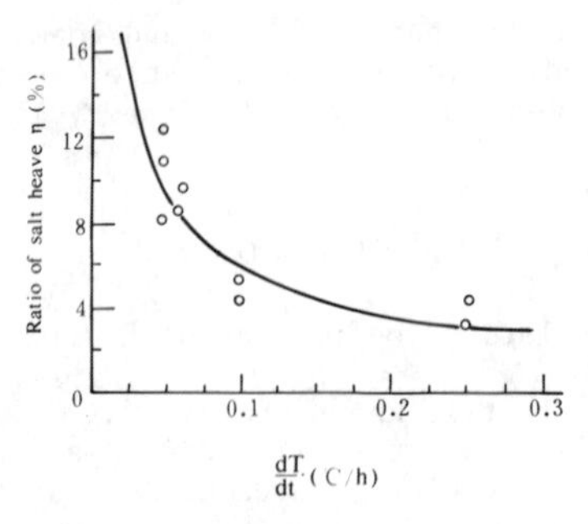

Fig.4 Ratio of salt heave (η) vs cooling rate (V_T)

4. Overburden surcharge: Overburden surcharge includes all kinds of overburden pressure, including the self-weight of soil acting on the subsurface of a soil layer. Of course, any subsurface within a subgrade is under a certain surcharge.

The relationship between salt heave ratio of loam, for 1.5 t/m³ in initial density and 17.7 % in water content, are shown in Fig.5 and Equation 4.

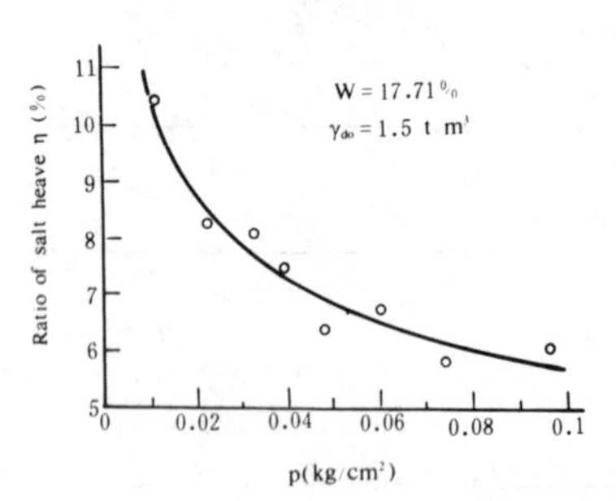

Fig.5 Salt heave ratio of loam vs overburden surcharge

$$\eta = 3.189\ P^{-0.255} \qquad (4)$$

Where, P--surcharge, kg/cm². The correlation coefficient of Equ.4 is 0.931.

It is obvious that the salt heave is strongly restrained by surcharge. An increase in surcharge will result in a sharp decrease of salt heave ratio. This is analogous to the case of frost heave retrained by surcharge (Chen, et al., 1983). Therefore, a study of salt heave of ground should be processed in steps: first, to evaluate the salt heave of each layer,then, to sum up the total possible value.

5. Water content: Experimental results show that as the cooling rate, dry density and surcharge stress are constant, the ratio of salt heave will reduce with the decrease of water content (Fig.6). This is because as the water content decreases the amount of dissolved solutes in pore water will decrease also; correspondingly the salt heave becomes weakened.

Fig.6 shows that there would be a critical value of water content, below which the soil will lose its ability to salt heave. Such a critical value can be defined as the critical water content for salt heave. In the loam case, for a dry

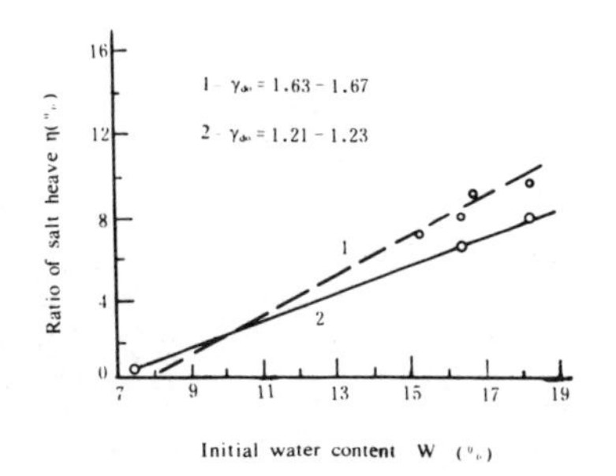

Fig.6 Ratio of salt heave of loam vs water content

density of 1.63 to 1.67 t/m^3, the critical water content for salt heave is about 8 %, for a dry density of 1.21 to 1.23 t/m^3, it is about 7 %.

4 CONCLUSIONS

As the saline soils cool, a series of changes will occur: parts of the solutes migrate towards the colder end; crystallization of sodium sulphate results in a salt heave with a nonreversible behavior, occurring in certain temperature intervals; the salt heave makes the soil reduce in compressive strength. All of these cause harmful effects to engineering construction.

The behavior of salt heave of a soil depends on a series of factors, namely, the temperature intervals, soil type, initial density, cooling rate, water content and surcharge. In this experiment, the salt heave in the loam case is one order higher than that in the clay case. There is a linear relationship between final dry density and the initial density. With the increase in cooling rate, the salt heave ratio attenuates following a power-function law. Salt heave will increase with water content. As the surcharge increases, the salt heave ratio attenuates following a power-function law also. An understanding of those laws is helpful to prepare measures to restrain the salt heave of soils.

(This program is supported by the research foundation from the Academia Sinica)

REFERENCES

Chamberlain, E.J., 1983. Frost heave of saline soils, Proc.4th International Conference on Permafrost, I:121-126.

Chen Xiaobai, et al., 1983. Influnence of frost penetration rate and surcharge stress on frost heaving, Proc.2nd National Conference on Permafrost, p.222-228.

Lo Weipu, 1980. Highway Engineering in saline Soil Regions, People's Transportation Pressing House. p.110.

Mahar, L.J., et al., 1983. Physical and numerical modeling of uniaxial freezing in a saline gravel, Proc. 4th International Conference on Permafrost, I:773-778.

5th International Symposium on Ground Freezing, Jones & Holden (eds)
 ISBN 90 6191 824 3

On a model for quasi-steady freezing processes of saturated porous media

E.Comparini
Istituto Matematico 'U.Dini', Università di Firenze, Italy

ABSTRACT: In a previous paper Fasano & Primicerio studied a modification of O'Neill-Miller's model for ground freezing, with the following aims:
(i) getting easy formulas for the location of the freezing front (i.e. the boundary of the so called frozen fringe) and for the water discharge in the quasy-steady approximation and planar geometry;
(ii) determining the location of possible ice lenses under the same circumstances;
(iii) studying the dependence of the physical quantities entering the model (e.g. water discharge) on the ratio between the density of the solid and the liquid phase. One of the advantages of this investigation is to estimate the error made in using the simpler formulas obtained setting the above ratio equal to one.

In this paper the same objectives are pursued, starting from a less crude approximation of O'Neill & Miller's model and proving that the results depend in a continuous way on the various parameters introduced.

INTRODUCTION

The mathematical model for freezing processes of saturated porous media proposed by O'Neill & Miller (1982) was based on the fundamental idea that the induced water flow is highly influenced by the thermodynamical structure of the so-called frozen fringe. The frozen fringe is a zone in which water and ice coexist in a proportion depending on temperature, pressure, and on some physical properties of the system, including an empirical relationship between the ice fraction and the average curvature of the ice-water interface (called the characteristic freezing curve).

In such a framework the water flow mechanism and the possible formation of ice lenses can be explained. For more references see Fasano & Primicerio (1978).

For sufficiently slow cooling processes quasi-steady approximations make sense (Black & Miller (1985), Fasano & Primicerio (1987), Holden, Piper &Jones (1985)).

In the paper of Fasano & Primicerio (1987) the scheme was reduced to a problem with one free boundary, across which the ice fraction jumps from zero to a given characteristic value. This was accomplished by introducing some changes with respect to O'Neill & Miller (1982), mainly acting on the characteristic freezing curve, and assuming approximations yielding explicit simple formulas for the main quantities involved in the phenomenon.

In particular the modified model provided a quick and reasonable qualitative description of the growth of ice lenses.

In the present paper we want to refine the model proposed by Fasano & Primicerio (1987), using less crude approximations and following the original scheme of O'Neill & Miller (1982) more closely. As a result two fronts (instead of just one) will have to be determined, besides the water velocity through the frozen fringe. Moreover we want to prove a continuous dependence of the unknowns on the main parameters entering the model, and finally we describe the growth of a sequence of ice lenses, following the outline of Fasano & Primicerio (1987).

As a conclusion we can say that the results obtained here confirm in most cases the validity of the simpler model used by Fasano & Primicerio (1987).

1 STATEMENT OF THE PROBLEM

Let us suppose that the freezing process has

plane horizontal symmetry. Then any dependent variable is a function of the vertical coordinate z, positive upwards, ($z=0$ at the bottom of the layer, $z=L$ at the top). Throughout the paper the layer thickness L will be supposed constant (prior to the formation of an ice lens), irrespective of the frost penetration.

According to O'Neill & Miller (1982), the equations describing the model are obtained from mass and energy conservation using the following physical assumptions:

(I) Darcy's law to express the volumetric water velocity v:

$$v=-\frac{\tilde{D}(w)}{\rho g}\left(\frac{\partial p}{\partial z}+\rho g\right), \quad g=9.8\,\mathrm{m/sec}^2, \tag{1.1}$$

where w denotes the fraction of pore volume occupied by water ($0\leq w\leq 1$), ρ the water density, p its pressure, $\tilde{D}(w)$ the hydraulic conductivity of the medium. Here we assume the same approximations as O'Neill & Miller (1982), i.e. $\tilde{D}(w)=D_0 w^7$, D_0 constant.

(II) Clapeyron's law in the region $0<w<1$

$$\frac{p}{\rho}-\frac{p_i}{\rho_i}=\frac{\lambda\theta}{T_0}, \qquad \theta=T-T_0, \tag{1.2}$$

where p_i and ρ_i are the ice pressure and density respectively, λ is the latent heat, T is the temperature and T_0 is the melting point (^{0}K) under standard conditions.

(III) An empirical law introduced by O'Neill & Miller (1982), expressing the pressure jump between the two phases

$$p_i-p=\omega\tilde{\psi}(w), \tag{1.3}$$

where ω is the ice-water surface tension and $\tilde{\psi}$ is an empirically determined function, related to the mean curvature of the ice-water interface.

The evolution of the system can be studied through a sequence of steady states, supposing that the thermal gradient changes sufficiently slowly in time, and noting that the perturbation induced by slowly growing lenses is generally very small. The differential equations which govern the freezing process are

$$\frac{\partial}{\partial z}\left\{-\frac{D_0}{\rho g}w^7\left(\frac{\partial p}{\partial z}+\rho g\right)\right\}=0, \tag{1.4}$$

$$\frac{\partial}{\partial z}\left\{K(w)\frac{\partial\theta}{\partial z}\right\}=0, \tag{1.5}$$

where K(w) is the thermal conductivity, with the boundary conditions

$$\theta=\theta_a<0,\ p_i=p_a,\ \text{at}\ z=L;$$
$$\theta=\theta_b>0,\ p=p_b=p_a+\rho g L,\ \text{at}\ z=0, \tag{1.6}$$

p_a being a constant reference pressure (e.g. atmospheric).

In order to solve the problem analytically, we consider an approximation ψ of the function $\tilde{\psi}$, more general than the one used by Fasano & Primicerio (1987), namely

$$\psi(w)=\begin{cases}\psi_0\left(\frac{w_0}{w}\right)^{m/n}, & 0<w<w_0,\\ \psi_0, & w_0<w<w_1,\\ \psi_1+(\psi_0-\psi_1)\left(\frac{1-w}{1-w_1}\right)^{v/u}, & w_1<w<1,\end{cases} \tag{1.7}$$

$w_0<<1,\ 1-w_1<<1,$

with m, n, u, v positive integers such that $m>n$, $u>v$, and $0<\psi_1<\psi_0$.

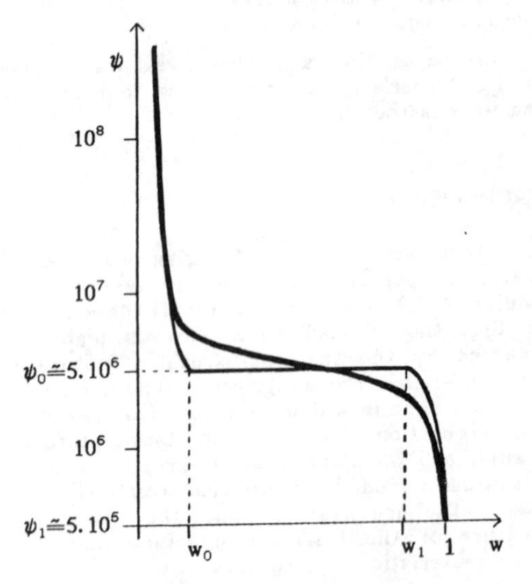

Fig 1. Freezing characteristic curve

On the basis of approximation (1.7), we can divide the layer into three parts, separated by two distinct fronts: a zone where only water is present ($0\leq z<\sigma$, $w\equiv 1$), a zone with a small amount of ice ($\sigma<z<s$, $w_1<w<1$) and a third one where the ice occupies almost all the volume ($s<z\leq L$, $0<w<w_0$). On the front $z=s$ the quantity w jumps from the value w_0 to the value w_1, while w is continuous across the front $z=\sigma$.

In the paper of Fasano & Primicerio (1987) heat conductivity was taken constant throughout the medium. Here, as a milder simplification, we will assume that the thermal conductivity K jumps across the front $z=s$ from a value K_0, assumed throughout the "frozen" zone $s<z\leq L$, to a value K_1, assumed throughout the remaining part of the medium. The constants K_0, K_1 depend on the structure of the porous medium (see De Vries (1963)).

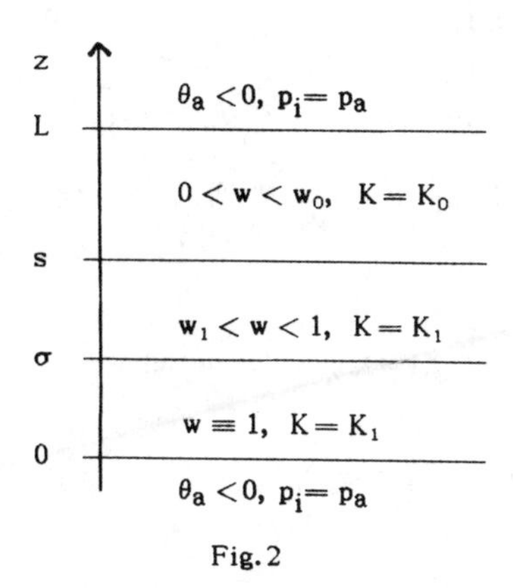

Fig. 2

Therefore we can integrate equation (1.5) to obtain:

$$\theta(z)=\theta_b-G_1 z, \qquad 0\leq z\leq s;$$
$$\theta(z)=\theta_a-G_0(z-L), \qquad s\leq z\leq L, \qquad (1.8)$$

$G_1=\frac{\theta_b-\theta_s}{s}$, $\quad G_0=\frac{\theta_s-\theta_a}{L-s}$, where θ_s represents the temperature on the still unknown freezing front, i.e.

$$\frac{\theta_s}{\theta_b}=\frac{1-s/L(1+\mathcal{K}(-\theta_a)/\theta_b)}{1+s/L(\mathcal{K}-1)}, \quad \mathcal{K}=\frac{K_0}{K_1}. \qquad (1.9)$$

So far this procedure is largely formal, since the problem of finding the front $z=s$ may not have a solution (this case is discussed in sec. 8). For the time being we go on assuming that s exists.

We shall denote by θ_σ the temperature at the front $z=\sigma$.

As to equation (1.4), it can be immediately integrated in the zone where there is no ice ($w\equiv 1$) to obtain

$$p(z)=p_b-(q+\rho g)z, \quad 0<z<\sigma, \qquad (1.10)$$

where $q\geq 0$ is an unknown constant ultimately determining the water velocity

$$v=\frac{D_0}{\rho g}q. \qquad (1.11)$$

Imposing the continuity of v throughout the medium, (1.4) yields

$$-w^7\left(\frac{dp}{dz}+\rho g\right)=q, \qquad \sigma<z<L. \qquad (1.12)$$

Following the methods of Fasano & Primicerio (1987), we introduce the quantity $\delta=\frac{\rho-\rho_i}{\rho}$, and we first study the problem in the case $\delta=0$. After that, we will derive expressions for the location of the fronts and the water discharge q as functions of δ, when $\delta\neq 0$.

We recall that (1.2), (1.3) can be used in the form:

$$p=-[\omega\psi(w)+C\theta]/\delta,$$
$$p_i=-[(1-\delta)\omega\psi(w)+C\theta]/\delta, \ \delta\neq 0,$$
$$0<w<1, \qquad (1.13)$$

where $C=\frac{\rho_i\lambda}{T_0}$, while in the case $\delta=0$ we obtain from (1.2), (1.3):

$$\theta=-\frac{\omega}{C}\psi(w), \quad 0<w<1. \qquad (1.14)$$

2 THE BASIC EQUATIONS

In order to solve equations (1.8), (1.10), (1.12) with boundary conditions (1.6), we make use of the continuity of the pressure p and of the volumetric water velocity v across the fronts. Moreover, considering approximation (1.7), we can integrate equation (1.12) separately in the zones $\sigma<z<s$ and $s<z<L$.

Neglecting the gravity term everywhere, which is orders of magnitude smaller than the other terms, the following system of equations is obtained:

$$\frac{\sigma}{L}=\frac{\theta_b-\theta_1+p_a\delta/C}{LG_1+qL\delta/C}, \qquad (2.1)$$

$$\frac{s}{L}-\frac{\sigma}{L}=v\frac{\theta_1-\theta_0}{LG_1}\left(\frac{w_a}{1-w_1}\right)^{v/u}\times$$
$$\times\int_0^{\left(\frac{1-w_1}{w_a}\right)^{1/u}}\frac{\left(\frac{1}{w_a}-t^u\right)^7 t^{v-1}}{\left(\frac{1}{w_a}-t^u\right)^7+\frac{q\delta}{CG_1}w_a^{-7}}dt, \qquad (2.2)$$

$$1-\frac{s}{L}=-\frac{m}{LG_0}\frac{p_a\delta/C+\theta_a}{1-\delta}\times$$

$$\times\int_1^{(\frac{w_0}{w_a})^{1/n}}\frac{t^{7n-m-1}}{t^{7n}+\frac{q\delta}{CG_0}w_a^{-7}}\,dt, \qquad (2.3)$$

where $\theta_0=-\frac{\omega}{C}\psi_0$, $\theta_1=-\frac{\omega}{C}\psi_1$, and

$$w_a=w\Big|_{z=L}=w_0\left|\frac{(1-\delta)\,\omega\psi_0/C}{p_a\delta/C+\theta_a}\right|^{n/m}.$$

Thus we have a system for the determination of the unknowns σ, s, q. Even if the integrals on the r.h.s. of (2.2), (2.3) can be calculated explicitly, solving (2.1)-(2.3) is difficult because G_0, G_1 depend on s, which was not the case of Fasano & Primicerio (1987).

Notice that (2.1) makes sense only if the r.h.s. is less than s/L, that is, in practical cases, if $\theta_1>\theta_0>\theta_a$.

3 THE CASE $\delta=0$

Performing the limit for $\delta\to 0$ in (2.1) and (1.9), setting $\sigma_0=\lim_{\delta\to 0}\sigma$, $s_0=\lim_{\delta\to 0}s$, and noticing that $\theta_{s_0}=\theta_0$, $\theta_{\sigma_0}=\theta_1$, we obtain:

$$\frac{\sigma_0}{L}=\frac{\theta_b-\theta_1}{\theta_b-\theta_0+\mathcal{K}(\theta_0-\theta_a)},$$

$$\frac{s_0}{L}=\frac{\theta_b-\theta_0}{\theta_b-\theta_0+\mathcal{K}(\theta_0-\theta_a)}. \qquad (3.1)$$

Concerning the water discharge, in the case $\delta=0$, one can immediately integrate (1.12) in the two zones $\sigma<z<s$ and $s<z<L$, recalling (1.3), (1.14) and (1.7), and using the continuity of water pressure throughout the medium, to obtain:

$$q=\frac{\omega\psi(w_a)}{\sigma_0+I_0-\frac{m}{7n+m}\frac{w_0^{-7}}{G_0^{0}}\theta_0^{\frac{-7n}{m}}(\theta_a^{\frac{7n}{m+1}}-\theta_0^{\frac{7n}{m+1}})}, \qquad (3.2)$$

where

$$I_0=(s_0-\sigma_0)^{\frac{7u}{v}}\int_{\sigma_0}^{s_0}[(s_0-\sigma_0)^{\frac{u}{v}}-(1-w_1)(z-\sigma_0)^{\frac{u}{v}}]^{-7}dz$$

and

$$G_0^{0}=G_0\Big|_{s=s_0}=\frac{\theta_b-\theta_0+\mathcal{K}(\theta_0-\theta_a)}{\mathcal{K}\,L},$$

$$s_0-\sigma_0<I_0<\frac{s_0-\sigma_0}{w_1^{-7}}.$$

In (3.2), we can neglect the two terms σ_0 and I_0 which are much smaller than the third one, and with the further hypothesis

$$(\frac{\theta_0}{\theta_a})^{7n/m+1}<<1, \qquad (3.3)$$

we obtain the value

$$q\doteq q_0=\frac{7n+m}{m}\,w_0^{7}G_0^{0}C\,(\frac{\theta_0}{\theta_a})^{7n/m}. \qquad (3.4)$$

4 CASE $\delta\neq 0$

Adding equations (2.1)-(2.3) and defining the new unknown

$$\eta=\frac{q\delta}{CG_0}w_a^{-7}=\frac{q\delta}{CG_1}\mathcal{K}\,w_a^{-7}, \qquad (4.1)$$

we obtain the equation

$$A(\eta,\delta)+C(\eta,\delta)-B(\eta,\delta)\equiv F(\eta,\delta)=0, \qquad (4.2)$$

where

$$A(\eta,\delta)=m\int_1^{(\frac{w_0}{w_a})^{1/n}}\frac{t^{7n-m-1}}{t^{7n}+\eta}\,dt,$$

$$C(\eta,\delta)=\frac{1-\delta}{p_a\delta/C+\theta_a}\,v\,(\theta_0-\theta_1)(\frac{w_a}{1-w_1})^{v/u}\times$$

$$\times\int_0^{(\frac{1-w_1}{w_a})^{1/u}}\frac{(\frac{1}{w_a}-t^u)^7\,t^{v-1}}{(\frac{1}{w_a}-t^u)^7+\frac{1}{\mathcal{K}}\eta}\,dt.$$

$$B(\eta,\delta)=\frac{1-\delta}{p_a\delta/C+\theta_a}\times$$

$$\times\Big[\frac{\theta_b-\theta_1+p_a\delta/C}{1+\frac{1}{\mathcal{K}}\eta\,w_a^{7}}-(\theta_b-\theta_a)\Big]. \qquad (4.3)$$

Notice that $F(0,\delta)=\delta$, and that (since $\theta_b-\theta_1+p_a\delta/C>0$)

$$\frac{\partial F}{\partial\eta}<0,\quad \lim_{\eta\to -1^+}F(\eta,\delta)=+\infty,$$

$$\lim_{\eta \to +\infty} F(\eta,\delta) = \frac{1-\delta}{p_a\delta/C+\theta_a}(\theta_b-\theta_a) < 0.$$

Therefore equation (4.2) can be solved obtaining the solution η as a function of δ ($\eta \in (-1,\infty)$, $-1<\delta<1$). The solution $\eta(\delta)$ is unique, it has the same sign as δ, and then the corresponding value of q is always positive.

We will look for such a solution, showing that the following approximations are valid in practical cases, provided we restrict the range of values of θ_a by means of assumption (3.3):

$$B(\eta,\delta) \doteq B(0,\delta) \equiv B_0,$$
$$C(\eta,\delta) \doteq C(0,\delta) \equiv C_0. \tag{4.4}$$

In fact we have that, if $\delta < 0$, then $-1 < \eta < 0$ and so $|\eta\, w_a^7| < w_0^7 << 1$.

If $\delta > 0$, we set $\tilde{\eta} = \frac{\delta}{B_0 - C_0}\left(\frac{w_0}{w_a}\right)^7$, with the hypothesis $B_0 - C_0 > 0$, corresponding to the following assumption (verified if (3.3) holds)

$$\left|\frac{\theta_0}{p_a\delta/C+\theta_a}\right| < 1. \tag{4.5}$$

Being $A(\tilde{\eta},\delta) - B_0 + C_0 < 0$, then

$$\eta\, w_a^7 < \tilde{\eta}\, w_a^7 = \frac{\delta}{1-\delta}\,\frac{\theta_a + p_a\delta/C}{p_a\delta/C+\theta_a-\theta_0}\, w_0^7,$$

which is negligible w.r.t. 1. As to the second of (4.4), we can show that $C_0 - C(\eta,\delta)$ is less than $\frac{1}{\mathcal{K}}\,\frac{\theta_0-\theta_1}{p_a\delta/C+\theta_a-\theta_0}\left(\frac{w_0}{w_1}\right)^7\delta$, which is negligible w.r.t. δ (recall $F(0,\delta)=\delta$).

From (4.4), we can study equation (4.3) in the form

$$A(\eta,\delta) = A_0 - \delta, \qquad A_0 \equiv A(0,\delta), \tag{4.6}$$

and apply the methods of Fasano & Primicerio (1987), recalling approximation (3.3).

Notice that the derivatives

$$D^i_\eta A(\eta,\delta)\Big|_{\eta=0} = (-1)^i \frac{m\, i!}{7ni+m}\left[1-\left(\frac{w_a}{w_0}\right)^{7i+m/n}\right]$$
$$\doteq q(-1)^i \frac{m\, i!}{7ni+m}, \quad i = 1, 2, \ldots$$

do not depend on δ. Then

$$A(\eta,\delta) = A_0 + m \sum_{i=1}^{\infty} (-1)^i \frac{\eta^i}{7ni+m}. \tag{4.7}$$

From (4.6) and (4.7), equation (4.3) is equivalent to

$$\delta = m \sum_{i=1}^{\infty} (-1)^{i-1} \frac{\eta^i}{7ni+m}, \tag{4.8}$$

where the series on the r.h.s. is convergent because $\eta \in (-1,1]$.

The tentative expansion

$$\eta(\delta) = \sum_{j=1}^{\infty} \lambda_j\, \delta^j \tag{4.9}$$

yields recursive formulas for the coefficients:

$$\lambda_1 = 7n/m + 1,$$
$$\lambda_2 = \frac{7n/m+1}{14n/m+1}\lambda_1 = a_2(n/m)\,\lambda_1^2, \ldots$$
$$\lambda_i = a_i(n/m)\lambda_1^{\,i} = a_i(n/m)\,(7n/m+1)^i, \ldots$$

Remark : The coefficients λ_j depend on the ratio n/m only.

Going back to (2.1), (2.2) and recalling (4.4), we obtain

$$\frac{\sigma}{L} \doteq \frac{\sigma_0}{L} + \frac{p_a\delta/C}{\theta_b-\theta_0+\mathcal{K}(\theta_0-\theta_a)}\left[1+(\mathcal{K}-1)\frac{\sigma_0}{L}\right], \tag{4.10}$$

$$\frac{s}{L} \doteq \frac{s_0}{L} + \frac{p_a\delta/C}{\theta_b-\theta_0+\mathcal{K}(\theta_0-\theta_a)}\left[1+(\mathcal{K}-1)\frac{s_0}{L}\right]. \tag{4.11}$$

Notice that the influence of δ amounts to an $O(\delta/C)$ perturbation.

Now we can express also q as a perturbation of the quantity q_0 found in (3.4). From (4.1), recalling expansion (4.9) and setting

$$Q(\delta) \equiv \frac{q-q_0}{q_0} \tag{4.12}$$

we have

$$Q(\delta) = \sum_{j=1}^{\infty} \frac{m}{7n+m}\Big\{\lambda_{j+1} - \frac{(\mathcal{K}-1)\,p_a\delta/C}{\theta_b-\theta_0+\mathcal{K}(\theta_0-\theta_a)}\lambda_j\Big\}\delta^j. \tag{4.13}$$

Notice that δ is an analytic function of η for $|\eta|<1$ in the complex plane and that (4.8) is equivalent to

$$\delta = 1 - m\int_1^{\infty} \frac{t^{7n-m-1}}{t^{7n}+\eta}\,dt. \tag{4.14}$$

Then, setting $\eta = \xi + i\zeta$, we can see that when η goes from -1 to 1 on the real axis, δ increases on the real axis from $-\infty$ to some

$\delta_0 \doteq 0.147$, and $\delta'(\eta) \neq 0$ in the circle of radius 1. Then the inverse function $\eta(\delta)$ is analytic in the image of $|\eta| < 1$. A representation of the image of $|\eta| < 1$ in the complex plane (obtained with numerical methods), shows that the series (4.13) is convergent for $|\delta| < \delta_0$.

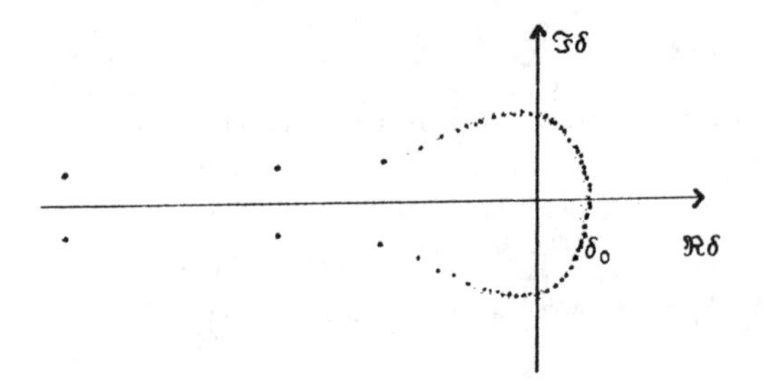

Fig.3 $\delta_0 \doteq 0.147$

5 CONTINUOUS DEPENDENCE ON THE DATA

In this section we are going to study how the solution of the set (2.1)-(2.3) depends on each of the data of the problem. In the following, both the varied parameters and the corresponding solutions will be indicated with a tilda.

We state the following results:

-(A) The most interesting case is a variation of the exponents m/n, v/u in (1.7). Let us set $\alpha = m/n$, $\beta = v/u$ for simplicity (recall $\alpha > 1$, $\beta < 1$).

Considering the case $\delta = 0$, we obtain that $\tilde{s}_0 = s_0$, $\tilde{\sigma}_0 = \sigma_0$ and $|\tilde{q}_0 - q_0| \leq \text{const}\ |\tilde{\alpha} - \alpha|$, that is q_0 is independent of β.

If $\delta \neq 0$, although rather formal, the best procedure is looking for the dependence of the coefficients λ_j in (4.9), being $\tilde{\eta} \neq \eta$. In fact following the methods of sec.4, we have

$$\tilde{\eta}(\delta) = \sum_{j=1}^{\infty} \tilde{\lambda}_j \delta^j, \quad \tilde{\lambda}_j = a_j(1/\tilde{\alpha}) \tilde{\lambda}_1^{\,j} = \tilde{\lambda}_j(1/\tilde{\alpha}).$$

Then

$$|\tilde{\lambda}_j - \lambda_j| \leq \text{const}\ |\tilde{\alpha} - \alpha|. \tag{5.1}$$

From the assumptions in sec.4 which led to (4.10),(4.11), we have that s and σ do not depend on α, β, then $\tilde{s} = s$, $\tilde{\sigma} = \sigma$. Recalling (4.1), the result obtained for η extends to q, namely

$$|\tilde{q} - q| \leq \text{const}\ |\tilde{\alpha} - \alpha|. \tag{5.2}$$

-(B) Varying parameters w_0, w_1 does not affect the location of the fronts, that is $s = \tilde{s}$, $\sigma = \tilde{\sigma}$. Concerning $\tilde{\eta}$, we note that it solves equation (4.2) (recall the approximations done in sec. 3 and notice that the ratio w_0/w_a does not depend on the parameters w_0, w_1). Therefore $\eta = \tilde{\eta}$ and finally

$$\begin{aligned} \tilde{q} - q &= (Q(\delta) + 1)(\tilde{q}_0 - q_0) \\ &= (Q(\delta)+1) \frac{7n+m}{m} G_0{}^0 C\left(\frac{\theta_0}{\theta_a}\right)^{7n/m} (\tilde{w}_0{}^7 - w_0{}^7), \end{aligned} \tag{5.3}$$

with $Q(\delta)$ defined in (4.13). Notice that the water discharge is larger if a larger fraction of pore volume is occupied by water for $z \to s_+$.

-(C) Varying ψ_0, ψ_1 (and consequently θ_0, θ_1) we obtain, in the case $\delta = 0$,

$$|\tilde{q}_0 - q_0| \leq \text{const}\ (|\tilde{\theta}_0 - \theta_0| + |\tilde{\theta}_1 - \theta_1|), \tag{5.4}$$

and analogous estimates for $|\tilde{s}_0 - s_0|$ and $|\tilde{\sigma}_0 - \sigma_0|$.

Now, if $\delta \neq 0$, with the methods used in the case (A), and recalling the Remark in sec.4, we obtain $\tilde{\eta} = \eta$. Moreover, (4.10) and (4.11) give

$$\begin{aligned} &\tilde{s} - s = \text{const}\ (\theta_0 - \tilde{\theta}_0), \\ &\tilde{\sigma} - \sigma = \text{const}\ [(\theta_0 - \tilde{\theta}_0) + (\theta_1 - \tilde{\theta}_1)]. \end{aligned} \tag{5.5}$$

Notice that if ψ_0 increases, then θ_0 decreases and s, σ increase; the variation of ψ_1 influences only the position of σ (σ increases with ψ_1). Finally

$$\tilde{q} - q = (Q(\delta) + 1)(\tilde{q}_0 - q_0), \tag{5.6}$$

and then, from (5.4), we obtain the continuous dependence of q on ψ_0, ψ_1.

-(D) Let us now consider perturbations of K_0, K_1. In the case $\delta = 0$ we have that the differences $(\tilde{s}_0 - s_0)$, $(\tilde{\sigma}_0 - \sigma_0)$, $(\tilde{q}_0 - q_0)$ are equal to some positive constant times the term $[(\tilde{K}_0 - K_0) + \mathcal{K}\,(K_1 - \tilde{K}_1)]$.

In the case $\delta \neq 0$, $\tilde{\eta} = \eta$ and the quantities $|\tilde{s} - s|$, $|\tilde{\sigma} - \sigma|$ are less than or equal to some constant times the term $(|\tilde{K}_0 - K_0| + |\tilde{K}_1 - K_1|)$; for $\tilde{q} - q$, an expression like (5.6) holds.

-(E) The variation of θ_a, θ_b yields that $(\tilde{s}_0 - s_0)$, $(\tilde{\sigma}_0 - \sigma_0)$can be written in the form const $[(\tilde\theta_a-\theta_a)+(\tilde\theta_b-\theta_b)]$, and

$$\left|\tilde q_0 - q_0\right| \le \text{const}\,(\left|\tilde\theta_a-\theta_a\right|+\left|\tilde\theta_b-\theta_b\right|).$$

In the case $\delta \neq 0$, $\tilde\eta = \eta$, then s, σ, q depend on θ_a, θ_b through expressions like those in the case $\delta=0$. It can be noticed that if θ_a increases (or θ_b decreases), then s is closer to L and the same for σ.

6 ICE PRESSURE . NEUTRAL STRESS

The study of the function $p_i(z)$ proceeds as in Fasano & Primicerio (1987), integrating (2.2) and (2.3) in order to determine $w(z)$ and then considering (1.13). The results concerning the zone $s<z<L$ are the same as in Fasano & Primicerio (1987) and we recall them :

$$\frac{d^2p_i}{dz^2}<0,\quad \frac{dp_i(s)}{dz}>\frac{dp_i(L)}{dz},\ p_i(s)>p_i(L)=p_a.$$

Then $p_i(z)$ has a maximum $\tilde p_i$ between $z=s$ and $z=L$ if

$$w_a<w_0\left(\frac{m}{7n+m}\,\frac{1}{1-\delta}\right)^{1/7},\ \delta\le 0;$$
$$w_a<w_0\left(\frac{7n\,m}{(7n+m)(14n+m)}\,\frac{1}{1-\delta}\right)^{1/7},\ 0<\delta<\delta_0, \qquad (6.1)$$

which is assumed for the value of w

$$\tilde w = w_a\left[\frac{7n+m}{m}\,(1+Q(\delta))(1-\delta)\right]^{1/7}, \qquad (6.2)$$

that is, for $\delta=0$, at the temperature

$$\tilde\theta = \theta_a\left(\frac{m}{7n+m}\right)^{m/7n}. \qquad (6.3)$$

If conditions (6.1) are not satisfied, then the function $p_i(z)$ decreases from the value $p_i(s)$ to p_a.

In the region $\sigma<z<s$, equation (2.2) with obvious modifications yields

$$\int_{\frac{w}{w_a}}^{\frac{1}{w_a}} \frac{\left(\frac{1}{w_a}-y\right)^{\frac{v}{u}-1} y^7}{y^7+\frac{1}{\mathcal{K}}\eta}\,dy = \frac{v}{u}\,\frac{G_1}{\theta_1-\theta_0}\left(\frac{1-w_1}{w_a}\right)^{\frac{v}{u}}(z-\sigma). \qquad (6.4)$$

From (6.4) one easily obtains that $w(z)$ is decreasing and then

$$\frac{d^2p_i}{dz^2}<0,\quad \frac{dp_i(s)}{dz}<\frac{dp_i(\sigma)}{dz},\quad p_i(s)>p_i(\sigma)>p_a.$$

A more careful study of the function $p_i(z)$ shows that, within the range of acceptable values for θ_a (recall (3.3)), $p_i(z)$ cannot have a maximum in (σ, s), and then it is always increasing there.

In order to study the possible growing of ice lenses, we introduce the function "neutral stress"

$$N=(1-\phi(w))\,p_i+\phi(w)\,p, \qquad (6.5)$$

where $\phi(w)$ is the stress partition parameter that we will consider equal to $w^{3/2}$ (see O'Neill & Miller (1982)). In the region $s<z<L$ the behaviour of the function $N(z)$ is almost the same as the behaviour of the function $p_i(z)$. Below the front $z=s$ one obtains that $N(z)$ starts from a value $N(\sigma)=p(\sigma)$ and decreases for a very short interval. After that it is always increasing up to the front $z=s$, where it approaches a value $N(s^-)>N(\sigma)$. Obviously $N(L)=p_a<N(s^-)<N(s^+)$.

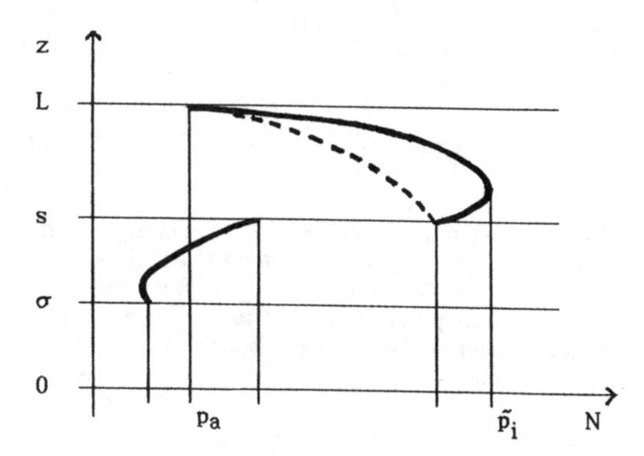

Fig.4 —— if (6.1) holds, ----- if (6.1) does not.

7 GROWING OF ICE LENSES

Generally it is assumed that no ice lens is formed as long as the neutral stress is less than some fixed value p^*, depending on the overburden and on the physical properties of the medium. Measured or estimated values for p^* have been given in Black & Miller (1985). Here we will assume that p^* is fixed and constant throughout the medium. So, comparing the function N with p^*, we have two possibilities :

(a) $p^* > p_a + \omega\psi_0 \doteq p_i(s)$

in this case, a lens can grow only if the function p_i assumes the value p^* somewhere in the region $s<z<L$, that is only if p_i has a

maximum $\tilde{p}_i$ with $\tilde{p}_i \geq p^*$. If we suppose that θ_a varies, in order to cool the medium, in a sufficiently slow way, θ_b being fixed, we obtain that an ice lens can start growing at $z=\tilde{z}$ (where p_i assumes its maximum) only if (6.1) are satisfied, that is, in terms of temperature, if

$$\theta_a < \theta_a^* = \theta_0 \tau^*, \tag{7.1}$$

where τ^* is the unique solution of the equation

$$p^* = \tilde{p}_i \equiv p_a + \omega \psi_0 \phi(\tau), \tag{7.2}$$

$$\tau = \frac{\theta_a}{\theta_0}, \quad \phi(\tau) = \frac{7n}{7n+m}\left(\frac{m}{7n+m}\right)^{\frac{m}{7n}} \tau + \tau^{-\frac{7n}{m}}.$$

The location of the base of the lens can be found from (1.8) and (6.3).

(b) $p_a < p^* < p_a + \omega \psi_0 \doteqdot p_i(s)$

An ice lens can start growing at $z=s$ if the temperature θ_a satisfies

$$\theta_a < \theta_0 < \theta_0^* = \frac{p_a - p^*}{C}. \tag{7.3}$$

Notice that neither θ_a^* nor θ_0^* depend on L, so that we can apply all analysis above to describe the formation of a sequence of ice lenses, setting the base of the new lens as the top of the new layer (see Fasano & Primicerio (1987), sec.8).

8. CASE $\theta_a > \theta_0$

If we suppose that the temperature θ_a at the top of the layer is such that $\theta_a > \theta_s$ (in practical cases $\theta_a > \theta_0$), then the front $z=s$ cannot appear. In this section we will study whether, under this hypothesis, the formation of ice lenses is possible. The thermal conductivity is now constant (K) throughout the medium and the function $\tilde{\psi}(w)$ is approximated by

$$\psi(w) = \psi_1 + (\psi_0 - \psi_1)\left(\frac{1-w}{1-w_1}\right)^{\frac{v}{u}}, \quad w_1 \leq w_a < w < 1. \tag{8.1}$$

The expression of the temperature is

$$\theta(z) = \theta_b - Gz, \qquad G = \frac{\theta_b - \theta_a}{L}. \tag{8.2}$$

With the same methods used above we obtain the localization of the front

$$\frac{\sigma}{L} = \frac{\sigma_0}{L} + \frac{\delta/C}{\theta_b - \theta_a}\left[p_b - (\rho g L + q L)\frac{\sigma_0}{L}\right];$$

$$\frac{\sigma_0}{L} = \frac{\theta_b - \theta_1}{\theta_b - \theta_a}, \text{for } \delta = 0. \tag{8.3}$$

Concerning the water discharge,in the case $\delta=0$, we get

$$q_0 = \frac{\omega \psi(w_a)}{\sigma_0 + I_a}, \tag{8.4}$$

$$I_a = (s_0 - \sigma_0)^{\frac{7u}{v}} \int_{\sigma_0}^{L} \left[(s_0 - \sigma_0)^{\frac{u}{v}} - (1-w_1)(z - \sigma_0)^{\frac{u}{v}}\right]^{-7} dz$$

with $s_0 = \dfrac{\theta_b - \theta_0}{\theta_b - \theta_a} L$.

For $\delta \neq 0$, we can try to follow the procedure of sec.4, defining $\eta = \frac{q\delta}{CG} w_a^{-7}$. The equation corresponding to (4.2) can be shown to possess a unique solution $\eta(\delta)$, with the same sign as δ; moreover $|\eta| < 1$.

If we suppose that $\psi(w_a) = \psi_0$, that is $w_a = w_1$, and $\theta_a = \theta_0 - \delta(\theta_0 + p_a/C) = \theta_s$, then the equation for η can be written in the form

$$v(\theta_1 - \theta_0)\left(\frac{w_1}{1-w_1}\right)^{v/u} \sum_{i=1}^{\infty} (-1)^{i-1} \eta^i I_i + (\theta_b - \theta_1$$

$$+ p_a \delta / C) \sum_{i=1}^{\infty} (-1)^{i-1} \eta^i w_1^{7i} = \omega \psi_0 \delta / C, \tag{8.5}$$

where $I_i = \displaystyle\int_0^{\left(\frac{1-w_1}{w_1}\right)^{1/u}} \frac{t^{v-1}}{\left(\frac{1}{w_1} - t^u\right)^{7i}} dt$ (which can be calculated explicitly) is independent of δ and $I_i < 1$ for any i. Then recursive formulas for the determination of the coefficients of an expansion like (4.9) can be found rather easily. The study of the function N(z) is analogous to the one done in sec.6:

$$\frac{dN(\sigma)}{dz} < 0 < \frac{dN(L)}{dz}, \quad N(L) < p_a, \quad N(\sigma) < p_a.$$

Of course, the hypothesis $p^* > p_a$ excludes the possibility of formation of ice lenses if $\theta_a > \theta_0$.

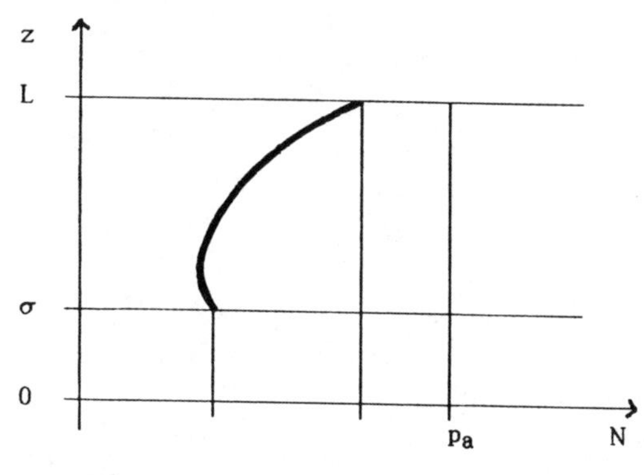

Fig. 5

9 CONCLUSIONS

The model presented here for freezing of saturated porous media is a refinement of the one proposed by Fasano & Primicerio in the spirit of the work of O'Neill & Miller. It is shown that such important quantities like water discharge, growth rate of the lowest ice lens, and the location of ice lenses can be calculated by means of explicit formulas. The influence of expansion or contraction of the freezing fluid is emphasized, showing in what cases it can be neglected, leading to even simpler formulas.

REFERENCES

Black, P.B. & Miller, R.D. 1985. A continuum app roach to modelling of frost heaving. Tech. Council on Cold Regions Engin. Monograph. M.Anderson-P.Williams Eds: 36-45.

De Vries, D.A. 1963. Thermal Properties of soils. Physics of plant environment. W.R.van Wijk Ed, North-Holland Publ. Co., Amsterdam (1963), 210-235.

Fasano, A. & Primicerio, M. 1978. Freezing of porous media: a review of mathematical models. Proc. German-Italian Symposium Applications of Mathematics in Technology. V.Boffi-H.Neunzert Eds, Teubner: 288-311.

Fasano A. & Primicerio M. Heat and mass transfer in quasi-steady ground freezing processes. 1987. Preprint

Holden, J.T. , Piper, D. & Jones, R.H. Some developments of a rigid-ice model of frost heave. 1985. Fourth Int. Symposium on Ground Freezing. Sapporo , S.Kinosita & M.Fukuda Eds, Balkema 93-99

O'Neill, K. & Miller, R.D. 1982. Numerical solutions for a rigid-ice model of secondary frost heave CRREL Rep.: 82-13

5th International Symposium on Ground Freezing, Jones & Holden (eds)
 ISBN 90 6191 824 3

Influence of freezing and thawing on the physical and chemical properties of swelling clays

K.A.Czurda & R.Schababerle
Department of Applied Geology, University of Karlsruhe, FR Germany

ABSTRACT: If a clay is frozen, at first the unbound water, which is contained in bigger pores, crystallizes. In case there is still unfrozen water available, ice lenses can be formed. For this purpose water is led from the nearer surrounding of the ice lenses as well as from the unfrozen zone.

During thawing one has to consider further structural changes of the clay. The melting of frozen water and the diffusion of melted water into the desiccated zone below the active front causes swelling. Both, the change of volume properties and the swelling of the clay will have influence on its texture.

These processes involve changes of the mechanical properties like plasticity, shear strength and permeability. Furthermore the mineralogical composition, the specific surface area and the cation exchange capacity are influenced by freezing and thawing. The migration of water during freezing as well as during thawing is responsible for the transport of ions.

To clarify the reasons and the extent of the above mentioned phenomena , laboratory studies were carried out. Temperature profiles, consumption of water and frost heaving were measured in a partially frozen clay. Before and after each freezing experiment the following were examined in detail:

cation exchange capacity
adsorbed cations
cations dissolved in pore water
specific surface area
water content
structural properties

Little is known about the extent of ion transport and the function of ions on the formation of soil fabric units. It has been one focus of our investigations, to gather more knowledge about these processes.

1 INTRODUCTION

If fine grained soils are cooled below the freezing point, first the unbound pore water freezes (in situ freezing). In this case, there is no water movement. If a suction force is caused by the growing ice, water is transported toward the ice. This process, which is called "segregation freezing" (TAKAGI 1979) must be regarded as the main cause of ice lense formation. In this case, water transport can also occur in the frozen zone. It is characteristic for clays, that they have a large number of small pores, which contain unfrozen water. This water remains unfrozen well below the freezing point (HORIGUCHI 1985). As driving force for the water movement the temperature gradient developing during freezing (HOEKSTRA 1967) as well as an electrical potential gradient must be regarded beside the suction force, which develops during ice lense formation.

The water which is needed for freezing is first taken from the nearest surrounding pore space. Later increasingly water from the unfrozen zone migrates towards the freezing front. The pore water contains free ions, which are carried also (CARY et al. 1979).

The formation of ice lenses on the one hand and the stripping of cations from the clay mineral interlayer interface on the other hand have an important influence on the soil structure. Structure itself controls the soil physical parameters such as permeability, pore space distribution etc. (CHAMBERLAIN & GOW 1979, JUMIKIS 1979, PUSCH 1979, NAKASAWA & UNEDA 1985). The soil mechanical behaviour changes as well (BOURBONNAIS & LADANYI 1985). The increase of volume after in situ freezing amounts up to 10 %. Ice lense formation leads to heavings up to 60 % of the original volume (CZURDA & WAGNER 1985).

2 TESTING PROCEDURE

First the frost tests were started with ten small plexiglass columns (r=2.5 cm; l=6 cm; fig. 1). The columns have a free water supply on the bottom side. On the top of the clay samples a freely moveable cold plate is located, whose movement can be registered by a displacement transducer.

Fig. 1 Photograph of the testing apparatus. On the left side the cooling device, on the rigth side the small plexiglass cylinders.

To minimize marginal effects and sample mean errors, a larger plexiglass column (r=7.5 cm; l=40 cm; fig. 2) was constructed. For the interpretation of the freezing phenomena it is very helpful to gain experience with the exact temperature distribution in the sample. For this purpose the sample was furnished with 8 temperature transducers. Water consumption and frost heaving could be registered with the help of displacement transducers. The values were recorded by a data acquisition control unit and transfered to a personal computer, where data were stored for later evaluation. The readings of the data were directed by a time switch and were carried out at a four hours interval.

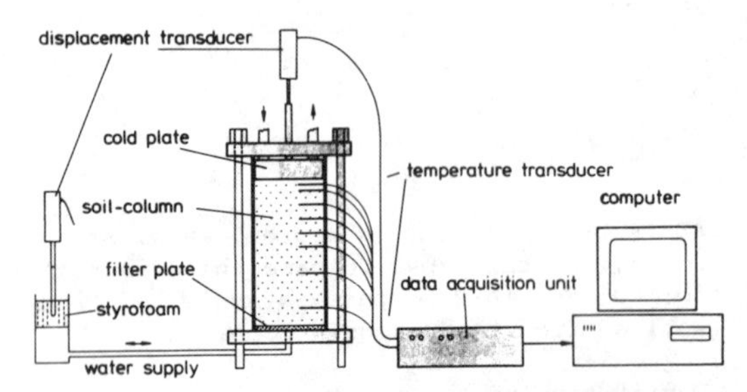

Fig. 2 Testing device for the determination of temperature course, frost heaving and water consumption of the soil column (r=7.5 cm, l=40 cm).

A quaternary varved clay from a drillhole near Ravensburg (SW-Germany) was used as test material. To compare the fabric changes with freezing, a Na-bentonite ("Tixoton") was used. The mineralogical composition of the investigated material is listed in table 1.

Tab. 1 Mineralogical composition and initial water content of the investigated samples. Values are in per cent.

	TIXOTON	RAVENSBURG CLAY		
type of columns	small	large	large	small
sample number	1-10	GS 1	GS 2	1-20
carbonate	6	38	38	25-38
quartz	11	18	18	18-25
montmorillonite	60	10	10	8-12
kaolinite	13	8	8	6-10
illite/muskovite	7	15	15	12-17
chlorite	-	5	5	3-7
others	3	6	6	-
water content	70	23	38	20-40

The installation of the sample columns within the small cylinders was carried out on disturbed as well as undisturbed samples. Disturbed samples were installed exclusively in the large columns. The big columns were cut in slices of 2 cm thickness after the tests were finished, to evaluate cation migrations within the samples.

3 RESULTS AND DISCUSSION

3.1 TEMPERATURE DISTRIBUTION

The temperature profiles in the soil column at the beginning of the test is shown in fig.3. As expected the upper part of the sample reacted very quickly as the temperature decreased. After 12 h the 0°C line was located 4 cm below sample top. This location did not change considerably during the test (fig. 4). After 24 h temperature equilibrium was reached in the whole sample. The heat flow could be regarded as almost stationary. Small disturbances during the test were effected by variation of room temperatures.

The temperature lines are divided into two parts. The upper part up to the zero centigrade line shows a almost linear profile, the lower part a steep, nonlinear course. This zoning has also been observed by YANAGISAWA & YAO (1985) and MIZOGUCHI & NAKANO (1985).

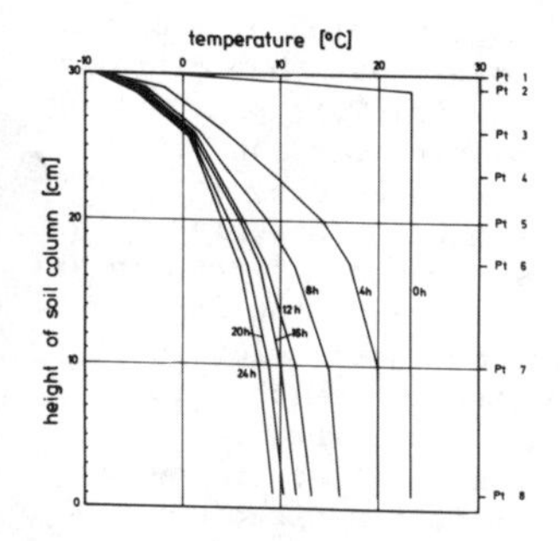

Fig. 3 Temperature course in column GS 2 in a four hours rhythm. After 24 hours a temperature equilibrium resulted.

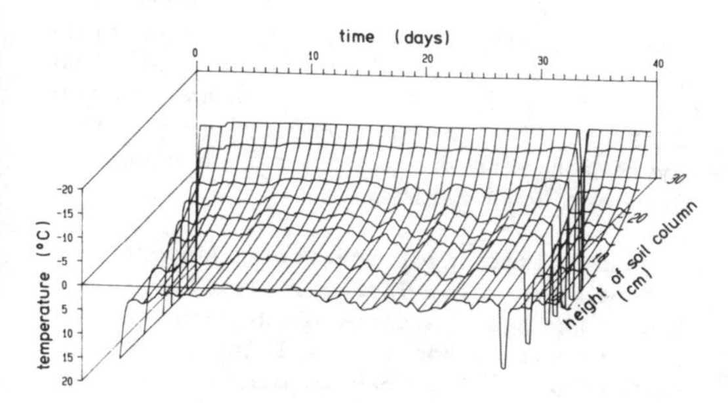

Fig. 4 Threedimensional presentation of the temperature profile in the soil column in dependence on time.

It is of importance, that the 0°C line corresponds to the frost line. Below this line ice lenses are not observed.

3.2 WATER CONTENT

Tests using both columns showed a good conformity in their water content curves. The unfrozen zone is characterized by a low water content, which reaches a minimum below the frozen zone. In the frozen zone the water content increases in proportion to the amount of ice. So an ice lens can be recognized clearly as a maximum of the water content curve of sample GS 2 (Fig. 7). During the second test with higher water content at the beginning, the differences in moisture contents proved to be much more distinct.

To reconstruct the water transport process during freezing, the following parameters were observed, namely the water content in the beginning, increase or decrease in the reservoir and the water content after the test was finished.

During test GS 2 water movement from bottom to top towards the frost front must have taken place. With column GS 1 the flow conditions were more complicated. The clay was oversaturated and gave off water to the storage reservoir. After thawing of the ice lenses, caused by electric power breakdown, an especially rapid water expulsion occured (Fig. 5). In total the column discharged about 54 ml of water. The ice lens formation in the upper part and the water discharge in the storage reservoir suggested two flows. The ice lens formation requires a flow of water towards the top. The originally water saturated unfrozen zone consolidated. The consolidation leads to a flow toward the bottom. The border of the two flow directions is difficult to determine.

3.3 FROST HEAVING

Fig. 6 shows frost heave curves of the small columns furnished with undisturbed samples. It can be recognized that the amount of heaving increases after each freeze-thaw cycle (CZURDA & WAGNER 1985). After thawing the sample top does not return to its original position. This is due to irreversible fabric changes in the soil. Different water contents in the beginning, differences in pore space distributions and presumably marginal effects might have caused the large differences in curve number 7.

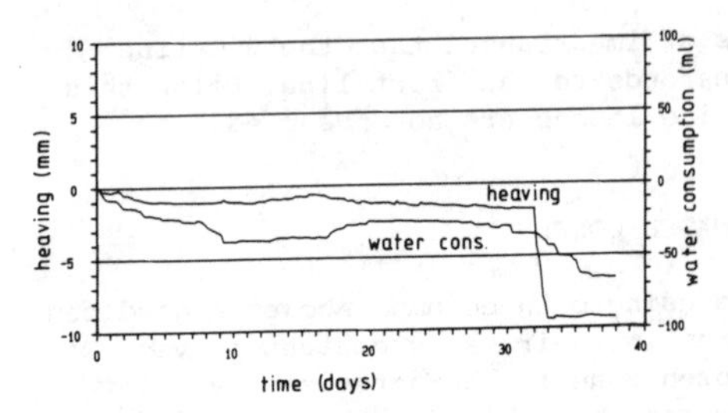

Fig. 5 Frost heaving and water consumption of the column GS 2. Due to the consolidation of the sample in the lower part water discharge and volume decrease of the upper part of the sample resulted.

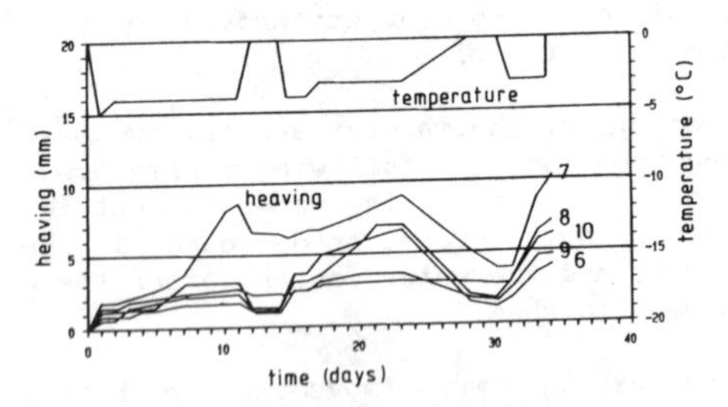

Fig. 6 Frost heaving and temperature course of five small columns (6-11). The upper part represents the temperature course, which is valid for all columns.

Sample GS 2, which had been installed as clay slurry, reacted very differently (fig. 5). Even though ice lens formation and length increase of the frozen part for 5 cm occured, this fact did not lead to a heave of the sample top, because of increased compaction of the unfrozen, unconsolidated part of the sample. The decrease of the pore space led to considerable water discharge at the sample bottom. The sample top was lowered for 9,4 mm. Compaction was supported by the generation of suction force below the ice lens. This caused a reorientation of the clay particles to a more compact, aggregated structure. CHAMBERLAIN & BLOUIN (1980) showed, that these phenomena are useful for compaction of clay muds.

3.4 SALT MIGRATION

Three different stages resulted from the investigation of the samples which had been prepared with distilled water. In the unfrozen zone the conductivity curve shows a relatively uniform course (fig.7). Below the ice lens a negative peak occurs, while the frozen range above reaches the same level as in the unfrozen zone. In comparison to the water content curve this curve shows a reciproke profile. As investigations on freezing of salt water shown (FARRAR & HAMILTON 1965), the dissolved salts are not absorbed into the crystal lattice of ice, but are concentrated at the water-ice interface. As result of the ice lense analyses, this mechanism must also be valid for freezing soil. This phenomenon is responsible for the increased conductivity below the ice lens. Field observations of CARY et al (1979) show, that a salt migration and accumulation below the frost front occurs not only under laboratory conditions. We know from different investigations that ice removes all particles disturbing the crystal lattice. Therefore mineral particles (RÖMKENS & MILLER 1972) or even submicroscopic inclusions of brines (JONES 1974) migrate towards the warm side. This explains the high purity of the observed ice lenses.

As the dissolved ions have different properties like ion radius, valence, hydration energy etc. we suppose, that they differ in their transport behavior. This presumption was enforced by AAS-investigations of sample slices of the frozen clay column, which had been shaken with destilled water. In fig. 7 it can be seen that first the Na+ ions increase from bottom to top continously, show a peak maximum just below the frost front and decrease from the frost front in a steep gradient. The K+ -curve shows the maximum some distance below the frost front, but the decrease becomes obvious in the frozen section according to the behaviour of Na+. Ca++ and Mg++ ions show a straight course in the lower part. Below the ice lens the concentration increases, however in the upper part it decreases again. But the differences in the course of these two ions are much less distinct. This indicates that monovalent ions are more mobile than the divalent cations. The comparatively large monovalent Na+ and K+ ions require only small hydration energy compared to the divalent Ca++ and Mg++ cations.

A quantification of the ion transport is not yet possible. Therefore further investigations must be carried out. Our research on this phenomenon is still in progress and further data will lead to more detailed information.

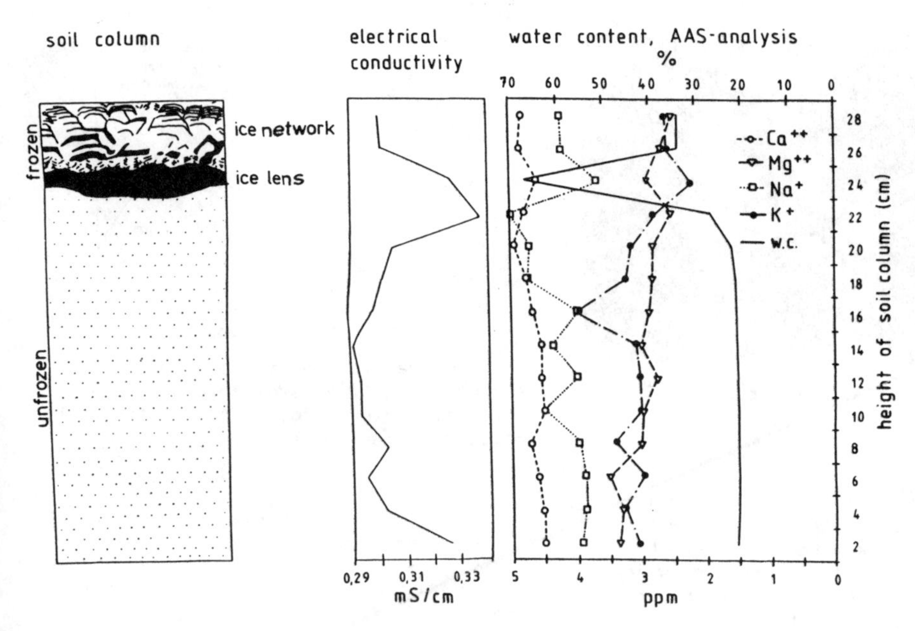

Fig. 7 Electrical conductivity, water content, and content of ions in the pore fluid.

3.5 SOIL STRUCTURE

Fig. 8 shows a photograph of the column GS 1. The lower limit of the frozen range form an ice lens of about 1,5 cm thickness. Next to it a zone is located, which shows fine ice sheets mostly inclined to 45°. The clay volume up to the sample top consists of more horizontal ice flakes than vertical ice structures. A polygonal pattern of contraction cracks filled by ice structures divide the clay by a pattern into single aggregates.

Fig. 8 Photo of the frost fabric of the soil column GS 2. The frozen zone is limited towards the bottom by an ice lens of 1,5 cm thickness.

Tixoton (Na-bentonite) showed a totally different behavior. Here no ice lenses had been observed at all. The frozen range was permeated by bundles of thin ice plates running mostly perpendicularly. These distinct differences of the two sample materials can be led back to different mineralogical compositions. The quarternary clay (Ravensburg) contains predominantly non swelling minerals with a high content of mobile water. Tixoton contains almost exclusively Na-montmorillonite, a three-layer mineral with very high specific surfaces. When saturated by water, these minerals adsorb water between the silicate layers. The extent of swelling and the degree of order of the water molecules depends predominantly on the cation packing within the interlattice space (GRIM 1952). The pore space is considerably diminished by the decrease of volume because of dewatering. Therefore water transport is not possible for ice lens formation (WAIBEL 1973). Indeed the water films of the swelling minerals are considerably thicker than those of the non swelling minerals, but the water molecules are bound much stronger due to distinct higher surface areas and the cation content. Because of this restricted mobility water was taken from the water hull. Water from the reservoir had not been taken.
The investigation of the clay with the help of x-ray diffractometry has proved, that

Fig. 9 Photo of a small frozen soil column (Tixoton)

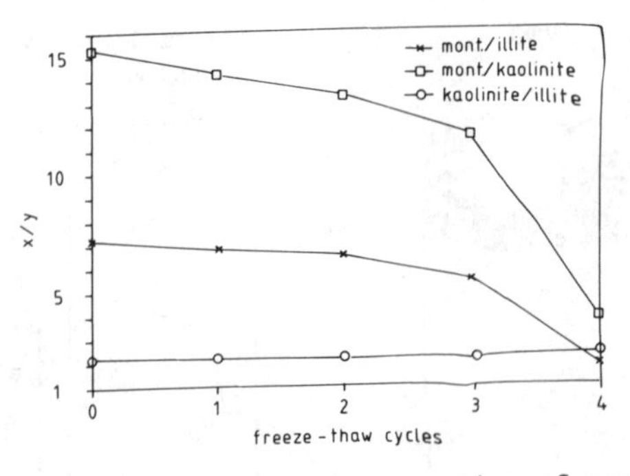

Fig. 10b Decrease of proportion of peak intensity (001) of swellable and non-swellable minerals in dependence of the number of freeze-thaw cycles.

frost has a great influence on the lattice distances within a stack of silicate layers. Before freezing the layers had a presumably uniform layer distance which could be observed by a sharp peak for all clay mineral layers. After each freeze-thaw cycle the layer distances for montmorillonite became more irregular. Consequently the peak height decreased, the peak basis became broader (fig. 10a).

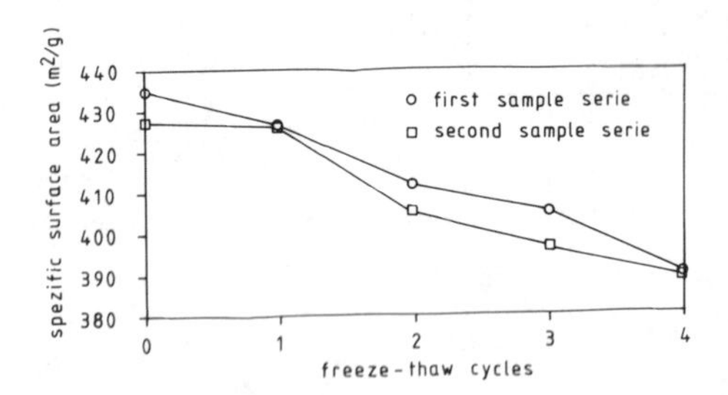

Fig. 11 Decrease of the specific surface area in dependence on number of the freeze-thaw cycles. The analysis was carried out with the methyleneblue adsorption method.

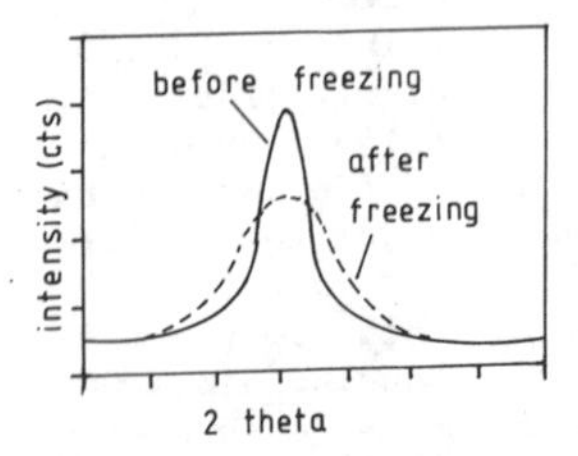

Fig. 10a Schematic representation of an x-ray diffractogram. After freezing the intensity of the peak (001) decreases as a result of a non-uniform layer distance. Yet the peak basis becomes broader.

The decrease of peak height after several freeze-thaw cycles is shown in fig. 10b.

YONG et al. (1985) could show, that the specific surface area and the CEC decreases due to aggregate formation through frost effects. A decrease of the specific surface area could only be proved by the calcigel-samples with the help of the methyleneblue method (fig. 11). The analysis of the clay from Ravensburg tended more to an increase of the specific surface area. A change of the CEC could not distinctly be proved.

4 CONCLUSIONS

The investigations have shown, that ice lenses in clay can be formed only, if there is a sufficient amount of mobile water. The mobility of water depends to a high degree on the mineralogical composition. Non swelling minerals like kaolinites have a thin water cover, but they are able to lead this water. These clays form ice lenses and three dimensional ice networks. Considerable frost heaving and a permanent change of the soil fabric results from it. Montmorillonites are able to bind much more water. This water is fixed strongly to the minerals. The type of absorbed cations strongly influences these properties. In case no ice lenses grow, frost heave and fabric changes are low.

Together with the water, the cations are carried towards the ice lens. These ions are not absorbed into the ice lattice, but remain on the border water-ice. This increase in salt concentration has important effects:

The ice forms a barrier for itself, because of salt concentrations below the active layer.This barrier can only be exceeded by decrease of temperature or dilution of the concentration.

The increased salt concentration causes an osmotic potential, which must be regarded as an additional force for water movements towards the ice lens.

Besides the discharge of water from the interlayers the increase of salt concentration leads to a decrease of the double layer (STERN-layer) of the clay minerals. Therefore the attracting forces predominate and an aggregate formation occurs.

Na+, K+, Ca++ and Mg++ ions are not transported by the water to the same extent. Obviously there are differences in ion mobility and a different affinity for the occlusion into the ice crystal lattice.

When judging the frost danger of clayey soils in the future, the mineralogical composition and the cation packing of the swellable minerals must be taken into account to a greater extent. The freezing of soils can be regarded as an effective mechanism to get fertilizer and agressive substances from the surface into the soil. The melting ice can effect a further shifting of those substances. The opposite direction, the transport of cations from lower levels up to the active freezing front could as well be evaluated. The latter is of extreme importance in case of clay covers for hazardous waste deposits.

5 ACKNOWLEDGEMENT

This study was supported by the DFG (German Research Group) through project number Cz 36/1-1.

REFERENCES*

BOURBONNAIS, J. & LADANYI,B., 1985: The mechanical behaviour of a frozen clay down to cryogenic temperatures. Proc. 4th Int. Sym. on Ground Freezing. Sapporo, Japan: Vol.2: 237-244.

CARY, J.W., PAPENDICK, R.I. & CAMPBELL, G.S., 1979: Water and salt movement in unsaturated frozen soil: Principles and field observations. Soil. Sci. Soc. Am. J., Vol. 43: 3-8.

CHAMBERLAIN, E.J. & GOW, A.J., 1979: Effect of freezing and thawing on the permeability and structure of soils. Eng. Geol., 13: 73-92.

CHAMBERLAIN, E.J. & BLOUIN,1978: Densification by freezing and thawing of fine material dredged from waterways. 3rd. Int. Conf. on Permafrost. Edmonton, Alberta, Canada: 622-628.

CZURDA, K.A. & WAGNER J.F., 1985: Frost heave and clay expansion in freshwater clays. Proc. 4th Int. Sym. on Ground Freezing. Sapporo, Japan: 129-136.

FARRAR, J. & HAMILTON, W.S., 1965: Off. Saline Water Res. Develop. Rept. No. 127.

GRIM, R.E., 1952: Relation of frost action to the clay mineral composition of soil materials. HRB Spec. Rep. 2: 167-172.

HOEKSTRA, P., 1966: Moisture movement in porous materials under temperature gradients with the temperture of the cold plate below freezing. Water Resources Res. 2, no 2: 241-250.

HORIGUCHI, K., 1985: Determination of unfrozen water content by DSC. Proc. 4th Int. Sym. Ground Freezing. Sapporo, Japan: 33-38.

JONES, D.R.H. 1974: Determination of the kinetics of ice-brine-interfaces from the shape of migrating droplets. J. Crystal Growth 26: 117-179.

JUMIKIS, A.R., 1979: Cryogenic texture and strength aspects of artificially frozen soils. Eng. Geol., 13: 157-162.

MIZOGUCHI, M. & NAKANO, M. 1985: Water content, electrical conductivity and temperature profiles in a partially frozen unsaturated soil. Proc. 4th Int. Sym. on Ground Freezing. Sapporo, Japan: 2: 47-52.

NAGASAWA, T. & UNEDA, Y., 1985: Effects of the freeze-thaw process on soil structure. Proc. 4th Int. Sym. on Ground Freezing. Sapporo, Japan: Vol.2: 219-224.

PUSCH, R., 1979: Unfrozen water as a function of clay microstructure. Eng. Geol., 13: 157-162.

RÖMKENS M.J.M. & MILLER, R.D. 1973: Migration of mineral particles in ice with a temperature gradient. J. Colloid Interface Sci., 42: 103-111.

SKARZYNSKA, K.M., 1985: Formation of soil structure under repeated freezing-thawing conditions: Proc. 4th Int. Sym. on Ground Freezing. Sapporo, Japan: Vol.2: 213-218.

TAKAGI, S., 1979: Segregation freezing as the cause of suction force for ice lens formation. Eng. Geol., 13: 93-100.

*See page X concerning ISGF papers

WAIBEL, P., 1973: Der Einfluß des Mineralbestandes auf die Frostsicherheit von Kiestrageschichten. Mittl. Inst. f. Grundbau u. Bodenmech. TH Wien. Vienna, Austria: 11-21.

YANAGISAWA, E. & YAO, Y.J., 1985: Moisture movement in freezing soils under constant temperature condition. Proc. 4th Int. Sym. on Ground Freezing. Sapporo, Japan: 85-91.

YONG, R.N., BOONSINSUK, P. & YIN. C.W.P., 1985: Alteration of soil behaviour after cyclic freezing and thawing. Proc. 4th. Int. Sym. Ground Freezing. Sapporo, Japan: 187-195.

5th International Symposium on Ground Freezing, Jones & Holden (eds)
© 1988 Balkema, Rotterdam. ISBN 90 6191 824 3

Field observations of water migration in unsaturated freezing soils with different ground water tables*

Gao Weiyue
Institute of Water Conservancy of Inner Mongolia Autonomous Region, Huhehaute, People's Republic of China
Xu Xiaozu
Lanzhou Institute of Glaciology and Geocryology, Academia Sinica, Lanzhou, People's Republic of China

ABSTRACT: Sandy silt and Silty clay with dry density of 1.38-1.40g/cm^3 were packed in steel cylinders. The cylinders were open at the top and sealed at the bottom and buried into ground. The soil columns in the cylinders had a water supply with constant groundwater tables and were frozen under natural conditions. The field observations lasted 5 years. It was found that the total amount of water intake flow during freezing was lowest and highest in the sandy silt when the groundwater table was 1.0 and 2.0 m, respectively, and was highest in the silty clay when the groundwater table was 1.5 m. The curve of the water intake flux vs. the elapsed time of freezing had a peak value. With the increase in groundwater table, the time when the peak occured was delayed. The relationship between the water intake flux during freezing and the elapsed time is expressed in the power form.

KEYWORDS: clay, groundwater, silts, thawing, water migration.

INTRODUCTION

In the northern part of China there exists a large area of seasonally frozen ground. One of the important factors affecting soil salinization is the water and salt migration during freezing, ground water with higher salinity being absorbed up to ground surface, resulting in the changes of soil properties. The buried depth of ground water directly influences the amount of water migration during soil freezing. Therefore, the study on the relationship between the amount of water migration during soil freezing and the groundwater table is not only of significance for the conservancy planning and the improvement of saline soils, but also of significance for protecting engineering constructions in cold regions from frost damage.

Xu Xiaozu (1982), Williams, P.J. (1984) have summarized the previous work on the subject of water migration in freezing soils, but a few of them have studied the influence of the ground water table based on the field observations. In combination with the study of frost heaving, Fukuda, M. (1985) and Knutsson, S. (1985) have reported their research results of frost heaving based on the field observations by using a constant and variable groundwater table, respectively, but they did not discuss the influence of the groundwater table on the frost heaving.

To understand the regularities of water and salt migration during soil freezing and to find the properly controlled depth of ground water which will not harm the crops, the Institute of Water conservancy of Inner Mongolia Autonomous Region has observed the water migration in unsaturated freezing soils under different groundwater tables since the field site was set up in 1982. Parts of the laboratory work have been carried out by Lanzhou Insti-

* Most of field observations were carried out by the following comrades: KangShuangyang, Wang Fene, Xing Chengcai, Wang Shufeng, Zhao Wengxue.

tute of Glaciology and Geocryology, Academia Sinica. The purpose of this paper is to describe the water migration in the unsaturated freezing soils in relation to the groundwater table based on the field observations.

SITE CONDITIONS

The test site is located in the middle part of the irrigation regions of the Great Bend of the Huanghe River. The terrain is smooth and the ground is seriously salified.

The soil samples used in the test are the sandy silt and the silty clay, which are widely distributed in this region. The physical properties of the soils are shown in Table 1.

Table 1. The physical properties of soils

Soil type	Grained size (%)			Specific gravity	Field capacity (%)	Salinity (%)
	0.1-0.05	0.05-0.005	<0.005			
Sandy silt	35.0	54.0	11.0	2.71	20.02	0.2
Silty clay	5.0	35.0	60.0	2.74	30.98	0.3

The soil samples with dry density of 1.38-1.40 g/cm³ are packed uniformly into steel cylinders. The size of the cylinders are 112.8 cm in diameter and 1 m² in area of the cross section and the cylinders are open at the top, sealed at the bottom and buried into ground. The upper end of the soil column in the cylinder is as high as the ground surface, and the lower end of the soil column has a reverse permeable layer made up of sandy gravel and is connected with the water supply apparatus with the constant water table of 0.6, 0.8, 1.0, L.3, 1.5, 1.8, 2.0, 2.5, 3.0 and 3.5 m, respectively (see Fig.1). The amount of intake or outlet flow of water migration is observed twice a day. During soil freezing and thawing, samples are taken by drilling from the soil columns in the cylinders to determine the layered water content of soils and the freezing o thawing depth.

RESULTS AND ANALYSES

Freezing and Thawing Processes of Ground

Observation data for the period of five years shows that the ground started freezing from the surface downward on about Nov. 10th each year and the maximal frost depth of 1.1-1.3 m is reached about on March 20th of the next year. The ground starts thawing from the middle ten days of March in two directions, downward and upward, and finishes thawing in the first ten days of May. Table 2 lists the values of the frost depth during 1984-1985 determined by drilling.

From Table 2 it can be seen that if the soil type and the date are the same, the values of the frost depth are very close to each other. It implies that the differences in frost depth caused by the groundwater tables are very small. The reason may be that the area of the

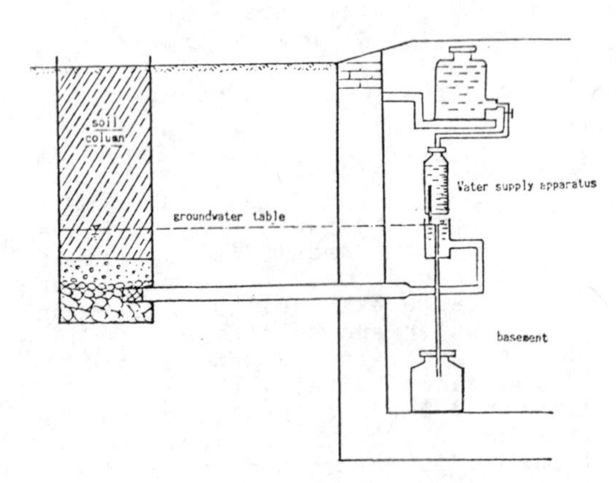

Fig.1 The scheme of water supply apparatus

Table 2, The values of the frost depth of soils, in cm

Soil type	Groundwater table (m)	Date							
		Nov.25	Dec.1	Dec.10	Dec.20	Dec.29	Jan.20	Feb.10	Mar. 20
Sandy silt	0.6	20	30	40	60				
	1.0	20	30	42	50	72			
	1.5	20	30	40	56	80	102	120	135
	2.0	20	30	40	55	71	100	115	125
	2.5	20	30	40	55	71	100	105	110
Silty clay	0.6	15	31	41	52				
	1.0	20	31	40	60	71			
	1.5	20	30	43	55	72	100	109	115
	2.0	20	30	40	60	71	101	110	122
	2.5	18	30	40	60	71	100	110	118

cross section of the soil column is less and the frost depth of soils in the cylinders is mainly controlled by the frost depth of natural ground. In this case the influence of the groundwater tables on the frost depth and freezing rate could be neglected when the amount of water migration during soil freezing in realtion to the groundwater table is analyzed.

The regression analysis indicates that there is a good correlative relationship between the frost depth (Hf) and the elapsed time (T), which could be expressed by

$$H_f = A\sqrt{T} + B \qquad (1)$$

where A and B are the empirical coefficients. Table 3 gives the values of A and B in equation (1) for different years.

Table 3 The values of A and B in equation (1)*

Period	Date starting freezing	Sandy silt				Silty clay			
		A	B	N	R	A	B	N	R
1982-1983	Nov.15	11.035	-10.662	43	0.9742	11.515	-9.327	45	0.9671
1983-1984	Nov.8	13.093	-26.104	46	0.9827	13.499	-31.062	52	0.9841
1984-1985	Nov.15	13.242	-20.769	43	0.9730	13.077	-19.979	44	0.9822
1985-1986	Nov.5	13.671	-24.262	45	0.9874	13.611	-28.379	42	0.9320

*Where N—the statistical number; R—the correlative coefficient

Total Amount of Water Intake Flow During Freezing

After the ground surface is frozen, the pore water pressure of soils is negative at the freezing front and is zero at the buried depth of ground water, so that there exists a pressure gradient, which is the driving force of water migration during soil freezing.

Figures 2 and 3 show the curves of the total amount of water intake flow during freezing vs. the elapsed time for the two types of soils, respectively. From Figs. 2 and 3 it can be seen that the total amount of water intake flow increases with increasing elapsed time no matter what the soil type is, but the curve shape depends on the groundwater table. When the buried depth of groundwater is less than the frost depth, there is an inflexion point on the curve, which means that once the frost depth reaches to the ground water table, the increasing rate of water intake flow suddenly drops. This situation is very clear in the sandy silt. When the buried depth of ground water is greater than the maximal frost depth, there is no inflexion point on the curve, but with the increase in the buried depth of ground water, the time occurring the water intake flow is gradually delayed. In other words, when the groundwater table is deep enough, the water intake flow is not occurred immediately at the beginning of soil freezing (see Fig.3). This situation is very clear in the silty clay.

The relationship between the total amount of water intake flow (Q) and the elapsed time (T) can be expressed as

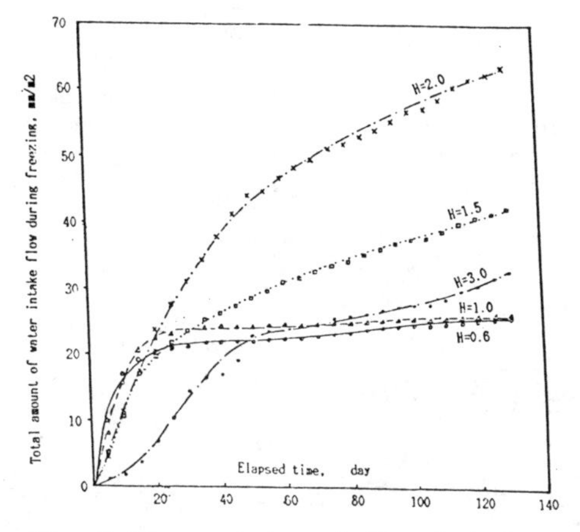

Fig.2 The curves of the total amount of water intake flow in the sandy silt during freezing vs. the elapsed time

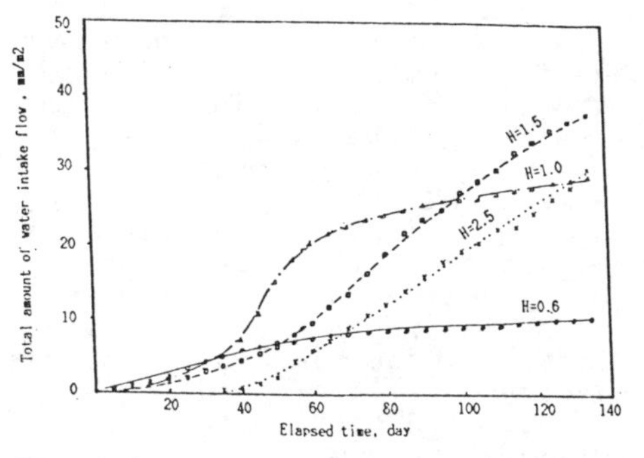

Fig.3 The curves of the total amount of water intake flow in the silty clay during freezing vs. the elapsed time

$$Q = A(T-D)^{(B-1)} \exp[-C(T-D)] \quad (2)$$

where A,B and C are empirical coefficients characterized by the soil type, freezing consition and hydrogeological consitions, D-the time when the water intake flow occurred.

Table 4 lists the values of the total amount of water intake flow for the two types of soils under different groundwater tables during freezing. From table 4 it can be seen that for the sandy silt, the total amount of water intake flow has a minimum and a maximum when the buried depth of ground water equals to 1.0 and 2.0 m, respectively, and for the silty clay, it is a maximum when the buried depth of ground water equals to 1.5 m.

Table 4, The values of the total amount of wate intake flow during soil freezing, in mm/m^2

Period	Soil type	Groundwater table (m)							
		0.6	1.0	1.5	1.8	2.0	2.5	3.0	3.5
1982-1983	Sandy silt	22.22	16.77	19.26	34.69	49.85			
1983-1984		32.27	30.59	39.19	38.84	60.14			
1984-1985		25.93	26.26	42.41	41.31	63.55			
1985-1986		29.01	16.15	47.12	47.56	38.28		32.94	
1982-1983	Silty clay	21.61	29.77	43.86		31.49	35.38		11.15
1983-1984		10.02	28.14	50.01					8.77
1984-1985		10.81	29.44	37.31			30.48		
1985-1986		19.77	37.84	48.56		24.96	26.45		

Water Intake Flux During Soil Freezing

Figures 4 and 5 show the curves of the water intake flux vs. the elapsed time for the two types of soils, respectively. From Figs.4 and 5 it can be seen that for the sandy silt, the curves are mainly decreasing. The peak value of water intake flux occurs earlier. The elapsed time corresponding to the peak value is usually in the range of 40 days when the ground water table is within the depth of 3.0m. For the silty clay, the elapsed time corresponding to the peak value is obviously delayed with the increase in the buried depth of ground water, for instance, the time when the peak value occurs is 30, 50 and 1o5 days when the buried depth of ground water equals to 0.6, 1.0 and 1.5 m, respectively. The curves are in the increasing shape when the buried depth of ground water is greater than 1.5 m. The relationship between

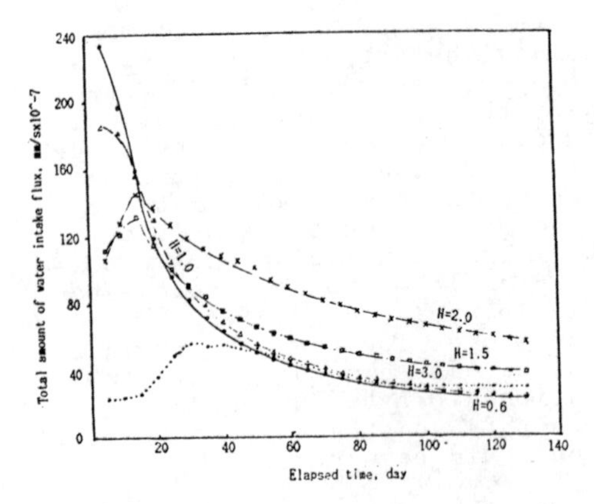

Fig.4 The curves of the water intake flux of the sandy silt during freezing vs. the elapsed time

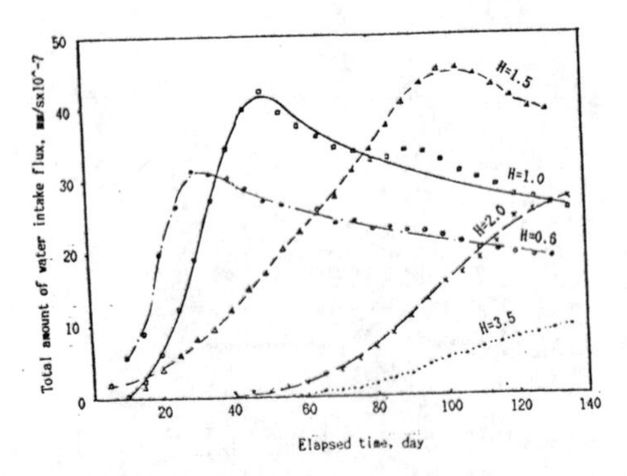

Fig.5 The curves of the water intake flux of the silty clay during freezing vs. the elapsed time

the water intake flux (q) and the elapsed time of freezing (T) could be expressed by

$$q = A(T-D)^{(B-1)} e^{[-C(T-D)]} \quad (3)$$

Table 5 and 6 list the values of the coefficients A B C D in equation (3) for the two types of soils, respectively. From Table 5 it can be seen that with the increase in the buried depth of ground water, A decreases and both B and C increase. But in Table 6 a regularity of the change in the empirical coefficients is not obvious.

Table 5. The values of A.B.C.D for the sandy silt in equation (3)

Year	Groundwatwr table (m)	A	B	C	D	R	N
	0.6	4.8936E-5	0.38015	1.95679E-3	0	0.993562	27
	0.8	4.2127E-5	0.41710	3.11205E-3	0	0.992647	27
Nor.,1982	1.0	2.9910E-5	0.56555	7.29674E-3	0	0.991154	27
	1.3	2.5524E-5	0.48395	3.08060E-3	0	0.985916	27
	1.5	1.2241E-5	0.80944	8.74520E-3	0	0.969271	27
Mar.,1983	1.8	2.2045E-5	0.80575	8.54337E-3	0	0.988138	27
	2.0	1.2484E-5	1.02332	9.75106E-3	0	0.978717	27
	3.0	1.1585E-8	2.92007	2.31314E-2	0	0.970844	27
	3.5	8.4040E-11	4.01623	2.40666E-2	0	0.872570	27
	0.6	4.3396E-5	0.60619	6.63584E-3	0	0.985392	28
	0.8	3.2915E-5	0.69069	8.68879E-3	0	0.980965	28
Nor.,1983	1.0	2.8218E-5	0.77052	9.88427E-3	0	0.983158	28
	1.3	2.6158E-5	0.48580	2.38131E-3	0	0.984888	28
	1.5	5.6864E-5	1.12646	9.27482E-3	0	0.961509	28
Mar.,1984	1.8	1.8180E-6	1.52657	1.60988E-2	0	0.926240	28
	3.0	4.5379E-12	5.27144	5.01986E-2	12	0.970728	26
	3.5	7.4274E-13	5.65417	5.10529E-2	17	0.977045	25
	0.6	8.3386E-5	0.33801	3.26973E-3	0	0.995364	26
	0.8	4.0062E-5	0.62169	6.61408E-3	0	0.977724	26
Nor.,1984	1.0	5.1408E-5	0.53916	7.24732E-3	0	0.991064	26
	1.3	3.5057E-5	0.67357	8.01736E-3	0	0.990430	26
Mar.,1985	1.5	1.8908E-5	0.82889	6.66332E-3	0	0.979404	26
	2.0	1.0957E-5	1.10293	9.63075E-3	0	0.976618	26
	2.5	1.4597E-6	1.78200	1.40006E-2	0	0.958267	26
	3.0	6.0389E-7	1.73581	1.65044E-2	0	0.870311	26
	0.6	5.7585E-5	0.48692	4.83220E-3	0	0.995516	29
Nor.,1985	1.0	4.0372E-6	0.95626	5.82667E-3	0	0.843356	29
	1.5	7.1419E-6	1.13433	9.77472E-3	0	0.956321	29
Mar.,1986	1.8	5.9217E-6	1.19419	1.04247E-2	0	0.954608	29
	2.0	6.8387E-7	1.86275	1.85057E-2	0	0.920463	29
	2.5	4.3807E-7	1.97528	1.26035E-2	0	0.843970	29

Table 6. The values of A.B.C.D for the silty clay in equation (3)

Year	Groundwater table (m)	A	B	C	D	R	N
	0.6	1.2832E-7	2.00153	1.73486E-2	0	0.853908	26
	0.8	1.6099E-9	3.41877	3.36417E-2	0	0.855276	27
Nor.,1982	1.0	2.6722E-11	4.66206	5.04303E-2	0	0.971065	26
	1.3	2.1555E-8	2.21967	6.25220E-3	0	0.950285	27
	1.5	2.8646E-8	2.00757	2.11434E-3	0	0.978512	27
Mar.,1983	1.8	1.1197E-8	1.86573	-1.60317E-2	0	0.941032	27
	2.0	1.0340E-12	3.93251	-5.17596E-3	15	0.980937	24
	2.5	2.0893E-13	5.15781	2.79960E-2	20	0.959863	23
	3.0	2.6680E-10	1.34439	-5.03775E-2	15	0.932633	24
	3.5	6.0752E-6	-0.73575	-5.19523E-2	25	0.977207	22
	0.6	2.8321E-11	4.31049	4.35434E-2	17	0.910736	25
	0.8	5.0841E-16	7.48560	7.44685E-2	17	0.968987	25
Nor.,1983	1.0	2.4722E-20	10.41777	1.05970E-1	17	0.957284	25
	1.3	1.8295E-13	5.09101	3.48750E-2	42	0.844443	20
	1.5	1.4818E-10	3.32546	6.61806E-3	7	0.972177	27
Mar.,1984	1.8	4.8700E-33	17.24432	1.42525E-1	57	0.969020	17
	2.0	5.5297E-19	6.55376	-4.21203E-3	62	0.982835	16
	0.6	1.4565E-8	2.53870	2.70802E-2	5	0.878626	26
	0.8	8.0662E-9	2.61118	1.94005E-2	0	0.955715	27
Nor.,1984	1.0	1.0241E-7	1.94385	9.42543E-3	0	0.872668	27
	1.3	4.4039E-7	1.26823	1.43580E-3	0	0.757169	27
Mar.,1985	1.5	8.8363E-7	1.00032	-1.15572E-2	0	0.951167	27
	2.0	1.7300E-6	0.47976	-2.17320E-2	0	0.977902	27
	2.5	2.5719E-14	5.73060	3.51482E-2	15	0.979254	24
	0.6	2.2045E-5	0.80575	8.54337E-3	0	0.988138	27
Nor.,1985	1.5	3.9333E-6	0.75103	-9.80816E-3	0	0.908985	29
Mar.,1986	1.8	6.2309E-6	-1.41901	-5.70119E-2	0	0.884371	28
	2.0	3.9387E-7	1.17966	-4.75564E-3	0	0.874922	29
	2.5	1.5499E-9	3.34449	-3.86010E-3	15	0.943345	26

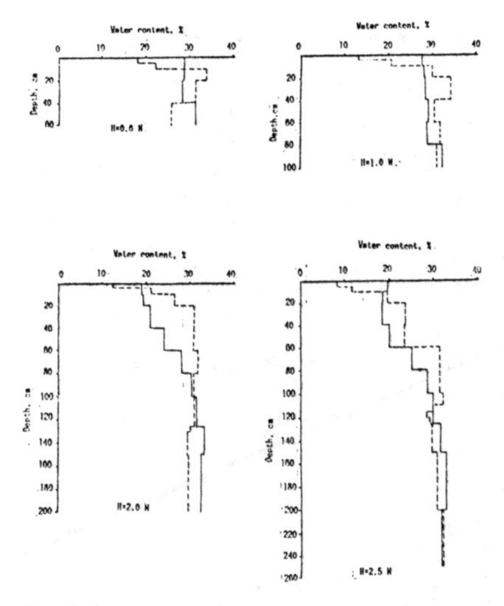

Fig.6 The water content profiles of the sandy silt before and after freezing, the real line-before freezing, the dashed line-after freezing

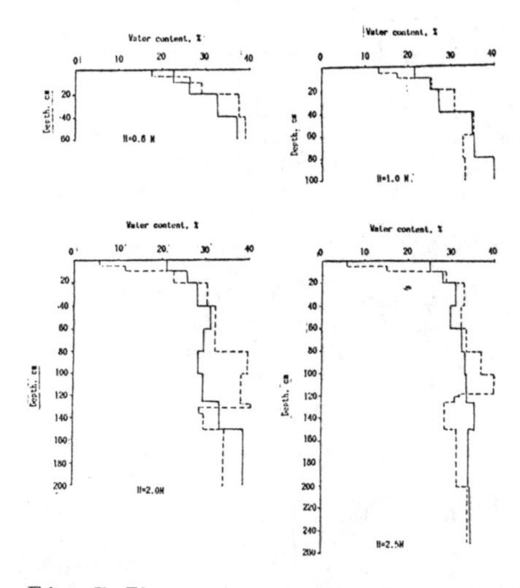

Fig.7 The water content profiles of the silty clay before and after freezing, the real line-before freezing, the dashed line-after freezing

Water Content Profiles of Soils Before and After Freezing

Figures 6 and 7 show the curves of the water content vs. the depth before and after freezing for the two types of soils, respectively. It can be seen from Figs. 6 and 7 that water migrates towards the freezing front and the water accumulated zone moves downward with the increase in the buried depth of ground water. The water content of soils in the range from the ground surface to the depth of 10 cm during freezing is less than that before freezing because of the ground surface evaporation. After freezing the water content of soils below the seasonally maximum frost depth is less than that before freezing, too, because the soils beneath the frost depth is denser due to the action of the frost heaving force and this phenomenon was proved by the laboratory tests (Xu Xiaozu, et al., 1986). After freezing the changing regularity of the water content profiles of soils under different groundwater tables is in good agreement with that of the total amount of water intake flow during freezing, as mentioned above.

CONCLUSIONS

1. The relationship between the total amount of water intake flow during freezing and the buried depth of ground water changes with soil types. The total amount of water intake flow during freezing is in the lowest and the highest when the buried depth of ground water equals to 1.0 and 2.0 m. Respectively, for the sandy silt, and is in the highest when the buried depth of ground water equals to 1.5 m for the silty clay.
2. The curve of the water intake flux vs. the elapsed time has a peak value. With the increase in the buried depth of ground water, the time when the peak value occurs is later.
3. The relationship between the water intake flux of soils and the elapsed time of freezing can be expressed in the power form.

ACKNOWLEDGEMENT

The authors wish to express sincere appreciation to Mr. Shen Zhongyan for his guidance in computer programing.

References*

1. Fukuda, M. & Nakagawa, S., 1985, Numerical analysis of frost heaving based on the coupled heat and water flow model, Proc. of 4th Intern'l. Sympo. on Ground Freezing, pp. 109-117.
2. Knutsson, S., Domaschuk, L. & Chandler, N., 1985, Analysis of large scale laboratory and in situ frost heave tests, Proc. of 4th Intern'l. Sympo. on Ground Freezing, pp.65-70.
3. Williams, P.J., 1984, Moisture migration in Frozen soils, Final Proc. of 4th Intern'l. Conf. on Permafrost, pp.64-66.
4. Xu Xiaozu, 1982, Research in the question of water migration in frozen soils abroad, Journal of Glaciology and Geocryology, Vol. 4, No.3, pp.98-103.
5. Xu Xiaozu, Deng Yousheng, Wang Jiacheng & Liu Jimin, 1986, A study on the behaviors of water migration in the saturated freezing soil, Proc. of 3rd National Conf. on Ground Freezing (in press).

*See page X concerning ISGF papers

5th International Symposium on Ground Freezing, Jones & Holden (eds)
© 1988 Balkema, Rotterdam. ISBN 90 6191 824 3

Experimental study of frost heaving of a saturated soil

Takeshi Ishizaki & Nobuaki Nishio
Tokyo Gas Co., Ltd, Research and Development Institute, Shibaura, Minato-ku, Tokyo, Japan

ABSTRACT: This paper reports an experimental study of the relationship between the water intake rate and the temperature conditions in the frozen zone of a soil sample. Both end plate temperatures were lowered at a constant rate, which allowed us to obtain both an approximately constant frost penetration rate and temperature gradient in the sample during the experiment. The experimental results showed that the water intake rate increased linearly with the temperature gradient in the frozen zone, and it did not change greatly with increasing frost penetration rate. The frost heave rate increased as the freezing front advanced and the length of the unfrozen zone decreased. These experimental results were utilized to test the frost heave model developed by Miller (1978).

1 INTRODUCTION

Since most of the inground LNG tanks are constructed around Tokyo Bay in Japan, the surrounding ground water level is very high and the soil is under saturated condition. The depths of the large tanks are more than 40 m, and therefore the surrounding ground is under the overburden load. Frost heave models which are applicable to these conditions have been proposed by Miller (1978), Hopke (1980), Takashi (1982) and others. It is assumed that the ice segregates at a plane where a criteria for ice segregation is satisfied in the frost heave models developed by Miller and Hopke. On the other hand, it is assumed that the ice segregation zone distributes widely in the frozen zone in the frost heave model developed by Takashi. Ishizaki et al. (1987) discussed the width of the ice segregation zone by the frost heave experiment using X-ray technique. The experimental results showed that the major portion of the heave was attributed to the ice lens growth nearest to the 0°C isotherm and supported the frost heave models developed by Miller and Hopke. In these frost heave models, the concept of the effective stress in the unfrozen unsaturated soil by Bishop et al. (1963) is applied for the frozen saturated soil, and the neutral stress of the pore contents (σn) is expressed by the pore water pressure (Uw) and the ice pressure (Ui) as follows

$$\sigma n = \chi Uw + (1 - \chi) Ui \quad (1)$$

where the quantity χ is a stress partition factor and a function of the unfrozen water content.

When the neutral stress (σn) surpasses the value of the overburden pressure (P), and the pore contents alone are able to support the overburden, a new ice lens is assumed to form in these models.

$$\sigma n > P \quad (2)$$

The objective of the study is to determine the frost heave amount and rate in the frozen zone under different temperature gradients and frost penetration rates by the frost heave experiments, and to compare the experimental results and the simulated ones to test the proposed frost heave models.

2 EXPERIMENTAL APPARATUS

The apparatus was similar to that reported by Ishizaki et al. (1985) and illustrated schematically in Fig. 1. The soil sample was contained in a plexiglass cylinder with a wall thickness of 2 cm and an inside diameter of 12.5 cm. Porous plates with a thickness of 1 cm were located directly between the end cooling plates and the soil sample, in order to obtain a steady heat flow condition at the start of soil freezing. The temperature of the end plates was controlled with thermoelectric cooling control systems. Radial heat flow to the soil sample was minimized with a 5 cm thick styrofoam around the soil container, placing the soil sample and styrofoam in a room with a constant temperature of 1°C. Eleven platinum resistance thermometers (3 mm in length and 0.6 mm in diameter) were mounted along the sample walls in contact with the sample. The seven nearest the cold end were spaced at 5 mm intervals, the rest were 10 mm apart. All platinum resistance thermometers were calibrated to 0.01°C to obtain precise temperature profiles. Small lead spheres (diameter 1.7 mm) were embedded in the sample at the intervals of about 5 mm in two vertical columns. The amount of frost heave in the frozen zone was measured by taking X-ray photos of the sample and locating the images of lead spheres. The source current of X-ray was 19 mA at 160 kV and the exposure time was 2.5 min. The total heave measurement was made with a dial gauge to a precision of ±0.01 mm. Data from the platinum resistance thermometers, the volume change indicator and the dial gauge with an electric output were collected periodically by a digital data acquisition system.

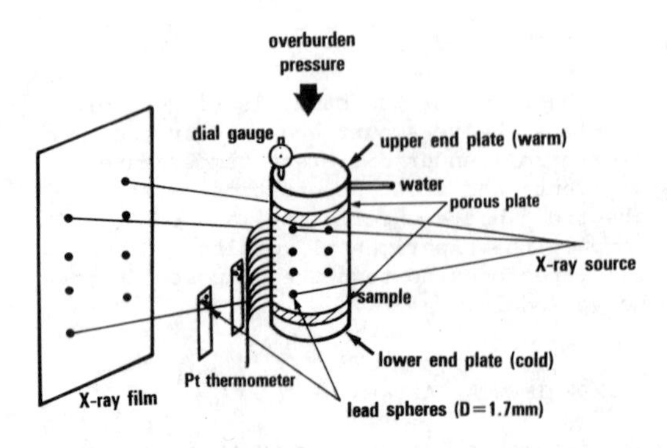

Fig. 1 Experimental apparatus

3 TEST PROCEDURE

An undisturbed clay sample with high frost susceptibility was used in the experiment. The various properties of the soil are summarized in Table 1. The initial sample length was 54 mm. Prior to the start of the experiment, both end plate temperatures were at 1°C. After equilibrating for 24 hours, the lower end plate was cooled down, and the lower porous plate started to freeze. Both end plate temperatures were then lowered at a constant rate to get approximately constant frost penetration rate and temperature gradients in sample. Ten experiments (A1-A7) were performed under different temperature conditions as shown in Table 2. The temperature change indicated by 10 - 0.4 t in the table shows that the initial temperature was 10°C and it was lowered at a constant rate of 0.4 (°C/h) from the start of the experiment. The applied load was 170 kPa and the back pressure was 50 kPa at the upper end of the sample.

Table 1 Properties of soil sample

	Unit	
silt (5 ~ 74 μm)	%	38
clay (< 5 μm)	%	62
porosity		0.487
bulk density	g/cm³	1.99
dry density	g/cm³	1.50
hydraulic conductivity	cm/s	$1 \sim 5 \times 10^{-8}$
specific surface area	m²/g	20.8

Table 2 Temperature conditions

Test	A1,A1'	A2,A2'	A3,A3'	A4	A5	A6	A7
Warm Plate Temperature(°C)	10−0.4t	10−0.2t	10−0.1t	5−0.1t	2.5−0.1t	2.5−0.05t	2.5−0.025t
Cold Plate Temperature(°C)	−0.4t	−0.2t	−0.1t	−0.1t	−0.1t	−0.05t	−0.025t

4 TEST RESULTS AND DISCUSSIONS

A typical example of temperature profiles (Test A2) is shown in Fig. 2. The nearly linear temperature profiles were obtained in both unfrozen and frozen zones in the soil sample. The movements of the 0°C isotherms which were obtained from the temperature profiles are shown in Fig. 3. The frost penetration rate was relatively constant for each experiment (Tests A1-A7). The average temperature gradients in the frozen zone, while the 0°C isotherms were located in the soil sample, are shown in Fig. 4. The temperature gradient showed a maximum value when the 0°C isotherm first penetrated into the soil sample and then decreased gradually with time. The frost penetration rate and the temperature gradient when the 0°C isotherm reached the upper end of the sample are all given in Table 3. The total heave curves, as a function of time, are shown in Fig. 5. The arrows show times when the upper porous plate started to freeze, which were determined by the water intake curves. When the total heave curves are compared among Tests A2, A4 and A6 of similar frost penetration rates but different temperature gradients, it can be seen that the total heave amount and heave rate are greatest for Test A2 and smallest for Test A6. This trend corresponds to the temperature gradient in the frozen zone, which is greatest for Test A2 and smallest for Test A6 as shown in Table 3. It is also found that the total heave amount is greatest for Test A3 and smallest for Test A1 among Tests A1, A2 and A3 which have similar temperature gradients in sample but different frost penetration rates. This corresponds to the frost penetration rate, which is lowest for Test A3 and highest for Test A1. The same trend is found among Tests A5, A6 and A7. The frost heave amounts after freezing of the upper porous plate are greater in Tests A5, A6 and A7 than in Tests A1, A2 and A3. This arises from the fact that the length of the frozen fringe is greater, and therefore, the unfrozen water content is higher in Tests A5, A6 and A7 than in Tests A1, A2 and A3.

The frost heave rates, as a function of time are shown in Fig. 6 for Tests A1, A2 and A3. The heave rate was calculated at ten minutes intervals by dividing the heave amount difference by the time interval. The fluctuation of the measured heave rate could be due to the limitation of the dial gauge resolution. The frost heave rates showed a similar value for the same frost depths in Tests A1, A2 and A3. The frost heave rate increased gradually with time in each test. This could be attributed to the decrease of the unfrozen zone length, therefore, the decrease of the total hydraulic resistance. The increasing heave rate will be discussed later by using the frost heave model developed by Miller (1978). Fig. 7 shows the frost heave rates for Tests A5, A6 and A7. The same trend was also observed for these tests. The frost heave rates for Tests A1, A2 and A3 are greater than those for Tests A5, A6 and A7.

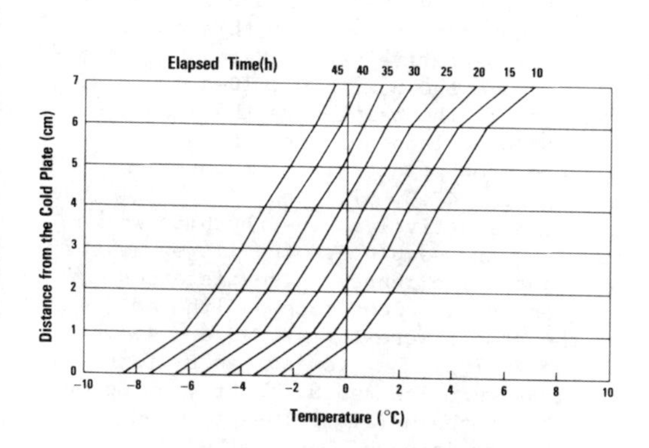

Fig. 2 Temperature profiles in sample

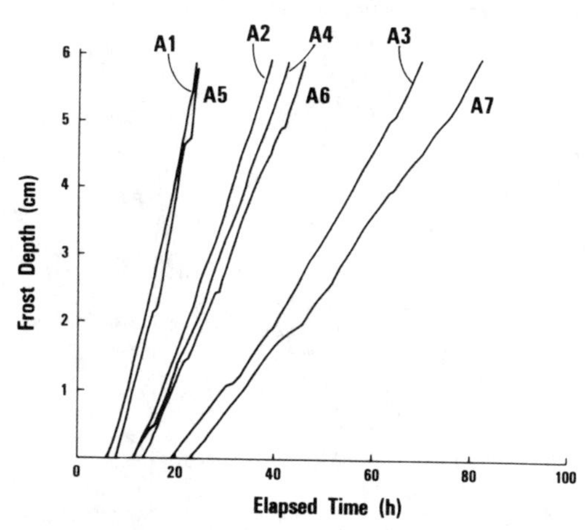

Fig. 3 Location of the freezing front (0°C line)

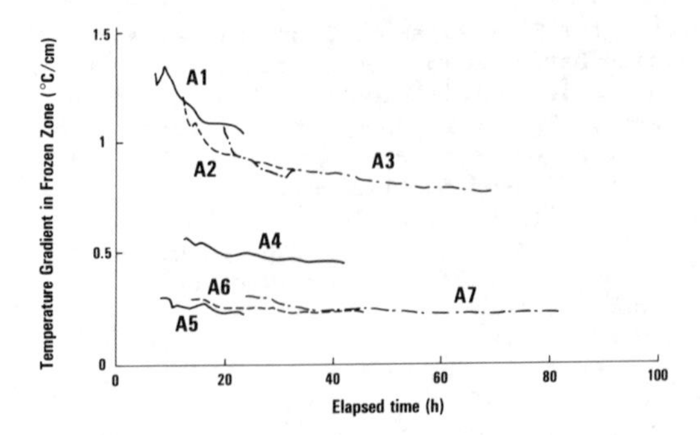

Fig. 4 Average temperature gradient in frozen zone

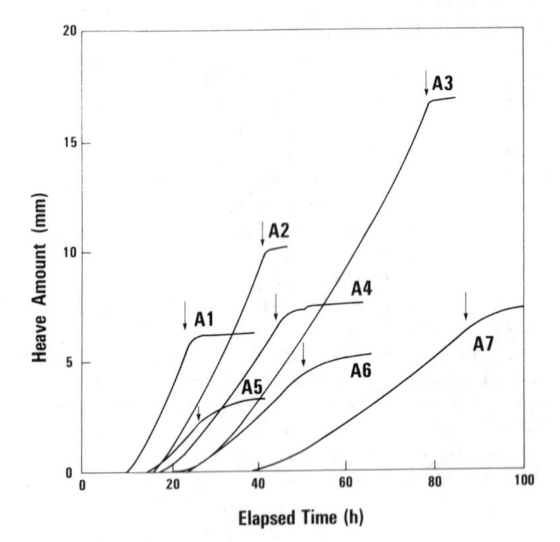

Fig. 5 Frost heave amount

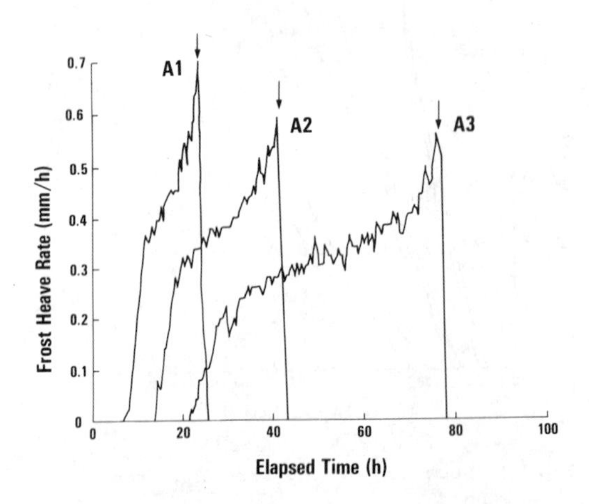

Fig. 6 Frost heave rate (Tests A1 - A3)

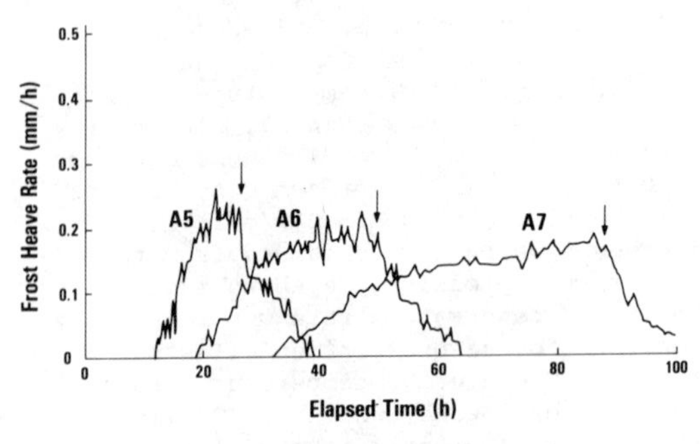

Fig. 7 Frost heave rate (Tests A5 - A7)

Table 3 Freezing rate and temperature gradient in frozen zone

Test	A1	A1'	A2	A2'	A3	A3'	A4	A5	A6	A7
Freezing Rate (mm/h)	3.84	3.65	2.31	2.18	1.30	1.21	2.09	4.01	2.09	1.07
Temperature Gradient in Frozen Zone(°C/cm)	1.01	1.05	0.86	0.94	0.78	0.86	0.46	0.23	0.24	0.22

The relationship between the water intake rate and the temperature gradient in the frozen zone when the 0°C isotherm reached the upper end of the sample is shown in Fig. 8. The water intake rate (Vw) increased linearly with increasing temperature gradient (αf) in the frozen zone. This corresponds to the results reported by Takashi (1982) and Takeda (1985). The relationship between the water intake rate and the frost penetration rate is shown in Fig. 9. This figure shows that there is not a direct relationship between the water intake rate and the frost penetration rate for this soil.

A representative set of X-ray photos for Test A3 are shown in Figs. 10-a to 10-d. The dots in the sample are the images of the embedded lead spheres. Figs. 10-a, b, c and d show prints of the X-ray films after 39.8, 47.9, 57.1 and 64.0 h from the start, respectively. The rhythmic white bands in the sample are the images of the segregated ice lenses. The distances between the ice lens face and the cold end of the sample were 16, 27, 40 and 49 mm, respectively. The growing faces of the ice lens were located at about 4 mm below the 0°C isotherm (Figs. 10-a to 10-d). The growing faces of the ice lens were relatively flat in all cases. This confirmed one-dimensional heat flow in the sample. Displacements of lead spheres in the sample for Test A3 are shown in Fig. 11. The curves, indicated by 1 cm - 4 cm, show the displacements of lead spheres, which were located at 1 cm to 4 cm from the cold end of the sample at the start of the experiment. The curve, indicated by DG shows the total heave amount measured by the dial gauge. The arrows in the figure show the times when the growing face of the ice lens reached the locations of these lead spheres. This figure shows that the displacement rates of lead spheres decreased rapidly just after the growing face of the ice lens passed their locations, which are indicated by the arrows. This shows that the major portion of the heave is attributed to the ice lens growth nearest to the 0°C isotherm. The result also indicates that heave in the frozen zone behind the growing ice lens has little to no effect on the total heave amount. This observation agrees with the results obtained by Mageau et al. (1980). The displacements of lead spheres are greater than the total heave amount just after the start of soil freezing. This shows that the unfrozen zone of the sample is slightly consolidated by about 1.5 mm after the start of soil freezing.

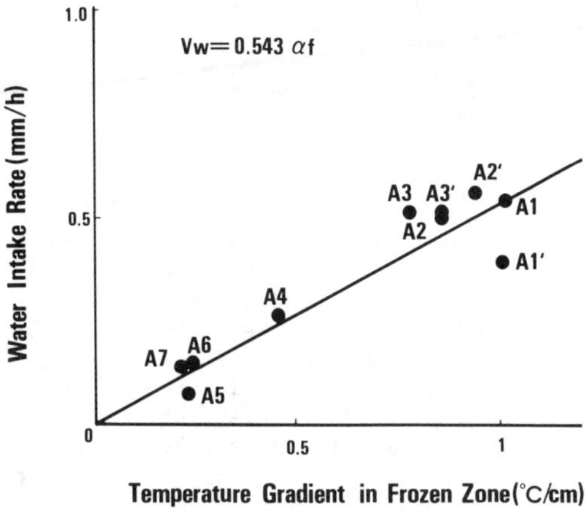

Fig. 8 Water intake rate vs temperature gradient in frozen zone

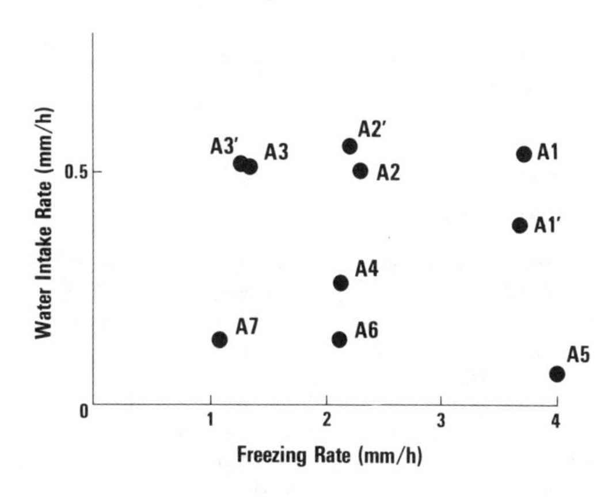

Fig. 9 Water intake rate vs freezing rate

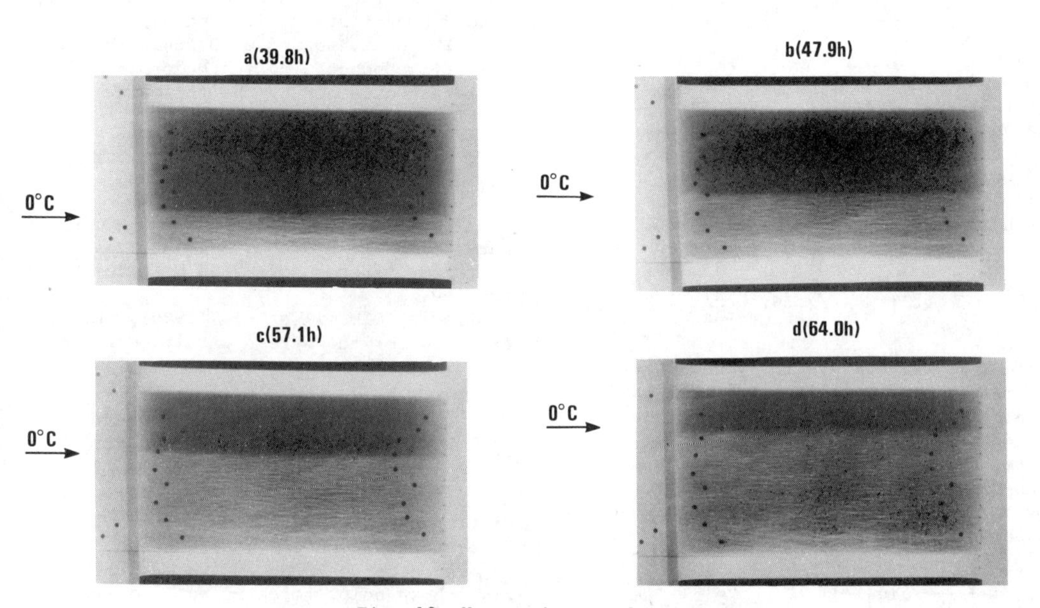

Fig. 10 X-ray photos of sample

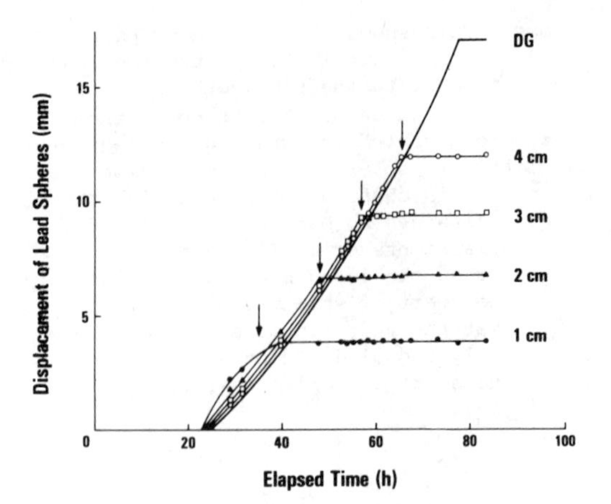

Fig. 11 Displacement of lead spheres in sample

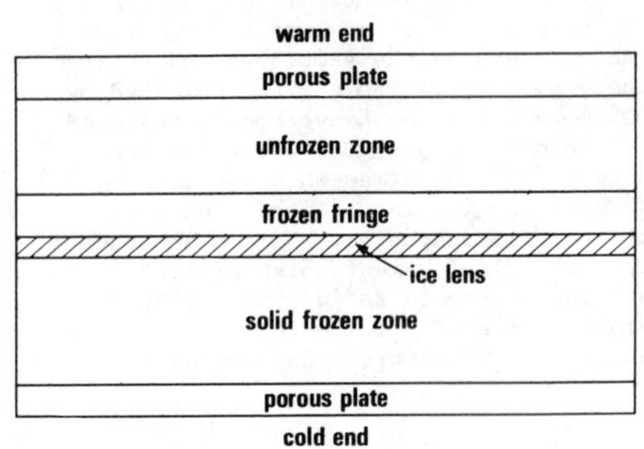

Fig. 12 Section through freezing soil sample

5 FROST HEAVE MODEL

The frost heave model, which was proposed by Miller (1978), was utilized to simulate the experimental results. When a column of soil is freezing, there are four distinct zones of soil. They are the unfrozen zone, frozen fringe, growing ice lens and solid frozen zone as shown in Fig. 12. A mass flow equation, which was developed following O'Neill et al. (1985), is shown in eq. 3.

$$(\rho i-\rho w)\frac{\partial I}{\partial t} - \frac{\partial}{\partial x}\left(\frac{k}{g}\cdot\frac{\partial Uw}{\partial x} - \rho i V_I I\right) = 0 \qquad (3)$$

where ρi and ρw are the density of ice and water, respectively, I is volumetric ice content, k is the hydraulic conductivity, g is the acceleration of gravity, Uw is the pore water pressure, and V_I is the ice velocity (heave rate).

A heat transfer equation is shown in eq. 4.

$$C\frac{\partial T}{\partial t} - \frac{\partial}{\partial x}\left(K\frac{\partial T}{\partial x}\right) - \rho i L\left(\frac{\partial I}{\partial t} + V_I\frac{\partial I}{\partial x}\right) = 0 \qquad (4)$$

where T is temperature (°C), C is the volumetric heat capacity, K is the thermal conductivity, and L is the latent head of fusion of water per unit mass.

The equation of mass conservation at the lens-fringe interface gives an expression for the ice velocity (heave rate), which is shown in eq. 5.

$$V_I = -k\frac{\partial Uw}{\partial x}/\rho ig(1-I) \qquad (5)$$

The pore water pressure profile (Uw) and the temperature (T) are obtained by solving the coupled head and mass transfer equations (eq. 3 and eq. 4). The volumetric unfrozen water content (Nu) is determined in the empirical formula shown in eq. 6 proposed by Anderson et al. (1972) using the specific surface area (S) data

$$Nu = \rho d\cdot Exp[a + b\ell nS + cS^{d}\ell n(-T)] \qquad (6)$$

where ρd is the soil dry density and a, b, c and d are constants i.e., a=0.2618, b=0.5519, c=-1.449 and d=-0.264. The hydraulic conductivity in the frozen fringe was obtained by the following equation.

$$k(Nu) = ko\left(\frac{Nu}{N}\right)^{5} \qquad (7)$$

where ko is the hydraulic conductivity of saturated unfrozen soil, and N is the porosity.

Following O'Neill et al. (1985), the form used for the stress partition factor χ is

$$\chi = \left(\frac{Nu}{N}\right)^{1.5} \qquad (8)$$

It is also necessary to obtain the relationship between the volumetric unfrozen water content (Nu) and the difference between ice and water pressure ($\emptyset iw$) shown below to carry out the simulation.

$$\phi iw \equiv Ui - Uw \quad (9)$$

where Ui is the ice pressure.

Since this relationship was not determined by experiment, the relationship was obtained by the following procedure. The integration of an appropriate form of the Clausius-Clapeyron equation gives

$$Uw - (\frac{\rho w}{\rho i})Ui = (\frac{\rho wL}{273})T \quad (10)$$

From eq. 9 and eq. 10, ϕiw can be expressed as follows, when Ui=0

$$\phi iw = -(\frac{\rho wL}{273})T \quad (11)$$

Elimination of T between eq. 6 and eq. 11 gives

$$Nu = \rho d \cdot Exp[a + b\ell nS + cS^{d}\ell n(\frac{273}{\rho wL}\phi iw)] \quad (12)$$

Although this equation was obtained for no ice pressure, it is assumed it is still valid when a small ice pressure exists. The other physical constants for the soil sample and porous plate used in the simulation are shown in Table 4.

Table 4 Physical constants of soil sample and porous plate

	Unit	
soil		
thermal conductivity in frozen zone	W/mK	1.74
〃 in unfrozen zone	W/mK	1.40
specific heat capacity of frozen zone	J/kgK	1350
〃 of unfrozen zone	J/kgK	2260
hydraulic conductivity	cm/s	3.0×10^{-8}
porous plate		
thermal conductivity in frozen zone	W/mK	0.708
〃 in unfrozen zone	W/mK	0.566
specific heat capacity of frozen zone	J/kgK	1730
〃 of unfrozen zone	J/kgK	2680
porosity		0.348

6 COMPUTATION AND RESULTS

The computer simulation was performed for the experiments (Tests A1 - A7). The calculated frost heave rate, frost heave amount and frost depth for Test A2 are shown in Figs. 13, 14 and 15 with the experimental results. The calculated heave amount is slightly greater than the measurement. This arises from the fact that the consolidation of the sample is not taken into account in this model (however the consolidation was observed at the start of the soil freezing as shown in Fig. 11). The calculated frost depths agreed with the measurements. The calculated frost heave rate increased with time. This trend, also, agreed well with the measurements. The calculated frost heave rate changes rhythmically with time. This cyclic behavior of heave rate was also observed by Holden et al. (1985). However, this rhythmic change of heave rate was not observed during the experiment. It can be said that the further work is required to improve the frost heave model to fully simulate the observed frost heave rate.

7 CONCLUSIONS

The frost heave experiments were performed to obtain the relationship between the water intake rate and the temperature condition when the both end plate temperatures were lowered at a constant rate. The experimental result showed that the water intake rate increased linearly with increasing temperature gradient in the frozen zone of the sample, and it did not change greatly with the frost penetration rate. The displacements of lead spheres showed that the major portion of the heave was attributed to the ice lens growth nearest to the 0°C isotherm. They also indicated that heave in the frozen zone behind the growing ice lens had little to no effect on the total heave amount. The frost heave rate increased as the freezing front advanced and the length of the unfrozen zone decreased. The experimental results were utilized to test the frost heave model developed by Miller (1978). The calculated frost heave rates showed the similar increase with advancing freezing front to the measured ones. However, the calculated heave rate fluctuated greatly with time, which was not observed during the experiment. It could be concluded that the further experimental work of high resolution is required to test and improve the frost heave models.

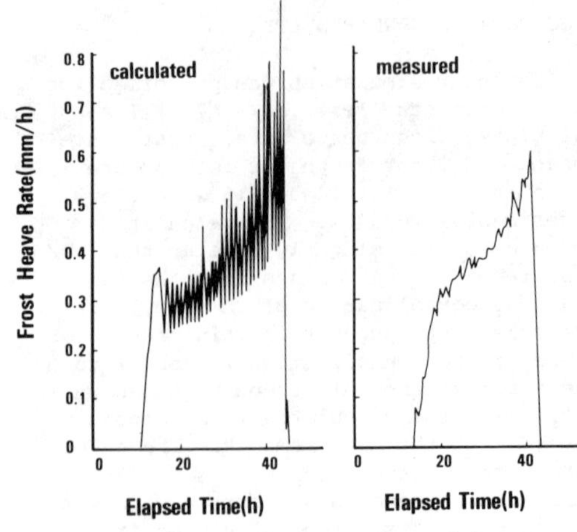

Fig. 13 Comparison of calculated frost heave rate with measurement

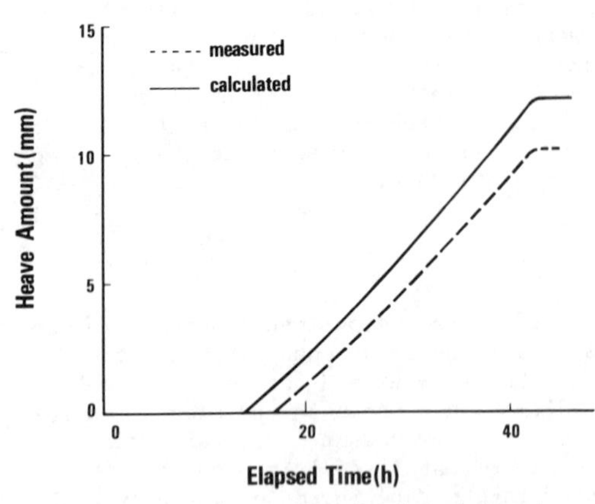

Fig. 14 Comparison of calculated frost heave amount with measurement

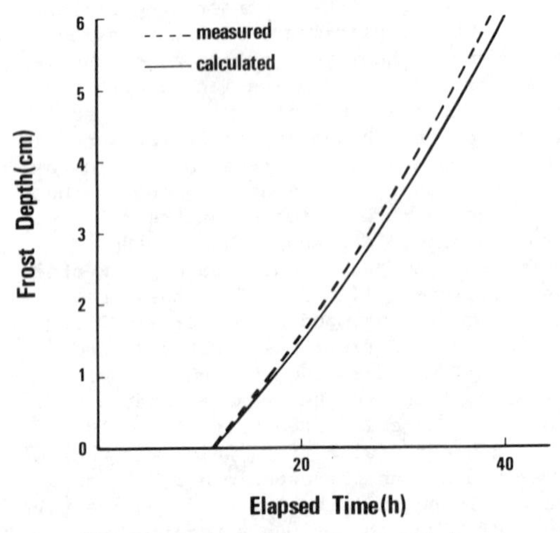

Fig. 15 Comparison of calculated frost depth with measurement

REFERENCES *

Anderson, D. M. and Tice A. R. 1972. Predicting unfrozen water contents in frozen soils from surface area measurements. Highway Research Record, no. 393 , p. 12-18.

Holden, J. T., Piper, D. and Jones, R. H. 1985. Some developments of a rigid-ice model of frost heave. Proc. 4th ISGF, Sapporo, Japan, vol. 1, p. 93-99.

Hopke, S.W. 1980. A model for frost heave including overburden. Cold Regions Science and Technology, vol. 3, p. 111-127.

Ishizaki, T. and Nishio N. 1985. Experimental study of final ice lens growth in partially frozen saturated soil. Proc. 4th ISGF, Sapporo, Japan, vol. 1, p. 71-78.

Ishizaki, T. and Nishio N., unpublished, presented at Meeting of Japanese Society of Snow and Ice, Oct. 1987.

Mageau, D. and Morgenstern, N.R. 1979. Observations on moisture migration in frozen soils. Canadian Geotechnical Journal, vol. 17, p. 54-60.

Miller, R. D. 1978. Frost heaving in non-colloidal soils, Proc. 3rd International Conference on Permafrost, Edmonton, Alberta, vol. 1, p. 707-714.

O'Neill, K. and Miller, R. D. 1985. Exploration of a rigid ice model of frost heave. Water Resources Research, vol. 21, no. 3, p. 281-296.

Takashi, T. 1982. Analysis of frost heave mechanism. Seiken Co. Ltd., Osaka, Japan.

Takeda, K., Nakazawa, J. and Kinoshita, S. 1985. Thermal condition for ice lens formation in soil freezing. Proc. 4th ISGF, Sapporo, Japan, vol. 2, p. 89-94.

*See page X concerning ISGF papers

5th International Symposium on Ground Freezing, Jones & Holden (eds)
© 1988 Balkema, Rotterdam. ISBN 90 6191 824 3

A finite element simulation of frost heave in soils

Roland W.Lewis & Wah K.Sze
University College of Swansea, University of Wales, UK

ABSTRACT: The flux equations in a freezing soil system are derived by using the principles of irreversible thermodynamics. The resulting coupled heat and mass transfer partial differential equations are similar in form to the Luikov equations. A numerical model is subsequently developed to depict the frost heave mechanism in the freezing soil system. The finite element method is applied in this model to solve the "partially nonlinear" form of these coupled differential equations.

1 INTRODUCTION

Every year, the freezing-induced mechanism in soil, referred to as "frost heave" causes a vast amount of damage to structures placed upon it. The structure may be the foundation of a building, underground services such as gas pipes etc., or the top of road or runway surfaces. The mechanism of frost heave is mainly due to the migration of water moisture from the source (usually the underground water table) toward a developing ice lens. The accumulation of water moisture and its consequent phase change at the ice lens generates an uplift force.

This paper concentrates on the different models which can describe the physical behaviour of wet soil subjected to freezing. Broadly speaking, it is possible to classify these into two major groups:

1. A *primary heave* model is characterized by the existence of a sharp freezing front which separates the frozen and the unfrozen zones of the soil system. Some typical papers relevant to this model are published by Taylor & Luthin [1976,1978], Outcalt [1976], and Holden et al [1982].

2. A *secondary heave* model extends the sharp freezing front into a region called the frozen "fringe", where the ice and liquid water are in a state of thermodynamic equilibrium. Some classical papers describing this model include those by Miller [1978,1980], O'Neill [1983], O'Neill & Miller [1985], and Holden et al [1985].

A detailed description of these two models, particularly the mathematical formulation, was presented in a previous paper by the authors [1986]. In that paper, a new model was proposed, based on Luikov's coupled heat and mass transfer partial differential equations, which depicted the migration of water moisture in the freezing soil system. The movement of water moisture ceased to propagate into the frozen region by the formation of an ice lens. The corresponding accumulation of water moisture and its consequent phase change will generate the heaving velocity. Below the ice lensing front is the region where water and ice co-exist in a state of thermodynamic equilibrium. It was assumed that water flux is the only contributing source to the calculation of the magnitude of frost heave. This differentiated our model from that of the secondary heave model presented by O'Neill & Miller [1985] who assumed that the "pore ice" and the "lens ice" were moving together as a rigid body.

The derived coupled heat and mass transfer equations are extremely complex, and any analytical solutions are restricted to simple one-dimensional forms, such as in the paper published by Mikhailov [1976]. Hence, a numerical method, such as the finite element technique, was used to solve the resulting equations.

Since the cross-coefficients in the coupled equations are not equal, the assembly of the whole stiffness matrix has to be considered in the solution scheme. This is an undesirable features which is expensive in terms of computer time. Therefore, in order to avoid this nonsymmetric stiffness matrix, certain parameters in the coupled equations are assumed to be constant. As a result, the coupled differential equations assume the form of a partially nonlinear system. The results achieved using this model will be discussed in the following section.

2 THE FREEZING SOIL SYSTEM

The microscopic features relating the migration of water moisture in a freezing soil system can be explained for most soil types in terms of the electrical charge which exists on the surface of the soil particles or the surface tension effects between the water and ice interface. As already known, the successful application of either of these theories depends mainly upon which is dominant in a particular soil type. The first theory makes use of a knowledge of the behaviour of the negatively-charged surface of the clay particles (see Beskow, 1935). When the soil undergoes freezing, the pore ice propagates towards the surface of the soil particle with a consequent rise in the concentration of cations in this region. This effect will create an osmotic pressure and water will tend to be absorbed from a region of lower concentration of cations (i.e. below the ice lens).

The surface tension effects on the freezing soil system draws on the analogy of the capillary action between an air/water interface and assumes the same action for the ice/water interface. The flow of water moisture toward the ice/water interface is due to the pressure difference across this surface. The analogy between freezing and drying for the porous media is discussed in a paper published by Koopmans and Miller [1965].

As mentioned earlier in this paper, the assumption made in the "rigid ice" model that the "pore ice" and the "lens ice" are moving together with the same heaving velocity is not adopted. In the present model, it is assumed that the pore ice velocity, if any, is small compared with the contribution from the accumulation of water flux. This assumption is well supported by most of the experimental evidence available. For instance, Taber's experiment [1930] showed that the amount of heave on the soil surface equals the total thickness of the segregated ice lens formed by drawing water from the source below the ice lensing front. In addition, a number of primary heave models have been proposed, e.g. Holden et al [1982] where the influence of the ice flux is not included in the heave rate calculation. However, the numerical results they obtained were similar to those obtained from actual experiments. The crucial point in the rigid ice model assumption is that although the ice flux exists, its velocity will be the same as the heaving velocity. However, in the present model it is assumed that the pore ice velocity is relatively smaller than the heaving velocity, hence its contribution towards heave can be neglected.

In our model the flow of water in the region of the partially frozen soil (sometimes termed the frozen fringe) is derived by using the principles of irreversible thermodynamics. A simple form of the Onsager system of water flux is given by the following equation, with a more detailed description presented by Kennedy and Lielmezs [1973]:

$$q_w = -\rho_o(D\nabla\theta_w + D\delta\nabla T_2) \qquad (1)$$

where q_w = water mass flux
D = moisture diffusivity
θ_w = mass-based water content
δ = thermo-gradient coefficient
T_2 = temperature in the partially frozen zone.

If the influence of temperature is ignored in equation (1), then the flow of water is governed by Darcy's law. Equation (1) is also similar to the flux equation derived by Luikov [1976] for the transportation of water moisture in a capillary-porous body for the drying problem.

The heat and mass transfer differential equations in the region of the partially frozen soil (region 2) are derived by considering the conservation of heat and mass flow through an elementary "control" volume. They can be written as :

$$\rho_o c_2 \frac{\partial T_2}{\partial \tau} = -\operatorname{div}(q_h) + \rho_o \lambda \frac{\partial \theta_i}{\partial \tau} \qquad (2)$$

and

$$\rho_o \frac{\partial \theta_i}{\partial \tau} + \frac{\partial \theta_w}{\partial \tau} = -\operatorname{div}(q_w) \qquad (3)$$

where the subscripts refer to
i = ice state
w = water state

and $\rho_o c_2$ = total heat capacity for the partially frozen region
τ = time
λ = specific latent heat
ρ_o = dry density
θ_i = ice content
q_h = heat flux

with
$$\varepsilon = \frac{\theta_i}{\theta_w + \theta_i} \tag{4}$$

On substituting equations (1) and (4) into (3) and assuming ε is constant, we obtain:

$$\rho_o\left(\frac{1}{1-\varepsilon}\right)\frac{\partial\theta_w}{\partial\tau} = \mathrm{div}\left\{\rho_o D(\delta\,\nabla T_2 + \nabla\theta_w)\right\} \tag{5}$$

The rate of change of ice content in equation (2) is converted into water content by using equations (4) and (5). The heat flux, q_h, is expressed by Fourier's law of heat conduction; hence, the heat transfer equation becomes:

$$\rho_o c_2 \frac{\partial T_2}{\partial \tau} = \mathrm{div}\left\{(K_2 + \lambda\,\rho_o \varepsilon D\delta)\nabla T_2 + \lambda\rho_o\,\varepsilon D\nabla\theta_w\right\} \tag{6}$$

Equations (5) and (6) are similar to the Luikov coupled heat and mass transfer differential equations. However, the cross-coefficients (i.e. $\lambda\,\rho_o\varepsilon D$ and $\rho_o D$) of these two equations are not equal if it is assumed that the parameters ρ_o, λ, ε, and δ are constants within the range of interest. On the other hand, if all the material properties defined in equations (5) and (6) were allowed to vary as functions of the prime variables, then the resulting system of equations would be non-symmetric. This is an undesirable feature, and therefore the form of equations (5) and (6) were referred to as "partially" nonlinear, and with some manipulation become:

$$\left.\begin{aligned} C_q \frac{\partial T_2}{\partial \tau} &= \mathrm{div}\left\{K_q \nabla T_2 + K\,\nabla\theta_w\right\} \\ C_m \frac{\partial \theta_w}{\partial \tau} &= \mathrm{div}\left\{K\,\nabla T_2 + K_m \nabla\theta_w\right\} \end{aligned}\right\} \tag{7}$$

where

$$C_q = \delta\,\rho_o\,c_2$$

$$C_m = \lambda\,\rho_o \frac{\varepsilon}{1-\varepsilon}$$

$$K_q = (K_2 + \lambda\,\rho_o\varepsilon D\delta)\delta$$

$$K = \lambda\rho_o\,\varepsilon D\,\delta$$

$$K_m = \lambda\rho_o\varepsilon\,D.$$

3 PARTIALLY NONLINEAR FORMULATION

The "partially" nonlinear formulations (i.e. equation 7) will be used to describe the moisture flow characteristics existing at the frozen fringe. The reason for using this method is that Luikov's coupled heat and mass transfer equations can be applied to both freezing and drying processes in capillary-porous bodies (Luikov, 1966).

In addition, Koopmans and Miller [1965] showed that the flow of water moisture toward the ice/water interface in the freezing soil system has a similar relationship to that which exists in the drying process. As a result of this analogy, the same method can be adopted to describe the migration of water in the frozen fringe.

For example, Thomas et al [1980] studied the results achieved from two different formulations by analysing the problem of drying in a timber section. In one case, all the material properties were allowed to vary as functions of the prime variables, hence a set of differential equations similar to those described in equations (5) and (6) were obtained which required a non-symmetric solver. In the second case, some of the material properties, e.g. ρ_o, λ, ε, and δ were held constant, and the resulting coupled heat and mass transfer equations, which allowed the use of a symmetric matrix solver.

Two of the parameters, ρ_o and λ, in both formulations (i.e. fully nonlinear and partially nonlinear) are usually constant. The influence of allowing two other parameters to vary, i.e. ε and δ, is illustrated in Table 1. There is virtually no difference in the results for both formulations for a particular timber drying problem (see Thomas et al [1980]). Therefore the same principles required to achieve a partially nonlinear formulation in the timber drying problem was adopted in the freezing soil system. It has already been explained that certain parameters assumed constant in the freezing case are preserved in the drying problem (i.e. ρ_o, λ, ε, and δ). These parameters do not have the same meaning in both the freezing and drying processes. In the drying situation, they are used to describe the phase change characteristics between liquid and vapour phases. On the other hand, in the freezing process, they describe the relationship of water which exists in the liquid and solid phases.

For both the freezing and drying cases, it is obvious that the parameters ρ_o and λ are usually constant. In addition, the effect of the thermogradient coefficient, δ, is quite small because the mass flux defined in equation (1) is mainly related to the gradient of moisture content. In other words, the assumption regarding a constant thermogradient coefficient, in both the freezing and drying processes, does not have much effect on the solution. The last parameter assumed to be a constant is the phase conversion factor, ε, which is a function dependent on the temperature.

According to Kennedy and Lielmezs [1973], ε exponentially increases w.r.t. the freezing temperature and the diagrammatic sketch of this parameter is illustrated in Fig.1. Since the temperature range within the frozen fringe is not far below 0°C, the phase conversion factor is usually within the range illustrated in Table 1. As already shown by Thomas et al [1980], a constant phase conversion factor does not greatly influence the results as compared with the fully nonlinear formulation. Therefore it would be expected that the results obtained by assuming a constant phase conversion factor, ε, for the freezing soil system will not be much different to thos obtained if the parameter was allowed to vary as a function of temperature.

4 THE INITIATION OF AN ICE LENS

In the region of the partially frozen soil, the total overburden pressure applied on the surface of the soil must be in equilibrium with the sum of the effective stress and the "total pore pressure" (sometimes referred to as neutral stress). This pore pressure comprises pore ice pressure and pore water pressure and are combined together by introducing a stress partition factor (ψ), which is believed to be a function of the unfrozen water content. A particular partition factor previously used by O'Neill & Miller [1985] is adopted in this paper:

$$\psi = \left(\frac{\theta_w}{\eta}\right)^{1.5} \tag{8}$$

where η is the soil porosity.

The relationship between the total overburden pressure (P_o) with the effective stress (σ') and the neutral stress (σ_n) can be expressed as:

$$P_o = \sigma' + \sigma_n \tag{9}$$

with

$$\sigma_n = \psi P_w + (1 - \psi) P_i \tag{10}$$

At a certain stage during the freezing process, the neutral stress below the former ice lens will reach a magnitude which can exceed the total overburden pressure. The whole soil mass above this level will then be supported by the initiation of a new ice lens.

The boundary condition along the ice lens assumes that the law of conservation of heat and mass transfer is obeyed. If the ice lens is moving in the direction of the y-axis, with the lensing front perpendicular to its heaving velocity, the heat balance equation gives :

$$- K_1 \frac{\partial T_1}{\partial y} = - K_2 \frac{\partial T_2}{\partial y} + \lambda q_w = \bar{q} \tag{11}$$

where subscript '1' represents the frozen region, and $\bar{q}$ is the equivalent heat flux prescribed at this boundary.

The mass balance equation with the heaving velocity V_i is written as :

$$\rho_i V_i = q_w \tag{12}$$

If the accumulation of water at the lensing front is known in advance, equation (12) gives the magnitude of the heaving velocity for the soil subject to freezing. The extraction of heat from the heart of the soil is dependent on the freezing condition on the surface. The heat flow through the frozen soil is governed by the equation :

$$\rho_o c_1 \frac{\partial T_1}{\partial \tau} = - \mathrm{div}\,(K_1 \nabla T_1) \tag{13}$$

where $\rho_o c_1$ = heat capacity in the frozen zone

K_1 = thermal conductivity in the frozen zone.

In order to solve this differential equation (i.e. equation 13), the boundary condition on the lensing front must be defined in advance. This is achieved by using equation (11) as the prescribed heat flux along the ice lens. It is not intended to describe the procedures for solving equation (13) by the finite element method as it has been the subject of many previous investigations, e.g. Wood & Lewis [1975].

The temperature along the ice lensing front is determined by solving the heat conduction equation (i.e. equation 13) in the frozen zone. The Clapeyron equation and the freezing soil characteristics (i.e. equations (14 and (15) defined in the following section) are used to check this temperature. The ice pressure is then replaced by the total overburden pressure. If the temperature does not satisfy equations (14) and (15), then the time-step is changed accordingly and the solution of equation (13) is repeated until the temperature along the ice lensing front satisfies the Clapeyron equation. This temperature then assumes the form of the prescribed boundary condition for the coupled heat and mass transfer equations in the partially frozen soil.

The second method of solving the unknown temperature along the ice lensing front requires much greater computing time. Instead of only using equation (13), the system is now "relaxed"by including the coupled heat and mass transfer equations (i.e. equation 7) with a prescribed flux boundary (see equation 11) which separates the frozen and partially frozen zones. All other procedures are identical to those of the first method which requires an equilibrium condition along with the Clapeyron equation.

These methods only approximate to the true boundary conditions along the lensing front. It is believed that this approximation is satisfactory in the freezing problem as most heat flux is generated by the latent heat effect due to phase change at the lensing front. In doing so, we avoid the expensive alternative of iteration to achieve an equilibrium criterion for the set of differential equations describing the physical behaviour in both the frozen and partially frozen soils. It also preserves the benefits of using the Lees three-level time-stepping scheme such that all the material properties (including the water flux reaching the ice lens) are calculated at the "central" time-step.

5 EVALUATION OF NEUTRAL STRESS

The calculation of neutral stress in the region of the partially frozen soil requires two independent equations relating the pore ice pressure and the pore water pressure. One of these is the Clapeyron equation which defines the state of equilibrium condition of water pressure and ice pressure in terms of temperature. This can be written as follows :

$$\frac{P_w}{\rho_w} - \frac{P_i}{\rho_i} = \left(\frac{\lambda}{k}\right) T_2 \qquad (14)$$

where k is the temperature in degrees Kelvin describing the change in state between water and ice.

The other equation is obtained by utilizing the surface tension effect and by comparing the processes of freezing and drying of particular soil types. From experimental evidence, Koopmans & Miller [1965] show that the relationship between water pressure and ice pressure in a frozen soil system has the same characteristics for the corresponding soil in the drying process. This shows that the difference in ice and water pressure at a particular temperature can be expressed as a function of unfrozen water content $\{F(\theta_w)\}$ in the following equation :

$$P_i - P_w = F(\theta_w) \qquad (15)$$

If the temperature and the moisture content are evaluated at each node in the finite element mesh, the corresponding pore ice pressure and pore water pressure can be calculated from the above equations.

6 THE FINITE ELEMENT DISCRETISATION

This section is concerned with applying the finite element technique to solve the partially nonlinear form of the coupled differential equations. The variation of the temperature and the water content throughout the domain of the frozen frings is approximated in terms of nodal values T_j and θ_j by the following equations :

$$T_2 \approx \hat{T}_2 = \sum_{j=1}^{M} N_j(x,y)\, T_j(\tau)$$

and

$$\theta_w \approx \hat{\theta}_w = \sum_{j=1}^{M} N_j(x,y)\, \theta_j(\tau) \qquad (16)$$

where N_j is the piecewise shape function, and M is the total number of nodes.

These approximation functions (equation 16) are substituted into the coupled differential equations (equation 7),and the corresponding residual is weighted over the domain by utilizing Galerkin's method. Green's theorem is then applied to reduce the order of the differential operators, and the resulting expression is often called the "weak form". The utilisation of the weighted residual statement is intended to minimize the error by using the above approximation functions. As a result of this partial disscretization, a set of ordinary differential equations is produced with respect to the time domain. They are written in matrix form as :

$$\mathbf{C}(\phi)\,\dot{\phi} + \mathbf{K}(\phi)\,\phi + \mathbf{F}(\phi) = 0 \qquad (17)$$

where

$$\mathbf{C} = \begin{bmatrix} C_q & 0 \\ 0 & C_m \end{bmatrix}$$

$$\mathbf{K} = \begin{bmatrix} K_q & K \\ K & K_m \end{bmatrix}$$

$$\phi = \begin{bmatrix} T_2 \\ \theta_w \end{bmatrix}$$

where $\mathbf{C}$ = mass matrix

$\mathbf{K}$ = stiffness matrix

and $\mathbf{F}$ = "forcing" term dependent on the boundary condition.

The numerical solution of equation (17) is solved in the time domain by introducing the Lees algorithm, viz :

$$C^m \left(\frac{\phi^{m+1} - \phi^{m-1}}{2\,\Delta t}\right) + K^m \left(\frac{\phi^{m+1} + \phi^m + \phi^{m-1}}{3}\right) + F^m = 0 \tag{18}$$

where m = time step number

Δt = time step.

This algorithm has been successfully applied to the coupled heat and mass transfer differential equations for the problem of timber drying by Thomas et al [1980].

7 NUMERICAL EXAMPLES

A soil column of 500 mm height was used to demonstrate our frost heave model. The geometry considered and the finite element mesh used are shown in Fig.1. The initial temperature is 8°C and the unfrozen water content is $0.3\, Kg_{moisture}/Kg_{dry\ soil}$. These values are assumed throughout the soil domain.

Dirichlet boundary conditions are assumed along AB and CD with the temperature on the surface AB held at -5°C, which is the average temperature chosen in the experiment (see Fig.3).

At CD the temperature and moisture content were maintained at the same value of the initial conditions. Non-conductive boundary conditions were applied to surfaces AC and BD. The material properties for the computer simulation were assumed to be constant and are shown in Table 2.

The formation of the first ice lens was assumed to begin at the soil surface. This assumption will not cause serious error because subsequent ice lenses close to the surface are formed within a short period of time. The time elapsed for the initiation of a new ice lens close to the soil surface is less than for those located deeper within the soil mass where a longer time is required to establish a new ice lens.

The curve shown in Fig.4 is the profile of the propagation of the ice lens versus time. The numerical model shows that the penetration of the ice lens is slower than that found for the experiment carried out at the British Gas Engineering Research Station to simulate a backfilled excavation crossing a gas pipe. The site is covered by insulated canopies within which the air temperature could be lowered to a value below the freezing point (see Fig.2). The slight diversion of the numerical curve from the measured values is due to the experimental results reporting the 0°C isotherm movement as measured by a thermocouple buried within the soil, as opposed to ice lens formation.

The heave for two different overburden pressures is shown in Fig.5. The heave indicated by these two curves has almost ceased after seven days, but the simulation of an overburden pressure of 90 KPa shows a slight increase in heave rate between points a and b. This is caused by a relatively large time-step value towards the end of the simulation. However, in the second simulation (i.e. 100 KPa overburden pressure) the maximum time-step attained is restricted to half the value previously chosen. As can be seen in Fig.5, the results achieved near the steady state are much better than those obtained in the previous simulation. In addition, both curves show that the heave rate gradually decreases from the onset of the freezing period. The decrease in heave caused by the increase in the total overburden pressure has similar characteristics to the secondary heave model introduced by Miller [1978].

8 CONCLUSIONS

The results given by the proposed model show a close relationship with the experimental evidence. In particular, the penetration of the ice lens, as given by the numerical model and the corresponding experimental measurements of the 0°C isotherm, have a similar degree of magnitude. Furthermore, the characteristics of the heaving profiles at different overburden pressures show a similar relationship with the secondary heave model proposed by Miller. Finally, it is important to stress that the material properties are known to a reasonable degree of accuracy if sensible predictions are to be made with any numerical model such as the one proposed.

Table 1 : Some numerical values for the timber drying problem.

Physical Parameter	Nonlinear	Constant
ε	1.0 at 12% m/c 0.1 at 30% m/c	0.3
δ	0.01/°C at 12% m/c 0.02/°C at 30% m/c	0.02/°C

Table 2 : Material properties used for the numerical simulation.

Physical Parameter	Magnitude
$K_1 = K_2$	2 W/mK
$\rho_o C_1 = \rho_o C_2$	3225 J/Kg K
D	2.5×10^{-10} m²/sec.
δ	0.002 /K
ε	0.1
λ	3.33×10^5 J/Kg
ρ_o	2000 Kg/m³

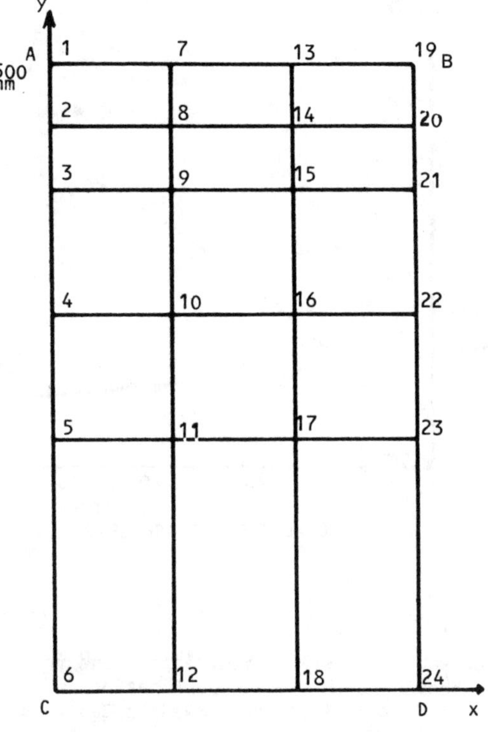

Legend:
AB - Soil surface
AC - Non-conducting boundaries
BD - Non-conducting boundaries
CD - Base of soil column with fixed temperature and moisture content

FIG.2 : FINITE ELEMENT MESH.

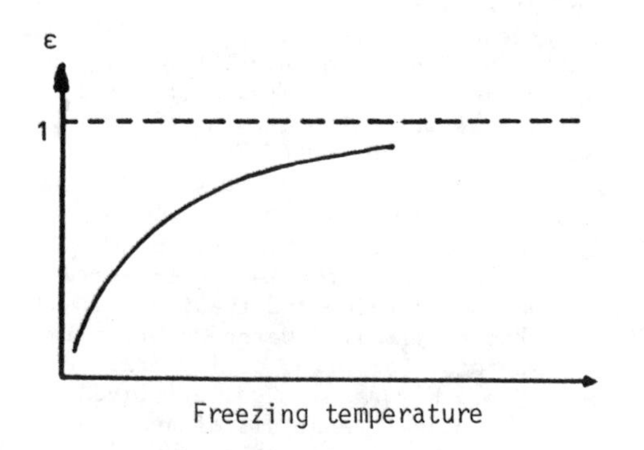

FIG.1 : VARIATION OF THE PHASE CONVERSION FACTOR.

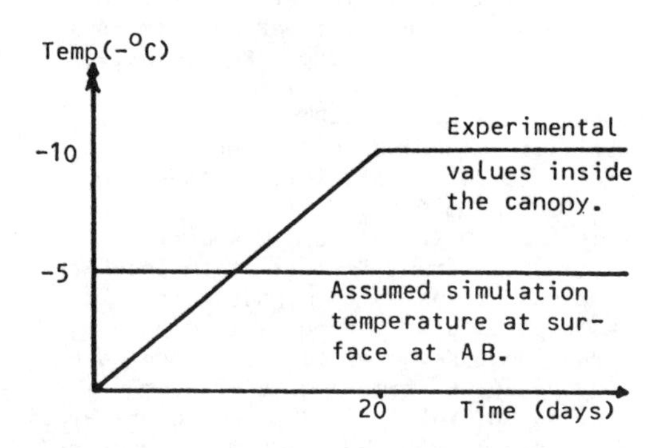

FIG.3 : EXPERIMENTAL AND ASSUMED NUMERICAL TEMPERATURE AT THE SOIL SURFACE.

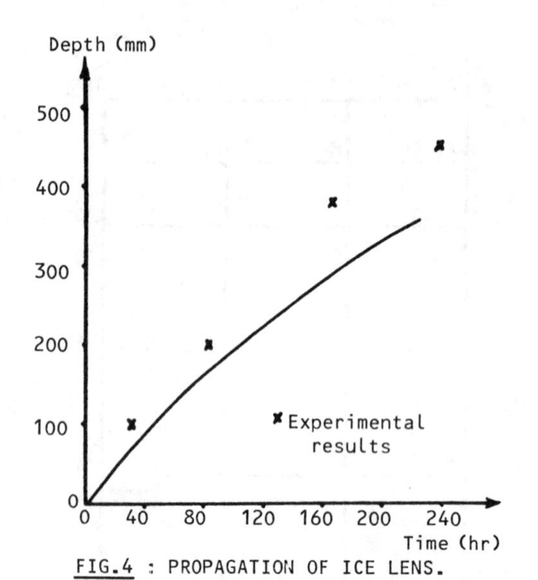

FIG.4 : PROPAGATION OF ICE LENS.

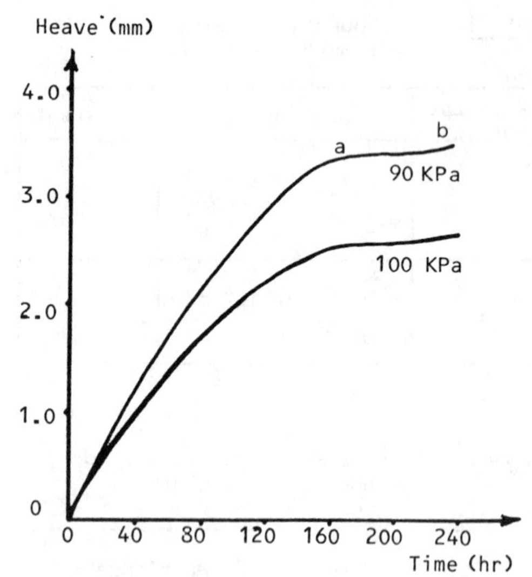

FIG.5 : ACCUMULATION OF HEAVE FOR DIFFERENT PRESSURES.

REFERENCES*

Beskow, G. 1935. Soil freezing and frost heaving with special application to roads and railroads. Swedish Geol.Soc., Series C, No.375.

Holden,J.T., R.H.Jones & S.J.M.Dudek. 1982. Heat and mass flow associated with a freezing front. Developments in Geotechnical Eng., Vol.28, Elsevier Scientific Publishing Company, pp.153-164.

Holden,J.T., D.Piper and R.H.Jones. 1985. Some developments of a rigid-ice model of frost heave. 4th Int.Symp. on Ground Freezing, pp.93-99.

Kennedy,G.F. and J.Lielmezs. 1973. Heat and mass trasfer of freezing water-soil system. Water Resources Res., Vol.9, pp.395-400.

Koopmans,R.W.R. and R.D.Miller. 1965. Soil freezing and soil water characteristic curves. Soil Sci.Soc.Amer.Proc., Vol.30, pp.680-685.

Lewis,R.W., W.K.Sze and P.M.Roberts. A finite element investigation into heat and mass transfer in porous materials with particular reference to ground freezing. Keynote lecture presented at Drying 86, Boston, August 1986. Drying 86, Vol.1, 22-29, Hemisphere Press.

Luikov,A.V. 1966. Heat and mass transfer in capillary porous bodies. Pergamon, Oxford.

Luikov,A.V. 1976. Systems of differential equations of heat and mass transfer in capillary-porous bodies (review). Int.J. Heat and Mass Transfer, Vol.18,pp.1-14.

*See page X concerning ISGF papers

Mikhailov,M.G. 1976. Exact solution for freezing of humid porous half-space. Int.J.Heat Mass Transfer, Vol.19, pp.651-655.

Miller,R.D. 1978. Frost heaving in non-colloidal soils. Proc. 3rd Int.Conf.on Permafrost, Edmonton, pp.708-713.

Miller,R.D. 1980. Freezing phenomena in soils- in 'Applications of Soil Physics' Chapter 11, Academic Press.

O'Neill,K. 1983. The physics of mathematical frost heave models : A review. Cold Regions Science and Technology, Vol.6, pp.275-291.

O'Neill,K. and R.D.Miller. 1985. Exploration of a rigid ice model of frost heave. Water Resources Res., Vol.21, pp.281-296.

Outcalt,S. 1976. A numerical model of ice lensing in freezing soils. Proc. 2nd Conf.Soil & Water Problems in Cold Regions, Edmonton, pp.63-74.

Taber,S. 1930. The mechanism of frost heaving. J.Geol., Vol.38, pp.303-317.

Taylor, G.S. and J.N.Luthin. 1976. Numeric results of coupled heat-mass flow during freezing and thawing. Proc. 2nd Conf.Soil & Water Problems in Cold Regions, Edmonton, pp.155-172.

Thomas, H.R., K.Morgan & R.W.Lewis.1980. A fully nonlinear analysis of heat and mass transfer problems in porous bodies. Int.J.Num.Meth.Engng., Vol.15, 1381-1393.

Wood, W.L. and R.W.Lewis. 1975. A comparison of time marching schemes for the transient heat conduction equation. Int.J.Num.Meth.Engng., Vol.9, 679-689.

5th International Symposium on Ground Freezing, Jones & Holden (eds)
© 1988 Balkema, Rotterdam. ISBN 90 6191 824 3

Thermal and hydraulic conductivity of unsaturated frozen sands

W.K.P.van Loon, I.A.van Haneghem & H.P.A.Boshoven
Department of Physics and Meteorology, Wageningen Agricultural University, Wageningen, Netherlands

ABSTRACT: The heat flow in unsaturated silver sand and river sand is determined by measuring thermal conductivities and temperatures at several depths in laboratory experiments. The thermal conductivity of the unfrozen sand is a function of the volume fraction of water. In the frozen sand however the thermal conductivity cannot be explained in terms of volume fractions of the components water, ice and sand. From the measured thermal conductivity and an assumed constant heat flow, the hydraulic conductivity of the frozen soil can be determined. The maximum value in river sand is 10^{-10} m/s and in silver sand $3*10^{-11}$ m/s. The hydraulic conductivity considered as a function of the temperature has the same characteristics as the unfrozen water content. The influence of the frozen water content in the unsaturated sands on the hydraulic water content is not significant.

List of symbols

symbol	description	unit
A	constant for unit consistency	m^4J^{-1}
C	volumetric heat capacity	$J\ m^{-3}K^{-1}$
c	specific heat	$J\ kg^{-1}K^{-1}$
d	grain diameter	m
h	pressure head of water	m
H	hydraulic head of water	m
k	hydraulic conductivity	$m\ s^{-1}$
L	heat of fusion of water	$J\ kg^{-1}$
p	pressure	$N\ m^{-2}$
Q	heat flow	$W\ m^{-2}$
S	specific surface area	m^2g^{-1}
T	temperature	K
t	time	s
v	velocity	$m\ s^{-1}$
X	position of the freezing front	m
z	vertical coordinate	m
Δ	difference operator	-
θ	volume fraction	-
λ	thermal conductivity	$W\ m^{-1}K^{-1}$
ρ	specific mass	$kg\ m^{-3}$
ϕ	porosity	-
∂	partial derivative operator	-
subscripts		
a	air	
f	frozen	
i	ice	
L	phase change water-ice, fusion	
s	sand	
u	unfrozen	
w	(liquid) water	
_	(underlined) measured value	

1 INTRODUCTION

Frost phenomena in soils are important in technical and agricultural engineering. Frost can damage pipelines, roads and buildings, but it can also improve the structure of the soil. Other fields of interest are artificial freezing (for instance of foods or human organs) and freeze- drying processes.

During the past few decades several theories have been developed to describe frost penetration as well as frost heave. It became clear that a coupling of heat and mass transfer takes place (eg Dirksen and Miller [1] or Kay and Groenevelt [2]). The coupling is very important to explain the secondary heave mechanism. A suction pressure was introduced to explain the water movement towards the freezing front and thermodynamic theories for equilibrium pressures were derived (Groenevelt and Kay [3]). Of course there was always a strong interaction between the development of those theories and laboratory experiments. In these experiments heave was measured, temperature was determined on several depths and often also the water content was followed.

However, up to now some important physical parameters were never determined in situ during a frost penetration test. For a good knowledge of the heat transport also the thermal conductivity has to be known as a function of time and place. Especially in unsaturated soil samples this physical property is hard to estimate. And for a better knowledge of the water transport through the frozen fringe the hydraulic conductivity has to be known in this area.

Our laboratory is specialized in thermal conductivity measurements with the non-steady state probe method [4,5]. Recent developments made it possible to perform these measurements in a short measuring time (less than 5 minutes). This allows us to follow quasi steady state processes. Besides, the averaging volume in which the effective thermal conductivity is determined has become smaller (a cylinder with a diameter of 3 cm). So, now a position dependent thermal conductivity can be measured. Theoretically it is also possible to obtain simultaneously a value for the volumetric heat capacity by this method [5]. From this heat capacity the water content can be determined. Theories of Kay et al [6] and confirmed by van Haneghem [7] allow us to estimate the hydraulic conductivity inside the frozen fringe from the measured heat conductivities (see also next sections).

In this paper the heat transport through two different sand samples is evaluated. From the heat transport other physical phenomena can be derived: eg water transport and hydraulic conductivity. Finally the results will be compared with experiments performed by others.

2 EXPERIMENTAL SET-UP

In real soils most of the transport phenomena caused by freezing occur in the vertical direction. Temperature is mainly a function of the distance to the earth surface and the water content is highly influenced by the distance to the water table. The different soil layers are also horizontally orientated. This allows us to use one- dimensional models both for the

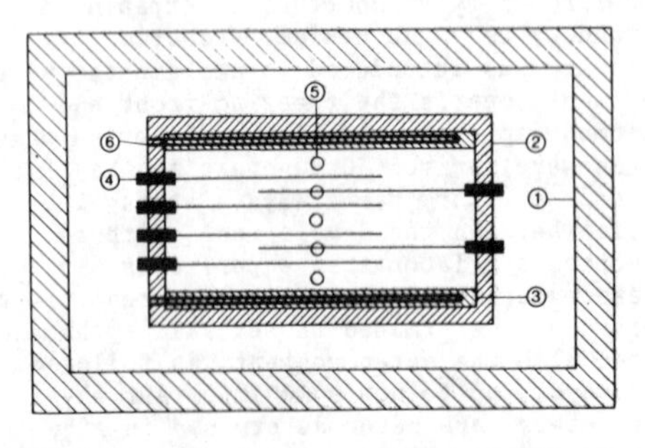

Fig 1 Schematic cross- section of the measuring box
1 insulated climate case, 2 insulated measuring box, 3 lower flat heat exchanger, 4 thermal conductivity probes, 5 thermocouples, 6 upper flat heat exchanger

theory and experiments. So the properties of the sand- water- ice- and air mixture are supposed to vary only in time and in the vertical dimension. To attain this a carefully isolated and closed measuring box was designed. This box has to contain the medium and has to provide the possibility to perform different kind of measurements (see figure 1).

The temperature range in which the experiments were performed is between -10°C and +10°C. For minimizing heat losses in horizontal direction the ambient temperature has to be in the middle of this range. We put therefore the isolated box into a climate case. For the ambient temperature +0.5°C was taken: freezing of the walls of the climate case had to be avoided. The overall vertical temperature difference was controlled by two flat heat exchangers. The heat exchangers were connected with thermostatic baths, so the temperature inside the heat exchangers remains constant until the temperature of the thermostatic bath is changed. Inside the measuring box with sand we placed thermocouples on several depths. Also six thermal conductivity probes were embedded in the sand. With these instruments it is possible to measure in situ the effective thermal conductivity of a sand- water- ice- air mixture.

The most important parts of the needle-shaped probe are a double folded constantan heating wire and a constantan-manganine thermocouple, both with a diameter of 0.1 mm and carefully fitted into a stainless steel envelope (see figure 2). The hot junction of the thermocouple is placed very close to the heating wire; the cold junction is placed at the end of the probe and is considered to stay at its original temperature. To fix the position of the wires in the cylindrical envelope and to prevent electrical contact, the remaining space is filled with a silicon rubber compound. The length of the probe is about 200 mm and its diameter is 2 mm.

In the nonsteady-state probe method the temperature response of the probe on a suddenly starting constant energy dissipation inside the probe is measured. At time t=0 a constant electrical heating current is switched on and the temperature response is recorded once per second. Then the thermal conductivity is inversely proportional to the measured temperature response.

Apart from these, other experiments were performed to characterise the sands more accurately. The following properties of the sand package were determined: hydraulic conductivity of the saturated sand, water retention curve (pF-curve), porosity, particle size distribution (sieve curve) and thermal conductivity as a function of the water content (for room temperature).

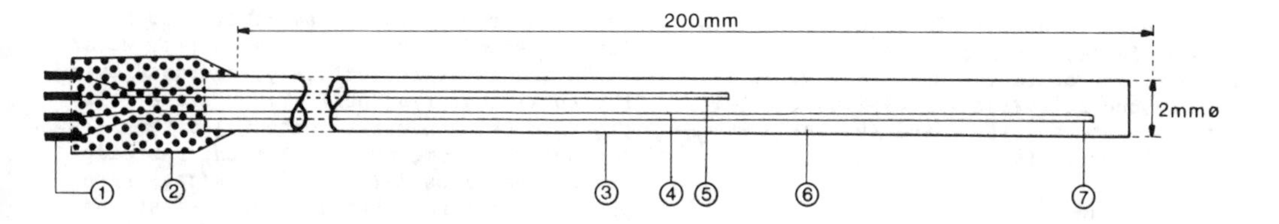

Fig 2 Cross- section of the needle-shaped thermal conductivity probe
1 contacting wires
2 polyvynyl chlorid protector
3 metalic tube
4 hot junction of thermocouple
5 heating wire
6 insulating material
7 cold junction of thermocouple

3 THEORETICAL APPROACH

The medium in which the experiments were performed consists of a solid matrix with pores. Inside the pores air, (liquid) water, vapor and, at temperatures below 0°C, also ice is present. In the latter temperature range both liquid water and ice exists in a thermodynamical equilibrium. At temperatures below the freezing point the liquid water consists of an adsorped film on the surface of the soil particles. In the temperature range below 10°C, in which the experiments were performed, heat transport by radiation can be ignored. The low temperatures cause also low vapor pressures, so heat transport in the vapor phase is also ignored. For our experiments carefully a one dimensional model was constructed: only changes along the vertical (or z-) axis are considered. This enables us to write the continuity equation for mass (ie water) in one dimension:

$$\rho_w \partial\theta_w/\partial t + \rho_i \partial\theta_i/\partial t = \rho_w v_w \quad (1)$$

with ρ the density, θ the volume fraction, t the time and v_w the water velocity. The left hand side of (1) accounts for accumulation of (liquid) water and ice, respectively, while the right hand side consists of transport terms for the liquid phase. The continuity equation for heat yields in this case:

$$C\,\partial T/\partial t + L\rho_w \partial\theta_w/\partial t = \partial/\partial z[\lambda(\partial T/\partial z) + v_w C_w T + v_w L \rho_w] \quad (2)$$

with C the effective heat capacity, T the temperature, L the heat of fusion, z the (vertical) position and λ the thermal conductivity. Here the left hand side accounts for an increase in temperature of the medium and for the latent heat to melt ice. The right hand side consists of the conduction term and the convection terms of sensitive heat and latent heat respectively (in the liquid water phase).

Most measurements however were performed under quasi steady state conditions, so the first term of equation (2) equals 0. The convection of sensible heat is ignored with respect to the conduction. The convection of latent heat might play a role near the freezing front. However, when no heave occurs (no water transport takes place), also this term can be neglected. Then (2) can be simplified into:

$$L\rho_w \partial\theta_w/\partial t = \partial/\partial z[\lambda(\partial T/\partial z)] \quad (3)$$

It is possible to integrate this equation over z. As a result a constant heat flow Q is obtained (see also [8]):

$$Q_L = Q_f - Q_u \quad (4)$$

in which Q_L is the flow of latent heat due to the moving freezing front (with position X), Q_u is the heat flow by conduction in the unfrozen zone and Q_f is the heat flow by conduction in the frozen zone.

The mathematical description of the heat flows yields:

$$Q_L = -\rho_w L \theta_w dX/dt \quad (5)$$

$$Q_f = -\lambda_f \partial T_f/\partial z \quad (6)$$

$$Q_u = -\lambda_u \partial T_u/\partial z \quad (7)$$

In determining the heat flow, the thermal conductivity λ is a very important parameter. To measure λ with the probe method, heat has to be dissipated. During this measurement in the temperature range below 0°C always a fraction of the ice will melt to establish a new thermodynamical equilibrium. But the model for the temperature response of the probe is based on the equation of Fourrier [5]. So:

$$\underline{C}\,\partial T/\partial t = \partial/\partial z[\underline{\lambda}\,\partial T/\partial z] \quad (8)$$

in which $\underline{C}$ and $\underline{\lambda}$ are the measured values.

The measured thermal parameters are not only determined by the material itself but also by the melting of ice, so we must write

(3) in the form of (8). Here the convection of sensible heat is neglected with respect to convection of latent heat. Kay et al [6] obtained the following expressions for the measured heat capacity $\underline{C}$ and the measured thermal conductivity $\underline{\lambda}$:

$$\underline{C} = C + L\rho_w(d\theta_w/dT) \qquad (9)$$

$$\underline{\lambda} = \lambda + L^2\rho_w{}^2Ak/T_0 \qquad (10)$$

The extra term for the heat capacity (9) reflects the latent heat which is needed to melt the small amount of ice which is necessary to obtain the new thermodynamical equilibrium between water and ice at the new (higher) temperature. The second term of the right hand side of (10) reflects the convection of latent heat, which is dissipated by the just mentioned melted ice. With these two equations substantial deviations of the measured $\underline{\lambda}$ and $\underline{C}$, with respect to the real values λ and C, could be explained. And even estimations for $\theta_w(T)$ and k(T) could be given in the subzero temperature range.

4 PROPERTIES OF THE USED SANDS

First we give a short characterisation of the two sands that are used (see table 1). The greyish 'silver' sand consists of only one sieve fraction (150-210 µm). The river sand is from the river Rhine and is well graded: bigger as well as smaller particles are present. This can be seen from the different diameters: eg 10 mass % of the particles has a bigger diameter than d10. Especially the presence of smaller particles (shown by the d90 and d99 diameters) influences the other physical properties of the river sand if it is compared with the silver sand. The capillary rise is bigger and the porosity ϕ is smaller. The water retention curves are shown in the appendix.

Table 1 Physical properties of the used sands

property	unit	silver sand	river sand
d10	µm	204	300
d50	µm	180	170
d90	µm	156	117
d99	µm	150	45
$k_{saturated}$	10^{-4}m/s	2	2
ϕ	volume %	51.7	37.9
θ_{quartz}/θ_s	volume %	99.99	99.9
$\theta_{organic}/\theta_s$	volume %	0	0
θ_w	volume %	7.6	12.3
S [1]	m^2/g	0.1-0.5	0.3-1.0

[1] The specific surface area is measured by isothermal nitrogen adsorption according to the BET theory [9].

Because the capillary rise of the sands is low, we added only a little amount of water (7.6% and 12.3% respectively, averaged over the total measuring box). To obtain a relationship between water content and thermal conductivity some experiments under steady state conditions ($\partial T/\partial t < 0.03$ K/day) were performed (at room temperature). First the thermal conductivity was measured at several depths. Immediately afterwards the water content was determined by taking sand samples at the same vertical position as the conductivity probes. The sand was weighed, dried in a stove and weighed again. From these simple measurements the water content could be calculated and compared with the measured thermal conductivities (see figure 3). Because this kind of water content determination is destructive these measurements were done after the freezing tests.

The measured relation between water content θ_w and thermal conductivity λ confirms the semi empirical equation of Johanson and Frivik [10]:

$$\lambda = \lambda_s + a(\lambda_{ws}-\lambda_s)\{{}^{10}\log(\theta_w/n)+1\} \qquad (11)$$

where λ_s is the thermal conductivity of the dry soil and λ_{ws} that of the water saturated soil. For silver sand they can also be calculated with empirical equations described in [10]: λ_s=0.13 W/mK and λ_{ws}=1.96 W/mK. The best fit was determined for a=0.53. For river sand these properties are: λ_s=0.26

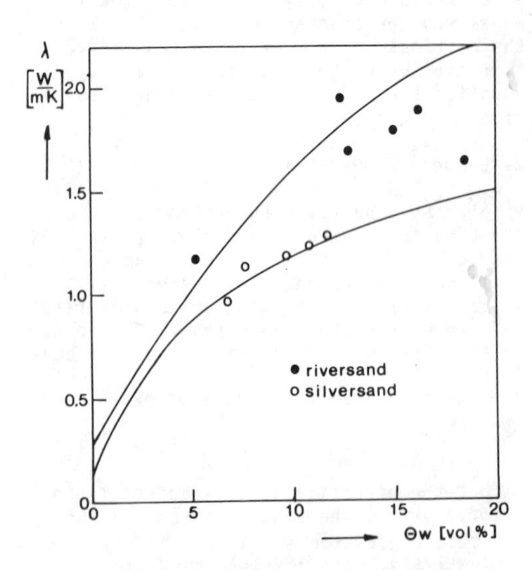

Fig 3 Thermal conductivity λ as function of volumetric water content θ_w in silver sand (outlined signs) and river sand (closed signs) under steady state conditions. The drawn lines are the best fits of the semi empirical Johanson and Frivik relation [10] (see equation 11).

W/mK, λ_{ws}=2.87 W/mK and a=0,83. This semi-empirical relation is for silver sand more accurate than for river sand (see figure 3). The average difference between measured and estimated λ is respectively 0.05 and 0.19 W/mK, which gives relative errors of 5% and 12%.

5 MEASURED HEAT FLOWS

The different non- isothermal experiments performed in the measuring box (indicated in figure 1) are overviewed in table 2. To test the experimental set- up a temperature difference was established for non freezing conditions. Under quasi steady state conditions the temperature gradient was determined by linear interpolation from the thermocouples: grad(T)=$\Delta T/\Delta z$. For position z of the thermal conductivity probe we find the measured heat flow $Q(z)=\underline{\lambda(z)grad(T)}$. For the five probes heat flows of respectively 7.3, 7.9, 6.9, 7.1 and 7.2 W/m^2 were determined. So by this method a constant heat flow of 7.3±0.6 W/m^2 is found. This means the measuring method for the heat flow by conduction is accurate to within 8% [2].
After the determination of the accuracy of the method (test 0), the freezing tests started. For different boundary temperatures the heat flow and the position of the freezing front was measured. Apart from the heat transport by conduction the heat absorbed or dissipated by the moving freezing front Q_L (5) had to be measured. The position of the freezing front itself is determined by linear interpolation of the measured temperatures. Its derivative to the time is calculated using the average displacement of the front during 12 hours. The only unknown quantity left is the water content of the volume element which is freezing. This quantity is estimated with the use of the measured thermal conductivity and equation (14). We admit that this can give errors in θ_w up to 30% (see figure 3: measured θ_w-λ-relations), but table 3 shows that Q_L itself is small compared with Q, so the error in $\underline{Q}$-Q_L remains relatively small.

In table 3 80% of the measurements of the heat flow has about the same value, even when frost penetration causes the largest value for Q_L: Q=$\underline{Q}$-Q_L= 51±3 W/m^2. However, the values on z=122 mm are significantly too big. This has to be caused by the measurement of the thermal conductivity on that depth. The position of the probe is just behind the freezing front (in the frozen area). In this area the liquid water content

[2] The accuracy will become less if bigger values of λ are measured, because the error in λ itself is proportional to λ^2.

Table 2 Overview of tests in silver sand (nr:0-3) and river sand (nr:11-14)

test nr	initial temp[°C] top on z=0	initial temp[°C] bottom z=202 mm	final temp[°C] top on z=0	final temp[°C] bottom z=202mm	freezing front [mm]
0	5.2	3.8	5.2	3.8	-
1	1.3	3.8	-9.0	3.2	125
2	-9.0	3.2	-7.2	3.3	114
3	-7.2	3.3	-5.0	3.2	96
11	10.0	10.1	-4.9	3.0	88
12	-4.9	3.0	-9.7	2.9	136
13	-9.7	2.9	-12.1	2.9	162
14	-6.4	1.6	-5.6	2.2	123

Table 3 Heat flow Q(z) [W/m^2] on several depths in the first freezing test in silver sand (horizontal heat flows were smaller than 0.5 W/m^2).

t[h]	$\underline{Q}_{71}$-Q_L	$\underline{Q}_{95}$-Q_L	$\underline{Q}_{122}$-Q_L	$\underline{Q}_{145}$	$\underline{Q}_{170}$	Q_L
42	-50	-50	-66	-53	-53	-2
48	-51	-52	-65	-52	-52	-1.5
66	-51	-52	-67	-49	-51	-0.5
90	-50	-50	-64	-48	-51	-0.2

is highly temperature dependent. During the measurement of $\underline{\lambda}$ a little heat is dissipated and a small amount of ice will melt. So, for this probe we must apply (10). The measured thermal conductivity $\underline{\lambda}$ consists of two parts: the real conductivity λ and an extra 'conductivity' $\Delta\lambda$ caused by the transport of latent heat:

$$\Delta\lambda = L^2 \rho_w^2 Ak/T_0 \quad (12)$$

If we now assume that also on z=122 mm the heat flow Q =-51 W/m^2, we can calculate the real thermal conductivity:

$$\lambda(122) = -(Q+Q_L)/grad(T_{122}) \quad (13)$$

and the extra 'conductivity':

$$\Delta\lambda = \underline{\lambda} - \lambda \quad (14)$$

Combining (12), (13) and (14) it also is possible to calculate the hydraulic conductivity:

$$k(z) = [T_0/L^2 \rho_w^2 A][\lambda(z) + (Q+Q_L)/(\Delta T(z)/\Delta z)] \quad (15)$$

We take the following values for the constants present in (15): T_0=273 K,

Table 4 Thermal and hydraulic conductivities on depth z=122 mm during the first freezing test in silver sand.

t[h]	$\underline{\lambda}$[W/mK]	λ[W/mK]	Δλ[W/mK]	k[m/s]	T[°C]
42	1.89	1.47	0.42	$1.0*10^{-11}$	-0.10
48	1.90	1.47	0.43	$1.0*10^{-11}$	-0.07
66	2.00	1.53	0.47	$1.1*10^{-11}$	-0.10
90	2.03	1.62	0.41	$1.0*10^{-11}$	0.00

L=334 kJ/kg and ρ_w= 1000 kg/m³. The results of the calculation are shown in table 4.

This kind of calculation has been done for each of the measured thermal conductivities in the frozen zone. Differences between the measured and the supposed thermal conductivity can go up to 5 W/mK (see table 8 in the appendix), which indicates a hydraulic conductivity k of more than 10^{-10}m/s.

Because the temperature is very important in the water- ice equilibrium, the calculated Δλ-values are presented as a function of the temperature. The different depths are indicated by using different signs (see figure 4).

From figure 4 some remarkable conclusions can be drawn:

- Δλ measured in river sand (filled signs) are bigger than those measured in silver sand (outlined signs).
- most points in figure 4 are found in two bands: one band for silver sand and one for river sand [3]. Such a band indicates a spreading in Δλ of ±0.4 W/mK or a spreading in T of ±0.2K.
- the depth of the measuring probe is not significant for the measured λ.
- some values of Δλ>0 are also found in the area T>0°C (only in silver sand). They are always measured very close to the freezing front. Knowing that the measured thermal conductivity is always a result of an averaged value over a finite region (a cylinder with a radius of about 15 mm), the top of this region can be frozen and then influences the measurement strongly. Of course the cylindrical symmetry around the measuring probe, which is necessary for the calculations of λ, is not present here, so these results are not reliable.

6 MEASURED THERMAL CONDUCTIVITIES

In the frozen zone an estimation for the real conductivity can be given (see previous

[3] Some points do not fit in these bands, therefore they are not indicated in figure 4. For these points the freezing front just passed their position, so the microscopical water- ice distribution might still be on its way to an equilibrium.

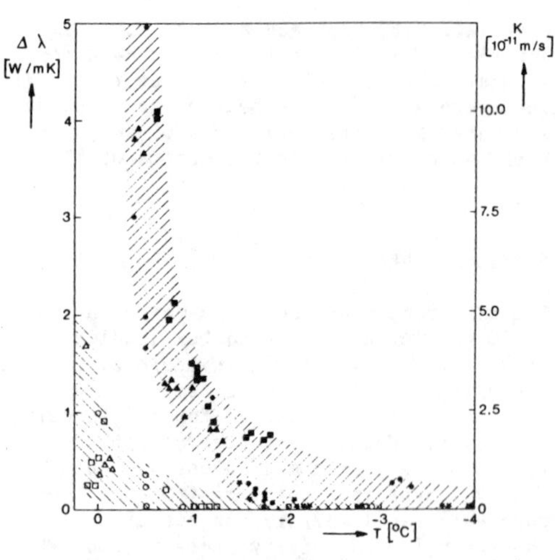

Fig 4. Calculated extra thermal conductivity Δλ and hydraulic conductivity k as a function of the temperature T. Closed signs are measured in river sand, outlined in silver sand, on several depths: * 46 mm, o 70 mm, □95 mm, Δ 122 mm.

section). But for lower temperatures, the measured thermal conductivity is very close to the real thermal conductivity. From figure 4 we see that in silver sand Δλ<0.03 W/mK for T<-1°C. So, in this temperature range a relationship between the ice content θ_i and thermal conductivity can be expected (see also [10]). This datum has been checked with some cylinders filled with silver sand with a given water content. We performed three succeeding freezing- thawing cycles. In a cycle the temperature varies from +10°C to -10°C and back in 24 steps. The temperature is adjusted with a thermostatic bath, in which the cylinder with the humid sand and thermal conductivity probe is placed. Again a very little amount of water was added to the sand because a homogeneous medium was needed. The results are presented in figure 5.

The first thing that is striking is that the measured thermal conductivities for T>0°C are well reproducable (for all cycles they can be found on one line). This means that during a cycle the average water content (on a macroscopical scale) around the probe does not change. So, during the freezing part of the cycle reversible water transport or no water transport at all took place. The reversible water transport can only take place if it is on a microscopical scale.

On contrary the measured thermal conductivities for T<0°C are 'cycle-dependent' (within one cycle the values are rather constant). This can not be caused by Δλ-effects (convection of latent heat), because the considered temperatures are too low (see

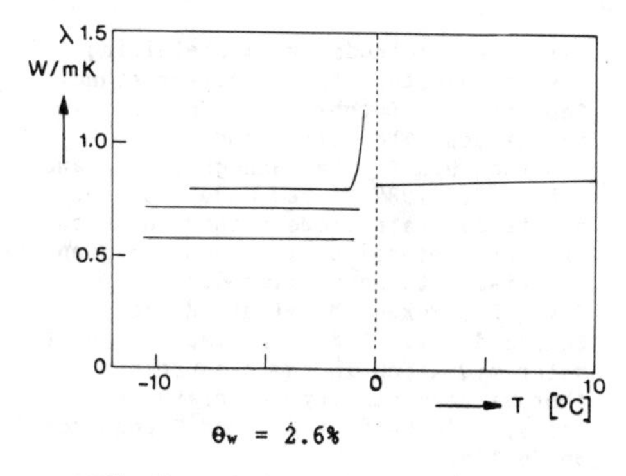

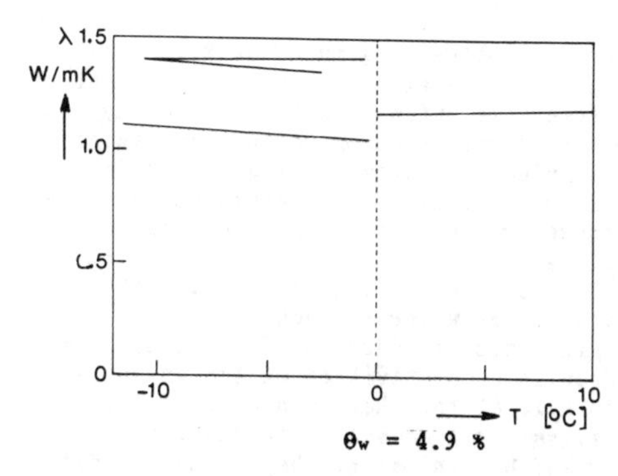

Fig 5. Measured thermal conductivities λ of humid silver sand as a function of temperature T during three freezing cycles.

also fig 4; the only $\Delta\lambda$-effect in figure 5 can be seen for θ_w=2.6% for $-1^0C<T<0^0C$, the upper line). Also the total water content $(\rho_i/\rho_w)\theta_i+\theta_w$ averaged over a bigger volume has to be a constant, because of the results above the freezing point. An explanation might be that the microscopical equilibrium between ice, water and sand is for each freezing cycle different. This could be caused by a different mutual configuration of the different particles. Thermal conductivities in granular materials are not only dependent on the volume fractions of the different components, but are also dependent on the way the solid particles contact each other [4].

7 DISCUSSION

When heat flow in porous media is evaluated, often the thermal conductivity is estimated. The composition of the multi- phase medium is used to calculate this physical property. In our experiments we found for $T>0^0C$ satisfying results, obtained by the model of Johanson and Frivik [8]. For frozen soils however this simple kind of relationship can not be applied: also the mutual configuration of the components might be important (see figure 5).

So the direct measurement of the thermal conductivity has become necessary. Here also some difficulties arise: the averaging area of the nonsteady state probe method is a cylinder with a diameter of about 3 cm. So, when the sand is not homogeneous (eg near the freezing front) it is not clear in which way the averaging exactly takes place. The averaging volume, which has to be as small as possible, is proportional to the temperature response of the probe. We adjusted it on only 0.15 K.

The heat capacity can be determined simultaneously with the nonsteady state probe method [5]. However with the very small temperature rise it is not possible to get accurate data for the heat capacity, because the determination of this quantity is only possible using higher order terms of the theoretical model for the nonsteady state probe method. The in situ determination could have given more precise information on the water content: $C(\theta_w)$-relations are more exact than $\lambda(\theta_w)$-relations. With the measured thermal conductivity and the assumption of a constant heat flow through the sand samples, the hydraulic conductivity k of frozen sand has been determined. The temperature dependence and 'kind of sand' dependence is clear. But there is no correlation between $\Delta\lambda$ (or k) and λ (see tables 7 and 8 where λ and $\Delta\lambda$ are given for different tests). Also the depth of the measuring probe has a minor influence on the calculated hydraulic conductivity. Because we suppose that during the tests the ice content on a certain depth is not changing much, except near the freezing front, we conclude now that the ice content θ_i has only minor influence on k. Because we performed our experiments in unsaturated sands, we must restrict this conclusion for unsaturated porous media: in ice lences very different values for k are probable.

To obtain a possible explanation of our k-T curve in the frozen zone we first look what is microscopical happening with the water transport. The hydraulic conductivity has to be influenced by the microscopical streamlines. The streamlines are situated in the unfrozen water layers, which are adsorbed on the surface of the sand particles. So in the frozen zone the water will flow easier if the adsorbed water layers are thicker and/or if there is more sand surface to flow along. Translating this back to the macroscopical level there must be a correlation between the hydraulic conductivity and the unfrozen water content and/or the specific surface. We compare this possible set up for a mechanistic theory with some other authors.

Horiguchi [11] measured unfrozen water contents for $T<0^0C$ with differential scanning calorimetry. The shape of his $\theta_w(T)$-

curve resambles very much our k(T)-curves (figure 4). He also remarks: 'the amount of nonfreezable unfrozen water is roughly equal to the amount of water which constitutes a mono-molecular layer coating all surfaces'. He finds a linear relationship between the unfrozen water content and the specific surface.

Also Ratke et al [12] looked at the transport of water through the frozen zone and the specific surface. They calculated hydraulic conductivities of frozen soils with help of the thermodynamics of the irreversible processes for Manaita-bridge clay and Negeshi silt. They found two different values for the diferent soils at the same temperature (-10°C): $3.5*10^{-14}$m/s and 1.4 to $1.7*10^{-14}$m/s for the clay and the silt respectively. They stated: 'This difference can be explained in terms of pore size of the two soils'. Next they related the pore size to the specific surface area S: higher values of k are correlating with higher values of S: S_{clay}=136.2 m²/g and S_{silt}=27.2 m²/g.

The specific surface area of river sand and silver sand are too small to be properly measured: S<1m²/g (see table 1). So we can not correlate these results with the measured hydraulic conductivities. Still we think that our measurements reinforce the conclusions of [11] and [12]. Therefore we continue our research in this mechanistic field of interest.

8 ACKNOWLEDGEMENTS

The authors wish to express their gratitude to A.E. Jansen, T. Jansen and W.H.M.C. Hillen of the workshop, who skillfully constructed the measuring box and the probes, to C. van Asselt who gave much electrotechnical advice, to P. van Espelo who accurately made the figures and to the students T. Allen (from Guelph, Canada) and W. Otten (Wageningen) who contributed to the measurements and conclusions.

REFERENCES*

1- Dirksen C. and Miller R.D., 1966, Closed-system freezing of unsaturated soil, Soil Sci. Soc. of Am. Proc., vol.30, no 2, pp 168-173.

2- Kay B.D. and Groenevelt P.H., 1974, On the interaction of water and heat transport in frozen and unfrozen soils, Soil Sci. Soc. Am. Proc., Vol.38, no 3, pp 395-404.

3- Groenevelt P.H. and Kay B.D., 1980, Pressure distribution and effective stress in frozen soils, Int. Sym. Ground Freezing 2 Proc., pp 597-610.

4- Van Haneghem I.A., 1981, Een niet stationaire naaldmethode (warmtegeleiding, warmtecapaciteit en contactweerstand), PhD. thesis (Dutch), Agr. Un. of Wageningen, The Netherlands.

5- Van Loon W.K.P., Van Haneghem I.A. and Schenk J., 1987, A new model for the nonsteady state probe method to measure thermal properties of porous and granular materials, to be published.

6- Kay B.D., Fekuda M., Izuta H. and Sheppard M.I., 1981, The importance of water migration in the measurement of the thermal conductivity of unsaturated soils, Cold Regions Sc. and Technology 5, pp 95-106.

7- Van Haneghem I.A. and Leij F.L., 1985, Thermal properties of moist porous media below 0°C, High Temp. High Press., vol.17, pp 611-621.

8- Holden J.T., Piper D. and Jones R.H., 1985, Some developments of a rigid- ice model of frost heave, Int. Sym. Ground Freezing 4 Proc., pp 93-99.

9- Gregg S.J. and Sing K.S.W., 1982, Adsorption, Surface area and Porosity, Academic Press, London UK.

10-Johanson O. and Frivik P.E., 1980, Thermal properties of soils and rock materials, Int. Sym. Ground Freezing 2 Proc., pp 427-453

11-Horiguchi K., 1985, Determination of unfrozen water content by DSC, Int. Sym. Ground Freezing 4 Proc., pp 33-38.

12-Ratkje S.K., Yamamoto H., Takashi T., Ohrai T. and Okamoto J., 1982, The hydraulic conductivity of soils during frost heave, Int.Sym. Ground Freezing 3 Proc., pp 131-135.

*See page X concerning ISGF papers

APPENDIX

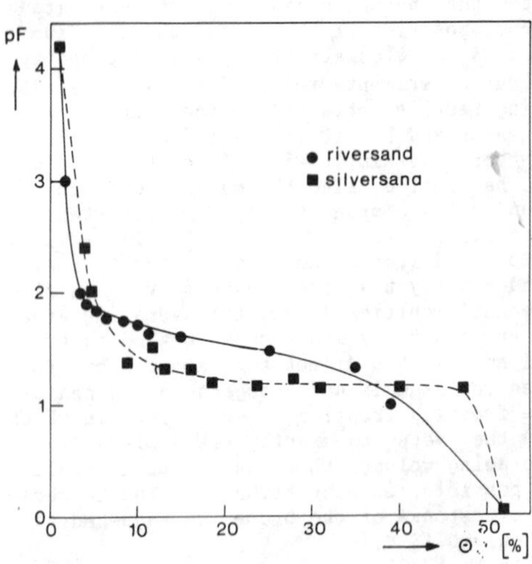

Fig 6 Water retention curves of river sand and silver sand. Logaritmic pore water pressure PF as a function of volumetric water content θ_w: $PF={}^{10}\log(-p_w/p_o)$ with $p_o \approx 1$ kPa (1 cm waterpressure).

Table 5 Second and third freezing test in silver sand, heat flow Q in W/m² *values in the frozen area; $Q=\underline{Q}-Q_L$ instead of $\underline{Q}$

test	t[h]	$\underline{Q_{71}}-Q_L$	$\underline{Q_{95}}-Q_L$	$\underline{Q_{122}}$	$\underline{Q_{145}}$	$\underline{Q_{170}}$	Q_L
2	18	-48*	-60*	-101*	-41	-50	10
3	19	-30*	-79*	-26	-39	-43	3
3	48	-30*	-52*	-16	-36	-39	1
3	144	-37*	-44*	-14	-34	-37	0.3
3	150	-37*	-51*	-13	-34	-37	0.2

Table 6 Freezing tests in river sand (test 11-14). Heat flow Q in W/m²· *values in the frozen area; $Q=\underline{Q}-Q_L$ instead of $\underline{Q}$

test	time[h]	$\underline{Q_{46}}-Q_L$	$\underline{Q_{71}}-Q_L$	$\underline{Q_{95}}$	$\underline{Q_{122}}$	$\underline{Q_{145}}$	$\underline{Q_{170}}$	Q_L
11	24	-35*	-47*	-39	-47	-41	-39	-5
	42	-39*	-93*	-40	-46	-42	-42	-1
	66	-39*	-103*	-41	-48	-43	-44	-0.5
	89	-46*	-136*	-41	-46	-42	-47	-0.3
	113	-50*	-198*	-41	-48	-43	-42	-0.1

test	time[h]	$\underline{Q_{46}}-Q_L$	$\underline{Q_{71}}-Q_L$	$\underline{Q_{95}}-Q_L$	$\underline{Q_{122}}-Q_L$	$\underline{Q_{145}}$	$\underline{Q_{170}}$	Q_L
12	15	-55*	-67*	-63*	-53	-54	-50	-16
	51	-51*	-69*	-119*	-169*	-64	-62	-8
	66	-53*	-56*	-116*	-155*	-66	-58	-2
13	22	-63*	-79*	-110*	-104*	-	-68	-6
	46	-66*	-89*	-107*	-89*	-51*	-77	-3
	75	-68*	-83*	-105*	-94*	-50*	-78	-2
	98	-72*	-94*	-108*	-90*	-104*	-71	-0.3

test	time[h]	$\underline{Q_{46}}-Q_L$	$\underline{Q_{71}}-Q_L$	$\underline{Q_{95}}-Q_L$	$\underline{Q_{122}}$	$\underline{Q_{145}}$	$\underline{Q_{170}}$	Q_L
14	23	-44*	-83*	-88*	-114*	-50	-47	10
	47	-44*	-57*	-83*	-80*	-45	-45	4
	71	-50*	-68*	-123*	-50	-55	-40	3
	96	-60*	- *	- *	-52	-43	-38	1

Table 7 Calculated thermal conductivities in silver sand [W/mK]. *values in the frozen zone

test	t	z=71 mm	z=95 mm		z=122 mm		z=145mm	z=170mm
nr	[h]	λ	λ	Δλ	λ	Δλ	λ	λ
1	42	0.80*	0.79*	0	1.47*	0.42	1.25	1.11
1	48	0.80*	0.79*	0	1.47*	0.43	1.24	1.10
1	66	0.76*	0.76*	0	1.53*	0.47	1.23	1.10
1	90	0.72*	0.72*	0	1.62*	0.41	1.22	1.09
2	18	0.70*	0.74*	0.18	1.39*	1.77	1.13	1.15
3	19	0.69*	0.78*	0.91	1.72		1.15	1.19
3	48	0.83*	1.38*	0.52	1.04		1.16	1.20
3	144	1.20*	1.62*	0.26	0.92		1.09	1.12
3	150	1.20*	1.58*	0.53	0.90		1.08	1.12

Table 8 Calculated thermal conductivities in river sand [W/mK]
*values in the frozen zone

test	z=46mm λ	Δλ	z=71mm λ	Δλ	z=95mm λ	Δλ	z=122mm λ	Δλ	z=145mm λ	Δλ	z=170mm λ
11	0.95*		1.66*	0.14	1.36		1.84		1.79		1.74
	1.00*		1.49*	1.71	1.36		1.76		1.74		1.81
	1.00*		1.43*	1.98	1.34		1.79		1.78		1.91
	1.14*		1.38*	3.00	1.33		1.72		1.74		1.98
	1.07*	0.17	1.40*	5.04	1.35		1.75		1.69		1.78
12	0.84*		1.94*	0.18	2.02*		1.85		1.67		1.60
	0.72*		1.77*	0.12	1.78*	1.54	2.41*	3.83	1.63		1.60
	0.69*		1.54*		1.66*	1.49	2.38*	3.65	1.66		1.47
13	0.69*		1.63*		1.66*	0.84	2.98*	1.27	-		1.46
	0.69*		1.62*	0.42	1.62*	0.78	2.83*	0.83	1.79*		1.48
	0.70*		1.60*	0.41	1.60*	0.75	2.91*	0.83	2.15*		1.41
	0.71*		1.55*	0.28	1.55*	0.81	2.85*	0.76	3.00*	1.41	1.33
14	0.71*		1.83*	1.88	1.85*	2.13	2.03*	3.86	1.63		1.52
	0.80*		1.89*	0.56	2.11*	1.89	1.62*	1.31	1.59		1.61
	0.93*		2.27*	1.18	2.30*	4.12	2.14		1.68		1.57
	0.85*	0.27	- *		- *		1.93		1.70		1.45

5th International Symposium on Ground Freezing, Jones & Holden (eds)
© 1988 Balkema, Rotterdam. ISBN 90 6191 824 3

A frost heave mechanism model based on energy equilibrium

Y.Miyata
Tokyo Gas Co., Ltd, Tokyo, Japan

ABSTRACT: A one dimensional, macroscopic frost heave mechanism model for saturated soil has been developed. The principle of this model is that there is a balance in the hydraulic energy of the unfrozen pore water and the released mechanical energy due to the freezing of subcooled water. The ice segregation rate is derived from this energy balance. The frost heave ratio is obtained by solving the equations for the mass balance, heat balance and for the ice segregation rate simultaneously. The numerical analysis method for this particular problem is simplified by the assumption of a quasi-steady state. The results of the numerical analysis agree with the experimental results on the dependence of the frost heave ratio on overburden pressure and on ice penetration rates. In addition, it has been shown that the frost heave ratio depends on the ratio of the temperature gradient in unfrozen soil to an ice penetration rate.

1 INTRODUCTION

The influence of frost heave on the structure of an inground or underground cryogenic storage will usually be significant. Therefore, the frost heave ratio, which is the ratio of frost heaved length to original length, is one of the most fundamental data in this field of engineering.

The method by Takashi, et al. (1978) is a convenient method for estimating the frost heave ratio. With this method, the frost heave ratio, ξ, is decided from the empirical formula below, the so-called Takashi's empirical formula, whose experimental constants, ξ_0, σ_0, U_0, are obtained as results of open-type frost heave tests with a constant ice penetration rate:

$$\xi = \xi_0 + \frac{\sigma_0}{\sigma_1}\left(1 + \sqrt{\frac{U_0}{U}}\right), \qquad (1)$$

where σ_1 is overburden pressure and U is an ice penetration rate.

Using this method, however, it is difficult to estimate the influence of the temperature gradient in the unfrozen soil on the frost heave ratio. Accordingly, a one dimensional, macroscopic frost heave mechanism model has been developed in order to improve the estimation accuracy of determining the frost heave ratio. The term "macroscopic" means that the model cannot deal with ice lensing.

Most researchers (e.g., Gilpin, 1980) of the frost heave mechanism create a frost heave model with unfrozen soil, and then calculate the frost heave using the model. However, it would be inconvenient to apply the model to the engineering field because there is often a great variety of conditions in the unfrozen ground - e.g. the distance between a freezing front and water supply source. In this paper, the frost heave capability without hydraulic resistance in unfrozen soil is called HEAVABILITY, and the term "frost heave ratio" means heavability. The dependence of heavability on overburden pressure, ice penetration rates, and the temperature gradient in the unfrozen soil will all be discussed using results obtained by simulation.

2 FUNDAMENTAL EQUATIONS

2.1 Outline

Why and how is unfrozen water flow produced in frozen soil? This question has not yet been agreed among experts, but a macroscopic solution will be derived from simple considerations of the energy balance.

Because the water in the soil flows, it

must receive mechanical energy as a driving source and must also compensate for the resulting fluid friction losses. From another point of view, the frozen soil does work to the water so as to draw it up, and this work is the mechanical energy which is equal to the friction loss of the water flow. This mechanical energy is converted from a portion of the thermal energy, which is conveyed through the frozen soil according to temperature difference, and it is totally re-converted to thermal energy by friction. This implies that the mechanical energy balance for the water flow is included in the heat balance, but that it does not affect the heat balance. Therefore, the mechanical energy balance and the heat balance are independent of each other and two equations are obtained from them.

The frost heave mechanism model is composed of three equations: mass balance, heat balance, and mechanical energy balance. Actually, the ice segregation rate is used instead of the mechanical energy balance. The derivation of the three equations will be described in detail later.

Five assumptions are adopted in order to simplify fundamental equations: (1) Darcy's law is applicable to the flow of unfrozen water in frozen soil. (2) The concept of effective stress is applicable to frozen soil. (3) The pore ice bears external forces, as well as the soil particles. (4) The soil particles are rigid. (5) The phase change is quasi-static.

2.2 Mass balance

Consider the water flow in an extremely small element in the frozen soil and assume that the flow rate per unit volume of soil decreases by ϕ in the element. Taking the coefficient of permeability in frozen soil, k, to be a function of temperature - i.e. a function of x, the equation below is obtained by balancing the flow rate in the element using Darcy's law:

$$\frac{\partial}{\partial x}\left(\frac{k}{\gamma_w}\frac{\partial P_w}{\partial x}\right) = \phi \, , \tag{2}$$

where P_w is the pore water pressure and γ_w is the specific weight of the water.

Here, the quantity of the ice segregation per unit volume of soil and per unit time, ψ, is called the ice segregation rate. The ice segregation decreases the water flow rate, and the decrease of the volumetric unfrozen water content, n, increases the water flow rate. Then the decreased flow rate ϕ is expressed below using the ratio of water specific volume to ice specific volume, μ;

$$\phi = \psi + (1 - \mu)\frac{\partial n}{\partial t} \, . \tag{3}$$

The relation between the effective stress, σ_e, the effective stress at the freezing front (i.e. overburden pressure) σ_1, and pore water pressure P_w, is given below;

$$\sigma_e = \sigma_1 - P_w \, . \tag{4}$$

The fundamental equation for the mass balance (5) is obtained by substituting Equation (3) and Equation (4) into equation (2):

$$\frac{\partial}{\partial x}\left(\frac{k}{\gamma_w}\frac{\partial \sigma_e}{\partial x}\right) = -\psi - (1 - \mu)\frac{\partial n}{\partial t} \, . \tag{5}$$

2.3 Released mechanical energy

According to thermodynamics, the quantity of energy per unit mass, Δf, is released when water subcooled by the temperature of $\Delta\Theta$, freezes. This is given by

$$\Delta f = \frac{L_w}{T_0}\Delta\Theta \, , \tag{6}$$

where L_w is the latent heat due to freezing and T_0 is the reference temperature.

Takashi, et al. (1981) have confirmed that the upper limit of heaving pressure, σ_u, and the temperature at the colder end of the sample, both conform to the modified Clausius-Clapeyron equation from the results of their closed type frost heave test with fixed temperature at both ends of a sample. The results imply that water and ice at temperature Θ are in phase equilibrium if the effective stress of the ice is

$$\sigma_u = -\frac{L_w}{v_i}\frac{\Theta}{T_0} \, , \tag{7}$$

where v_i is the specific volume of the ice.

If the actual effective stress at this point is σ_e and the equilibrium temperature is Θ_e, then

$$\sigma_e = -\frac{L_w}{v_i}\frac{\Theta_e}{T_0} \, . \tag{8}$$

The sub-cooled temperature $\Delta\Theta$ is $\Theta - \Theta_e$, and substituting Equation (7) and Equation (8) into Equation (6), the released energy Δf is

$$\Delta f = v_i (\sigma_u - \sigma_e) . \quad (9)$$

2.4 Mechanical energy balance

Consider the mechanical energy balance of the water flow in the extremely small element of length, dx, and the area, dA, for a short time, dt. In the unfrozen water flow, the kinetic and potential energy can be neglected when compared with other kinds of mechanical energy.

When the unfrozen water, with a velocity V flows into the element at the pore water pressure P_w, the difference between the energy, E_i, brought into the element and the energy, E_o, taken out by the flow is

$$E_o - E_i = (P_w \frac{\partial V}{\partial x} + V \frac{\partial P_w}{\partial x}) \, dx \, dA \, dt . \quad (10)$$

The energy, E_1, lost by the pressure drop in the element is

$$E_1 = -V \frac{\partial P_w}{\partial x} \, dx \, dA \, dt . \quad (11)$$

When the quantity of water per unit volume of soil and per unit time, ψ', freezes in the element, then the released energy, E_f, which is due to the freezing of the sub-cooled water, is obtained from Equation (9) :

$$E_f = \frac{\sigma_u - \sigma_e}{\mu} \psi' \, dx \, dA \, dt . \quad (12)$$

The quantity of the freezing water ψ', is the sum of the ice segregation rate ψ and the unfrozen water content decreasing rate $-\partial n/\partial t$; namely;

$$\psi' = \psi - \frac{\partial n}{\partial t} . \quad (13)$$

Substituting Equation (13) into Equation (12) yields

$$E_f = \frac{\sigma_u - \sigma_e}{\mu} (\psi - \frac{\partial n}{\partial t}) \, dx \, dA \, dt . \quad (14)$$

The surplus water due to the decrease of the water content must be absorbed into the unfrozen water flow. Equation (15) gives the energy, E_s, necessary for this:

$$E_s = P_w (1 - \mu) \frac{\partial n}{\partial t} \, dx \, dA \, dt . \quad (15)$$

The energy balance of the water flow in this element is

$$E_o - E_i = E_f - E_1 - E_s . \quad (16)$$

Substituting Equations (10), (11), (14) and (15) into this equation and rearranging it by using Equation (2) and (3) gives the equation for the mechanical energy balance (17):

$$(P_w + \frac{\sigma_u - \sigma_e}{\mu}) \psi = \frac{\sigma_u - \sigma_e}{\mu} \frac{\partial n}{\partial t} . \quad (17)$$

This equation represents the situation where the released mechanical energy, due to the so called in-situ freeze, compensates for the shortage of mechanical energy, which is the difference between the energy necessary to draw the quantity of the freezing water and the released energy due to its freezing.

2.5 Ice segregation rate

Substituting Equation (4) into Equation (17) and rearranging gives the following equation:

$$\{\sigma_u - (1 + \mu) \sigma_e + \mu\sigma_1\} \psi = (\sigma_u - \sigma_e) \frac{\partial n}{\partial t} . \quad (18)$$

In the case of $\sigma_u - (1 + \mu) \sigma_e + \mu\sigma_1 \neq 0$, the ice segregation rate ψ, is given by

$$\psi = \frac{\sigma_u - \sigma_e}{\sigma_u - (1 + \mu) \sigma_e + \mu\sigma_1} \frac{\partial n}{\partial t} . \quad (19)$$

A positive ice segregation rate means frost heave will occur. Considering, that the term of $\partial n/\partial t$ is negative, the necessary conditions for this rate to occur are as follows:

$$\sigma_e < \sigma_u, \text{ and } \sigma_e > \frac{\sigma_u + \mu\sigma_1}{1 + \mu} . \quad (20)$$

On the other hand, in the case of $\sigma_u - (1 + \mu) \sigma_e + \mu\sigma_1 = 0$, the mechanical energy necessary for intake of the quantity of freezing water is balanced by the released energy due to its freezing. The system does not need compensation by the released energy due to the in-situ freeze. The ice segregation rate ψ, may have a positive value even if $\partial n/\partial t = 0$. This condition is the necessary condition of the so-called "Perfect Frost Heave Phenomenon", if the temperature distribution between this point and the freezing front is static.

2.6 Heat balance

The last factor in the frost heave mechanism is heat transfer. Assume that the temperature of the element is Θ and that the quantity of water ψ' given by Equation (13) freezes in the element and that the properties, such as the thermal conductivity, λ_f, etc., are constant with temperature. Omitting small heat transfer factors such as the quantity of heat brought by unfrozen water flow, the fundamental heat balance equation in the element is given by

$$\frac{\partial^2\Theta}{\partial x^2} = \frac{1}{\kappa}\frac{\partial\Theta}{\partial t} - \frac{L_w\rho_w}{\lambda_f}\left(\psi - \frac{\partial n}{\partial t}\right) . \tag{21}$$

where κ is the termal diffusivity and ρ_w is the density of the water.

3 FROST HEAVE MECHANISM MODEL

3.1 Frost heave mechanism

Figure 1 shows the concept of the frost heave mechanism which is described by the fundamental equations. The features of the mechanism are such that the coefficient of permeability in the unfrozen soil is infinite, that the partially frozen soil where the unfrozen water exits is divided into the super-heated zone and the sub-cooled zone, and that the ice segregation occurs in the sub-cooled zone between the line of upper limit of heaving pressure and the mechanical energy equilibrium line, which is shown as the dotted line in Figure 1.

3.2 Quasi-steady state model

The fundamental equations for the frost heave mechanism model are Equations (5), (19), and (21). It is not practical to obtain unsteady solutions directly from these equations because solving an unsteady and non-linear simultaneous boundary value problem which includes an inner boundary, makes the computation complicated and time-consuming. Because of this the frost heave mechanism model is converted from the unsteady state to the quasi-steady state.

Consider the ideal situation where the ice penetration rate U is constant as well as the temperature gradient at the freezing front in the unfrozen soil, α_u, and the velocity of the pore water flow in the unfrozen soil, V_o. Then, also adopt the coordinate ζ, whose origin is the freezing front. The equations for the coordinate

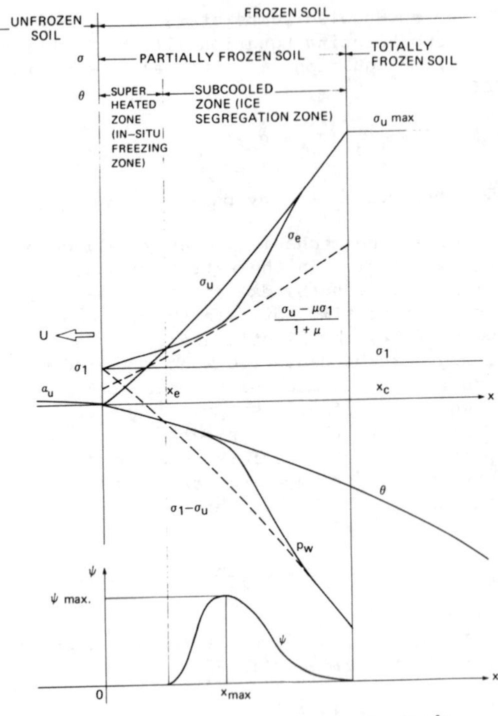

Fig.1 Examination of frost heave mechanism

transformation as described below will hold approximately. Thus:

$$\frac{\partial}{\partial t} = U\frac{d}{d\zeta} , \tag{22}$$

$$\frac{\partial}{\partial x} = \frac{d}{d\zeta} . \tag{23}$$

Rearranging Equations (5), (21) and (19), by substituting the equations above, gives the fundamental equations for the quasi-steady state.

For the mass transfer;

$$\frac{d}{d\zeta}\left(\frac{k}{\gamma_w}\frac{d\sigma_e}{d\zeta}\right) = -\psi - U(1-\mu)\frac{dn}{d\zeta} . \tag{24}$$

For the heat transfer;

$$\frac{d^2\Theta}{d\zeta^2} = \frac{U}{\kappa}\frac{d\Theta}{d\zeta} - \frac{L_w\rho_w}{\lambda_f}\left(\psi - U\frac{dn}{d\zeta}\right) . \tag{25}$$

For the ice segregation rate;

$$\psi = \frac{\sigma_u - \sigma_e}{\sigma_u - (1+\mu)\sigma_e + \mu\sigma_1} U \frac{dn}{d\zeta}$$

where, $\sigma_u > \sigma_e$, (26)

$$\psi = 0 \qquad \text{where, } \sigma_u \leq \sigma_e. \qquad (27)$$

The boundary conditions at the freezing front ($\zeta = 0$) are:

For the mass transfer;

$$\sigma_e = \sigma_1 \ , \qquad (28)$$

$$V_0 = \int_0^{\zeta_c} \psi d\zeta - U\,(1 - \mu)\,(n_u - n_c). \qquad (29)$$

For the heat transfer;

$$\Theta = 0 \ , \qquad (30)$$

$$\alpha_f = \frac{\lambda_u}{\lambda_f}\alpha_u - U\frac{L\rho_w w}{\lambda_f}(n_u - n_f) \qquad (31)$$

The subscripts u, f, and c express the unfrozen side, the frozen side at the freezing front, and the colder end of the partially frozen soil respectively.

3.3 Numerical analysis method

The quasi-steady state model composed of Equation (24) to Equation (31) is still complicated, even though it has been simplified.

Two points are highlighted below. The first point is that both the integration and the Runge-Kutta methods are used in the super-heated zone and the Adams-Bashford method is used in the sub-cooled zone, after Equation (24) and Equation (25) are changed into first-order differential equations by integration. The second point is that the simultaneous solutions in the sub-cooled zone and the flow rate at the freezing front are obtained by iteration.

3.4 Estimation methods for properties

Both the unfrozen water content and the coefficient of permeability are important in the frost heave mechanism, though their availability are restricted. In this model, the unfrozen water content, w, is determined by an equation of the form, $f(w + w_c, \Theta + \Theta_o) = 0$, which is modified from the experimental formula by Anderson et al. (1972) whose form is $f(w, \Theta) = 0$. W_c and Θ_o in the equation are calculated from the unfrozen water contents at the freezing front and at the colder end of the partial frozen soil.

The coefficient of permeability is set to be proportional to the 2.5 th power of the ratio of the unfrozen water content to the reference content.

4 RESULTS

The results of numerical analysis on three kinds of soil, whose properties have been published by Takashi et al. (1976), and are shown in Table 1, will be compared with the results of experiments.

Referring to Takashi's empirical formula (1), the results of the numerical analysis are shown in Figure 2 concerning the dependence of the frost heave ratio on the overburden pressure, and in Figure 3 on the dependence of the frost heave ratio on the ice penetration rate. The results of both agree well over a wide range, though the main properties are estimated mechanically.

Table 1. Main properties of sample soil

Item	Unit	Sample 1	Sample 2	Sample 3
Soil	-	Silt	Silt	Clay
ρ_u	Mg/m³	1.53	1.70	1.36
n_u	-	0.569	0.448	0.707
S	m²/g	16.3	27.2	136
ξ_o	%	0.55	0.3	0.555
σ_o	kPa	2.5	6.54	7.59
U_o	mm/h	2.97	1.70	67.1

ρ_u; density of unfrozen soil
S; specific surface

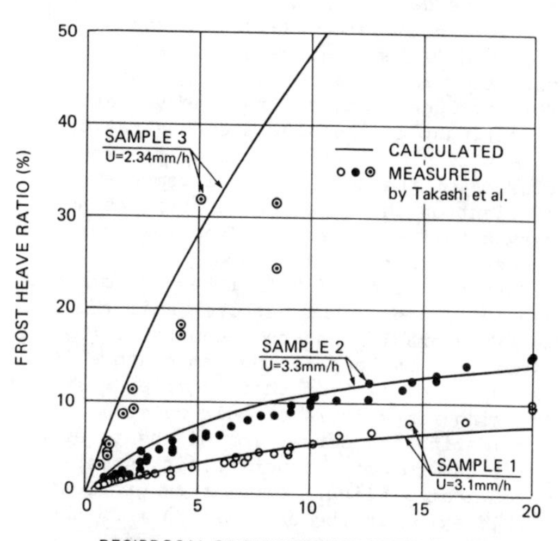

Fig.2 Comparison of calculated results with experimental results on overburden pressure dependence.

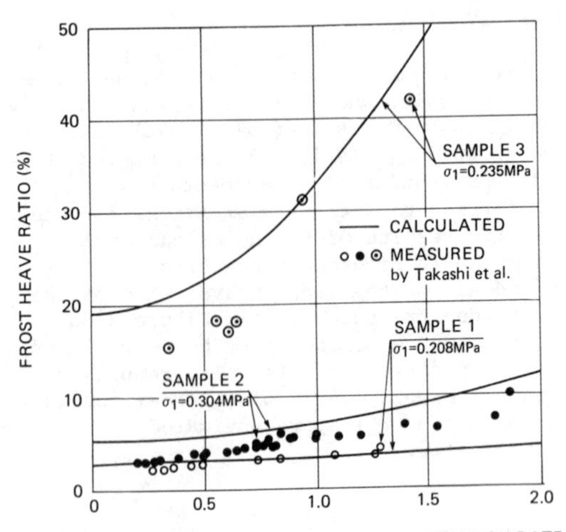

Fig.3 Comparison of calculated results with experimental results on ice penetration rate dependence.

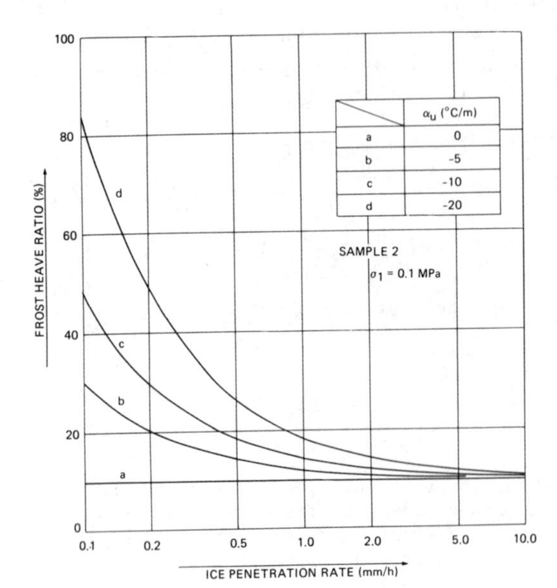

Fig.4 Influence of temperature gradient in unfrozen soil

5 DISCUSSION

5.1 Temperature gradient in unfrozen soil

Figure 4 is one of the simulation results, and shows the relationship between the frost heave ratio and the ice penetration rate, using the temperature gradient in the unfrozen soil α_u as a parameter. It is clear from the figure that the frost heave ratio increases as the ice penetration rate decreases only if $\alpha_u \neq 0$. The greater the absolute value for the unfrozen soil temperature gradient α_u, the more the frost heave ratio increases. This influence on the frost heave ratio is significant in the slow ice penetration rate zone. On the other hand, the influence of the unfrozen soil temperature gradient on the frost heave ratio is not very significant, even in the low overburden pressure zone.

Takashi, et al. (1971) intended to avoid the influence of the unfrozen soil temperature gradient on the results of their experiment by keeping the temperature in the unfrozen part of their samples at the freezing temperature of bulk water. Considering the depression of the freezing point due to soil structure, it is estimated that a slight temperature gradient should exist in the unfrozen part of the samples of about 30 mm thick. This is the reason why Takashi's empirical formula (1) has the term which represents the influence of the ice penetration rate on the frost heave ratio.

In order to clarify how to amend Takashi's empirical formula, the influence of the term α_u/U on the frost heave ratio is shown in Figure 5. From this figure, it is clear that the values for the frost heave ratio are almost the same, if the values for the term α_u/U are held constant.

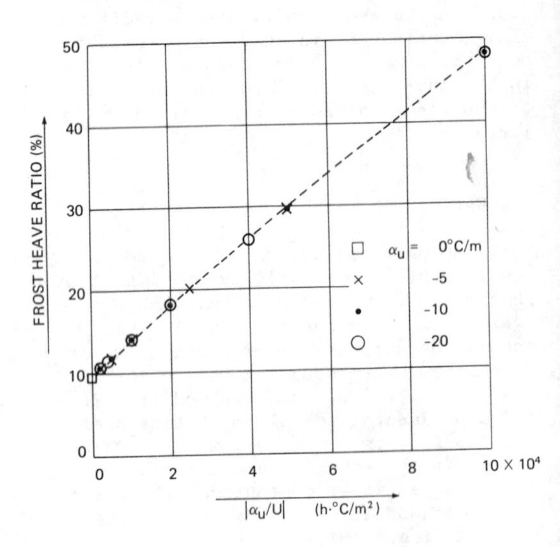

Fig.5 Influence of temperature gradient in unfrozen soil (Sample 2, σ_1=0.1 MPa)

5.2 Overburden pressure

The relationship between the frost heave ratio and the overburden pressure is shown in Figure 6. In this figure, both coordinate axes are on a log scale and the ice penetration rate is unified to 5 mm/h, so as to make it easy to compare the results of the numerical analysis and Takashi's empirical formula.

Consider the reason why the frost heave ratio estimated from the frost heave mechanism model decreases more than that given by Takashi's empirical formula as the overburden pressure of σ_1 increases in the high overburden pressure region ($\sigma_1 >$ 1 MPa).

In the frost heave test by Takashi, et al., the colder end of the sample is cooled to a temperature of -10°C after the temperature at the warmer end of the sample reaches the freezing point of the bulk water. At that time, the bulk water just outside the sample freezes and prevents the flow of the unfrozen water in the sample to the outside. Then, the retained unfrozen water at the temperature of -10°C and greater in the sample expands due to the phase change. The measured value in the frost heave test by Takashi, et al. includes this expansion quantity due to the phase change.

According to the mechanism model, the reason why the increasing rate of the frost heave ratio decreases as the overburden pressure decreases in the low overburden pressure region (σ_1 < 0.2 MPa), is that the increase of the unfrozen water flow rate in the frozen soil, which is due to the increase of the frost heave quantity, makes the pore water pressure decrease and the effective stress increase.

Figure 2 shows the agreement between both results in this region. In this figure it is observed that the model underestimates the frost heave ratio as the soil particles become large. It implies there must be some room for improving the estimation method for determining properties.

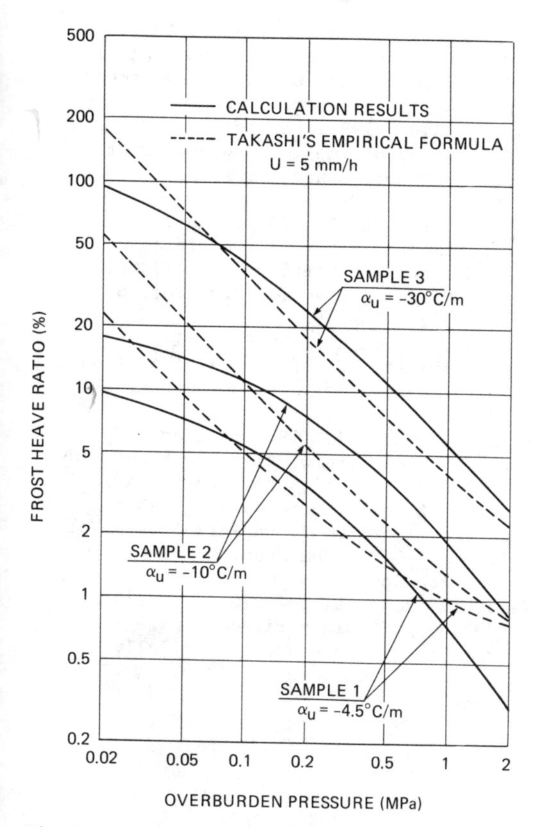

Fig.6 Calculated results on overburden pressure dependence.

5.3 Ice penetration rate

The relationship between the frost heave ratio and the ice penetration rate is shown in Figure 7, unifying the overburden pressure to 0.1 MPa.

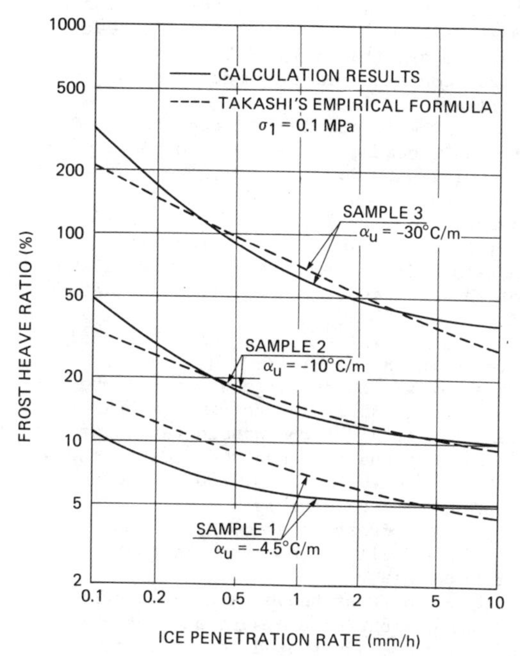

Fig.7 Calculated results on ice penetration rate dependence

Consider the cause of the difference between both results by the model and Takashi's empirical formula in the fast ice penetration rate region ($U > 1$ mm/h). The faster the ice penetration rate U, the greater the hydraulic resistance in the unfrozen soil. This makes the effective stress at the freezing front, i.e. the overburden pressure σ_1, increase. The real frost heave ratio will be underestimated when the effective stress at the warmer end of the sample is used for the arrangement of measured data. This is the reason why the increasing rate of the frost heave ratio by the frost heave mechanism model is different from Takashi's empirical formula in the fast ice penetration rate region.

Consider the cause of the difference between both results in the slow ice penetration rate region ($U < 0.2$ mm/h). In this region, the frost heave does not reach its final expected level in some parts of the sample due to a mild temperature gradient, therefore, the frost heave ratio by the frost heave test is smaller than that given by the frost heave mechanism model.

6 CONCLUSIONS

1. The frost heave mechanism model estimates the frost heave ratio by solving the ice segregation rate, the equation for mass balance and the equation for heat balance simultaneously. The novel features contained in the model are as follows:

(a) The released mechanical energy due to the freezing of the sub-cooled water, quantified by Equation (9), draws water into the frozen soil and causes ice segregation.

(b) The ice segregation rate is obtained from a mechanical energy balance of the flow and is quantified as Equation (19).

2. The results of the numerical analysis using the mechanism model in the quasi-steady state agree well with the results of published experiments by Takashi, et al. From consideration of the details of the agreements and differences between the two methods, the following conclusions are derived:

(a) The frost heave mechanism model simulates well the macroscopic phenomena of frost heave .

(b) The frost heave ratio is a function of the overburden pressure and the ratio of the temperature gradient at a freezing front in unfrozen soil to an ice penetration rate.

3. The frost heave mechanism model is applicable for the estimation of the frost heave ratio of ground for the engineering of cryogenic underground structures. It is hoped that the accuracy of estimation by the frost heave mechanism model will soon be improved by the use of measured properties, such as the coefficient of permeability in frozen soil, which are not available at present.

Acknowledgements

The author would like to express his appreciation to Thornton Research Center of Shell Research Ltd., especially Dr M.H. Collins and Dr B.J. Wadie, for supporting and assisting his research during his stay there.

The author also would like to thank Seiken Co., Ltd., which kindly offered the experimental data compiled by Takashi et al.

REFERENCES

Anderson, D.M. and Tice, A.R., 1972. Predicting unfrozen water contents in frozen soils from surface measurements. Highway res. record. 393:12-18.

Gilpin, R.R., 1980. A model for the prediction of ice lensing and frost heave in soil. Water resources res.. 16, No. 5: 918-930.

Takashi, T., et al.. 1971. An experimental study of the effect of loads on frost heaving and soil moisture migration. Seppyo (J. Jpn. Soc. Snow Ice). 33: 115-125.

Takashi, T., et al.. 1976. Influence of permeability of unfrozen soil on frost heave. Seppyo (J. Jpn Soc. Snow Ice). 38:1-10.

Takashi, T. et al.. 1978. Effect of penetration rate of freezing front and confining stress on the frost heave ratio of soil. Proceedings of 3rd International Conf. on Permafrost, Vol. 1. p.737-742. Edmonton: National Research Council of Canada.

Takashi, T., et al.. 1981. Upper limit of heaving pressure derived by pore water pressure measurements of partially frozen soil. Engineering geology. 18: 245-257.

5th International Symposium on Ground Freezing, Jones & Holden (eds)
© 1988 Balkema, Rotterdam. ISBN 90 6191 824 3

Sensitivity of a thaw simulation to model parameters

Sorab Panday & M.Yavuz Corapcioglu
Department of Civil & Environmental Engineering, Washington State University, Pullman, Wash., USA

ABSTRACT: Construction and maintenance of engineering structures in arctic regions demand an understanding of and the ability to cope with environmental problems produced by permafrost. Where thawing of the permafrost cannot be prevented, the range, amount, and extent of degradation are factors that must be taken into account in the design of engineering structures. Predictions of these factors are facilitated by the use of a mathematical model designed to describe permafrost thaw consolidation based upon a complete formulation of the problem. This problem is one of multiphase (air, ice, water, and solid) transport and deformation in a porous medium. The conservation of mass equations for liquid water, ice with phase change, and deforming soil solids are developed. Darcy's law, which expresses the specific discharge of water relative to the moving solids, is extended to include moisture movement due to thermal gradients. The ice and soil solids are assumed to settle at the same rate of deformation. The quasi-static equilibrium equations and stress-strain relations which assume a perfectly elastic solid matrix are also employed. Simultaneous variations of moisture retention and phase composition curves with temperature and pore pressure are introduced. The energy conservation equation which includes terms due to viscous dissipation and compression effects, along with the other equations expressed above represents the significant processes occurring in thawing permafrost soils. A numerical scheme has been used to solve these governing equations, and the model, once calibrated, can be used to predict the thaw settlement behavior of a frozen soil. However, values of material and physical parameters are required for such a model, and under field conditions, the variability of parameters makes it difficult to determine these parameters accurately. A sensitivity analysis helps to determine to which parameters the model is sensitive and hence we can spend time, effort, and money to determine these parameters more accurately. On the other hand, values borrowed from handbooks could be used for the general type of soil under consideration, for parameters which do not affect significantly the model's performance, and we need not perform elaborate studies to determine the exact values of those parameters for the particular soil under consideration.

This paper introduces a sensitivity analysis to some crucial parameters involved in modeling the thaw consolidation of a one-dimensional soil column.

1 GOVERNING EQUATIONS

The model describing permafrost thaw settlement can be developed from the basic heat and mass conservation equations and other parametric and constitutive relations. A detailed treatment of the derivation and underlying assumptions is given by Corapcioglu and Panday (1988a). Only the basic ingredients used to develop the model will be presented here along with the final form of the equations. The conservation of mass equation for the unfrozen water phase is given by

$$\nabla \cdot \rho_f \underset{\sim}{q} + \frac{\partial}{\partial t} (\rho_f \theta_w) - R_a = 0 \tag{1}$$

where, for a deforming porous medium, the flux of water relative to the moving solid particles due to soil matrix deformation is expressed by a modified Darcy's Law

$$\begin{aligned} \underset{\sim}{q}_r &= \theta_w (\underset{\sim}{V}_f - \underset{\sim}{V}_s) = \underset{\sim}{q} - \theta_w \underset{\sim}{V}_s \\ &= -\underset{\approx}{K} \left(\frac{1}{\rho_f g} \nabla p - \nabla z \right) - \underset{\approx}{D}_{MT} \nabla T \end{aligned} \tag{2}$$

A definition of the symbols can be found in the notation list.

The conservation of mass for the soil solids is expressed as

$$\nabla\cdot[(1-n)\underset{\sim}{V}_s] + \frac{\partial}{\partial t}(1-n) = 0 \qquad (3)$$

where the velocity of the soil solids is related to the strain rate as

$$\frac{d_s\varepsilon}{dt} = \frac{\partial(\varepsilon)}{\partial t} + \underset{\sim}{V}_s\cdot\nabla(\varepsilon)$$

$$= \nabla\cdot\underset{\sim}{V}_s = \frac{1}{(1-n)}\frac{d_s n}{dt} \qquad (4)$$

For a one-dimensional vertical case, the strains are stress and temperature related, and we can write

$$\frac{\partial\varepsilon_{ij}}{\partial t} = \left.\frac{\partial\varepsilon_{ij}}{\partial\sigma'}\right|_T \frac{\partial\sigma'}{\partial t} + \left.\frac{\partial\varepsilon_{ij}}{\partial T}\right|_p \frac{\partial T}{\partial t}$$

$$= \alpha_p\frac{\partial}{\partial t}(S_w p) + \alpha_T\frac{\partial T}{\partial t} \qquad (5)$$

for a constant total stress.

The conservation of mass equation for the melting ice in the porous medium is given by

$$\frac{\partial}{\partial t}(\rho_i\theta_i) + \nabla\cdot(\rho_i\theta_i\underset{\sim}{V}_s) + R_a = 0 \qquad (6)$$

and the energy conservation equation is written as

$$\frac{\partial}{\partial t}[(\rho C)_m T] - L R_a + \nabla\cdot[\{\theta_w\rho_f C_v\underset{\sim}{V}_f$$

$$+ \theta_i\rho_i C_i\underset{\sim}{V}_s + (1-n)\rho_s C_\varepsilon\underset{\sim}{V}_s\}T]$$

$$- \nabla\cdot[\underset{\sim}{\Lambda}_m\nabla T] - p\nabla\cdot(\theta_w\underset{\sim}{V}_f) \qquad (7)$$

$$+ [\theta_i + (1-n)]T\gamma\frac{\partial\varepsilon}{\partial t} = 0$$

We incorporate equations (2), (3), (4), (5), and (6) into equations (1) and (7) to try and obtain solution for the variables p, and T. However, saturations, S_w, and the thermal compressibility of the soil, α_T, are functions of the pore pressure p and temperature, T, and need to be expressed in terms of these basic variables.

The variation of saturation with pore pressures is expressed by the moisture retention curve, $S_{w1}(p)$ vs. p, and with respect to temperatures (below freezing) by the phase composition curve $S_{w2}(T)$ vs. T. The simultaneous variation of moisture content with p and T is the equation of the surface $\theta_w = \theta_w(p,T) = n[S_{w1}(p)S_{w2}(T)]$ which is plotted in Figure 1, and we can write

$$\frac{\partial S_w}{\partial t} = \left.\frac{\partial S_w}{\partial p}\right|_T \frac{\partial p}{\partial t} + \left.\frac{\partial S_w}{\partial T}\right|_p \frac{\partial T}{\partial t}$$

$$= \xi_p\frac{\partial p}{\partial t} + \xi_T\frac{\partial T}{\partial t} \qquad (8)$$

where the slopes, ξ_p and ξ_T, can be obtained from the equation of the surface generated. We can also express the amount of ice mass on this surface. For example (see Figure 1), the ice mass at point D is equal to the mass of water at point A minus the mass of water at point D. Note that point A is on the θ_w vs. p plane, i.e., T = °C. Thus, the volumetric ice fraction, θ_i can be expressed as $\theta_i = n(\rho_f/\rho_i)\,S_{w1}(p)[1 - S_{w2}(T)]$ where we neglect any mass transfer into the air phase from the water or the ice phase, (i.e., condensation, vaporization, and sublimation). This can be used to plot the variation of ice content with pressure and temperature, as in Figure 2, and the temporal variation of ice content θ_i can be written as

$$\frac{\partial\theta_i}{\partial t} = \left.\frac{\partial\theta_i}{\partial p}\right|_T \frac{\partial p}{\partial t} + \left.\frac{\partial\theta_i}{\partial T}\right|_p \frac{\partial T}{\partial t}$$

$$= \xi_{ip}\frac{\partial p}{\partial t} + \xi_{iT}\frac{\partial T}{\partial t} \qquad (9)$$

where the slopes ξ_{ip} and ξ_{iT} can be calculated from the surface of Figure 2.

The compressibility coefficient due to thermal changes, α_T, is a parameter not readily found in literature. However, it can be determined from the thaw settlement parameter, A_o. For complete thawing of a fully frozen, fully saturated medium, the thaw settlement parameter A_o as defined by Sykes, et al. (1974) represents the thaw settlement due to the volume change of ice on melting and release of pore water. Watson, et al. (1973) have performed experiments to determine the value of A_o. By definition, for a saturated medium, $\theta_w = n$. Thus, initially, if all of the pore spaces are ice filled (i.e., $\theta_i = n_o$), then after total melting (i.e., $\theta_i = 0$),

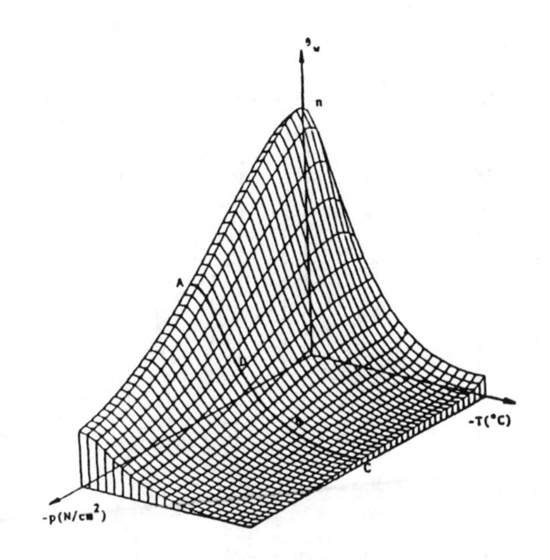

Fig. 1 Variation of liquid water content with temperature and pressure.

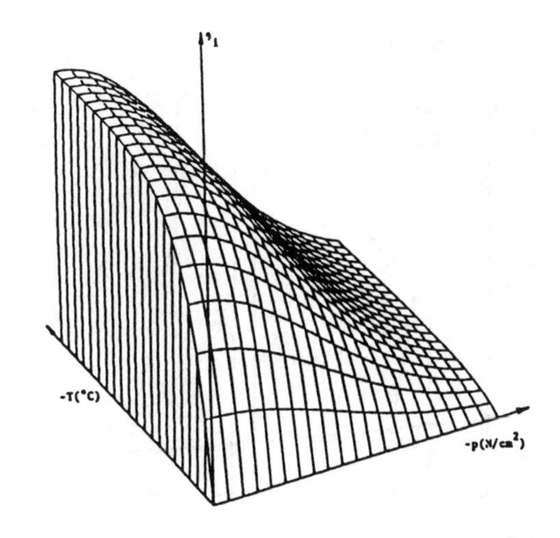

Fig. 2 Variation of ice content with temperature and pressure.

incremental change in ice content is $\partial\theta_i = n_o - 0 = n_o$. In this case, $\partial\varepsilon|_p = A_o$, where A_o is the thaw settlement parameter based on the definition of Sykes, et al. (1974). Sykes, et al. (1974) showed that for a fully saturated and frozen non-clay soil

$$A_o = \frac{\Delta e}{1+e}\Big|_p = \frac{\Delta n}{1-n}\Big| \equiv \partial\varepsilon|_p \tag{10}$$

Now, assuming a linear relationship for a change of ice content, $\partial\theta_i$, the strain would then be

$$\partial\varepsilon|_p = A_o \frac{\partial\theta_i}{n_o} \tag{11}$$

Further, in (5), α_T has been defined such that $\partial\varepsilon|_p = \alpha_T\,\partial T$ and hence, we get

$$\alpha_T = \frac{A_o}{n_o}\frac{\partial\theta_i}{\partial T} = \frac{A_o}{n_o}\xi_{iT} \tag{12}$$

Also, parameters like the thermal liquid diffusivity are a function of the moisture content, and curves for the relationships have been supplied, for example, in Farouki (1981, p. 49) after Higashi (1958). Least squares fit expressions can be obtained from such curves for the purpose of numerical solutions.

The flow equation (1) with q_r represented as in (2), and coupled by R_a term to the ice mass conservation equation is finally expressed in one-dimensional (vertical) form as

$$\begin{aligned}\frac{\partial}{\partial z}&\left[-\underline{\underline{K}}\left\{\frac{1}{\rho_f g}\frac{\partial p}{\partial z} - 1\right\} - \underline{\underline{D}}_{MT}\frac{\partial T}{\partial z}\right]\\
&+\Big[S_w p\alpha_p\xi_p + \alpha_p S_w^2\\
&\quad + n\xi_p + (\rho_i/\rho_f)\xi_{ip}\\
&\quad + (\rho_i/\rho_f)\theta_i\alpha_p\left\{p\xi_p + S_w\right\}\Big]\frac{\partial p}{\partial t}\\
&+\Big[S_w p\alpha_p\xi_T + S_w\alpha_T + n\xi_T\\
&\quad + (\rho_i/\rho_f)\xi_{iT}\\
&+ (\rho_i/\rho_f)\theta_i\left\{p\alpha_p\xi_T + \alpha_T\right\}\Big]\frac{\partial T}{\partial t} = 0\end{aligned} \tag{13}$$

and the energy equation (7) in its one-dimensional form, coupled again with the ice mass conservation equation is given by

$$\begin{aligned}&[(\rho C)_m + pn\xi_T + \{p\alpha_p\xi_T + \alpha_T\}\\
&\{(\theta_i+1-n)T\gamma + (1-n)pS_w\}\\
&- \{\rho_i\xi_{iT} + \rho_i\theta_i(p\alpha_p\xi_T + \alpha_T)\}\\
&\{T(C_v - C_i) - L - p/\rho_f\}]\,\frac{\partial T}{\partial t}\end{aligned}$$

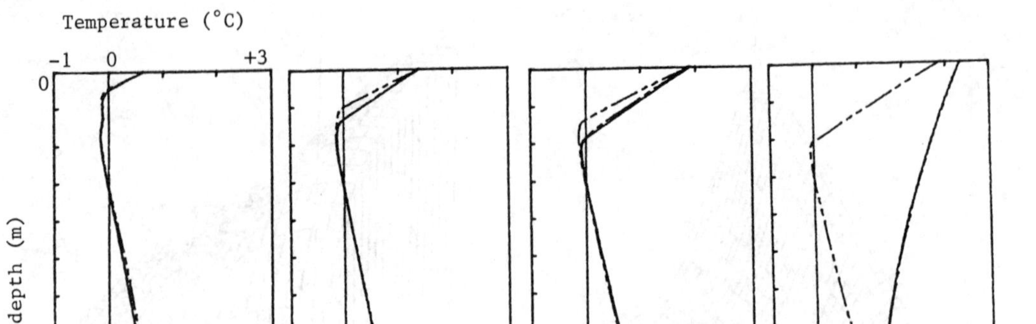

Fig. 3 Sensitivity of predicted temperature profiles to thermal parameters.

——— Simulation results with λ_i = 0.7106 J/m/s/°C , λ_s = 2.926 J/m/s/°C
C_ε = 1445.7 J/kg/°C, C_i = 908.7 J/kg/°C

—·—·— Sensitivity results with λ_i = 0.1338 J/m/s/°C, λ_s = 2.926 J/m/s/°C
C_ε = 2153.2 J/kg/°C, C_i = 908.7 J/kg/°C

— — — Sensitivity results with λ_i = 0.92 J/m/s/°C, λ_s = 3.3 J/m/s/°C
C_ε = 1445.7 J/kg/°C, C_i = 908.7 J/kg/°C

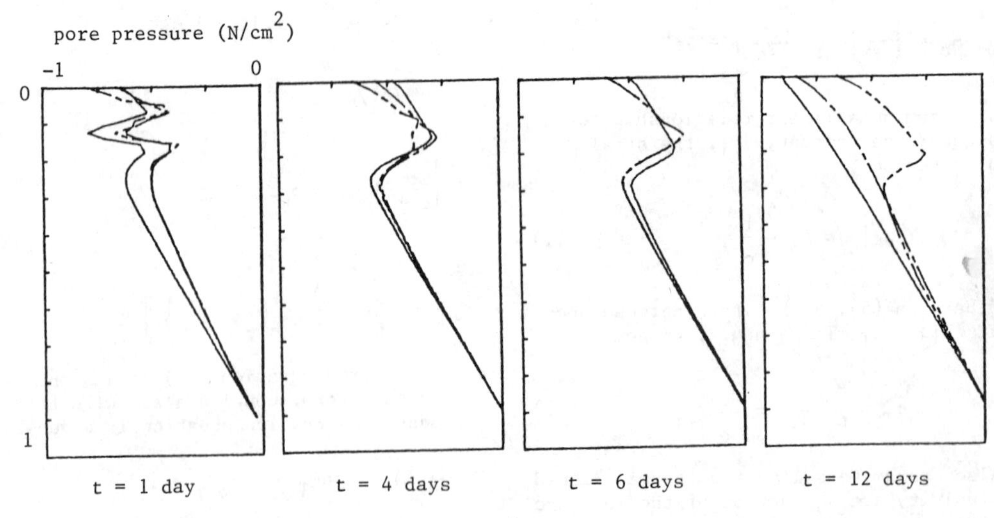

Fig. 4 Sensitivity of predicted pore pressure profiles to thermal parameters.

——— Simulation results with λ_i = 0.7106 J/m/s/°C , λ_s = 2.926 J/m/s/°C
C_ε = 1445.7 J/kg/°C, C_i = 908.7 J/kg/°C

—·—·— Sensitivity results with λ_i = 0.1338 J/m/s/°C, λ_s = 2.926 J/m/s/°C
C_ε = 2153.2 J/kg/°C, C_i = 908.7 J/kg/°C

— — — Sensitivity results with λ_i = 0.92 J/m/s/°C, λ_s = 3.3 J/m/s/°C
C_ε = 1445.7 J/kg/°C, C_i = 908.7 J/kg/°C

$$+ [pn\xi_p + \{p\xi_p + S_w\}\{(1-n)T\gamma\alpha_p$$

$$+ (1-n)S_w p\alpha_p\} - \{\rho_i\xi_{ip}$$

$$+ \rho_i\theta_i\alpha_p(p\xi_p + S_w)\} \qquad (14)$$

$$\{T(C_v - C_i) - L - p/\rho_f\}] \frac{\partial p}{\partial t}$$

$$+ C_v\rho_f q_r \frac{\partial T}{\partial z} - \frac{\partial}{\partial z}(\Lambda_m \frac{\partial T}{\partial z}) = 0$$

A numerical solution scheme is applied to solve for the basic variables p, and T for a vertical one-dimensional soil column, and the water and ice contents can be calculated from the (θw, p, T) and (θi, p, T) curves given in Figures 1 and 2, respectively, and porosity changes, and settlements can be determined from equation (4). The model was tested against experimental results obtained at the U.S. Army CRREL (Guymon et al, 1985), and the results compared very favorably (Corapcioglu and Panday, 1988b).

2 SENSITIVITY ANALYSIS

To deduce the effect of variability of certain parameters in the system to the overall simulation, a sensitivity analysis was carried out for certain parameters. The simulation was conducted over a 12-day period, to match experiments conducted at CRREL.

First, the general behavior of the simulation, with respect to a variation in the thermal parameters is tested. The thermal conductivity values of ice and soil (silty sand) were reduced. At the same time, the heat capacity of ice was lowered, and that of the soil, increased. For a second case, the heat capacity values remained unchanged, but the thermal conductivities of ice and soil were raised. All sensitivity tests were performed within a realistic range of values for soil and ice, as supplied by Van Wijk and De Vries (1963). The temperature and pressure curves obtained for these two cases are compared with the original results in Figures 3 and 4, respectively. The second case, where the thermal conductivities were increased, with the heat capacities unchanged, gave very close results to the actual simulation, showing that a variation such as this does not influence the results much. Thus, in analyzing the first case, we should consider the heat capacity changes only. We can conclude that the heat capacity of the soil being increased greatly decreases the ability of the system to change temperatures rapidly, as can be seen from Figure 3. For this case, the ice had not yet melted after 12 days of simulation, which is also reflected by the buildup of water above, and slightly into more impermeable ice region as shown in the pressure diagram of Figure 4. The total settlements noticed for this case were about 6 cm less than for the actual simulation run. The effect of lowering the heat capacity of ice was expected to be noticed within the frozen zone of Figures 3 and 4, but this was not so, probably due to the small fraction of ice present, as compared to the soil which will give a much larger contribution to the total heat capacity of the system. The heat capacity effects of liquid water, towards the combined heat capacity of the system, will also be quite small, due to the smaller fraction of water present, and due also to the much smaller value of heat capacity of water as compared to that of soil.

The model was relatively insensitive to the variation of hydraulic conductivity with moisture content. Varying the degree of the phase composition curve affected the rate of melting, but the forms of the temperature and pressure profiles were not affected greatly, moving towards similar equilibrium conditions once all the ice had melted. The coefficient of compressibility, α_p is also an insensitive parameter on this model. A value equal to half of that used in the simulation provided a final settlement difference of only 0.1 cm for a 12-day simulation period.

The next parameter on which a sensitivity test is performed, is the thaw settlement parameter A_o. The value was first decreased to 0.20, and then increased to 0.32. These values are within the range of experimentally determined values supplied by Watson, et al. (1973). A = 0.28 is the value used for the actual simulation. The p and T profiles generated by employing these values, are not affected to a noticeable extent. However, settlements are extremely sensitive to this parameter. The thaw settlement curve presented in Figure 5 compares these results. A_o = 0.32 produced a final settlement of 2.2 cm, A_o = 0.20 produced 1.2 cm of final settlement, while the simulation value of 0.28 provided a 1.8 cm settlement.

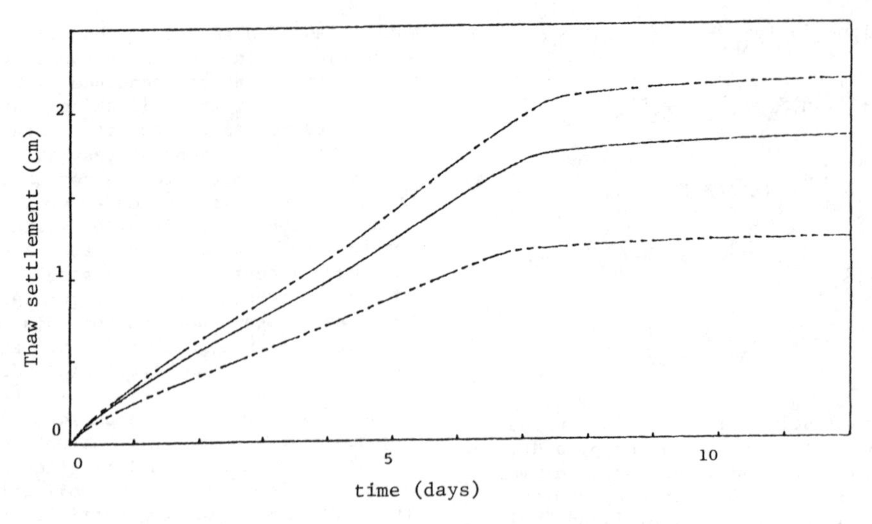

Fig. 5 Sensitivity of predicted thaw settlement variation with time to thaw settlement parameter, A_o (——— A_o = 0.28, —— - —— A_o = 0.32, ——-·-—— A_o = 0.20)

A trial run was also performed in which the moisture retention curve, and the phase composition curve were combined in a dynamic way wherein the change of saturation due to temperature affected the pore pressure directly. This was done as follows. If the ice in the pore spaces is considered to reduce the flow paths, the pore pressures will be affected by this, and considering the ice to behave as interstitial solids (modifying the flow regime), we can write the moisture content as $\theta w = (n - \theta_i)S_w^*$ $[= n\ Sw_1\ (p)]$, where $S_w^* = f_1(p)$ is defined by the moisture retention curve. A phase composition curve representing the temperature variation of the liquid water content as a percentage of the total (ice + water) water content is expressed by $f_2(T) = \theta_w/(\theta_i+\theta_w)$, and hence $\theta_i = (1 - f_2(T)/f_2(T))\theta_w$. Solving for these two curves simultaneously gives

$$\theta_w = nS_w = \frac{nf_1(P)\ f_2(T)}{f_2(T)+f_1(p)-f_1(P)f_2(T)} \tag{15}$$

and

$$\theta_i = \frac{n[f_1(p) - f_1(p)f_2(T)]}{f_2(T)+f_1(p)-f_1(p)f_2(T)} \tag{16}$$

In such a representation, the pore pressures directly reflect the liquid water saturation. This is not the case for a representation like Figure 1 and Figure 2. For example, on Figure 1 we can move in a direction parallel to the T axis along ABC, such that the same pore pressure represents many different saturation values. Figures 6 and 7 for the temperature and pore pressure profiles show the results to be very similar indicating that a static way of interpreting the combination of phase composition and moisture retention curves can produce good results, in slow processes such as groundwater freeze-thaw and flow, hence affording greater simplicity.

3 SUMMARY AND CONCLUSIONS

A mathematical model describing the thaw settlement processes in permafrost soils, as developed in Corapcioglu and Panday (1988a) and verified in Panday and Corapcioglu (1988b) can be used to predict the extent and rate of thawing of permafrost soils. The ability to use such a model successfully, is enhanced if the importance of parametric data used in the simulation is studied through a sensitivity analysis. The settlements are found to be very sensitive to the thaw settlement parameter Ao, and a good estimate is required for its value for the soil under consideration, if settlement results are important. Pore pressure and temperature profiles, and settlement

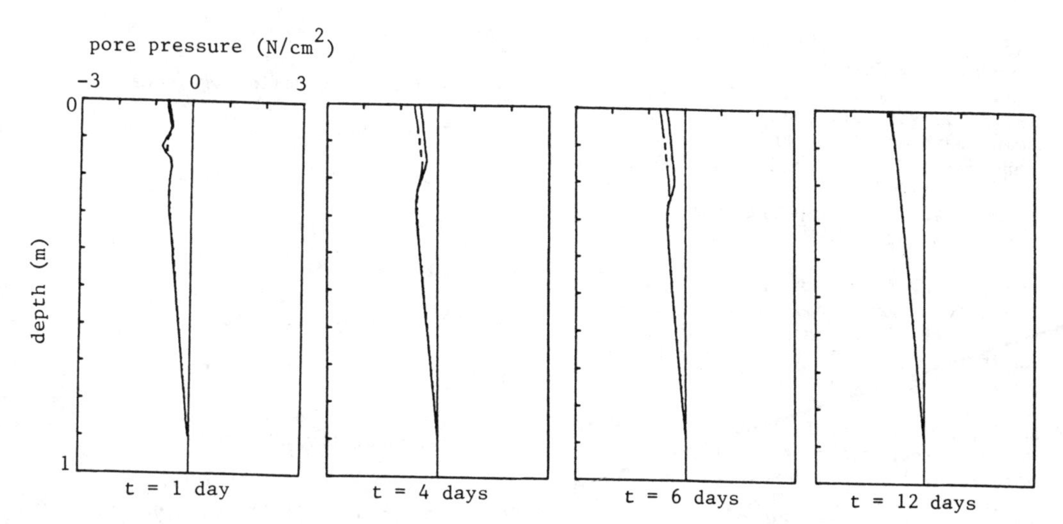

Fig. 6 Senstitivity of predicted temperature profiles to θ_w vs. p and T relation (——— Using the curves of Figures 1 and 2, — - - — Using Eq. 15 and 16)

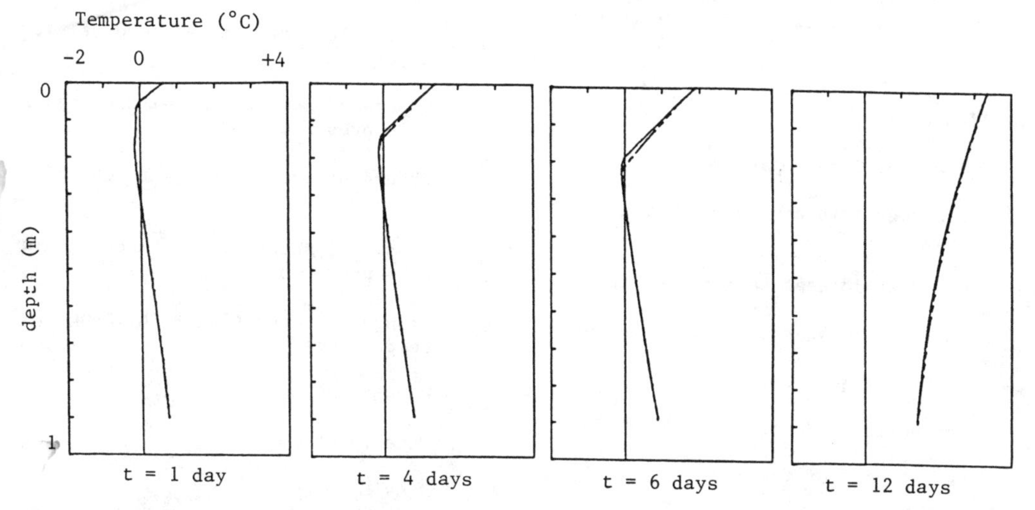

Fig. 7 Sensitivity of predicted pore pressure profile to θ_w vs. p and T relation (——— Using the curves of Figures 1 and 2, — - - — Using Eqs. 15 and 16)

variations are sensitive to the heat capacity values used for ice and soil grains. The model was relatively indifferent to variations in the values of the coefficient of compressibility, α_p, and to the hydraulic conductivity and phase composition relations used. This study also confirmed that phase composition and moisture retention curves, although determined experimentally under static conditions, can be used in a static way, for slow transient processes such as groundwater freeze-thaw and flow.

REFERENCES

Corapcioglu, M.Y., and S. Panday, 1988a. Multiphase approach to thawing of unsaturated frozen soils: Equation development (submitted for publication).

Corapcioglu, M.Y., and S. Panday 1988b, Thawing in permafrost--Simulation and verification, in Permafrost, 5th Int. Conference Proceedings, National Academy Press, Washington, DC.

Farouki, O.T., 1981, Thermal properties of soils, Cold Regions Research and Engi-

neering Laboratory Monograph 81-1, Hanover, New Hampshire.
Guymon, G.L., R.L. Berg, and J. Ingersoll, 1985, Partial verification of a thaw settlement model, in Freezing and Thawing of Soil Water Systems, ASCE, New York, pp. 18-25.
Higashi, A., 1958, Experimental study of frost heaving, SIPRE Research Report 45, AD214672.
Sykes, J.F., W.C. Lennox, and Unny, T.E., 1974, Two-dimensional heated pipeline in permafrost, American Society of Civil Engineers, Journal of the Geotechnical Engineering Division, v. 100, pp. 1203-1214.
Van Wijk, W.R., and De Vries, D.A., 1963, Periodic temperature variations in a homogenous soil, in Physics of Plant Environment edited by W.R. Van Wijk, Amsterdam, North Holland Publishing Company, pp. 102-143.
Watson, G.H., W.A. Slusarchuk and R.K. Rowley, 1973, Determination of some frozen and thawed properties of permafrost soils, Canadian Geotechnical Journal, v. 10, pp. 592-606.

NOTATION LIST

A_o Thaw settlement parameter

C_ε Gravemetric heat capacity of soil solids

C_i Gravemetric heat capacity of ice

C_v Gravemetric heat capacity of water

$\underset{\approx}{D}_{MT}$ Thermal liquid diffusivity

$\underset{\approx}{K}$ Hydraulic conductivity

L Latent heat of fusion

n Porosity

p Pore water pressure

$\underset{\sim}{q}$ Specific discharge of water

$\underset{\sim}{q}_r$ Specific discharge of water relative to the moving solid

R_a Rate of melting of ice mass

S_w Degree of saturation

T Temperature

t Time

$\underset{\sim}{V}_f$ Velocity of flowing water

$\underset{\sim}{V}_S$ Velocity of solid particles

z Verticle coordinate

α_p Compressibility of soil

α_T Thermal compressibility of soil

ε Strain

θ_i Volumetric ice content

θ_w Volumetric water content

$\underset{\approx}{\Lambda}_m$ Coefficient of thermal conductivity of porous medium

λ_f Coefficient of thermal conductivity of water

λ_i Coefficient of thermal conductivity of ice

λ_s Coefficient of thermal conductivity of soil

ξ_{ip} Change of ice content with change in pore pressure

ξ_{iT} Change of ice content with change in temperature

ξ_p Change of saturation with change in pore pressure

ξ_T Change of saturation with change in temperature

ρ_f Density of water

ρ_i Density of ice

ρ_s Density of soil

$(\rho C)_m$ Heat capacity of the porous medium

σ' Effective stress

ACKNOWLEDGMENTS

This research has been supported by a grant from National Science Foundation ECE-8602849.

5th International Symposium on Ground Freezing, Jones & Holden (eds)
© 1988 Balkema, Rotterdam. ISBN 90 6191 824 3

Investigations of the frost heave of colliery spoil

Maria Porębska & Krystyna M.Skarżyńska
University of Agriculture, Kraków, Poland

ABSTRACT: The utilization of colliery spoil has in recent years been the subject of extensive investigations in various countries. Among other applications, it was used in Poland in the construction of river embankments, impoundment dams for the fuel ash storage, and as fill material. One of the problems with colliery spoil is that it can undergo degradation through weathering caused by moisture changes and cycles of freezing and thawing. The experiments reported in this paper describe laboratory studies on unburnt colliery spoil from Upper Silesia for which observations were made of the moisture changes and temperature distribution produced during freezing. The results were compared with frost susceptible silty clay tested under similar conditions. Conclusions are drawn from the laboratory results concerning the possible behaviour of the colliery spoil in the field.

1 INTRODUCTION

The problem of the utilization of various wastes, "artificial" or so called man-made soils, for engineering structures began some years ago and provided the incentive for much research into their properties from soil mechanics point of view.

Only some of this work, however, delt with the behaviour of mentioned soils subjected to freezing. In principle they are connected with standard tests for general classification as frost susceptible or non-frost susceptible material, mainly for road structures. More significant references are discussed below.

The investigations of Kettle (1973), Kettle and Williams (1976) concerned the measurements of frost heave, heaving pressure, and frost susceptibility of colliery shales. The tests were performed on specimens of unbound and cement stabilized colliery spoil from different coalmines, both unburnt and burnt shales being studied. The general conclusion is that burnt shales are frost susceptible but the behaviour of unburnt shales changes from non-susceptible to highly susceptible. Among the unbound shales, the finer grained ones developed the highest heaving pressure. Cement stabilization did not significantly affect the heaving pressure developed by the coarser grained shales but, with the finer grained shales, it reduced the pressure developed. Rainbow (1987) presented the result of TRRL tests for the unburnt colliery spoils from all parts of the United Kingdom. The author stated that material was usually within the permissible limits of frost heave (13 mm to 16 mm) and therefore generally non-frost susceptible.

Croney and Jacobs (1967) discussed the results of a large number of burnt colliery shale samples. No clear correlation of heaving value with saturation or percentage of fines was found. It was suggested that the non-uniformity of material properties causes significant heaving variations ranging from 0.2 to 84 mm. The tests on burnt colliery shales were also performed by Soczawa (1981, 1984) who found the correlation between the percentage of fines and frost heaving values.

The investigations of the heaving pressure and the frost susceptibility of "artificial" soils were also performed by McCabe and Kettle (1982, 1984). For those tests they mixed the basic matrix sand and chalk with three types of selected coarse particles (slag, basalt, limestone) to change the particle size distribution of the original material. They found that the frost susceptibility of soil was reduced by mechanical stabilization. It is more important, however, to note that not only the amount of finer particles, but also the quantity and type of coarse particles have significant effect on frost heave and heaving pressure.

Quite a number of existing references are connected with investigations of fly ash. In a paper of Katsuyuki and Takashi (1985) are also shown the results of experiments on fly and volcanic ash. There it is said that frost heave of volcanic ash is much higher than of non-volcanic ash and the magnitude of this frost heave decreases with decreasing dry bulk density. Frost susceptibility of fly ash was investigated by Croney and Jacobs (1967). The results showed that the tested material heaved from marginally to satisfactorily. Grading was clearly not the only factor determining the heave, and as in the case of limestones and other porous aggregates, the porosity of the individual particles played a part. Stabilization of ash with cement reduced the frost heave significantly. Sutherland and Gaskin (1970) investigated ashes in their plain state and also after stabilization with ordinary Portland cement and calcitic hydrated lime. They stated, as did other authors, that the use of additives was the successful method of reducing frost heave. The results allowed classification of stabilized ashes as non-frost susceptible.

From this short review of selected references describing the behaviour of coalmine wastes, man-made and artificial soils under frost action it may be seen that it is necessary to carry out more tests to obtain representative results. It is also important to carry out separate investigations for each colliery spoil type if we want to use them as an embankment or fill material. This is why our paper tries to show some results of the essential properties of unburnt colliery spoil.

2 MATERIAL AND METHODS

Unburnt colliery spoil is a mixture of rock fragments such as claystone, shales, siltstones, carbon shales, and sandstones with a small admixture of coal. This material can undergo degradation through weathering mainly caused by moisture changes and cyclic freezing and thawing. Susceptibility to weathering depends also on its duration, mineralogical composition, type of rocks and its structure.

In earlier work, the present authors carried out experiments on the disintegration of colliery spoil subjected to freezing (Skarżyńska et al. 1987, Porębska 1987). Tests on the frost resistance of fresh colliery shale were carried out on fully saturated samples with $\phi < 100$ mm, the freezing was followed by thawing in water, in 25 cycles at -20°C. The selected material of ϕ < 10 mm with established moisture content was also subjected to freezing in 16 cycles at -5°C. Both experiments confirmed significant disintegration of the material. In the first case this consisted mainly in reduction of boulder and gravel sized particles and increase of sand particles. This strongly affected the angle of internal friction which decreased from 46° to 24°, and other material properties. In the second case the increase of sand as well as silt and clay also was significant.

Site investigation results of river embankments (made of colliery spoil from different sources) showed that the disintegration of material is the quickest and most intensive on the embankment surface and declines with depth. A very significant increase (10-25%) of particles finer than 2 mm was found at a depth of 0-0.5 m and a smaller one (6-8%) at a depth of 0.5-1.1 m (Skarżyńska et al., 1987). It was desided then to investigate the colliery spoil material which may be denominated as a far product of disintegration and the other designat-

Table 1. Characteristics of the material tested

Material	Optimum m/c %	Max.dry density g/cm³	Organic matter %
Colliery Spoil ø < 10 mm	11.6	1.81	18.9*
Colliery Spoil ø < 0.06 mm	22.8	1.52	18.9*
Silty Clay	14.0	1.82	1.8

*organic matter, mainly pure coal

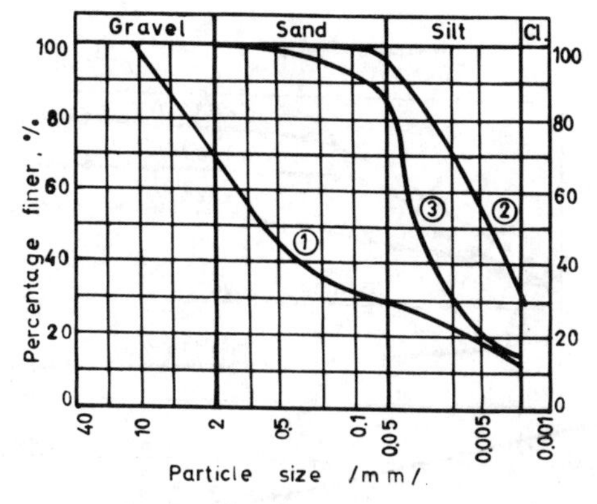

Fig.1. Particle size distribution curves of the investigated soils, 1.colliery spoil ø < 10 mm. 2. colliery spoil ø < 0.06 mm. 3. silty clay

ed as a transition one.

For the tests, unburnt colliery spoil from the Brzeszcze coalmine - Silesia Region and natural soil-silty clay, for comparison, were used. In result of mineralogical analysis of colliery spoil the clay minerals amounting to 40% of the material, represented chiefly by kaolinite (25%), illite (10%), and montmorillonite (5%) were identified. Moreover, calcite (4%) and considerable quantities of organic matter were found. Silty clay consists mainly of quartz (60-70%), feldspar (15%), and calcite (7%). Clay minerals (10%) represented by illite and kaolinite were also evaluated.

The colliery spoil material was sorted into two groups, that with a diameter of ø < 10 mm and ø < 0.06 mm being taken for testing. Using terminology accepted for mineral soils, colliery spoil material can be classified as gravelly clay and clay, respetively. In Table 1 and Figure 1 are compiled the parameters characterizing the selected material.

After applying suitable initial moisture, specimens were compacted in 4 layers in plexiglass cylinders 11 cm in diameter and 16 cm high. Then they were left for 24 hrs until the conditions stabilized. The experiments were carried out in an air-conditioned chamber with a temperature monitoring system. Specimens, insulated from all sides were subjected to single cooling (+16°C to -5°C) from the top surface in the open system. There were also carried out single and repeated changes of temperature in the same range in the closed system. The sub-zero temperature was maintained until its stabilization occured. In general two different moisture contents (see Table 2) were established for tested material. The investigations contained measurements of temperature, frost heaving and moisture distribution. Scanning electron micrographs were also taken.

3. RESULTS

3.1. Frost heaving

In the closed system for the colliery spoil samples of ø < 10 mm with an initial moisture content of 11.5% a decrease in sample height was observed, both after single (- 0.61 mm) and cyclic freezing (- 1.56 mm). When the moisture content was 16.7% the magnitude of heaving was 0.52 mm following single cooling and 2.57 mm after 16 freezing-thawing cycles (Table 2, Figure 2a). It should be noticed that after single and cyclic cooling the frost heaving of the colliery spoil of ø < 0.06 mm was much higher than that for material of ø < 10 mm. The magnitude, for assumed initial moisture

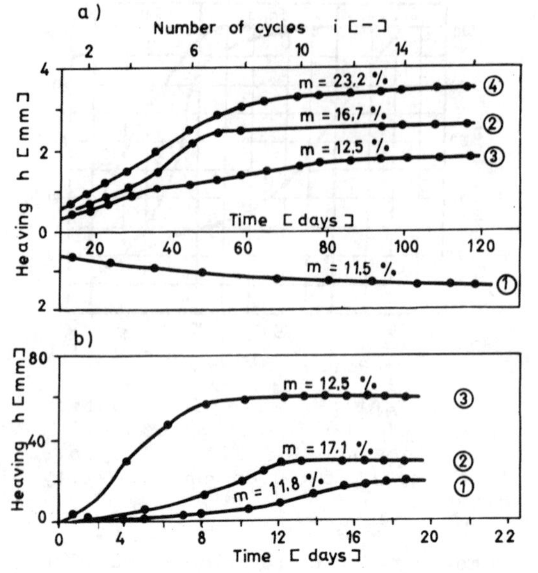

Fig.2 The course of frost heaving a. closed system, b. open system, 1,2 colliery spoil $\phi < 10$ mm, 3,4 colliery spoil $\phi < 0.06$ mm, m - initial moisture content

content increased, but did not exceed 3.4 mm.

In the open system, during single cooling, the same general regularity was observed. The heaving values were three times higher for the material of $\phi < 0.06$ mm than for $\phi < 10$ mm, i.e. 60 mm and 19.2 mm, respectively. They were also much higher than achieved in a closed system for the same initial moisture content (Figure 2b).

In colliery spoils, similarly to the natural soils, repeated freezing and thawing conditions caused more intensive processes of water migration and consequently higher value of heave (Kukiełka 1983, Skarżyńska 1985, Porębska 1987), as well as the availability of water significantly increased the magnitude of heave. Both processes were confirmed by the results achieved for silty clay tested under the same conditions.

3.2. Moisture distribution

The distribution of moisture, in general comparable with the heave values and the intensity of ice

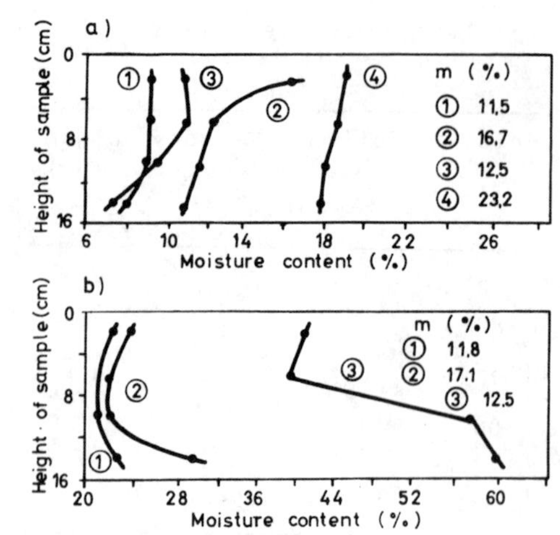

Fig.3 Distribution of moisture. a. closed system, b. open system, 1,2 colliery spoil $\phi < 10$ mm, 3,4 colliery spoil $\phi < 0.06$ mm, m - initial moisture content

lens segregation. In the closed system under repeating freezing a decrease in moisture content of 2% to 5% in the moisture section of all the samples took place (see Table 2). This was caused partly by horizontal migration of water to the sample sides what suggested insufficient side insulation of specimens and also water evaporation during the long-lasting test. Nevertheless, it may b seen from the graphs (Figure 3a) that the bottom zone of samples dried out more owing migration of water to the freezing front for both colliery shale materials, irrespective of the initial water content. In the open system the final moisture content showed the easily recognized increase of water content along the whole profile of colliery spoil samples (Figure 3b). Moreover the moisture distribution along the profile was more varried than in the close system with significant accumulation of water in the sample parts close to the free excess of water. The moisture distribution in silty clay and the increase in moisture content is similar to that in colliery spoil of $\phi < 0.06$ mm.

3.3. Course of temperature

Analysis of the temperature curves

Table 2. Freezing test results

Material	Moisture content (%)			Frost heaving (mm)	
	before freezing	after 1 cycle	after 16 cycles	after 1 cycle	after 16 cycles
	Closed system				
Colliery spoil $\phi < 10$ mm	11.5	10.9	9.0	-0.61	-1.56
	16.7	12.7	12.9	0.52	2.57
Colliery spoil $\phi < 0.06$ mm	12.5	-	10.3	0.50	1.8
	23.2	-	18.2	0.61	3.4
Silty clay	16.7	-	14.0	1.20	3.6
	Open system				
Colliery spoil $\phi < 10$ mm	11.8	20.5		19.2	
	17.9	21.7		28.9	
Colliery spoil $\phi < 0.06$ mm	12.6	44.0		60.0	
Silty clay	11.8	44.0		35.2	
	17.24	24.0		40.2	

recorded in the open and closed system permits their classification as classic ones (Figure 4). In the open system, the temperature of the whole sample decreases in the first stage. For an initial moisture 12% and wastes $\phi < 10$ mm and $\phi < 0.06$ mm and silty clay this period lasts for about 42, 30, and 36 hours, respectively. In the second period, corresponding to that of ground freezing, the temperature stabilizes at about 0°C to ±0.2°C, this being connected with the intensive formation of crystallization nuclei and ice lenses in the lower zone, and small, diffuse centres of crystallization in the middle zone. This period lasts for 24 h at $\phi < 10$ mm, 3 h at $\phi < 0.06$ mm and 42 h for silty clay. During the third period a distinct linear decrease takes place in the temperature, corresponding to its lowering in the already frozen soil-water medium, causing further freezing of loosely bound water. At a moisture content of about 12% this process lasts for about 322 hours for $\phi < 10$ mm, 400 h for $\phi < 0.06$ mm, and 268 h for silty clay. With higher initial moisture content (17%), the processes of cooling and subsequent freezing take place similarly but last much longer. In the case of a closed system, however, the time required for this is very short as compared with that in the open system, this being associated with the amount of water migrating to the freezing zone.

3.3. Structure observation

Macroscopic and microscopic observation was carried out for frozen samples in the open system. As the macroscopic description of sample vertical cross-sections shows the magnitude and the arrangement of ice lenses is different for colliery spoil and silty clay. There are numerous large, 5 mm thick horizontal ice lenses at the bottom of colliery spoil samples. In the central and upper part only fine and dispersed ice can be seen. The arrangement of ice lenses for silty clay is similar at the sample bottom but their thickness is smaller, around 3 mm. In the upper part, besides horizontal lenses there are also vertical ones (parallel to the direction of freezing) but these are thin, about 1 mm.

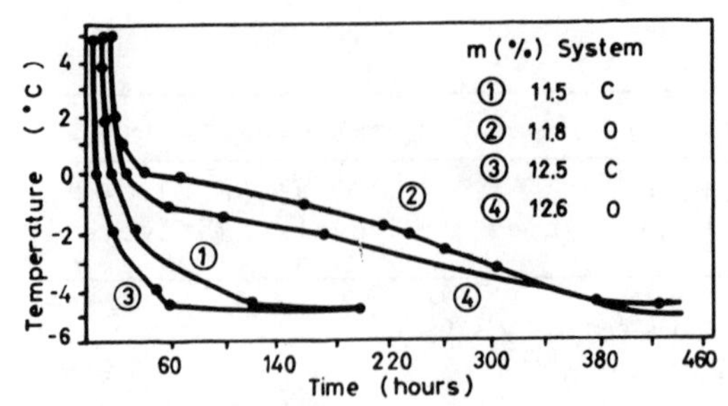

Fig.4 The course of temperature in closed and open system, 1,2 colliery spoil ϕ <10 mm, 3,4 colliery spoil ϕ <0.06 mm

Typical of the central part are numerous ice lenses in the shape of fine needles and grains.

SEM photographs were taken for colliery spoil ϕ < 10 mm and for silty clay before and after freezing. In the case of colliery spoil the micro-ice lenses greatly exceed 1200 μm. Their distribution is fairly regular. In silty clay, however, the distribution of micro-ice lenses is much more varried with respect to the sizes and respective positions than for colliery spoil. Their magnitude for silty clay varies mainly in the range from 200 to 600 um reaching about 1100 μm.

Analyzing the micro-structure it may be noted that original colliery spoil material consists mainly of flaky grains with irregular shapes and arrangements. They can adhere to each other on three ways: (EE) edge-to-edge, (EF) edge-to-face, (FF) face-to-face (Grabowska-Olszewska 1980). After freezing reconstruction of structure took place, evidenced by partial smoothing of the aggregate surface close to the formed ice lenses. It may be seen that aggregates are composed of regularly ordered domains: these consist of parallelly adhering flaky particles (Figure 5a,b). The original structure of silty clay is completely different. The oval--shaped grains create a spatial structure so that contacts between them are mainly EE or EF type. After freezing it is easy to separate the single aggregates as well as large ones with highly differentiated morphology of the surface (Figure 6a,b).

The freezing process significantly affected the arrangement of the particles and aggregates in a degree dependent on the mineralogical composition of the material, mainly on clay minerals content which is much higher in colliery spoils (40%) than in silty clay (10%).

4. DISCUSSION AND CONCLUSIONS

When the suitability of unburnt colliery spoil as a fill material is considered, the behaviour of this material should be taken into account; it undergoes disintegration caused by climatic condition as well as degradation during incorporation in the structure. As was concluded by Rainbow (1987) colliery spoils are generally not frost susceptible, which is reasonable as his results represented the values for a sample of natural grading of ϕ < 80 mm, Kettle and Williams (1976) reported that the behaviour of unburnt colliery shales changed from not susceptible to highly susceptible. They tested materials of ϕ <40 mm with different amounts of smaller particles.

Analysing the conditions under which this particular material is working when incorporated in a structure, it should be said that as a result of moisture and temperature changes the upper layer (0-1 m thick, especially between 0-0.5 m) yields to disintegration relatively quickly. The material grading passes to silt and clay with the amount of gravel particles increasing with depth. The deeper layers (> 1 m) are not subjected to disintegration caused by weathering processes and frost action.

On the basis of our experiments it may be concluded that:
- colliery spoils of ϕ <0.06 mm correspond to the surface layer (0-0.3 mm) in terms of complete disintegration. They are highly frost susceptible, especially when access of infiltration water is considered.
- colliery spoils of ϕ < 10 mm correspond to conditions of the layer at a depth of 0.3-1.0 m. They are moderately frost susceptible when they become saturated with moisture.
- repeated heaving and thawing of the upper layer bears upon a new structure and micro-structure

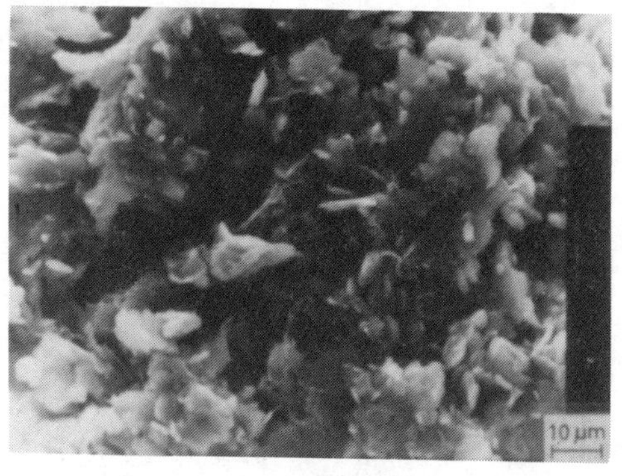

a)

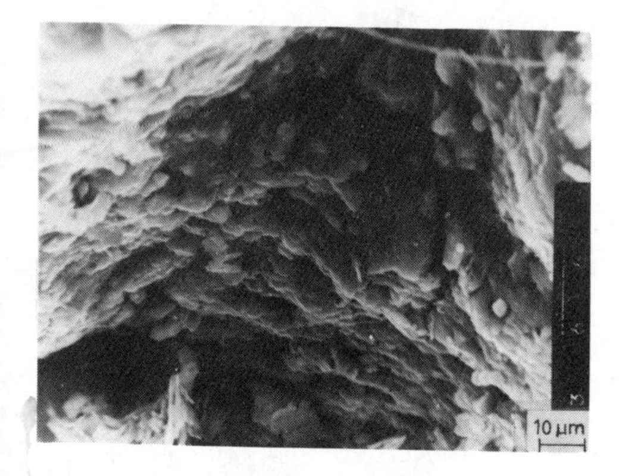

b)

Fig.5 Scanning electron photomicrographs of colliery spoil ∅ < 10 mm, a. original, b. after freezing in the open system

arrangement in both the colliery spoils tested.

This heaving and thawing also contribute significantly to decreasing of strength parameters and also leads to liquefaction of the material. The last was often observed in the embankments and dump slopes. The experiments testify moreover that heaving processes and their effects become more intensive in the case of a shallow ground water table.

REFERENCES*

Croney, D., Jacobs, J.C. 1967. The frost susceptibility of soils

*See page X concerning ISGF papers

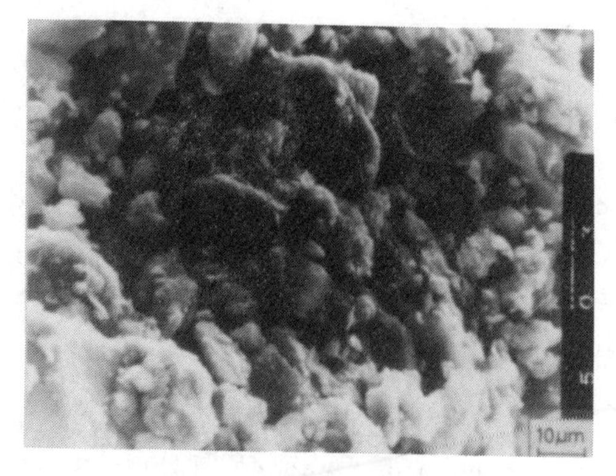

a)

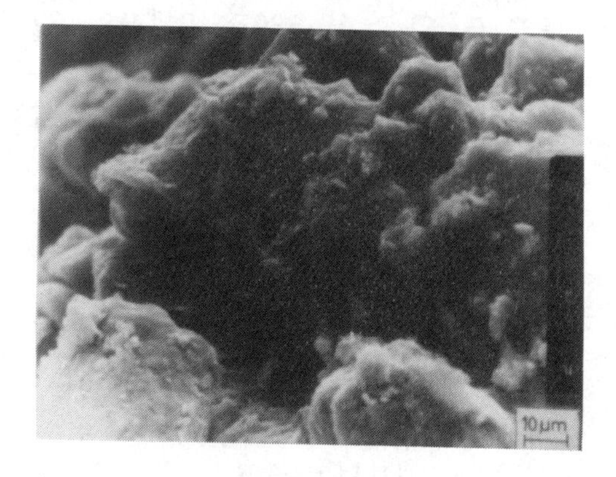

b)

Fig.6 Scanning electron photomicrographs of silty clay, a. original, b. after freezing in the open system

and road materials. Road Res. Lab. Report, LR 90 Crowthorne. pp. 68

Grabowska-Olszewska, B. 1980. Metody badań gruntów spoistych, Warszawa: Wyd.Geol.

Katsuyuki, S., Takashi, M. 1985. Frost heaving of volcanic ash soils. Fourth ISGF, Sapporo. Vol.2: 163-166

Kettle, R.J. 1973. Freezing behaviour in colliery shale. Ph.D. thesis, Univ. Surrey. Guildford, England. pp.225

Kettle, R.J., Williams, R.I.T. 1976. Frost heave and heaving pressure measurements in colliery shales. Can.Geotech.J., Vol.13: 127-138

Kettle, R.J., McCabe, E.Y. 1984. The control of frost action in unbound materials. Third Int. Cold Regions Engineering Speciality Conference, Edmonton, Alberta, pp.20

Kukiełka, J. 1983. Laboratory experiments on loess frost susceptibility. Drogownictwo Vol. 10: 285-290

McCabe, E.Y., Kettle, R.J. 1982. Heaving pressures and frost susceptibility. Proc. of Third ISGF. Army Corps of Engineers. CRRL Hanover, NH. Special Report 82-16: 285-294

Porębska, M. 1987. The influence of repeated freezing-thawing conditions on the heaving of coal mining wastes. Symposium Polish-French. Paris, in print

Rainbow, A.K.M. 1987. An investigation of some factors influencing the suitability of minestone fill in reinforced earth structures. British Coal. London, pp.562

Skarżyńska, K.M. 1985. Formation of soil structure under repeated freezing-thawing conditions. Fourth ISGF, Sapporo. Vol.2: 213-218

Skarżynska, K.M., Burda, H., Kozielska-Sroka, E., Michalski, P. 1987. Laboratory and site investigations on weathering of coal mining wastes as a fill material in earth structures. Proc. of Second Conf. of Coal Mining Wastes. Nottingham, p.179-195

Soczawa, A. 1981. Zdolność odpadów kopalnianych do mrozowego pęcznienia. Drogownictwo Nr 9: 227-282

Soczawa, A. 1984. Akumulacja wody w przemarzających odpadach kopalnianych. Zeszyty Naukowe P.Śl. seria: Budownictwo z.59: 123-135

Sutherland, H.B., Gaskin, P.N. 1970. Factors effecting the frost susceptibility characteristics of pulverized fuel ash. Can. Geotech. J. Vol.7.No.1: 69-78

5th International Symposium on Ground Freezing, Jones & Holden (eds)
© 1988 Balkema, Rotterdam. ISBN 90 6191 824 3

Frost expansion pressure and displacement of saturated soil analyzed with coupled heat and water flows

Kimitoshi Ryokai
Shimizu Corporation, Tokyo, Japan

Fujio Tuchiya
Obihiro University of Agriculture and Veterinary Medicine, Japan

Masataka Mochizuki
Fujikura Ltd, Japan

ABSTRACT:Heat diffusivity and frost expansion analyses were carried out simultaneously in order to estimate the amount of frost heave and its pressure of a saturated soil. Through these analyses, a frost heave model which couples the heat and water flow where the entire earth pressure changes is developed. This report presents an applicability of the model to cylindrical form of heat and moisture flows by comparing the changes of the freezing front, displacement in soil and earth pressure with time. One of the conclusions is that the model simulate the behavior of soft clayey soil.

NOTATION

C_c, C_s :Index of consolidation and swelling
C_v :Consolidation coefficient
E :Young's modulus
K_o :Coefficient of earth pressure at rest
L_s, L_w :Latent heat of soil and water
R :Distance from the center of tank to boundary where displacement is zero
S_f :Suction force due to frost heave
T_c, T_w :Temperature of cooling and warming
T_f, T_u :Temperature distribution of frozen and unfrozen soil
V, V_w :Rate of freezing and water intake
Z :Coordination indicated depth
a, b :Radius of tank and freezing front
e :Void ratio
k :Permeability
n_f :Volumetric moisture content
r :Radial coordination
t :Time
u_r :Radial displacement
u_w :Pore water pressure
u_{wo} :Initial pore water pressure
α :Thermal transfer coefficient
Γ :Frost expansion ratio of water
γ_s, γ_w :Unit density of soil and water
δ :Amountof frost expansion
δ_c :Amount of consolidation
d_e :Amount of elastically deformation
ε_r :Radial strain
ζ :Frost heave constants
κ_1, k_2 :Thermal diffusivity of frozen and unfrozen soil
λ_1, λ_2 :Thermal conductivity of frozen and unfrozen soil
μ :Poisson's ratio
ξ, ξ_w :Ratio of frost heave and water intake
ξ_o, σ_o, V_o :Frost heave constants
σ, σ' :Frost expansion and effective pressure
σ_r, σ_θ :Radial and tangential stress

1 INTRODUCTION

In the ground surrounding an in-ground storage tank for Liquified Natural Gas (LNG), it is required to precisely estimate changes in the freezing front with the elapse of time, amount of frost expansion of the ground and the resultant frost expansion pressure when the soil is frozen.

When the soil is frozen, an ice layer called an ice lens develops. This is due to a suction force drawing water towards the freezing front. The analytical model that the total pressure do not change during the freezing ,was presented by Ryokai(1985). When the soil is frozen at the cylindrical form, the frost expansion pressure increases. When the suction force and frost expansion pressure appear, the effective pressure increases. To express such a freezing phenomenon, an analysis method has been developed and compared to the results of model tank experiments.

2 FROST HEAVE ANALYSIS COUPLED THE HEAT AND WATER FLOW

2.1 Assumption for analysis

Following assumptions are made for the analyses:

(1)The soil is homogeneous and perfectly water-saturated.

(2)The frost heave ratio, ξ, is obtained by the equation of Takashi(1977):

$$\xi=\xi_0+(1+\sqrt{V_0/V})\,\sigma_0/\sigma' \tag{1}$$

(3)The unfrozen soil behaves elastically when subjected to heave.

(4)The displacement, u_r, is zero at R from the center of the tank shown in Fig.1.

(5)The insulation layer of zero thickness with a heat transfer coefficient exists on the internal surface.

2.2 Suction force due to frost heave

As the laboratory model is of cylindrical form, as shown in Fig.1, the movement of heat inside the frozen soil and unfrozen soil and that of water inside the unfrozen soil are obtained respectively from the following equations:

$$\partial T_f/\partial t=\kappa_1(\partial^2 T_f/\partial r^2+1/r\ \partial T_f/\partial r) \tag{2}$$

$$\partial T_u/\partial t=\kappa_2(\partial^2 T_u/\partial r^2+1/r\ \partial T_u/\partial r) \tag{3}$$

$$\partial u_w/\partial t=C_v(\partial^2 u_w/\partial r^2+1/r\ \partial u_w/\partial r) \tag{4}$$

At the internal surface of tank (r=a) continuity of heat flow is required:

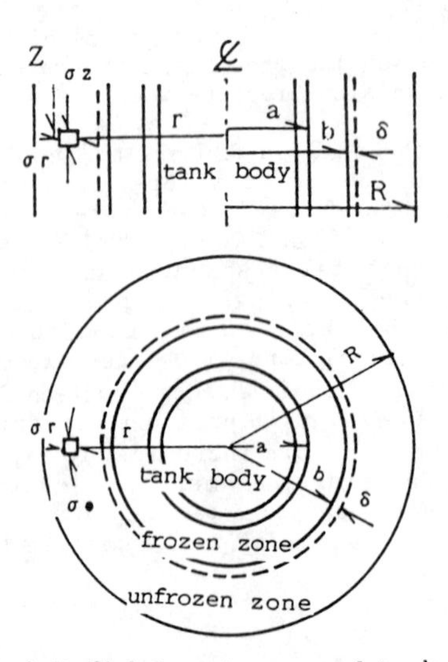

Fig.1 Radial freezing around tank

$$\lambda_1\partial T_f/\partial r=\alpha\ (T_f-T_c) \tag{5}$$

Moreover, the eq.(6) should be satisfied at the boundary between frozen and unfrozen soil:

$$\lambda_1\partial T_f/\partial r-\lambda_2\partial T_u/\partial r=L_s\gamma_s V+L_w\gamma_w V_w \tag{6}$$

Where Darcy's law is applicable for unfrozen soil, eq.(7) is satisfied:

$$V_w=k/\gamma_w\ \partial u_w/\partial r \tag{7}$$

While the frost expansion ratio of a saturated soil is given by eq.(1), the ratio of water intake/discharge, g_w, is expressed according to water balance:

$$\xi=\xi_0+n_f\Gamma+(1+\Gamma)\,\xi_v \tag{8}$$

V_w at the freezing front is given by the following equation:

$$V_w=\frac{(V+\sqrt{V_0V})\,\sigma_0}{(\sigma-u_{w0}-S_f)\ (1+\Gamma)}-\frac{V n_f\Gamma}{(1+\Gamma)} \tag{9}$$

When the pore water pressure at time j-1 is used instead of the pressure at time j, V_w becomes:

$$V_w=k\,\{u_w(i+1,j-1)-(S_f+u_{w0})\}/(dr\,\gamma_w) \tag{10}$$

When the freezing front grows at the earth pressure, S, and initial pore water pressure, $K.\gamma z$, the suction force due to frost heave in a freezing front, S_f, is generated in Fig.2 and is given as:

$$S_f=\sigma-u_{w0}-\sigma' \tag{11}$$

Using eqs.(9), (10) and (11), the suction force constituting a source of water absorption force can be obtained from the following equation:

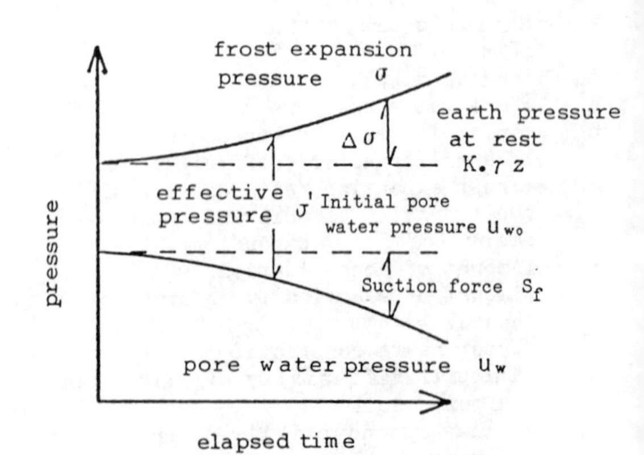

Fig.2 Pressure versus time

$$S_f=\frac{1}{2C}\Big[B+CD - \sqrt{(B+CD)^2-4C\,[(\sigma-u_{wo})\{B+C(u_w(i+1,j+1)-u_{wo})-A\}]}\,\Big]$$

$$\text{where } A=(V+\sqrt{V_o V})\sigma_o/(1+\Gamma)$$
$$B=V n_f \Gamma/(1+\Gamma)$$
$$C=k/\gamma_w/dr \qquad (12)$$
$$D=\sigma+u_w(i+1,j-1)-2u_{wo}$$

2.3 Displacement of frost expansion and consolidation by freezing

The eqs.(2) and (3) are iterated by the time when eq.(6) is satisfied. And the temperature distribution, the amount of absorbed water, and the freezing rate can be obtained at a specific time.

The distribution of pore water pressure due to the frost heave pressure can be obtained from eq.(4). Since the effective pressure is given from the pore water pressure, the change of void ratio and displacement of consolidation, δ_c, is obtained from e-log(σ') curve as a function of effective pressure.

When the freezing front has advanced from b-dr to b (Fig.3), the displacement of frost expansion is given:

$$\delta=\{\xi_o+(1+\sqrt{V_o/V})\sigma_o/\sigma'\}dr \qquad (13)$$

2.4 Frost expansion pressure

The consolidated deformation appears due to the suction force. This deformation, however, does not contribute to the increase in the frost expansion pressure (total earth pressure at the freezing front). The frost expansion pressure is obtained from the forced deformation, δ_e, equivalent to the following:

$$\delta_e=\delta-\delta_c \qquad (14)$$

As shown in Fig.1, the stress in the ground prior to the freezing of cylindrical form is given by the following equations:

$$\sigma_r=\sigma_\theta=K_o\gamma Z \qquad (15)$$
$$\sigma_z=\gamma Z \qquad (16)$$

The principal stresses at position, r, in the unfrozen soil is obtained by solving the following equation in the form of cylindrical freezing:

$$\partial\sigma_r/\partial r+(\sigma_r+\sigma_\theta)/r=0 \qquad (17)$$

When the freezing front reaches a nodal point i-1 at the time j-1 and moves to the nodal point i at time j, the change of

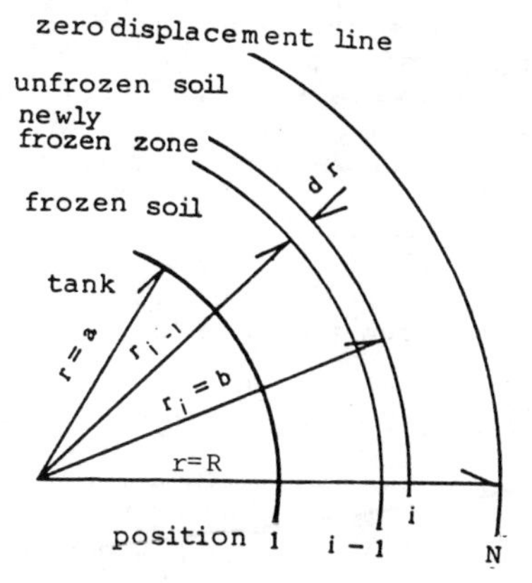

Fig.3 Analysis model of ground around tank

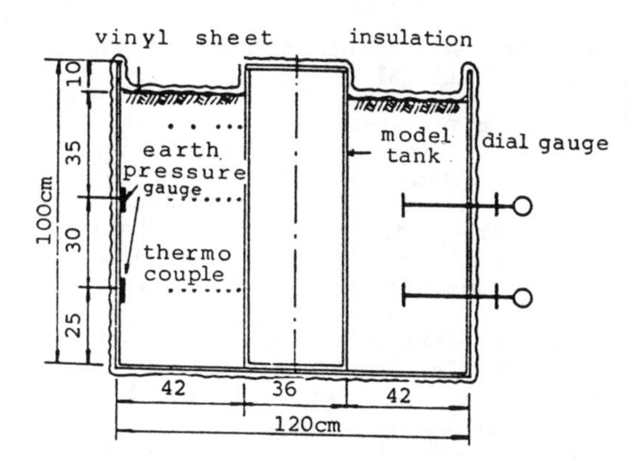

Fig.4 Model tank used in the laboratory experiment

Table 1 Physical properties of soil

ITEM	loam
specific gravity	2.810
porosity	0.65
consistency	
liquid limit w_L(%)	97.0
plastic limit w_p(%)	78.0
plastic index I_p	19.0
gradation	
sand	48.0
silt	38.0
clay	14.0
D_{50}(mm)	0.062
permiability k(m /h)	8.0E-4
consolidation coef.Cv(m^2/h)	1.2E-3
consodidation index Cc	0.193
swelling index Cs	0.015
ξ_o	0.0002
σ_o(kgf/cm²)	0.01
V_o(mm/h)	3364
n_f	0.439

stresses are obtained based on the boundary conditions.

The following equation shows the relationship between δ_e and the displacement in the unfrozen soil:

$$2\pi b \delta_e = \int_b^R (2\pi r \varepsilon_r)\, dr \tag{18}$$

Consequently, the frost expansion pressure is obtained from the following equation:

$$P(i,j) = P(i,j-1) + \frac{\delta_e E}{(R-b)\left[(1-\mu) + \dfrac{(1+\mu)b/R-(1-\mu)}{1+\dfrac{1+\mu}{1-\mu}\left(\dfrac{b}{R}\right)^2}\right]} \tag{19}$$

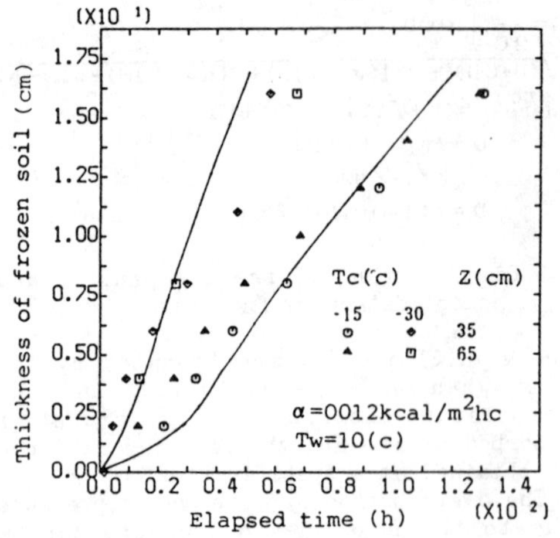

Fig.5 Growth curve of freezing front for two rates of cooling and at two depths

3 COMPARISON WITH THE RESULTS OF MODEL TANK EXPERIMENT

3.1 Method of experiment

The model tank made of steel for this experiment has 36 cm diameter and 100cm height in a soil vessel, 120cm diameter and 100cm height, as shown in Fig.4. The physical properties of the soil are listed in Table 1.

This experiment was carried out until the thickness of frozen soil around the tank had grown to about 16cm under the condition that the temperatures of the tank were -15℃ and -30℃. The earth pressures acting on the inner surface of soil vessel, displacement and temperature were measured.

3.2 Results of measurements

The relationship between the cooling time and thickness of frozen soil is shown in Fig.5. Comparison between the measurements and calculated values of the earth pressure at the outer boundary and of displacement at r=38cm are shown in Figs.6 and 7 respectively. It shows that the calculated values obtained from the model and the measured values are in excellent agreement.

The frost heave constants indicated in Table 2 are values computed from the displacement and earth pressure measurements without counting the displacement of consolidation. ξ_0 is equivalent to the amount of secondary frost heave (wherein adsorbed water around soil particles is frozen and expanded after freezing of soil due to a reduction of soil temperature) and is

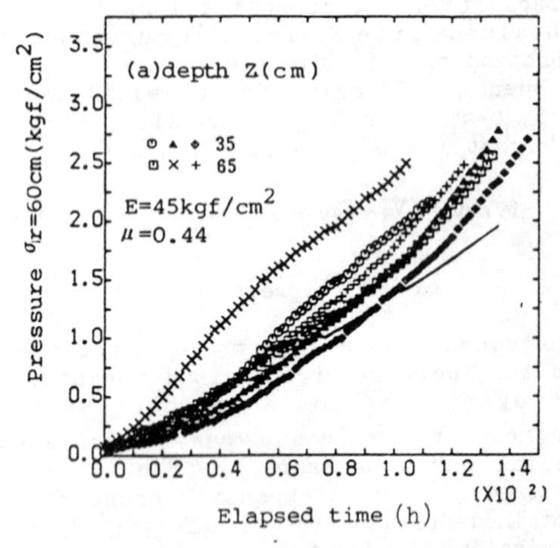

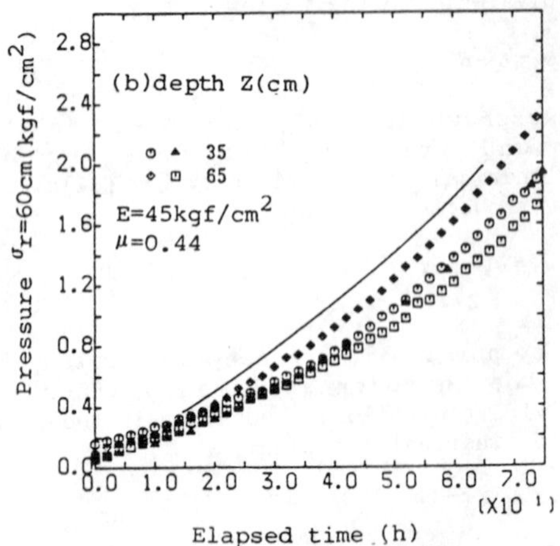

Fig.6 Measured and caluculated pressures at the outer wall versus time for two cooling temperatures,(a)Tc=-15℃,(b)Tc=-30℃

normally not more than 0.01. This calculated values are significant. Furthermore, these values differ according to the cooling temperature twice as much.

These calculated values include the amount of deformation of unfrozen soil associated with the suction force. The lower the cooling temperature, the greater the suction force, as shown in Fig.8(a).

Meanwhile, the frost heave force for the case where u_w is constant at $r=R$, and the resultant distribution of pore water pressure are presented in Fig.8(b). Even in the case of a lower cooling temperature, the suction force is greater.

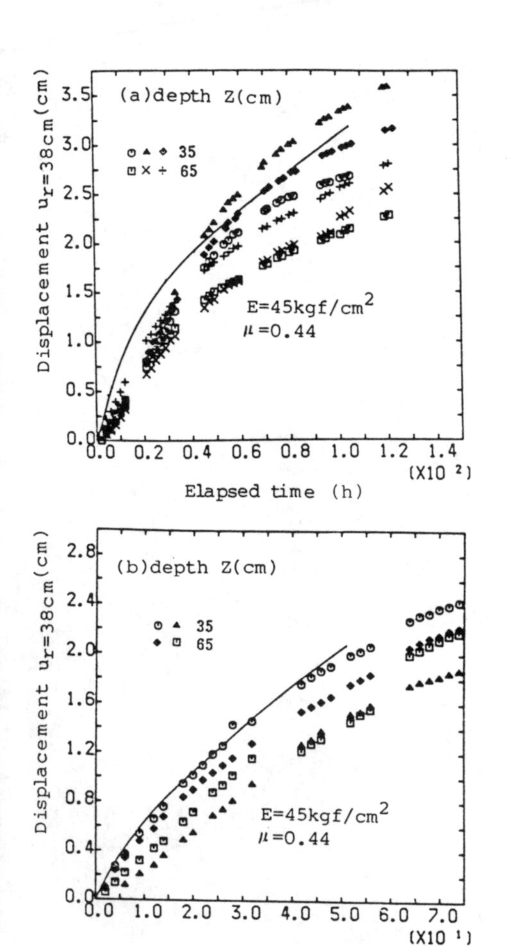

Fig.7 Measured and calculated displacements at r=38cm versus time for two cooling temperature, (a)Tc=-15℃, (b)Tc=-30℃

4 CONCLUSION

We have developed a frost heave model that couples the flows of heat and water where the entire earth pressure changes. Namely, where the freezing expansion pressure changes moment by moment while taking into consideration deformation of unfrozen soil associated with freezing.

For comparison with the model results, comparative studies were conducted by

Table 2 Frost heave constants computed from the displacement and earth pressure measurements

cooling temp.(-15℃)	ξ_0	0.321
	ζ (kgf/cm²)	0.032
cooling temp.(-30℃)	ξ_0	0.161
	ζ (kgf/cm²)	0.098

where $\xi = \xi_0 + \zeta / \sigma$

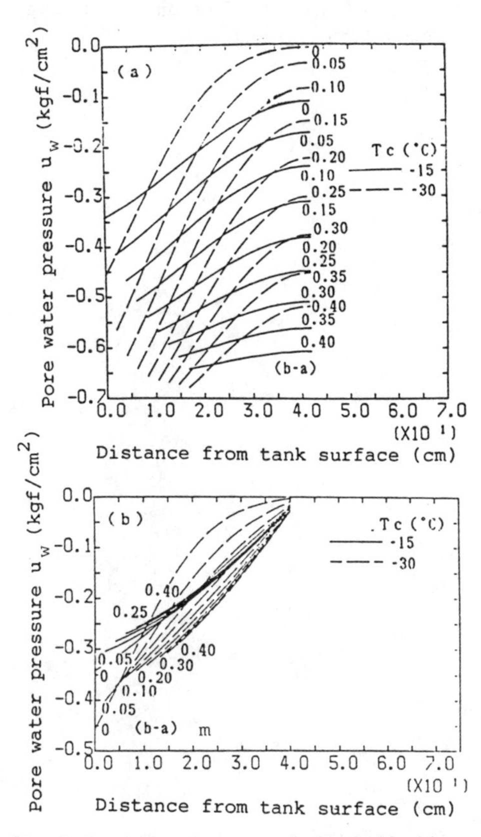

Fig.8 Suction force and distribution of pore water pressure for two cooling temperature, (a) $\left.\frac{\partial u_w}{\partial r}\right|_{r=R}=0$ (b) $u_w|_{r=R}=0$

executing a laboratory model experiment. The following two major conclusion were reached:
(1) Frost heave analysis that includes the coupling of the flows of heat and water is applicable in the case when total pressure increases.
(2) The frost expansion pressure rises due to the amount which is subtract the amount of frost heave from the consolidated deformation.

References *

1)T.Takashi, T.Ohrai & H.Yamamoto;Frost expansion of soils in frozen ground working method, Soil and Foundation, 1977
2)K.Ryokai;Frost heave theory of saturated soil coupling water/heat flow and its application, 4th ISGF, 1985 p. 101-108.
3)M.Fukuda & K.Ryokai;Frozen soil Reclamation with Heat Pipe and its Utilization,2nd Cold Region Technology Conference,1986

*See page X concerning ISGF papers

5th International Symposium on Ground Freezing, Jones & Holden (eds)
© 1988 Balkema, Rotterdam. ISBN 90 6191 824 3

Calculation of the stress field in soils during freezing

M.Shen & B.Ladanyi
Northern Engineering Centre, Ecole Polytechnique, Montreal, Quebec, Canada

ABSTRACT: A conceptual model for calculating stresses and strains associated with soil freezing, on the basis of coupled heat, moisture and stress field, is presented. The model takes into account the non-homogeneity of the freezing zone due to temperature variation, as well as the effects of frozen soil creep on stress distribution. Using this model, a numerical simulation of a saturated cylindrical sample under unidirectional freezing was carried out. The results of simulation of both temperature field and the amount of heave are in a good agreement with the experimental results published by Penner (1986). However, in addition, the simulation furnished also the complete stress field during freezing, and it enables us to take into account the effect of external loading on the amount of frost heave.

1 INTRODUCTION

During the past two decades, many numerical models for simulating the heat and moisture transfer during frost heaving process have been developed (Harlan, 1973; Guymon and Luthin, 1974; Taylor and Luthin, 1978; Sheppard, Kay and Loch, 1978, Jame and Norum, 1980). However, all of these models were based only on the coupling of heat and moisture transfer, and did not consider the external loading. Since the 1980's, some mathematical models of frost heaving that include applied loading have, in fact, also been proposed (Hopke, 1980; Gilpin, 1980; O'Neill and Miller, 1985), but in all of them the applied external loading was only considered as a factor affecting frost heaving, while the resulting deformation and stress fields were not considered. However, from the practical viewpoint, the prediction of stress and deformation fields during frost heaving is considered to be of a great importance for the prediction of stability of structures in cold regions. In this paper, a model for calculating stresses and deformations during soil freezing on the basis of coupled heat, moisture and stress fields, is presented. Using this model, a numerical simulation of a saturated cylindrical sample under unidirectional freezing is carried out.

2 OUTLINE OF THE MODEL

During freezing, the transfer of heat, moisture and the variation of stresses in the soil are all interrelated, so the analysis must deal with the coupling of all these effects. The mathematical model requires three systems of equations for expressing the interrelationship among the laws of heat and moisture-transfer and a varying stress field.

For the case in which the convective heat transfer by thermal vapor is negligible, the heat transport equation may be written as:

$$C \cdot \frac{\partial T}{\partial \tau} = \nabla(\lambda \nabla T) + L \cdot \rho_i \cdot \frac{\partial \Theta_i}{\partial \tau} \tag{1}$$

where C — heat capacity of soil (J/m^3°C); λ — thermal conductivity of soil (W/m°C); T — temperature (°C); L — latent heat of fusion (J/Kg); τ — time (sec).

For the moisture transfer, we only consider the liquid water flow driven by a gradient in unfrozen water pressure, while the air phase and vapor transfer is neglected. In most cases, the effect of gravitational potential on the water flow in frozen soil is negligible, so the moisture transfer equation for steady or unsteady flow during freezing can be written as (Sheppard, et. , 1978):

$$\frac{\partial}{\partial \tau}\left[\Theta_l + \frac{\rho_i \Theta_i}{\rho_l}\right] = \nabla(K \cdot \nabla P_l) \tag{2}$$

where ρ_l, ρ_i — densities of liquid water and ice (kg/m^3); Θ_l, Θ_i — volumetric fractions of liquid water and ice; K — hydraulic conductivity ($m^2 sec^{-1} Pa^{-1}$); P_l — liquid water pressure (Pa).

The pressure of liquid water P_l in frozen soil can be described by the Clapeyron equation (Kay and Groenevelt, 1974):

$$\frac{P_l}{\rho_l} - \frac{P_i}{\rho_i} = L \cdot \ln \frac{T_K}{T_o} \qquad (3)$$

here T_K is the absolute temperature of soil (°K); T_o is the absolute freezing temperature of pure water (°K). Because the role of the ice pressure is poorly understood, some simplifying assumptions are proposed (Fig.1), in order to use the Clapeyron equation. The ice pressure P_i at the freezing front is assumed to be zero, and that at the coldest side of the freezing fringe is assumed to be equal to the local mean pressure. The frost heaving is assumed to start when the ice content exceeds its critical value equal to 85% of the soil porosity, regardless of the unfrozen water content (Taylor and Luthin, 1978).

Once the ice pressure P_i is defined, substituting (3) and (2) into (1), and rearranging gives:

$$\bar{C} \cdot \frac{\partial T}{\partial \tau} = \nabla(\bar{\lambda} \cdot \nabla T) + \rho_l \cdot L \cdot \nabla(K \cdot \nabla P_i) \qquad (4)$$

with

$$\bar{C} = C + L \cdot \rho_l \cdot \frac{\partial \Theta_l}{\partial T} \qquad (5.A)$$

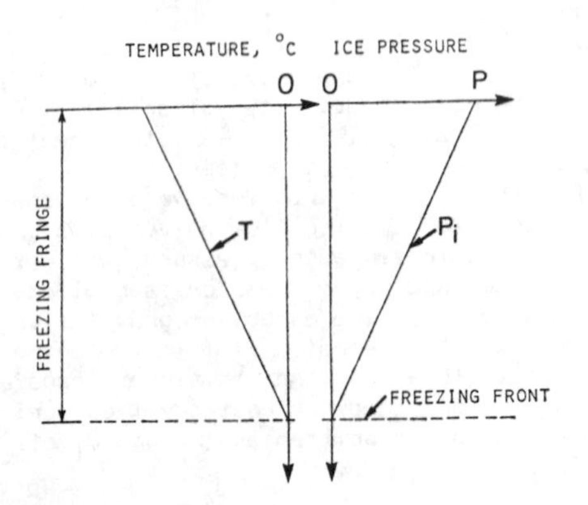

Fig.1. Idealized distribution of ice pressure and temperature within freezing fringe adopted in the calculation.

$$\bar{\lambda} = \lambda + \frac{K \rho_l^2 L^2}{T_K} \qquad (5.B)$$

Since the effect of stress on heat transfer is very small and may be neglected, eqn.(4) can be rewritten as

$$\bar{C} \cdot \frac{\partial T}{\partial \tau} = \nabla(\bar{\lambda} \cdot \nabla T) \qquad (6)$$

In the above equation, the relationship between the liquid water content in frozen soil and temperature must be determined experimentally.

If the volume of soil particles is assumed to remain constant during freezing and thawing processes, the volumetric expansion strain ϵ^V caused by the moisture transfer and by the fact that water in soil is changed into ice, can be given by

$$\epsilon^V = 0.09(\Theta_o + \Delta\Theta - \Theta_l) + \Delta\Theta + (\Theta_o - n) \qquad (7)$$

where Θ_o — initial water content (m^3/m^3); $\Delta\Theta$ — increment of water content (m^3/m^3); n — porosity of the soil.

Because the soil is considered to be isotropic, the normal expansion strain in each direction is equal, and the shear strain due to the expansion is zero:

$$\epsilon_{ij}^V = \frac{1}{3} \cdot \delta_{ij} \cdot \epsilon^V \qquad (8)$$

here the δ_{ij} is Kronecker delta.

The stress-strain relationship of the soil is given by the following incremental form:

$$d\{\sigma\} = [D](d\{\epsilon\} - d\{\epsilon^C\} - d\{\epsilon^V\}) \qquad (9)$$

where $d\{\sigma\}$ — stress increment vector; [D] — matrix of elastic constants, related to the temperature of frozen soil; $d\{\epsilon\}$, $d\{\epsilon^C\}$, $d\{\epsilon^V\}$ — strain increment vectors, creep increment vector and expansion strain increment vector, respectively.

In eqn.(9), the creep strain increment vector is defined by the Prandtl-Reuss law

$$d\{\epsilon^C\} = d\bar{\epsilon}^C \cdot \frac{\partial \bar{\sigma}}{\partial \{\sigma\}} \qquad (10)$$

where $\bar{\epsilon}$ and $\bar{\sigma}$ are equivalent creep strain and equivalent stress.

In multiaxial states of stress, the empirical power creep law for the uniaxial case can be generalized as follows (Ladanyi, 1975):

$$\dot{\bar{\epsilon}}^c = \left[\frac{\bar{\sigma}}{\sigma_{CT}}\right]^n \left[\frac{\dot{\epsilon}^c \cdot \gamma}{b}\right]^b \tag{11.A}$$

with

$$\sigma_{CT} = \sigma_{co} \left[1 + \frac{T}{T_r}\right]^\omega \tag{11.B}$$

where n, b, ω are experimental coefficients, and $\dot{\epsilon}^c$, σ_{co} and T_r are reference values of strain rate, stress and temperature.

From eqs. (9), (10) and (11), the equivalent nodal forces increment vector $\Delta\{R\}$ due to creep and volumetric expansion in soil can be written as

$$\Delta\{R\} = \iiint [B]^T [D] \Delta\{\epsilon^v\} dv + \iiint [B]^T [D] \cdot b \cdot \left[\frac{\bar{\sigma}}{\sigma_{CT}}\right]^n \left[\frac{\dot{\epsilon}^c}{b}\right]^b \gamma^{b-1} \frac{\partial \bar{\sigma}}{\partial\{\sigma\}} \cdot \Delta\gamma \cdot dv \tag{12}$$

The above equations constitute the system of equations for calculating stress and deformation increments in each time step by the finite element method.

3 NUMERICAL SIMULATION

Using this model, the numerical simulation for the frost heaving experiment published by Penner (1986) was carried out. Penner measured the frost heaving, ice lens growth and frost penetration during open system freezing of a saturated soil under 50kPa overburden pressure. Because the freezing and the external water supply were both unidirectional, and the test samples were cylinders, the heat and moisture transfer eqns. (2) and (6) may be simplified to 1-D equations under the following condition:

$$\frac{\partial T}{\partial r} = \frac{\partial \Theta_i}{\partial r} = \frac{\partial \Theta_l}{\partial r} = 0 \tag{13}$$

where r is the coordinate in the cylindrical coordinate system. Therefore, the heat and moisture transfer is a one-dimensional problem, but the mechanical equations are a two-dimensional axial symmetry problem.

The soil heat capacity and thermal conductivity are calculated as follows:

$$C = \Sigma C_j \Theta_j \tag{14.A}$$

$$\lambda = \Pi(\lambda_j)^{\Theta_j} \tag{14.B}$$

where C_j, λ_j, Θ_j are heat capacity, thermal conductivity and volumetric fraction of each phase in the soil, respectively. The constants used in this simulation and given in Table(1), have been selected according to Harlan and Nixon (1978).

TABLE 1. Thermal constants used

C & λ	soil grains	water	ice
C ($MJm^{-3\circ}C^{-1}$)	2.2	4.18	1.93
λ ($Wm^{-1\circ}C^{-1}$)	1.95	0.602	2.22

The relationship between hydraulic conductivity and temperature was deduced from experimental data published by Horiguchi and Miller(1983), using a regression analysis:

$$K = \begin{cases} 3.072\times10^{-11} e^{13.438T} & -0.3<T<Tf \\ 5.453\times10^{-13} & T<=-0.3 \end{cases} \quad \text{(m/sec)} \tag{15}$$

The experimental parameters in the creep eqns. (11.A) and (11.B) shown in Table(2) originate from Vyalov (1962) and Eckardt (1982).

TABLE 2. Coefficients in Creep Equation

State*	b	n	ω	σ_{co}(MPa)	Tr(°C)	$\dot{\epsilon}^c(h^{-1})$
C	.45	2.50	.97	0.18	-1	10^{-5}
T	.44	2.33	1	1.83	-1	10^{-5}

*(C — Compression; T — Tension)

The Young's modulus of unfrozen soil is considered to be constant and equal to 11.2 MPa (Lambe and Whitman, 1969). The stress strain curve and the relationship between the initial tangent modulus of frozen soil E_o and the temperature follow the results published by Zhu Yuanlin and Carbee (1984). The Poisson's ratio of both unfrozen and frozen soils is assumed to be 0.3.

For reducing the amount of calculation, the heat and moisture transport eqns. (6) and (7) were solved by the finite difference method, but the mechanical equations were still calculated by FEM. The two grid systems with the vertical coordinate of each grid point made to coincide, were set up for calculating by both the finite difference method and the finite element method.

As for the heat transfer eqn. (6), it can be approximated by the predictor-corrector implicit scheme:

$$- a\bar{\lambda}(j-\tfrac{1}{2},n)T(j-1,n+\tfrac{1}{2}) + [\bar{C}(j,n) + a\bar{\lambda}(j-\tfrac{1}{2},n) + b\bar{\lambda}(j+\tfrac{1}{2},n)]T(j,n+\tfrac{1}{2}) - b\bar{\lambda}(j+\tfrac{1}{2},n)T(j+1,n+\tfrac{1}{2}) = \bar{C}(j,n)T(j,n) \quad (16.A)$$

$$- a\bar{\lambda}(j-\tfrac{1}{2},n+\tfrac{1}{2})T(j-1,n+1) + [\bar{C}(j,n+\tfrac{1}{2}) + a\cdot\bar{\lambda}(j-\tfrac{1}{2},n+\tfrac{1}{2}) + b\bar{\lambda}(j+\tfrac{1}{2},n+\tfrac{1}{2})]T(j,n+1) - b\cdot\bar{\lambda}(j+\tfrac{1}{2},n+\tfrac{1}{2})T(j+1,n+1) = \bar{C}(j,n+\tfrac{1}{2})T(j,n) + a\bar{\lambda}(j-\tfrac{1}{2},n+\tfrac{1}{2})T(j-1,n) - [a\bar{\lambda}(j-\tfrac{1}{2},n+\tfrac{1}{2}) + b\bar{\lambda}(j+\tfrac{1}{2},n+\tfrac{1}{2})]T(j,n) + b\bar{\lambda}(j+\tfrac{1}{2},n+\tfrac{1}{2})T(j+1,n) \quad (16.B)$$

with

$$a = \Delta\tau/[\Delta z^2(j)+\Delta z(j)\Delta z(j+1)] \quad (16.C)$$

$$b = \Delta\tau/[\Delta z(j)\Delta z(j+1)+\Delta z^2(j+1)] \quad (16.D)$$

where: j — the spatial node (vertical); n — the time increment; Δz — the positional node spacing; $\Delta\tau$ — the time step interval.

As for the moisture transfer equation(2); it may be expressed in finite difference form by

$$\Theta_i(j,n+1) = \Theta_i(j,n) + \frac{\rho_l}{\rho_i}\{aK(j-\tfrac{1}{2},n+1)P_l(j-1,n+1) - [aK(j-\tfrac{1}{2},n+1) + bK(j+\tfrac{1}{2},n+1)]P_l(j,n+1) + bK(j+\tfrac{1}{2},n+1)P_l(j+1,n+1) - \Theta_l(j,n+1) + \Theta_l(j,n)\} \quad (17)$$

In each time step, first the temperature in each node is calculated, then the moisture content is defined according to the temperature and ice pressure at each node. Once the heat and moisture fields are determined, the volumetric strain increment can be defined from the variation of moisture field within the time step, and the creep strain increment is determined on the basis of stress level and soil temperature at the time step by Prandtl-Reuss eqn.(10). The displacement increment field in the time step can be calculated according to the incremental initial strain method in FEM analysis. Once a nodal displacement increment is defined, the nodal stresses may be calculated by the constitutive law. Then the nodal average stress can be determined for calculating heat and moisture transport in the next time step.

4 RESULTS AND DISCUSSION

In this paper, a simulation of Penner's experiments of frost heaving (Penner, 1986, Soil 1,Run 1) was carried out. In addition, the prediction of frost heaving and stress fields for other loadings was also undertaken.

Figs. (2) and (3) compare the simulated frost penetration and heaving under 50 KPa applied loading with the experimental curves published by Penner(1986).The calculated values are seen to be slightly lower than experimental data, but the precision of simulation becomes better after early stages of the test.

The calculated moisture distributions in the sample under 50 KPa and 300 KPa applied loading are shown in Figs. (4) and (5). As the dry density of tested soils was not given in the Penner's(1986) paper, the dry density was assumed to be 1750 Kg/m^3 according to other experiments under similar conditions published by Penner and Ueda (1977). According to our simulation, the moisture increase near the cold side is not very large, because the frost penetration progresses very quickly in the initial freezing stage. Later, the total moisture content smoothly increases to a maximum value behind the advancing front, when the applied loading is small. For the case of a large applied loading, the total moisture content slightly decreases behind the advancing front, and the total migrated moisture content is much smaller than under a the small applied loading. Discrete ice lenses cannot be predicted by the proposed model.

The stress fields in the soil during freezing under 50KPa and 300KPa loading are shown in Figs. (6)-(8). Before the soil is frozen, the initial stress field defined by FEM according to the elastic theory, is found to be uniform. When the soil is partially frozen, the soil water changes into ice and the moisture migrates. This causes a volumetric variation in frozen soil,and results in a change in stresses. This phenomenon is very similar to the thermal stresses in structures. When the applied loading is small, the volumetric expansion in the frozen part is large because a lot of moisture is transferred from unfrozen to frozen soil, causing a change in the state of stress from initial compression to tension stresses in the major portion of the frozen zone. In the portion of soil under tension,the resulting stress field produces creep strains which actually increase the total heaving.A comparison of stress fields and moisture fields, shows that the maximum values of the expansion stresses appear at the same level as the maximum moisture contents. At the cold side, the stress concentration is caused by the confinement of the bottom of the chamber in the tests .

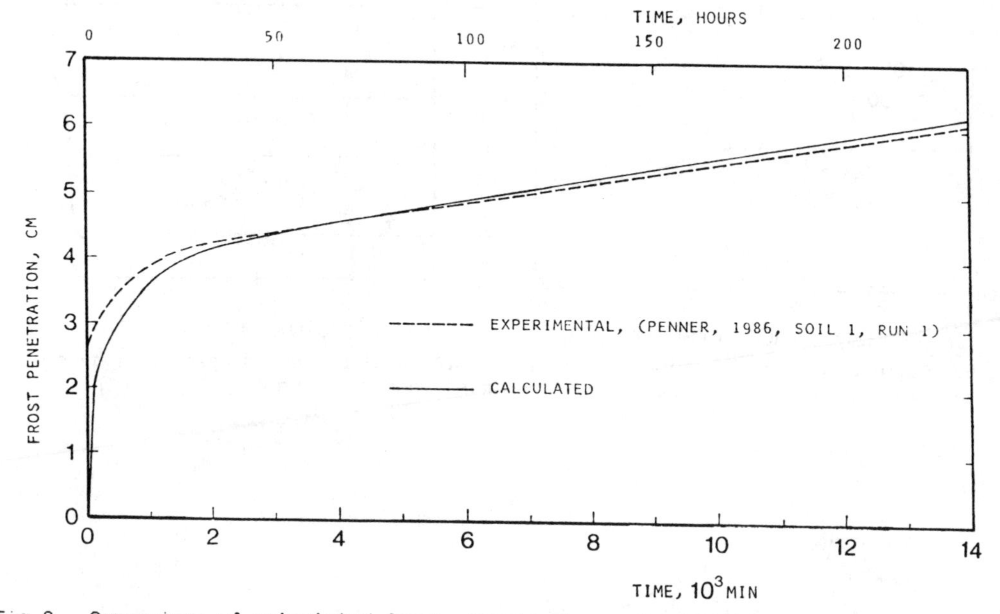

Fig.2. Comparison of calculated frost penetration under 50 kPa with that measured by Penner (1986).

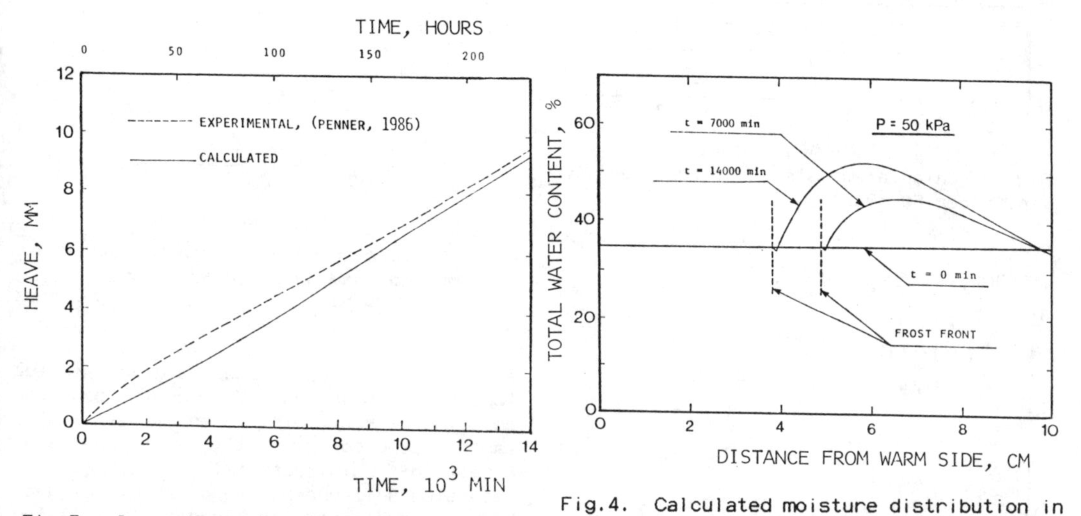

Fig.3. Comparison of calculated heave with experimental data by Penner(1986).

Fig.4. Calculated moisture distribution in the sample after 7000 and 14000 min, under 50 kPa.

Figs. (9) and (10) show the calculated relationship between the frost heaving, the heaving rate and the applied loading. The predicted results are seen to simulate very well the often observed effect of applied loading on frost heaving, showing that, when the applied loading increases, the heaving decreases rapidly.

5 CONCLUSION

A model that makes it possible to calculate the deformation and stress field in the

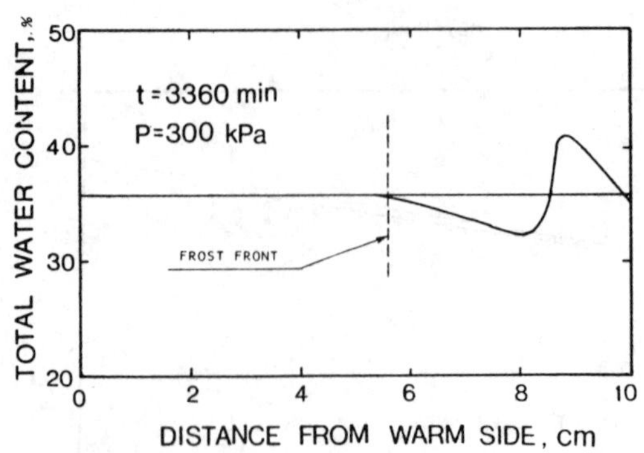

Fig.5. Calculated moisture distribution in the sample after 3360 min, under 300 kPa.

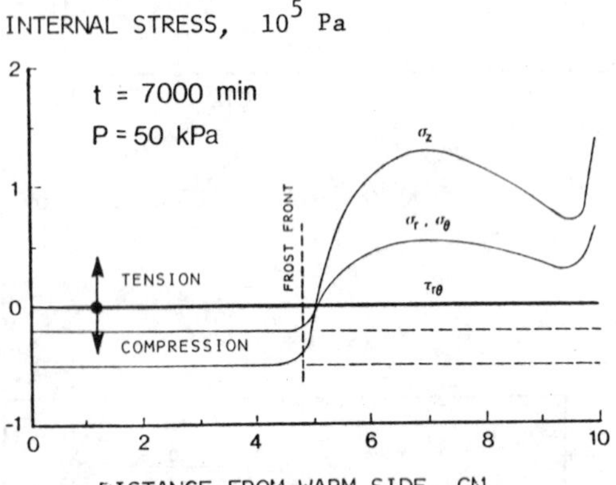

Fig.6. Calculated stress field after 7000 min, under 50 kPa.

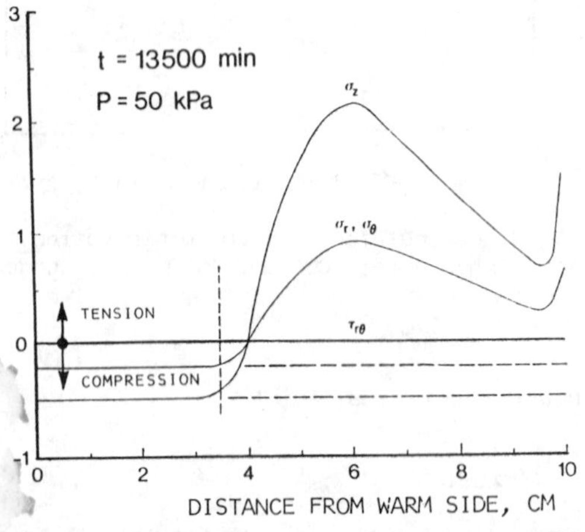

Fig.7. Calculated stress field after 13500 min, under 50 kPa.

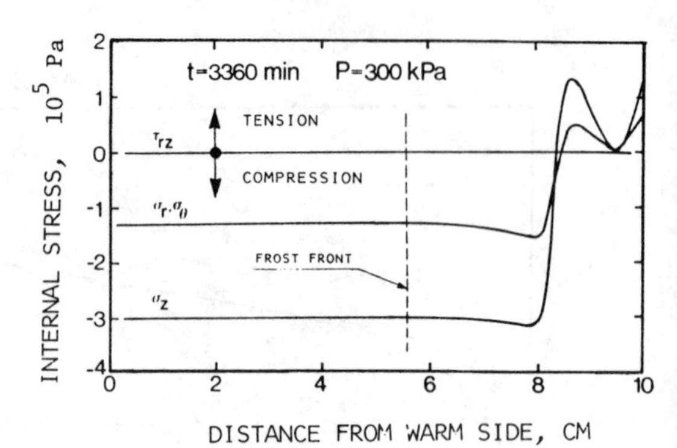

Fig.8. Calculated stress field after 3360 min, under 300 kPa.

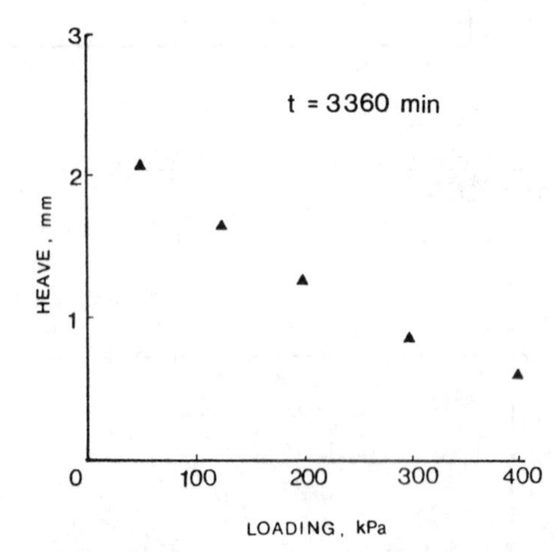

Fig.9.Calculated relationship between heave and applied loading after 3360 min.

soil during freezing under a given applied loading is proposed. The model, based on coupling of heat,moisture and stress fields is then used for simulating the results of a test of unidirectional freezing in a saturated soil under an applied loading of 50 kPa, published by Penner (1986). The calculated temperature field and heaving are found to be in a good agreement with the experimental results.

ACKNOWLEDGEMENTS

Financial support provided by the Lanzhou Institute of Glaciology and Geocryology, Academia Sinica for M.Shen, and by the National Science and Engineering Council of

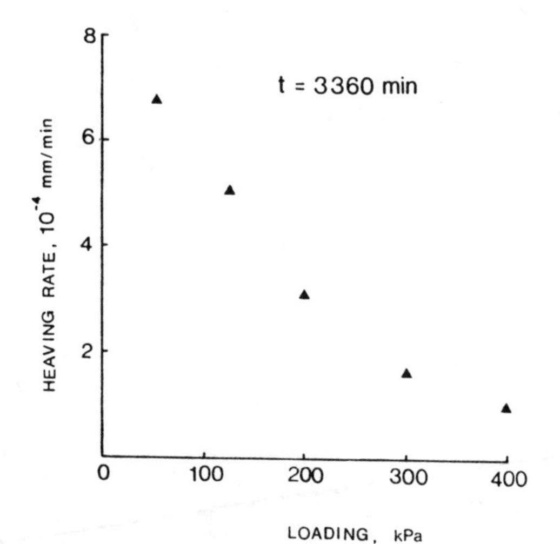

Fig.10.Calculated relationship between heaving rate and applied loading after 3360min.

Canada,for B.Ladanyi, is gratefully acknowledged.

REFERENCES*

Eckardt, H. 1982. Creep tests with frozen soil under uniaxial tension and uniaxial compression. Proc. of the 4th Canadian Permafrost Conf., PP.394-405.

Gilpin,R.R. 1980. A Model for the prediction of Ice Lensing and Frost Heave in Soils. Water Resources Research, 16(5), pp.918-930.

Guymon,G.L. & Luthin,J.N. 1974. A coupled heat and moisture transport model for Arctic soil. Water Resources Research, 9(5), PP.1314-1323.

Harlan,R.L. 1973. Analysis of coupled heat-fluid transport in partially frozen soil. Water Resources Research, 9(5), pp. 1314-1323.

Harlan, R.L. & Nixon, J.F. 1978. Ground thermal regime.In "Geotechnical Engineering for Cold Regions". Edited by O.B. Andersland and D.M. Anderson. McGraw-Hill Book Company. pp.103-150.

Hoekstra, P. 1966. Moisture movement in soils under temperature gradients with the cold-side temperature below freezing. Water Resources Research,2(2),pp.241-250.

Hopke, S.W. 1980. A model for frost heave including overburden. Cold Regions Sci. & Technol., 3(1980), pp.111-127.

Horiguchi,K. & Miller,R.D. 1983. Hydraulic conductivity functions of frozen materials. Proc. of the 4th Int. Conf. on Permafrost, pp.504-508.

Jame,Y.W. & Norum,D.I. 1980. Heat and mass transfer in a freezing unsaturated porous medium. Water Resources Research, 16(4), pp.811- 819.

Kay, B.D. & Groenevelt, P.H. 1974. On the interaction of water and heat transport in frozen and unfrozen soils: I. Basic theory; The vapor phase. Soil Sci. Soc. Amer. Proc., Vol.38, pp.395-400.

Ladanyi,B. 1975. Bearing capacity of strip footings in frozen soils. Can. Geotech. J.,12,pp.393-407.

Lambe,T.W. & Whitman,R.V. 1969. Soil machanics. John Wiley & Sons, Inc., New York.

Miller,R.D. 1972. Freezing and Heaving of Saturated and Unsaturated Soils. Highway Research Record, No.393, pp.1-11.

O'Neill,K. & Miller,R.D. 1985. Exploration of a Rigid Ice Model of Frost Heave,Water Resources Research, 21(3), pp.281-296.

Penner,E. & Ueda,T. 1977. The dependence of frost heaving on load application - preliminary results. Proc. Int. Symp. Frost Action in Soils, Lulea, Vol.1, pp. 92-101.

Penner,E. 1986. Aspects of ice lens growth in soils. Cold Regions Sci. & Tech., 13, pp.91-100.

Shen, M. & Ladanyi, B. 1987. Modelling of coupled heat,moisture and stress field in freezing soil. Cold Regions Sci. & Tech., Vol.14, pp.237-246.

Sheppard,M.I., Kay,B.D. & Loch,J.P.G. 1978. Development and testing of a computer model for heat and mass flow in freezing soils. Proc. of the 3rd Int. Conf. on Permafrost, pp.76-81.

Taylor,G.S. & Luthin,J.N. 1978. A model for coupled heat and moisture transfer during soil freezing. Can. Geotech. J., Vol.15, pp.548-555.

Vyalov,S.S. 1962. The strength and creep of frozen soils and calculations for ice -soil retaining structures. Transl. 76, CRREL, 1965

Xu Xiaozu, Oliphant,J.L. & Tice,A.R. 1985. Prediction of unfrozen water contents in frozen soils by a two-points or one-point method. Proc. of the 4th Int. Symp. on Ground Freezing, Vol.2, pp.83-87.

Zhu Yuanlin & Carbee, D.L. 1984. Uniaxial compressive strength of frozen silt under constant deformation rates. Cold Regions Sci. & Technol., 9(1984), pp.3-15.

*See page X concerning ISGF papers

5th International Symposium on Ground Freezing, Jones & Holden (eds)
© 1988 Balkema, Rotterdam. ISBN 90 6191 824 3

The effect of surcharge on the frost heaving of shallow foundations

Tong Changjiang & Yu Chongyun
Lanzhou Institute of Glaciology and Geocryology, Academia Sinica, People's Republic of China

ABSTRACT: In order to meet the requirement for the design of shallow foundations of light buildings in seasonal frost areas, the authors, studied the effect of light pressures (<200 kPa) on frost heaving through in-situ tests. The test results show that the total amount of frost heaving decreases with increasing pressures. It can be reduced by 44--52% in the high frost susceptible soils when the pressures is equal to 200 kPa. The effective depth of the frost heaving beneath foundations increases with increasing pressures. It is about 80 cm when the pressure is equal to 150-200kPa. According to the test results, the influence coefficient of frost heaving for various frost susceptibilities of soils and pressures (0-200 kPa), which can be used to estimate the attenuation of frost heaving of subsoils, were presented.

KEYWORDS: clay, foundation, frost heave, insitu testing, seasonal frost.

INTRODUCTION

Investigations (Penner and Walton, 1979; Chen et al., 1983; Tong and Guan, 1985) show that frost heave ratio decreases with the increase in external load. Even though under a large surcharge, there is still frost heave (Ishizaki and Kinosita, 1979). To meet the demand for the shallow burial of light foundations in seasonal frozen ground regions (i,e. the depth of the foundations is less than the maximum frost depth), the authors have studied the effect of a surcharge (P<200 kPa) on the repression of frost heave for foundation soils under field conditions. The test results proved that the above-mentioned conclusion is correct. Our test results also showed that the thickness of the define-frost heave repression layer beneath a foundation increased with the increase in applied load. Results showed that a certain thickness of frozen soil could be maintainal beneath the foundations with light surcharge in seasonal frost regions, for which both the frost deformation and thawing settlement could not damage the structures. From tests results, the influence coefficients of different loads on the frost heave of various foundation soils have been presented. From this, the decrease of frost heave beneath foundation can be calculated.

TEST SITE CONDITIONS AND METHODS

Tests were carried out at three sites. The soils at the sites are silty clay. Their grain-size curves and physical properties are shown in Fig.1 and Table I, respectively. The average maximum freezing depths at the three sites were 194, 160 and 205 cm and their average frost heave ratios were 6%, 19.5% and 5%, respectively. These soils being classified as either frost or strongly frost susceptible soils. The groundwater levels before freezing (in October) and after completely frozen (in middle March of next year) were 100/230, 30/190 and 100/ >250 cm for the three test sites.

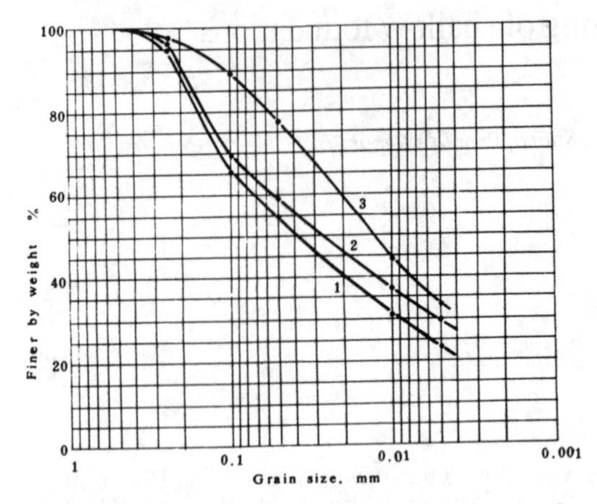

Fig.1, Grain-size curves for site 1 (1), site 2 (2) and site 3 (3).

Table I. Physical properties of the soil at the test sites

Test site	1	2	3
plastic limit Wp (%)	16.4	17.7	17.9
liquid limit WL (%)	26.9	30.9	31.8
plasticity index Ip(%)	10.5	13.2	13.9

The test set-up is shown in Fig.2. The test foundation was a concrete pile with the area of 50x50 cm^2. In order to eliminate the tangential frost heaving force on the lateral surface of the foundation, the foundation surface was coated with 2-mm-thick grease, and wrapped with two layers of asphalt felt. Four deformation observation points were installed at the middle points of the four sides of the concrete loading platform. Frost deformation was measured at regular intervals with a highly precise leveling instrument. On each test site, the following data were measured: freezing depth, groundwater level, temperature and frost heave at ground surface and other depths. The buried depths of the test foundations were about 0.4, 0.7, 1.0 and 1.3m. For each buried depth, there are four test foundations, for which the applied loads on the foundation bottom were 50, 100, 150 and 200 kPa.

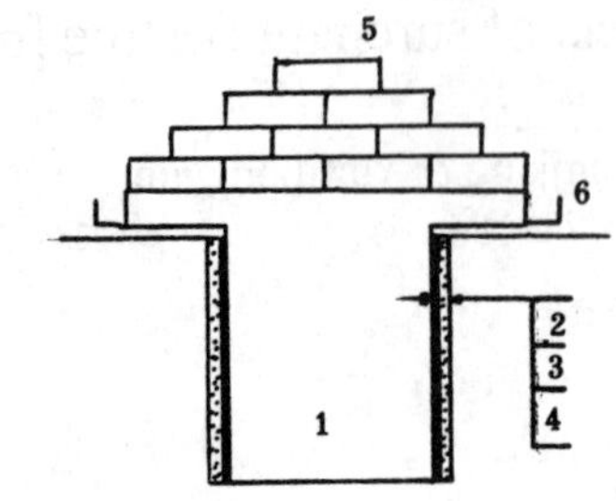

Fig.2, The test set-up
1--50x50 cm^2 concrete foundation
2--grease layer,
3--asphalt felt layers,
4--sand,
5--dead weights, and
6--deformation measurement points

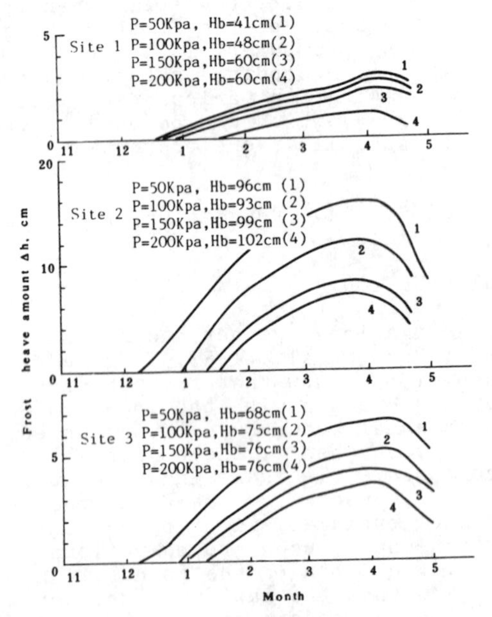

Fig.3, Frost heave amount vs time under different loads and buried depths

TEST RESULTS AND ANALYSIS

The curves of frost heave amount vs time for the test foundations with various buried depths (Hb) and applied load (P) at the three sites are shown in Fig.3.

From this figure it can be seen that the onset time of the initial frost heave of the foundations increase

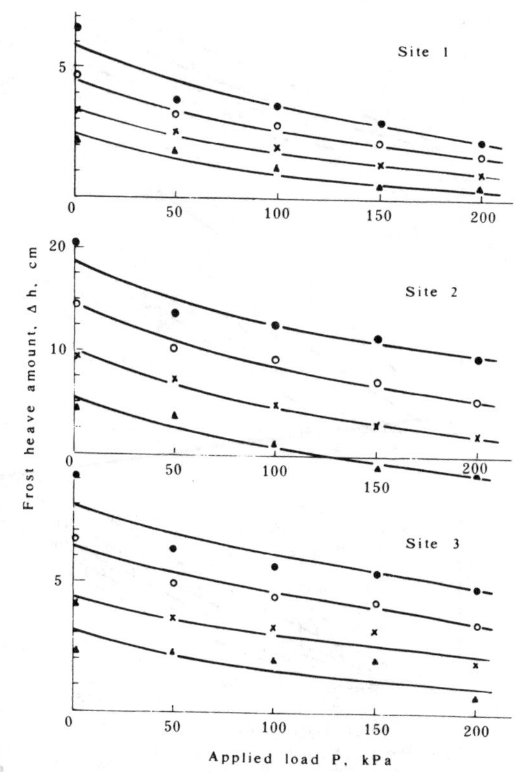

Fig.4, Frost heave amount vs applied load for various buried depths: •---40cm, ∘---70cm, ×---100cm, ▲---130cm.

with both increasing burial depths of the foundation and applied loads. That is to say, the frost heave of foundation with a certain load occurs only when the freezing depth attained a certain magnitude.

The test results showed that frost heave amount (Δh) decreased with an increase in the applied load (P) for the same burial depth (see Fig. 4).

The degree of repression was different for different burial depths. But, it can be described by a power equation:

$$\Delta h = a\ e^{-bP} \qquad (1)$$

where a and b are empirical parameters dependent upon foundation burial depth and the frost susceptibility of the soil.

According to test results, when the load is equal to 200 kPa, the frost heave amount of a very strong frost susceptible soil can be reduc-

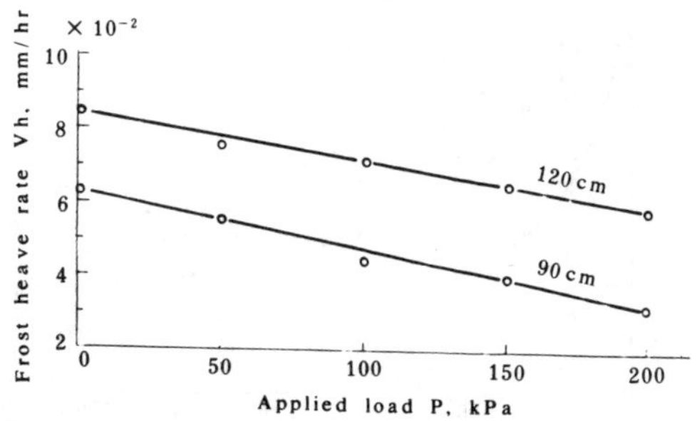

Fig.5, Frost heave rate vs applied load for buried depths of 120cm (1) and 90cm (2).

ed by 44-52%.

The frost heave rate as a function of applied load for site 2 is shown in Fig.5, at which the ground temperature gradient was 4.17×10^{-2}°C/mm before the maximum frost heave appeared. The frost heave rate decreases linearly with the increasing load. If the curves were extended, the pressures under which no frost heave occurs would be obtained. It varies with the thickness of frozen soil beneath foundation. From Fig.5, it is known that they are equal to 700 and 400 kPa when the frozen soil thickness are equal to 120 and 90 cm, respectively. It also varied with the the temperature gradient of the foundation soil. Frost heave suppression by external loading can be evaluated by an influence coefficient, i.e.,

$$\lambda = \frac{\Delta h_o - \Delta h_p}{\Delta h_o} \qquad (2)$$

where Δh_o and Δh_p are the frost heave amounts without and with external loads. Obviously, the influence coefficient (λ) decreases with an increase in the thickness of frozen soil beneath foundation (δ_f) (see Fig.6). The maximum decrease appeared at site 2 and the minimum decrease at site 3. When the thickness of frozen soil under the foundation is less than 150 cm, the relationship between them can be described by

$$\lambda = \alpha\delta_f^{-\beta} \qquad (3)$$

where α and β are the empirical parameters dependent on the load and the ground temperature. The average values for the three sites are given in Table II.

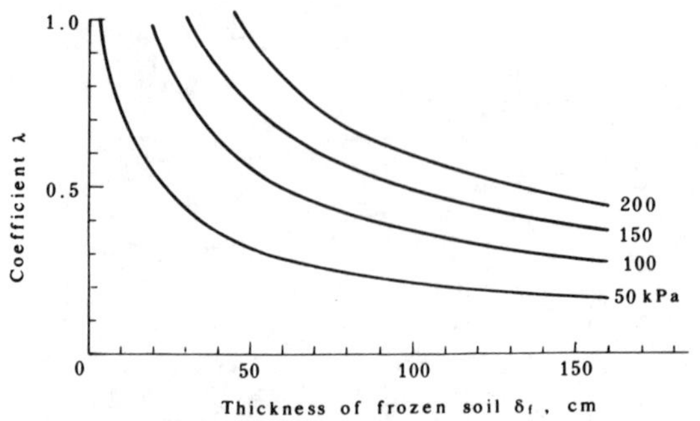

Fig.6, Influence coefficient λ as a function of the thickness of frozen soil under different loads

TABLE II Values of α and β in eq(3)

load, kPa	50	100	150	200
α	2.60	5.49	7.06	10.13
β	0.53	0.58	0.58	0.61

Fig.7 shows the variation of frost heave amount (Δh) with the thickness of frozen soil (δ_f) beneath a foundation under various applied loads, which can be described by

$$\Delta h = m\delta_f^n \qquad (4)$$

where m and n are empirical parameters dependent upon soil properties, load and the ground temperature gradient.

From Fig.7, it is seen that under the action of an external load, the frost heave tends to zero within a certain depth beneath the foundation, which is called the influence zone of the load on frost heave. This depends mainly on the load and the properties of the foundation soil (soil type, water content and density). Under the same geological condition, the influence zone increases with increasing load. If the influence of frost deformation on the stability of structures can be ignored when frost heave ratio is less than 1%, the influence zone can reach about 80 cm when the load is up to 150-200kPa.

In summary when structures are built on the strongly frost susceptible foundation soils, a certain thickness of frozen soil beneath the foundation can exist. This thickness will depend upon the load level and the ground temperature gradient. About 1,400,000 m² of residential quarters built in the test area has proven that this method was practicable.

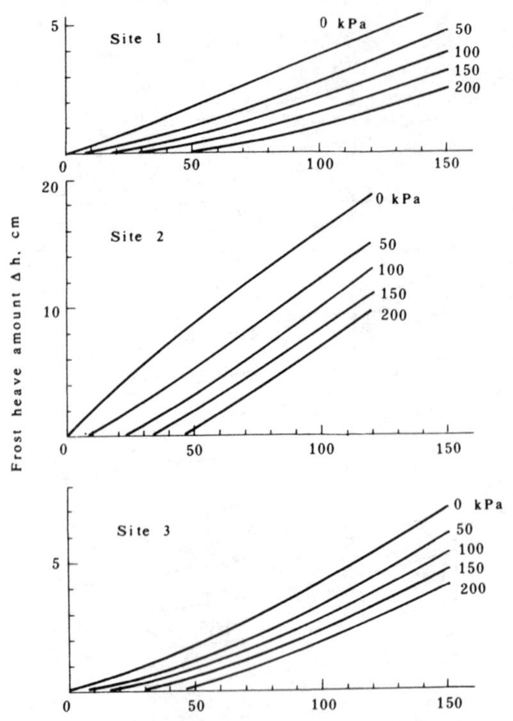

Fig.7, Frost heave amount vs the thickness of frozen soil under various applied loads

CONCLUSIONS

The test results have shown that if the load increases, frost heave rate will decrease rapidly, and the influence zone (i.e., the no frost heave zone) will increase. This is due to the fact that under the action of a load, freezing point of the foundation soil is reduced and moisture migration decreases. The influence coefficient increases with the increasing applied load and decreasing frozen soil thickness under the foundation. Therefore, in engineering practice the foundations can be shallowly buried under certain conditions.

ACKNOWLEDGEMENTS

The authors are grateful to their colleages of Daqing Oil-field Construction Design and Reseasch Institute, Heilongjiang Province for their technical assistance. The assistance received from Mr. Me Heizheng and Prof. Zhu Yuanlin is also sincerely appreciated.

REFERENCES

Chen Xiaobai, Wang Yaqing and Jiang Ping, 1983, Influence of frost penetration rate and surcharge stress on frost heaving. In Proceedings of the Second National Confeerence on Permafrost, Lanzhou (1981), Gansu People's Publishing House.

Ishizaki, T. and Kinosita, S., 1979, Frost heaving under overburden pressure. Low Temperature Science. Ser. A, 38.

Penner, E. and Walton, T., 1979, Effects of temperature and pressure on frost heaving. Engineering Geology J., Vol. 13 Nos. 1-4.

Tong Changjiang and Guan Fongnian, 1985, Frost heaving and prevention of frost damage to constructions. China Water Conservancy and Power Press.

5th International Symposium on Ground Freezing, Jones & Holden (eds)
© 1988 Balkema, Rotterdam. ISBN 90 6191 824 3

Thermal characteristics of fine-grained soils

E.D.Yershov, I.A.Komarov, N.N.Smirnova, R.G.Motenko & Ye.N.Barkovskaya
Department of Geocryology, Moscow State University, Moscow, USSR

ABSTRACT: Modern methods for determining thermal properties of fine-grained soils are analysed and their dependence on the composition and texture of such soils as well as on the thermal conditions is discussed. The data obtained through long-term experimental work have been generalized.

1 INTRODUCTION

Over many years, many Soviet and foreign specialists have been investigating the processes of heat transfer through the use of parameters that are determined both theoretically and experimentally. Theoretical studies are aimed at developing a heat transfer mechanism based on the presentation of texture and structure of real soils in the form of their model analogs. However, the error of calculations, made with the help of models representing a soil as a totality of particles of predetermined shape and size with a definite system of arrangement, is still great. Therefore, the thermophysical properties of soils are mainly investigated experimentally today.

The development of heat transfer theory, coupled with a noticeable enhancement of measuring and microprocessing techniques, has resulted in sufficiently efficient integrated methods for determining thermophysical properties based on constant-rate heating, periodic heating, etc., and in the elaboration of noncontact, medium undisturbing method for determining properties using, for instance, radiation sources of energy (Ivanov(ed) 1970; Lykov(ed) 1973; Yershov(ed) 1984).

The unsteady-state method, utilising quasisteady and especially regular regimes of heat transfer (Chudnovsky et al.1962), are used to investigate a broad spectrum of solid, rigidly cemented, and fine-grained soils. It should be noted that the term "regular regime", which is widely used in Soviet literature, implies that the relative rate of heating (cooling) is constant for all points of a body and does not depend on time. The above-mentioned tests are made predominantly in laboratory conditions for a wide range of temperatures except the temperature range from 0^0 to-5^0C.

When used for soil investigation, the unsteady-state methods proper are represented most extensively by various modifications of probe method employed both in laboratory and, particularly, in field conditions. They differ by the technique of temperature generation, heat flow and by their design.

The situation is somewhat different when the thermophysical characteristics of freezing and fine-grained soils are determined in the range of temperatures corresponding to the main phase transitions, where the application of unsteady-state method is impeded because phase transformations of moisture are accompanied by heat liberation or absorption. Even though unsteady-state methods may undoubtedly be applied, the experimental data should however be processed in accordance with the calculation formulas obtained from the solution of inverse problems of heat transfer with phase transitions in the range of temperatures. The solution of even direct problems of this type is associated with certain difficulties, whereas to use approximate solutions in data processing is not always permissible due to incorrectness of inverse problems. Apart from mathematical problems, the need to take into account the phase transitions of moisture makes it mandatory to use in the calculations a number of parameters that should also be determined indirectly. This problem is aggravated even more by the fact that the characteristic times of water crystallisation in rocks are frequently commensurable with those of making an experiment by unsteady-state methods (Danielian & Yanitsky 1979). The above-given considerations show that steady-state methods are most reliable for investigating the thermophysical properties of moist fine-grained soils in the range of temperatures of the main phase transitions. It should be mentioned that these methods can more expediently be used to investigate anisotropic soils. The effect of heat transfer with migrating unfrozen water can usually be minimised by small temperature gradients.

2 METHODS USED IN INVESTIGATIONS

Different modifications of the above-mentioned methods have been used at the Department of Geocryology, Moscow State University, to study the relationships governing the changes of the thermal properties of soils depending on their composition and texture. For instance, the thermophysical characteristics of moist fine-grained soils are determined by an integrated method combining both methods of regular regime, used to determine thermal diffusivity, and of steady-state regime to determine thermal conductivity (Yershov, et al. 1984). Theoretically this method is based on the solution of a heat conduction equation for an infinite plate with a continuous heat source inside the plate. Estimation of the temperature fields for an infinite and a finite plate showed that the temperature difference in the plate's centre is no more than 2% if its diameter exceeds the thickness by 3 to 4 times (Lykov(ed) 1973). This method allows us to make measurements in the range of temperature from -25^{0}C to +20^{0}C. Temperatures can be maintained at 0.03^{0}C and recorded with an accuracy of 0.01^{0}C.

The regular regime of the 1st kind in the modification of flat or cylindrical layer (Yershov(ed) 1984) was used to make urgent mass determinations. In this case, the mean value of thermal diffusivity was determined for a certain interval of positive or negative temperatures (usually from 3^0 to 10^{0}C). The temperature range, where most phase transitions of moisture occur, was excluded.

The above-stated temperature limits restrict also the applicability of the utilised method of period heating (Ananian, et al, 1974). From the data of one experiment the method allows us to determine two characteristics at a time: thermal diffusivity and coefficient of heat capacity. The first variant of this method is convenient for a broad spectrum of solid rocks and loose soils. The need to work with relatively thin soil samples is one of the disadvantages of this technique.

When this method is applied, temperature fluctuations on one of the plate surfaces are created by a periodic source of heat. The temperature "wave" propagates through the plate and is recorded on its reverse side. The mathematical solution for the steady-state temperature regime makes it possible to relate the thermophysical characteristics of the plate material to the temperature fluctuations on the opposite surface with respect to the heater. The phase shift between periodical surges of power on the heater surface and temperature fluctuations on the reverse side of the plate is associated with the thermal diffusivity of a sample, whereas the amplitude of fluctuations depends on its heat capacity. By varying the period of fluctuations of heating and the sample thickness, it is possible to achieve conditions under which the thermophysical characteristics are determined independently from one another with approximately the same error. To make a calculation it is usually sufficient to record the change of temperature and power of heating with time during 3-5 periods.

When themophysical characteristics are determined in a laboratory, it is expedient to select methods that can ensure sufficient approximation to the natural conditions of rock occurence. One such condition is external load. For instance, heat conduction of a sandstone mass changes under the effect of load from 2.9 W/m.K at a depth of 1300m to 4.4 W/m.K at a depth of 2350m. The effect of external load on the thermophysical characteristics of soils manifests itself by the change of their texture. Failure to take this factor into account may result in appreciable methodic errors.

The stress fields must be taken into account to explain the nature of some methodic errors in determining the thermophysical characteristics of frozen soils. These errors are associated with the difference in the rate of heating and cooling of samples both in the course of experiment and in the process of sample preparation for the test (Chudnovsky 1962). Such errors are caused by the sign-variable stresses in soils that are stipulated by an anomalously high value of the coefficient of volumetric expansions of frozen fine-grained soils as compared to its analogous value for the materials usually used to manufacture sample holders. Irrespective of the direction of temperature variation (heating or cooling of samples), this error usually results in decreasing the absolute value of heat conduction either due to the appearance of an air gap between the soil sample and its holder or due to formation of fractures in this sample. This necessitates strict standards in the methods used to carry out particular experiments and those employed in sample preparation. It seems expedient to freeze cylindrical samples of soils outside the holder and to conduct an experiment in the cycle of temperature increase.

It is more rational to determine the thermophysical characteristics of soils and rocks in their natural occurrence with a minimum disturbance of their structure, heat and moisture regime. This can most successfully be made by the probing methods that allow us to investigate rocks in natural conditions and relatively large volumes.

Under field conditions, the thermal properties of soils and rocks were investigated by cylindrical and flat probes with continuous power source which disturbed insignificantly the structure and temperature regime of a rock mass. The principle of their operation is based on the relationships governing the temperature field variation in a medium inside or on the surface of which there is a constant or continuous source of heat. Probe overheating, relative to the medium, does not exceed 0.5^0 -1^0C, whereas the accuracy of temperature measurement is 0.005^0C. Thermal conductivity is determined both in the process of probe heating and during its cooling (when heating is stopped). The radius of testing amounts to 10-15 cm. In thawed and frozen soils this characteristic is obtained over about 20-30 minutes. Although this method is theoretically not applicable in the range of temperatures from 0^0 to-5^0C wherein intensive phase transitions occur, we have managed to expand the temperature limits for the application of the cylindrical probe to -1^0C by decreasing the power of heating and increasing the measurement time by 50-80%.

3 RELATIONSHIPS GOVERNING THE CHANGES OF THERMAL PROPERTIES

The thermophysical properties of soils and rocks depend on many geological-geographic factors. Rock temperature as well as direction and rate of its variation are the most important parameters characterising the influence of external conditions on heat conduction. Temperature decrease in solid rocks brings about an increase in the heat conduction of crystalline rocks and an analogous decrease in the amorphous ones. In contrast to solid and loose rocks, moist fine-grained soils are characterised by a nonlinear and, as a rule, extreme nature of the relationship $\lambda=f(T)$ (Fig.1) due to soil texture transformation and phase transitions of moisture in the range of negative temperatures (Barkovskaya et al, 1983).

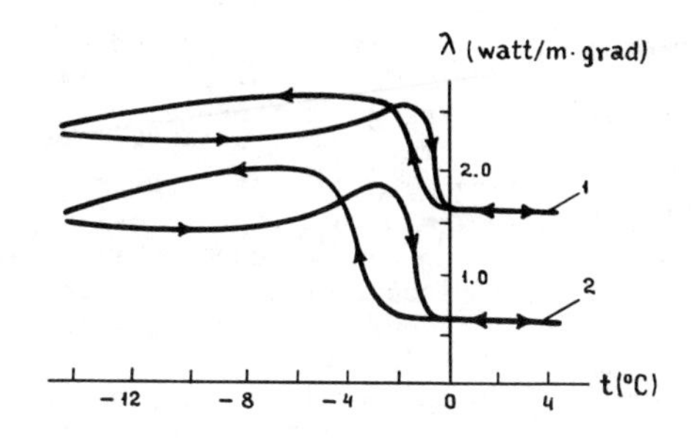

Fig.1 Temperature relationship of diffusivity for water saturated fine-grained soild of different particle-size distribution and mineral composition
1 - Heavy loamy quartz sand
2 - Montmorillonite clay

Moreover, fine-grained soils without rigid bonds are notable for the hysteretic pattern of the thermal conductivity dependence on the direction of temperature variation (heating or cooling). The hysteresis is not expressed in the rocks with rigid bonds.

The heat conduction of fine-grained soils depends appreciably on the size of contacting aggregates, their number and mutual arrangement, which determine the pore space structure and correlation of free and bound water that occurs in the soil in solid and liquid phases at negative temperature. Heat transfer is also influenced by the heat conduction of the soil-forming structural elements, composition and concentration of pore fluid, and presence of organic admixtures.

The experimental studies of fine-grained rocks and soils with a disturbed or natural texture have shown a decline in their heat conduction with a decrease in particle size in the following sequence: coarse-fragmental rock-sand-loamy sand-less-loam-clay. An increase in soils fineness is accompanied by that of water-absorbing capacity which determine the phase composition of frozen soils. It should be noted that the percentage of unfrozen water is increasing in the above-presented sequence.

The relationship between the heat conduction of soils and fineness of their particles is observed in the entire range of the investigated temperatures including those corresponding to the phase transition of moisture. This trend is valid for soils with a different extent of moisture saturation.

The heat-conducting properties of rocks and soils are affected appreciably by their density and any disturbance of continuity. Experiments with fine-grained soils of different composition have proved that thermal conductivity is higher in denser, less porous soi s. This relationship is valid for a broad range of temperatures including that of intensive phase transitions. It is interesting to note that in a dry, fine-grained system with a high porosity (more than50%), the heat conduction is practically independent of the properties of the skeleton material. Thus, similar heat conduction of 0.04 W/m.K has been obtained for two fine-grained systems with a porosity of 62.5% consisting, respectively, of iron balls and rock and differing in the heat conduction of material almost 20 fold (58 and 3.2 W/m.K) (Chudnovsky 1962). The proximity of thermal conductivity values of dry material of various composition may be explained by the dominating influence of porosity. Soil moistening increases the values of heat conduction and augments the differentiation of the thermal properties of soils depending on their composition and texture. There is a sufficiently wide spectrum of changes in the heat conduction of soils depending on their moisture content at different values of density. The dependence of heat conduction on humidity is more intricate in the unfrozen than in frozen soil. The effect of soil density on the relationship of heat conduction and degree of moisture saturation is observed only in unfrozen soils. The heat conduction of frozen soil can be assumed to depend only on the degree of moisture because practically it does not depend on their density.

Fig. 2 presents the relationship between thermal conductivity and the degree of moisture saturation for several soil varieties.

It is seen that not only the value but also the shape of relationship $\lambda = f(q)$ depends appreciably on the particle-size distribution. Thus an increase in humidity up to the values $q \simeq 0.15$ changes the heat conduction of sands 5-10-fold and practically does not change that of loams. A further increase in soil humidity up to complete water saturation increases the heat conduction of sands about 1.5-2 times and loams

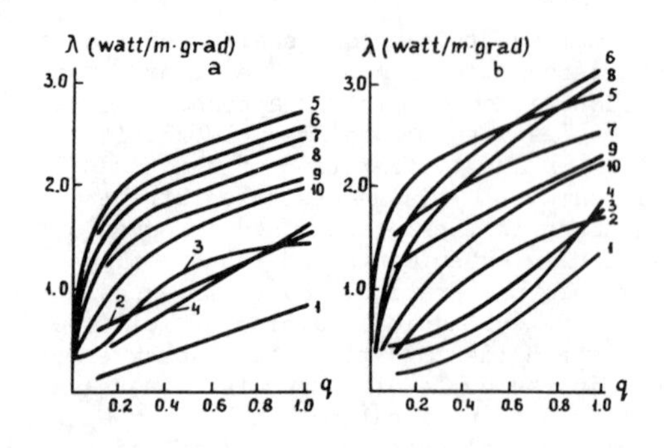

Fig. 2 Relationship between thermal conductivity of soils of different composition and degree of moisture saturation in the unfrozen (a) and frozen (b) state:
1 - Montmorillonite clay; 2 - Kaolinite clay; 3 - Heavy silty loam; 4 - Heavy loam; 5 - Quartz sand (0.25-0.5 mm); 6 - Pure medium and fine-grained quartz sand; 7 - Quartz sand (0.1-0.25mm); 8 - Medium and fine-grained silty quartz sand; 9 - Polymineral sand (0.1-0.5 mm); 10 - Fine-grained polymineral silty sand.

and clays 3-5 times.

A decrease in the amount of clay and silt fractions changes the shape of relationship $\lambda = f(q)$ from concave (montmorillonite clay) to convex (quartz sand). Such differences in the shape of the heat conduction dependence on the degree of humidity are explained not only by different mechanisms of humidification, but also by the presence of unfrozen water. The relationship between heat conduction of rocks and their humidity is also affected by the differences in their mineral composition. It should be noted that this influence cannot be attributed only to water receptivity. The shape of relationship $\lambda = f(q)$ is practically identical for quartz and polymineral sands. However, there is a difference (up to 30%) in the absolute values of thermal conductivity at the equal values of humidity and density (see Fig. 2). This divergence may be explained by different values of heat conduction of the mineral skeleton. The heat conduction of kaolinite clay exceeds that of montmorillonite clay almost twofold at the same values of water saturation. It is hardly possible to explain this difference only by the divergence of the quantitative values of density (0.8 g/cm^3 for montmorillonite and 1.2 g/cm^3 for kaolinite clays). One should take into account the character of water bonding with clay particles. Thus, in kaolinite, hydration takes place only on the outer cleaves and chips of mineral par-

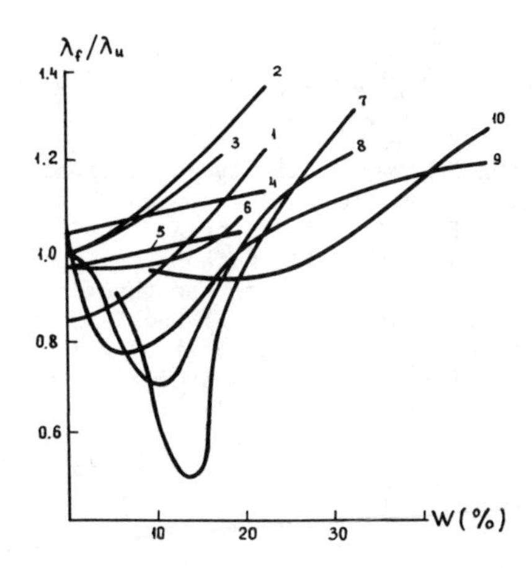

Fig.3 The λ_f/λ_u=f(w) relationship dependence on the humidity of rocks and soils of different composition:
1 - Fine-grained polymineral silty sand;
2 - Medium and fine-grained silty quartz sand; 3 - Pure medium and fine-grained quartz sand; 4 - Quartz sand (0.25-0.5 mm); 5 - Quartz sand (0.1-0.25 mm); 6 - Polymineral sand (0.1-0.5 mm); 7 - Heavy loam; 8 - Heavy solty loam; 9 - Kaolinite clay; 10 - Montmorillonite clay.

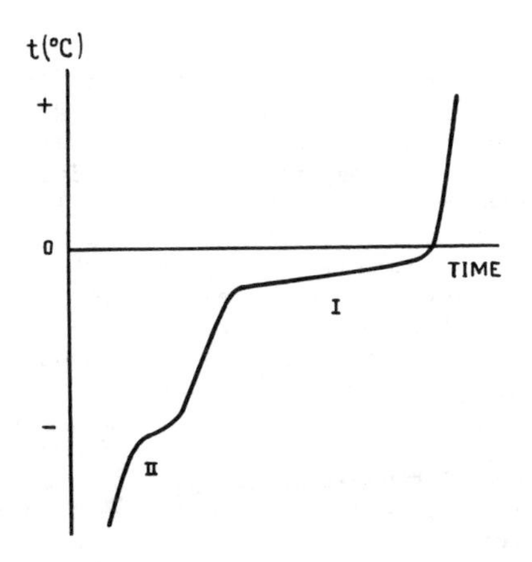

Fig.4 Typical thermogram of saline soil thawing

ticles, whereas in montmorillonite water penetrates into the interlayer space of the crystalline lattice which results in the separation of particles and swelling of this mineral.

The λ_f/λ_u ratio depends primarily on soil humidity. The higher soil humidity the greater is this ratio. The typical curves of the λ_f/λ_u=f(w) relationship for certain soil and rock varieties are presented in Fig.3.

This relationship is influenced by the particle-size distribution and mineral composition of soils and rocks. It can be assumed as linear for sands whose fractions vary from 0.1 to 0.5 mm. The change of soil humidity from 0 to 20% increases the λ_f/λ_u ratio by 7-8%. For sands of natural composition the curve is steeper and the change of humidity in the range discussed is about 30%. The curve for loams and clays runs through the minimum whose depth depends on soil variety. The value of λ_f/λ_u ratio may differ by 1.5-3 times.

The λ_f/λ_u ratio depends on the mineral compostiion of earth materials. For quartz and polymineral sands of close particle-size distribution, the difference can amount to 10-15% at the same values of humidity.

One of the most important factors influencing the formation of thermal properties of rocks is their salinity which appreciably affects the phase state of moisture and the pattern of texture transformation. In the majority of cases, the heat-conducting properties are decreased by the presence of salts. The frozen state is characterised by a complicated dependence of the heat-conducting properties which is primarily associated with the specific features of the dynamics of change of the phase composition and concentration of pore fluid caused by a temperature decrease. Thus, investigation into the freezing-thawing temperatures, carried out in moisture saturated saline samples of rocks of different composition and texture, has allowed us to distinguish in the range of negative temperatures on the thermograms (Fig.4) two narrow intervals (I and II) with a nearly constant temperature which is indicative of the heat effect of the phase transition type.

The calorimetric studies have proved heat release at these sites. Site (I) located near zero is interpreted as the temperature at the beginning of freezing (thawing). The second interval, situated near the temperature of eutectics of the water solution of salt in a free volume, is characterised by a considerably smaller thermal effect which degenerates with a decrease in rock humidity and salinity.

The limit of applicability of methods used to determine the thermophysical characteristics of frozen soils is rather significant since there is a real possibility of additional heat release due to the change in the pore fluid concentration, and lack of data on the kinetics of salt redistribution in the temperature field. While evaluating the temperature limits of the applicability of the regular regime method, it has been revealed that apart from the area of intensive phase transitions near zero, it is necessary to exclude the interval close to the temperatures of eutectics of a salt solution in a free volume. In the interval of temperatures that are higher and lower than a eutectic one, it is possible to process the experimental curves with an error which does not exceed that for non-saline soils.

Our investigations have made it possible to develop a new model of the freezing-thawing processes occurring in saline fine-grained soils. It is distinguished by the presence of several phase transition fronts and by the dependence of the temperature and heat of phase transition on the concentration of pore fluid.

4 SUMMARY

The progress in experimental studies of the thermophysical properties is associated primarily with an increase in the experiment's information capacity and urgency. Most challenging in this respect is further development of the unsteady-state and integrated methods.

It is expedient to use the unsteady-state methods in the cycle of heating due to minimal methodic error and insignificant texture transformation.

During experimental determination of heat transfer coefficients it is frequently necessary to take into account the hysteretic pattern of their dependence on temperature as well as the presence of several temperature ranges in saline soils wherein additional heat release is possible similarly to phase transitions.

REFERENCES

Ananian, A.A., R.G. Motenko & N.N. Smirnova 1974. On the possible use of the periodic heating method for determining the thermophysical characteristics, Issue XIV. 175-178. Moscow

Barkovskaya, Ye.N., E.D. Yershov, I.A. Komarov & V.G. Cheverev 1983. The mechanisms and relationships governing the changes in the heat conduction of soils during freezing and thawing. Proc. IV Intern. conf. on permafrost, Fairbanks, USA.

Chudnovsky, A.F. 1962. The thermophysical characteristics of fine-grained materials. Moscow: Fizmat Publishers.

Danielian, Yu.S. & P.A. Yanitsky 1979. In: Izv. Siberian Branch, USSR Acad. Sci., tech.sci.series, issue 3, 13: 89-92.

Ivanov, I.I. (ed) 1970. Methods for determining the thermal properties of rocks. Moscow: Nauka.

Lykov, A.V. (ed) 1973. Methods for determining heat conductivity and thermal diffusivity. Moscow: Energiya.

Yershov, E.D., I.A. Komarov & N.N. Smirnova 1984. An integrated method for experimental determination of the thermophysical properties of soils. In: Problems of engineering geology associated with industrial and civil construction and development of mineral deposits, vol. 3, p.136-139. Sverdlovsk.

Yershov, E.D. (ed) 1984. The thermophysical properties of soils. Moscow: MSU Publ. House.

2. Mechanical properties

5th International Symposium on Ground Freezing, Jones & Holden (eds)
© 1988 Balkema, Rotterdam. ISBN 90 6191 824 3

State of the art: Mechanical properties of frozen soil

Francis H.Sayles
US Army Cold Regions Research and Engineering Laboratory, Hanover, N.H., USA

1. INTRODUCTION

Since the last State-of-the-Art Report was presented by Prof. Jessberger in 1980 at the Second International Symposium on Ground Freezing, much activity has taken place in the areas of testing and modeling the behavior of frozen soil in response to load and temperature. Accordingly, progress has been made in understanding the fundamental physics that govern the behavior of frozen soil. As Ladanyi and Sayles (1979) have pointed out, our understanding of the mechanical properties, strength, and constitutive relationships that are being developed for permafrost can also be used for artificially frozen ground. However, important differences between permafrost and artificially frozen ground must be considered. In permafrost, the temperature below the active zone is nearly constant at values above -15°C and, except for polygonal soil where ice wedges exist, the ground ice is dominated by ice lenses nearly parallel to the ground surface. In contrast, the temperature of artificially frozen ground usually is characterized by steep temperature gradients with temperatures as low as liquid nitrogen temperatures near the freezing elements and varying to 0°C at the freezing front. Since the orientation of the ice lenses is usually controlled by the direction of freezing, the lenses are generally parallel to the freezing elements. However, ice inclusions nearly parallel to the freezing directions have been reported (Radd and Wolfe (1978), Chamberlain and Gow (1978)). Notwithstanding the differences between naturally and artificially frozen soils, much knowledge can be gained by exchange of information between investigators in the two fields. Such an exchange of information is one of the purposes of this Symposium.

The purpose of this paper is to summarize the state of understanding of the mechanical behavior of frozen soil due to changes in applied stresses and thermal conditions. Dynamic properties are not covered in this report. The focus is on the most recent advances, although some earlier available papers that reflect progress in the past will also be covered. There are many fine papers in the literature that have advanced our understanding of the mechanical behavior of frozen soil, but space does not permit the literature citation to be exhaustive. This is regrettable since the past decade has produced a large number of contributions, each worthy of a more extensive review than is possible in this space.

2. PHYSICAL NATURE OF FROZEN SOIL

It is commonly recognized that frozen soil consists of four components: namely, the soil grains, the structured unfrozen water at the interface between the ice and the surface of the soil grains (Anderson et al. (1978)), ice within the pores between the grains, and air within the pore ice. The response of frozen soil to applied stress and temperature changes is determined by the combined reaction of these individual components and their interactions with each other.

The mechanical properties of the soil skeleton contribute to the behavior of the frozen soil through the sharing of stress with the ice (Ladanyi (1985)) and through a major contribution of the frictional resistance of the soil. Evidence indicates that the long-term strength of frozen soil can be traced to the cohesion and frictional resistance of the soil particles (Miller (1963), Sayles (1973), Vrachev et al. (1984)). This resistance is determined by the same factors that influence the strength of unfrozen soils, which include the soil mineral compositions, particle shape, soil particle distribution, the unit weight (or void ratio or porosity), cohesion, and soil structure.

The ice matrix of the frozen soil provides a major portion of the resistance to stresses that are suddenly applied to the frozen soil mass. The amount of ice in the soil (Sayles and Carbee (1981), Kuribayashi et al. (1985)), the temperature of the ice (Vialov et al. (1963)), and the orientation of ice crystals with respect to the direction of an applied stress (Vialov et al. (1963), Chamberlain (1985)) influences the behavior of the frozen soil due to changes in stress. When the applied constant stresses are less than the failure-producing stress, the ice gradually deforms and stress is transferred to the soil structure to the limit of its frictional resistance (Ladanyi (1981, 1985)). If this limit is exceeded, the soil mass continues to deform despite the frictional resistance until failure eventually occurs.

The structured unfrozen water in the frozen soil is in equilibrium with the ice, and the amount of water is dependent upon the kind of soil minerals, the specific surface area of the soil, the temperature, and the concentration of impurities such as salt in the soil. Although the unfrozen water can resist very little shear, it can withstand tension and compression by virtue of its molecular bond to the soil particles. However, the resistance to stress and deformation of the frozen soil is reduced with an increase in unfrozen water content.

Air present in frozen soil contributes negligible resistance to its applied stress or deformation. However, much of the nonrecoverable compressibility (i.e. consolidation) can be attributed to the presence of air in frozen soil (Tsytovich (1975)).

Stress-Strain Strength of Frozen Soil

The strength of frozen soil is interrelated with its deformational characteristics. When strength is considered, the term "failure" must be defined. Soil that is fractured by stressing clearly has reached its strength limit or failure, but when plastic flow or creep occurs the definition of failure often becomes arbitrary as to the amount of strain that is considered to represent failure. In practical cases where structures are dependent upon frozen ground for support or stability, failure would be represented by deformations that are large enough to prevent the structure from functioning properly. For the purpose of this discussion three terms are used to describe strength: namely, (a) maximum (or peak) stress, which is observed when a soil specimen is subject to an applied increasing stress (Figure 1), (b) residual strength, which is the resistance the soil can sustain at large strains, (i.e. usually greater than about 5%), and (c) creep strength, which is the point of inflection between primary creep and tertiary creep (i.e. the point that represents secondary creep). Other criteria for creep failure have been suggested. For instance, Parameswaran (1987) proposes use of the tangent point on the creep curve that is determined by drawing a line through the origin. Assur (1979) suggests that strain failure is related to the product of the time to reach the minimum strain rate and the minimum strain rate.

The stress-strain strength behavior of a given frozen soil at a constant temperature depends upon the method of loading and the rate at which loads are applied, as well as the initial three-dimensional stress state. Laboratory methods for applying stress to frozen soil include (a) creep tests where a constant deviator stress or load is applied to the test specimen and held constant, (b) a constant rate of either deformation, stress, or load is applied to the test specimen in one direction, and (c) the relaxation test where an initial deformation is held constant and the reduction or relaxation of the stress is observed with time. At the present time the most common tests are the uniaxial compression or tension and the triaxial (confined) compression.

Both the strength of frozen soil and the shape of the stress vs. strain curve from laboratory tests are influenced by the rate of applied stress or strain. As the rate of the applied strain increases, the resistance of the frozen soil increases and the type of failure changes from plastic to brittle. This general trend is illustrated for uniaxial compression tests in Figure 1. A purely brittle failure occurs with an abrupt rupture; in contrast, the purely plastic failure occurs by deforming continuously under a constant stress. As can be seen in Figure 1, the stress-strain curves for frozen soil lie between these two extremes at the temperature used for those tests (-3°C). It should be noticed that yield (initial sharp bending of the curve) occurs at less than 1% strain even though the maximum

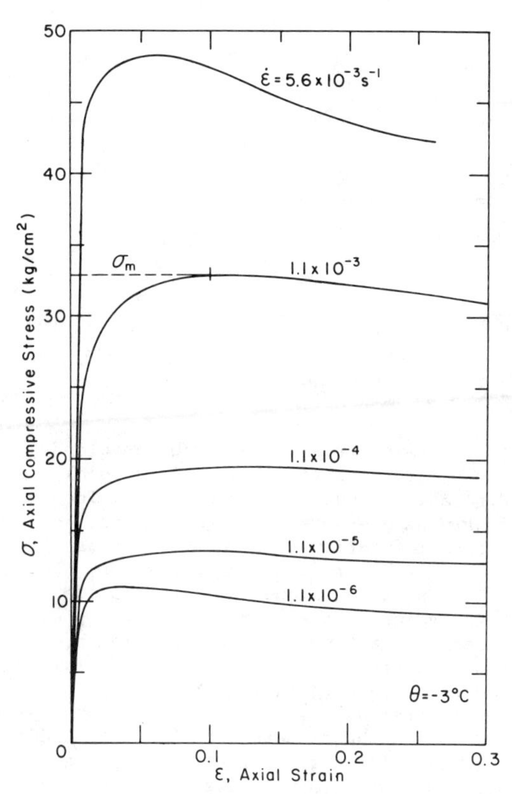

Fig. 1. Stress versus strain curves for uniaxial compression. (After Zhu and Carbee 1984.)

stress is not reached. There is evidence to suggest that initial cracking of the ice matrix occurs at this point (Sayles and Carbee (1981), Hawkes and Mellor (1972)) and that for the plastic failure the peak stress is the result of further cracking of the ice along with friction and dilatancy of the soil structure. The relatively constant stress beyond the maximum stress can be attributed to friction. When the failure is of the brittle type, the strength tends to be independent of the rate of applied load (Figure 2) (Sayles and Epauchin (1966), Zhu et al. (1988)). However, some investigators have reported a slight decrease in strength as the rate increases in the brittle range (Bragg and Andersland (1980)). The strain rate at which the transition between ductile and brittle type failure occurs depends on the type of soil (i.e. clay, silt, sand, gravel) and the testing temperature. In general, for similar temperatures the transition for clays and silts occurs at higher strain rates than for clean sands or gravels. In addition, at higher temperatures the transition zone occurs at a higher rate of strain for all soils. The higher strain rates for the finer grained soil at higher temperatures have been attributed to their greater unfrozen water content (Parameswaran 1980)).

Tensile tests on frozen soils show a behavior similar to the unconfined compression tests except that the tensile strength is typically less and the failure occurs at a smaller strain (Zhu and Carbee (1985, 1987), Haynes et al. (1975)).

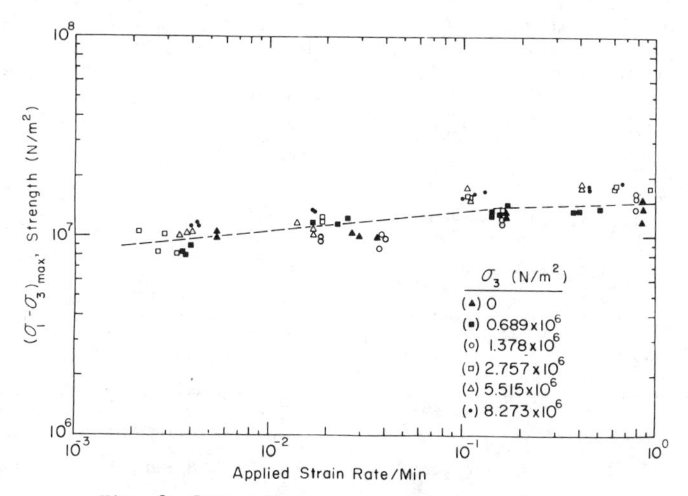

Fig. 2. Strength versus applied strain rate.

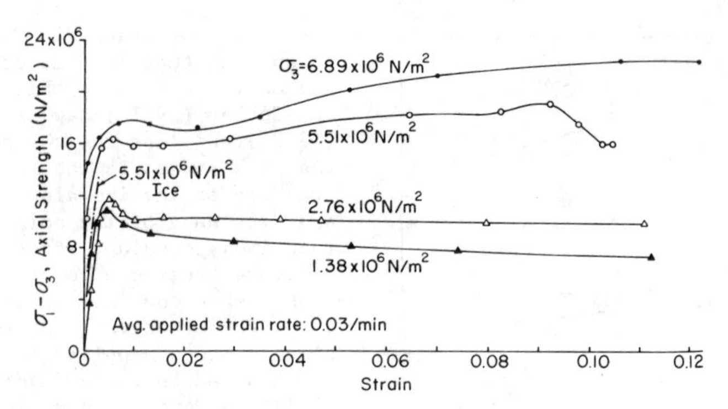

Fig. 3. Stress versus strain curves for triaxial compression.

The effect of confining pressure on the stress-strain behavior of saturated frozen Ottawa sand is illustrated in Figure 3. Two peak stresses occur when the confining pressure on the Ottawa sand (Sayles (1973) is above about 2.76 MPa. The first peak, or initial yield, occurs at a strain of less than 1% and the second peak occurs at strains in the range of 10%. The strain at the initial yield occurs close to the failure strain for columnar-grained ice shown in Figure 3. This observation suggests that the first peak (initial yield) is a consequence of the ice matrix strength, and the second peak (associated with residual stresses) is the result of internal friction developing between the grains of sand and/or ice crystals.

That the initial yield represents the domination of the ice matrix cohesive component while the residual or second peak represents the frictional strength is reinforced by the shape of the two Mohrs' strength envelopes shown in Figure 4. The envelope in Figure 4b is nearly a straight line with an angle of internal friction of about 31°, a value near to that obtained for unfrozen Ottawa sand. The envelope for the initial yield (i.e. first peak) in Figure 4a is curved at the lower values of confining pressures, indicating that some internal friction is involved at these lower levels of confining pressures. This "first peak" envelope approaches a constant value at the higher confining pressures, suggesting that ice cohesion provides nearly all the initial yield resistance at these confining pressures and small strains.

Goughnour and Andersland (1968) tested sand-ice mixtures in triaxial compression starting out with only ice and gradually increasing the concentration of sand. They observed that at low sand volume concentrations, the shear strength increased slightly with increased proportions of sand. Upon reaching a sand volume concentration of about 42%, a rapid increase in shear strength occurred. They attributed this increase to friction between sand particles and the initiation of dilatancy. Jones and Parameswaran (1983) also triaxially tested mixtures of ice with from 10 to 60% concentrations of sand using confining pressures from 0.1 to

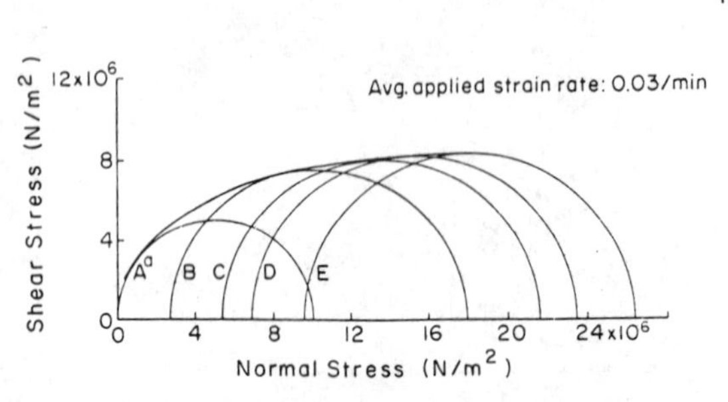

a. Ottawa sand first resistance peak.

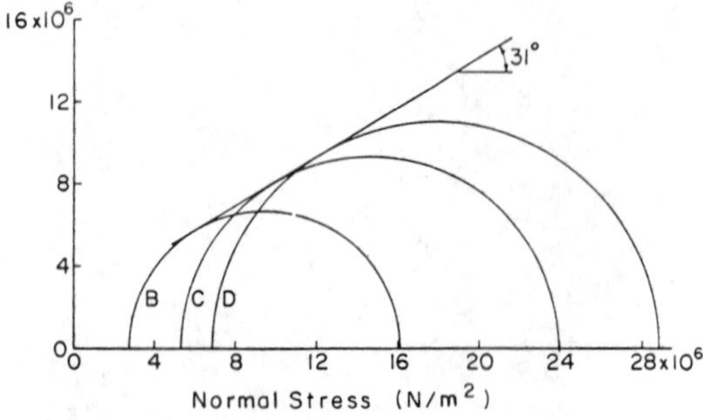

b. Ottawa sand second resistance peak.

Fig. 4. Mohr's envelopes for constant applied strain rate.

85 MPa. They found that as confining pressure was increased the strength increased to a maximum value and then decreased with a continued increase in confining pressure. This maximum was about 30 MPa for polycrystalline ice, 40 MPa for saturated sand, and 25 MPa for the ice-soil mixtures. Chamberlain et al. (1972) and its discussion by Andersland (1972) concluded that at high confining pressures (about 69 MPa) dilatancy was suppressed and that pressure melting of the ice developed with an accompanying decrease in shear strength because of excess pore pressure buildup. Since air in the water is more compressible than sand grains, it is presumed that at even higher confining pressures (above 110 MPa) the frictional component of the shear strength increases in a manner similar to that for an undrained test on saturated unfrozen soils. Parameswaran and Jones (1981) tested an Ottawa fine sand using confining pressures up to 75.7 MPa, confirming the findings of Chamberlain et al. (1972). For confining pressures varying from 0 to 8.3 MPa, Sayles (1973) reported that the strength of Ottawa sand is related empirically to the rate of applied strain by:

$$\sigma_{max} = A\,\dot{\epsilon}^{b}$$

where σ_{max} is maximum stress or strength, $\dot{\epsilon}$ is the rate of applied strain and A and b are experimentally determined parameters. The value of b was found to be about 0.1. This relationship was confirmed by Parameswaran and Jones (1981) for a confining pressure of 5 MPa and supported by data from Shibata et al. (1985). In contrast to the response of frozen sands to confining pressure, Ouvry (1985) found that undisturbed, artificially frozen clay showed a slight strength decrease with an increase in confining pressure. Chen (1988) found that failure stress for clay increased slightly with an increase in confining pressure. It is not clear why there is a difference in the results of the tests on clay, but small differences in ice content or rates of loading could account for the different conclusions.

In testing frozen sand in triaxial compression, Youssef (1984) found that the volume remained nearly constant when the test specimen was not enclosed in a rubber membrane. When this sand with the same initial void ratio was tested under the same conditions with a membrane, it showed an initial volume decrease and then the volume increased continuously with increasing strain after the initial yield occurred. This preliminary result indicates that the volume increases occurring in triaxial test results at low confining pressures should be attributed to crack formation and crack opening in addition to dilatancy of the soil grains.

The initial slope of the stress vs. strain curve is of interest where small stresses and strains are important. Since the frozen soil is usually inelastic even at very small strains, the term "initial tangent modulus" is used rather than "modulus of elasticity." From compression tests on sand the initial tangent modulus tends to increase with an increasing strain rate and with a decreasing temperature (Bragg and Andersland (1980)). For frozen silt, Haynes et al. (1975) found that the modulus for both tension and compression increased slightly with increase in strain rate. Studies by Haynes and Karalius (1977) and Zhu and Carbee (1984, 1987) agree with these findings for silt except that they found the initial modulus for tension is nearly independent of strain rate at about 10 GPa at -5°C.

Values of Poisson's ratio have been measured for a few frozen soils under a limited number of temperature and stress conditions. Tsytovich (1975) reported on Poisson's ratio values for a frozen sand, silt, and clay at temperatures in the range from 0 to -5°C. In general, the values decrease with decreasing temperature (e.g. for clay the Poisson's ratio reduced from 0.45 at 0.5°C to 0.26 at -5°C). He noted that the values approach 0.5 as the temperature approaches 0°C. Akagawa (1980) computed values of Poisson's ratio from the measured axial and radial deformations of a frozen sand during uniaxial creep tests. His findings show that the Poisson's ratio increases as the temperature of the soil increases, which is the same trend noticed in the values published by Tsytovich (1975).

3. CREEP

The term "creep" as used in this paper is the irrecoverable time-dependent deformation that occurs at a constant volume. This definition distinguishes creep from consolidation where a permanent decrease in volume occurs. Vialov and Tsytovich (1955) explain the physical process of creep in frozen soils as a time-dependent process where an applied constant load concentrates the stress between the soil particles at their points of contact with ice, causing pressure-melting of the ice. Differences in stress within the soil mass cause the melted ice to move to regions of lower stress where it refreezes. This

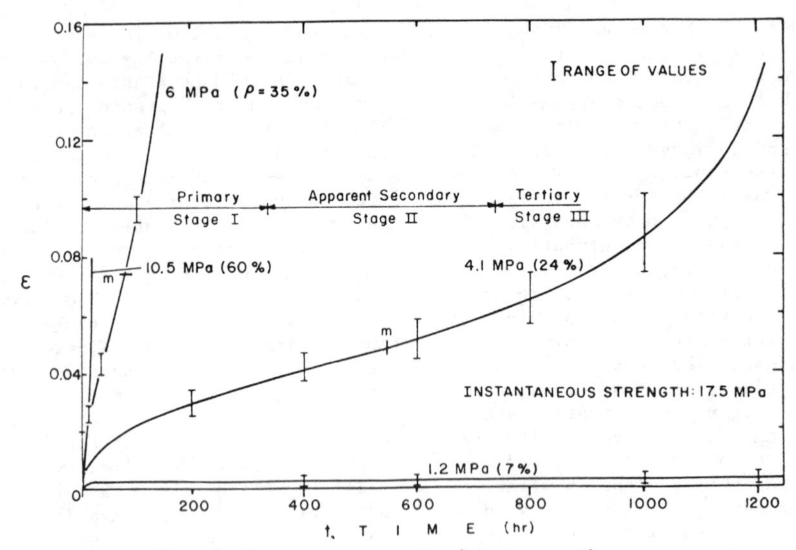

Fig. 5. Creep curves--strain versus time.

process is accompanied by a breakdown of the ice and the structural bonds between the soil grains, plastic deformation of the pore ice, and a readjustment in the soil particle arrangement. This structural deformation leads to a denser packing of the soil particles, which in turn causes an increase in strength of the material due to the increased number of firm contacts between grains and hence an increase in internal friction between the grains. Counter to this process there is a weakening of the structural cohesion and possibly an increase in the amount of unfrozen water. If the applied load does not exceed the ultimate long-term strength of the frozen soil, then the weakening process is compensated by the gain in strength. The rate of deformation decreases with time (i.e. damped deformation); this condition is known as primary creep. However, if the applied load exceeds the long-term strength of the soil, the breakdown of internal bonds is not completely compensated by the strengthening process. The rate of deformation increases with time, resulting in plastic flow and cracking and causing a breakdown of the soil-ice matrix structure (i.e. undamped deformation); this condition is known as tertiary creep.

When the applied constant stress exceeds the ultimate long-term strength of the frozen soil, a plot of the observed deformation vs. time reveals a form similar to that of classical creep curves for metals (Figure 5). These curves include: an initial instantaneous deformation consisting of both an elastic and a plastic irrecoverable component; a primary creep stage characterized by a decreasing rate of deformation; an "apparent secondary stage" with a point of minimum creep rate; and a tertiary stage where the creep rate is increasing, ending in rupture. When the logarithm of the deformation rates or strain rates is plotted against the logarithm of time (or strain), curves similar to those shown on Figure 6 result. The shape of this log-log curve shows that the secondary creep stage, shown as a constant strain rate on Figure 5, is in reality a point of inflection separating primary from tertiary creep. This inflection point at which the creep rate reaches a minimum has been chosen to represent creep failure (Fish (1980), Martin et al. (1981), Ting (1981), Zhu (1984)). Although the inflection point may be defined as creep failure, many ice-rich soils have an apparent long-duration "secondary creep stage" because the rates of change in the primary and tertiary creep stages can be quite small near the inflection point. This portion of the creep curve for such soils can be represented by a straight line for many engineering purposes (Thompson and Sayles (1972)). Ladanyi (1972) takes advantage of this feature to represent an engineering theory of creep of frozen soils.

When the applied stress is less than the ultimate long-term strength of the soil, the rate of deformation attenuates continuously and approaches a constant value asymptotically (Vialov et al. (1963), Sayles (1968)). The design of permanent structures that depend upon the support of

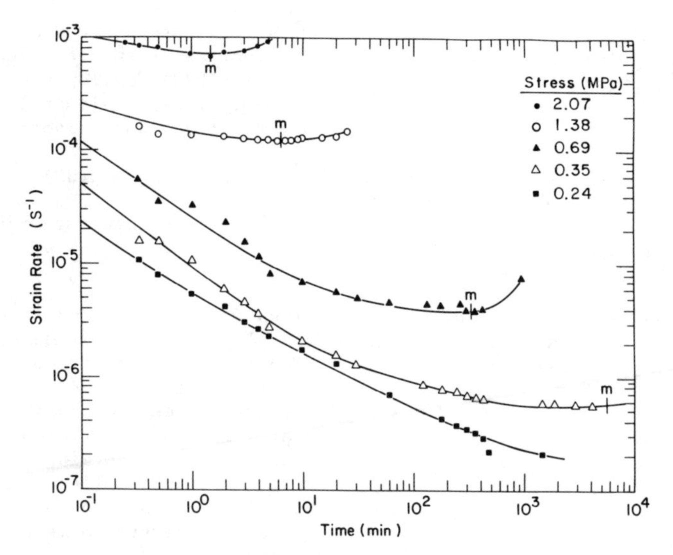

Fig. 6. Creep curves--log (strain rate) versus log (time).

frozen soil with minor deformations must be based on allowable stresses less than the long-term strength.

The shape of the creep curve is influenced by temperature, the magnitude of the applied stress, the type of soil, ice content, the anisotropic nature of the soil-ice matrix, and impurities in the pore ice. Creep rate increases with an increase in temperature for a given soil under constant stress conditions (Vialov et al. (1963), Sanger and Kaplar (1963)). Creep data from tests on frozen soil suggest that thermal activation is involved in the deformation process (Andersland and AlNouri (1970)) and the temperature-dependence of the creep rate can be represented by:

$$\dot{\epsilon} = A \exp\left(-\frac{U}{RT}\right)$$

where U is the apparent activation energy, R is the universal gas constant, T is the absolute temperature, and A is an experimentally determined stress-dependent soil parameter. Goughnour and Andersland (1968) offer

$$\dot{\epsilon} = \text{constant}\ \frac{1}{|T|^n}$$

as a close approximation to the exponential form, where T is in degrees Celsius and n is an experimentally determined constant.

Typical time vs. strain curves are shown on Figure 5 for different magnitudes of applied uniaxial stresses. The percent values shown on each curve refer to the percent of "instantaneous strength" determined by uniaxial compression test at an applied strain rate of 3.3% per minute. The curve at the 7% stress level is below the ultimate long-term strength of this frozen sand.

Models for Creep Behavior

Plots from creep tests of strain rate vs. time on logarithm coordinates, shown in Figure 6 for an undisturbed silt, provide a convenient representation for use in analyses and the development of models to describe the creep process. Fitting curves to data similar to that shown in Figures 5 and 6 are the basis for the development of empirical expressions for use in the design and analysis of frozen soil structures (Vialov (1959, 1973), Vialov et al. (1963), Goughnour and Andersland (1968), Sayles (1968, 1973), Andersland and AlNouri (1970), Ladanyi (1972), Jessberger (1977), Eckardt (1979), Fish 1980), Takagawa et al. (1979), Klein (1979), Ting and Martin (1979), Pusch (1980), Ting (1981), Gardner et al. (1984), Zhu and Carbee (1984), Berggren and Furuberg (1985), and others). In general, the expressions were developed from observations of uniaxial compression creep tests. Of the many expressions available in the literature, because of space limitations, only a few will be presented here to illustrate the salient features of creep behavior. One of the earliest equations to describe primary creep was developed by Vialov et al. (1963). It is recognized that the total strain should be represented by

$$\epsilon = \epsilon_o + \epsilon_d + \epsilon_c$$

where

ϵ_o = initial elastic component of the strain
ϵ_d = delayed elastic effect
ϵ_c = irrecoverable creep component.

Usually ϵ_o is small in comparison with ϵ_c and often is neglected. An expression for the creep component was developed using the power function

$$\sigma = B\,\epsilon_c^m \;;\quad B = f(t,\theta)$$

$$B = C(\theta)\, t^{-\lambda}$$

then

$$\epsilon = \left[\frac{\sigma\, t^{\lambda}}{C(\theta)}\right]^{1/m}$$

$$\epsilon = A(\theta)\sigma^{1/m} t^{\lambda}$$

where σ is applied stress; m and λ are is considered basic characteristics of the soil, while $A(\theta)$ is a function of temperature below freezing (θ); and t is time. The final expression by Vialov et al. (1963) with a modification by Assur (1963) to make the temperature dimension consistent is

$$\epsilon_c = \left[\frac{\sigma\, t^{\lambda}}{\omega\left(\frac{\theta}{\theta_o} + 1\right)^{k}}\right]^{1/m}$$

where λ, m, and k are dimensionless constants that are characteristic of the material. ω has dimensions of temperature, and θ_o is an arbitrarily selected reference temperature.

A similar equation for strain was developed by Sayles (1968, 1973) for primary creep by fitting a straight line to data on logarithmic plots of strain rate vs. time values. This equation and Vialov's expression have been used successfully in analyses for the primary creep settlement of footings on frozen soil. Takagawa et al. (1979) suggest that two equations that are modifications of the Vialov-type creep strain equation be used for describing creep strains for stresses below the initial yield stress. One equation is suitable for creep behavior without ice lenses in the soil and the other is suggested for use when ice lenses are present.

Using creep data from low confining pressure (less than 0.4 MPa) triaxial differential creep tests, Andersland and AlNouri (1970) developed the following exponential expression relating strain rate to stress and temperature:

$$\dot{\epsilon} = A \exp(-\frac{L}{T}) \exp(ND) \exp(-m\sigma_m)$$

where

T = absolute temperature
$D = (\sigma_1 - \sigma_3) - \sigma_m$
$\sigma_m = 1/3(\sigma_1 + \sigma_2 + \sigma_3)$

Values for A, L, N, and m are determined from plots of the test data.

The investigators noted that this equation is similar to the form indicated by the Rate Process Theory shown by Mitchell et al. (1968) for unfrozen soil.

Ladanyi (1972) developed an engineering creep theory using linear creep curves (Hult (1966)) and a pseudoinstantaneous strain. Total strain is represented by the expression:

$$\epsilon = \frac{\sigma}{E} + \epsilon_k\left(\frac{\sigma}{\sigma_k}\right)^k + t\dot{\epsilon}_c\left(\frac{\sigma}{\sigma_c}\right)^n$$

where

σ = applied stress
t = time
$(\frac{\sigma}{E})$ = instantaneous elastic strain, E = fictitious Young's modulus that contains the delayed elastic effect
$\epsilon_k(\frac{\sigma}{\sigma_k})$ = instantaneous plastic strain
σ_k = temperature-dependent deformation modulus.
ϵ_k = arbitrary small reference strain introduced for convenience in calculation and plotting data

σ_k and k are obtained from logarithmic plots of instantaneous plastic strain vs. applied stress (see Andersland et al. (1978)).

As mentioned earlier, this expression is based on the assumption that strain or deformation can be represented by a secondary or constant rate creep. This assumption is reasonable for some ice-rich soils.

Ting and Martin (1979) examined the validity and reliability of the Andrade equation as a predictive tool to determine the secondary strain rate from short-term tests on ice and frozen soil. The empirical equation formulated by Andrade (1910) for primary creep took the form:

$$\ell = \ell_o(1 + \beta t^{1/3}) \exp(kt)$$

where

ℓ = length at any time
ℓ_o = original or initial length of the test specimen

t = time

β and k = constants for curve fitting

By converting this equation to true strains, it becomes:

$$\epsilon = \ell_o + \ln(1+\beta t^{1/3}) + kt$$

ϵ = total strain

They found this equation provided a good curve to fit the strain vs. time data for those creep tests exhibiting an initial slope of less than 0.7 on a logarithm plot of time vs. strain rate. (See Figure 6.) However, even when the initial slope is less than 0.7, the equation does not provide a reliable prediction of the secondary strain rate or the time that occurs.

Berggren and Furuberg (1985) propose an empirical model that utilizes the reciprocal of the slope (R) of the creep curve as a function to describe the creep behavior of frozen soil. Since R is a function of time, it can be represented by

$$R = \frac{dt}{d\epsilon} = r(t-t_t)$$

Then for primary creep:

$$\epsilon = \epsilon_o + \frac{1}{r}\ln t \; ; \quad \text{for} \quad t < t_p$$

where

ϵ_o = instantaneous strain

$r = r_f(f)^{-i}$

f = degree of mobilization $= \frac{\sigma}{\sigma_\theta}$

σ = applied stress

σ_θ = temperature-dependent reference strength from constant strain rate tests

also:

$$\sigma_\theta = \sigma_u\left(\frac{\omega_s}{\omega_u}\right)^u \; ; \text{ for } \omega_u > 0$$

ω_u = unfrozen water content

ω_s = water content at 100% saturation

If all the water is unfrozen, $\omega_u = \omega_s$.

t_p = time to secondary creep $= t_{pf}(f)^{-j}$

i, j, γ_f, t_{pf}, u, and σ_u are parameters determined from creep and unfrozen water content data.

For secondary creep:

$$\dot{\epsilon}_s = \frac{d\epsilon}{dt} = \text{constant and } \epsilon_s = \frac{1}{rt_p}(t-t_p);$$

$t \geq t_p$

Tertiary creep is not included in the model.

It should be noted that the effect of temperature is expressed in the σ_θ term, which reflects the temperature of the ice as well as the amount of unfrozen water content in the soil when σ_θ is determined by a constant strain rate test. But if σ_θ is determined by evaluating the unfrozen water content only by the expression

$$\sigma_\theta = \sigma_u\left(\frac{\omega_s}{\omega_\mu}\right)^u ,$$

then the effect of temperature on the creep of pore ice is not included. Berggren and Furuberg (1985) suggest that the model be verified for other soils in addition to the clay that was used in their investigation.

Using the Rate Process Theory (Glasstone et al. (1941), Frenkel (1946)) and thermodynamic considerations, Fish (1976, 1980, 1982a,b, 1983a,b) proposed a creep model that describes all three stages of creep, which unifies the process of deformation and failure. This expression is

$$\dot{\epsilon} = \tilde{C}\,\frac{kT}{h}\exp\left(-\frac{E}{RT}\right)\left(\exp\frac{\Delta s}{k}\right)\left(\frac{\sigma}{\sigma_o}\right)^{n+m}$$

where $\tilde{C}$, $n \geq 0$, and $m \geq 1$ are dimensionless parameters independent of temperature;

σ_o = temperature-dependent, the ultimate instantaneous strength of the frozen soil

E = activation energy

k = Boltzmann's constant

h = Planck's constant

R = universal gas constant

T = absolute temperature, °K

$\frac{kT}{h}$ = frequency of vibration of elementary particles around their equilibrium position

Δs = change in entropy

$\frac{\Delta s}{k} = \frac{\Delta s(T)}{k} = \delta f(\tau)$

δ = a dimensionless parameter independent of temperature

$f(\tau)$ = a function of normalized time

$f(\tau) = \tau - \ln\tau - 1;$

and $\tau = \frac{t}{t_m}$

t_m = failure time, the time to reach the point of inflection in the creep curve (See point "m" on Figure 5 and 6)

The minimum strain rate can be determined by setting $\Delta s = 0$, then

$$\dot{\epsilon} = \dot{\epsilon}_m\,\tilde{C}\,\frac{kT}{h}\exp\left(-\frac{E}{RT}\right)\cdot\left(\frac{\sigma}{\sigma_o}\right)^{n+m}$$

By substituting back and rewriting the expression

$$\dot{\epsilon} = \dot{\epsilon}_m e^{\delta f(\tau)}$$

by integration (Fish (1983b), Fish and Assur (1984)) the expression for creep strain becomes

$$\epsilon = \dot{\epsilon}_m \psi \tau \exp(\delta f(\tau))$$

where ψ is the integration factor.

Although this series of equations by Fish (1976, 1983b) contain empirical components, the general form of the equation can be tied to the Rate Process Theory. These equations like many others seem to fit creep data reasonably well for ice-rich soil at stresses well above the ultimate long-term strength level. However, at creep stress levels approaching the long-term strength of the soil (i.e. where the interaction between soil grains apparently dominate the soil behavior (Sayles (1973)), the equations do not represent the creep behavior well.

Ting (1983) presented an empirical model that he named a Tertiary Creep Model for frozen sands in the following form:

$$\dot{\epsilon} = A\, t^{-m} \exp(\beta t)$$

where A, m, and β are experimentally determined constants.

This expression also describes the entire curve for creep. In their discussion, Fish and Assur (1984) point out that this expression belongs to a general family of equations proposed by Zaretsky and Vialov (1971), Assur (1979), and Fish (1976, 1980, 1982b).

Hampton et al. (1985) recently compared the creep equations proposed by Fish (1983b), Ting (1983), Gardner et al. (1984) and the equation by Vialov et al. (1963) for primary creep that was used by Klein (1979). Vialov's model for primary creep at constant temperature is

$$\epsilon_c = A\, \sigma^{1/m}\, t^{\lambda} = A\, \sigma^{B}\, t^{c}$$

where ϵ_c is creep strain, σ is stress, and t is time. A, m, and λ are parameters for the soil at constant temperature, $B = 1/m$, and $c = \lambda$.

An equation for describing the entire creep curve developed by Gardener et al. (1984) is

$$\frac{\epsilon}{\epsilon_m} = \left(\frac{t}{t_m}\right)^c \exp\left[(c^{1/2}-c)\left(\frac{t}{t_m} - 1\right)\right]$$

where ϵ_m = creep strain at the point of inflection on the creep curve (see point m on Figures 5 and 6)

t_m = time to the point of inflection on the creep curve to correspond to

$$c = \left(\frac{\dot{\epsilon}_m t_m}{\epsilon - \epsilon_o}\right)^2$$

ϵ_o = initial strain

This expression also belong to the same family of equations as that of Zaretskiy and Vialov (1971), Fish (1976, 1980), Assur (1979), and Ting (1983).

The models developed by Fish (1983b) and Ting (1983) have been described earlier. A closeness of fit comparison by Hampton et al. (1985) showed that the Ting model was close to the experimental data near the start of the test but underpredicted the strain for most of curve. Although the Vialov's model slightly underpredicts strain, it gives the best fit to the data in the primary creep stage but is not applicable beyond the point of inflection. Both the Gardner and Fish models greatly overestimated the creep strain in the early portion of the curve but closely fit the data from a point at $t/t_m = 0.7$ to rupture. It should be noted that the Fish, Ting, and Gardner equations are the same basic family. Klein's equation is of the same form for primary creep as those developed by Vialov et al. (1963) and Sayles (1968).

Ashby and Duval (1985) have made progress in developing a physically-based model for the deformation of polycrystalline ice. Some of their concepts may have application to frozen soils.

Determining Creep Parameters from Other Types of Tests

Because long-term creep tests on frozen soil are time-consuming and require special attention to the control of temperature and applied stresses for long periods of time, test procedures such as a series of constant step loads applied to a single specimen of soil or constant applied strain rate tests have been suggested. It is reasonable to assume that the mechanical properties of the soil should be independent of loading conditions. One difficulty, however, often lies in the way tests must be conducted (i.e. by applying boundary conditions which reduce some types of tests to index tests). Even though index tests have their limitations, they are quite useful for comparison purposes and are used extensively in evaluating materials; however, they have limited value in advancing the state of the art. A second difficulty with using rapid types of tests to determine creep parameters is that the physical arrangement of the soil components changes during the loading period. Therefore, the stress as well as the temperature history of the soil sample could influence the value of the parameters. To date, little quantitative evidence has been published in the English literature on the influence of stress or temperature history on creep of frozen soils.

A procedure using a series of step loads where each successive step applies a constant stress higher or lower than the preceding one has been used. This procedure has the advantage of testing the same soil specimen at different stress levels, thus avoiding the need for obtaining nearly identical samples for testing at each of the different stress levels. Andersland and AlNouri (1970) used this procedure in studying the effect of changing the confining pressure and deviator stresses on frozen sand in triaxial tests. Ladanyi (1972) suggests a procedure by summation of strains, which accounts for the complete loading and temperature changes. These studies did not show a comparison of results with those from a single-step creep test. However, Eckardt (1979) used step-load tests to compute creep curves for a single stress. The comparison with measured values from a test using a single stress creep was good for the high stress levels. It should be noted that when long-term strength is to be determined by a series of step loads, it is uncertain how long the stress of each step is to be held to ensure that final steps will yield the desired long-term strength value.

Perkins and Ruedrich (1974) and Ladanyi (1981) have suggested that the creep stress and the strain at the point of inflection on the creep curve corresponds to the peak stress and strain at the peak stress on a curve of stress vs. strain for a constant rate of strain. However, Fish (1983b) and Rein (1985) suggest equivalency for the entire curve. Fish used a thermodynamic model along with normalized stresses, strain rate, and times to demonstrate the equivalency of the curves. Rein constructed a creep curve from a series of stress vs. strain curves obtained for a variety of strain rates. His assumption is that at a given stress level, the strain rate is the same for both types of loading. Both methods show good equivalency for high stress levels but poor correspondence for low stress levels. This lack of agreement at low stress levels could be related to the friction component of the soil grains, which can dominate frozen granular soil behavior at low stresses (Miller (1963), Sayles (1973)).

Isochronous stress-strain curves have been developed from a family of constant stress creep curves for frozen soil at a constant temperature (Vialov et al. (1963), Rein et al. (1975)) for use in the analysis of engineering structures. This type of curve is obtained by selecting a specified time, then using the values of strain and the creep stress that produced this strain at the specified time. On logarithmic plots, these curves have been represented as straight lines even though it is recognized that the data indicated a curve is appropriate (Vialov et al (1963), Sayles (1968), Sayles and Haines (1974)). Rein (1975) suggests that these curves are best represented by two straight lines. The initial line starting at zero with a steep slope would be intercepted by a line with a flatter slope. Data for frozen sand indicates that the point of interception is near the yield point (i.e. at a strain of less than 1%) which supports the concept that the initial fracturing of the ice at the yield point changes the characteristics of the deformation process.

Time-Dependent Creep Strength

The creep strength of frozen soil is defined as the stress level that can be resisted up to a finite time at which instability occurs (Andersland et al. (1978) or creep failure is reached. As mentioned above, creep failure is defined in this paper as the creep at the point of inflection on the creep curve. (See point m on Figure 5 and 6.) Relationships for creep failure stress, failure time, and failure strain have been proposed by different investigators (Vialov (1959), Ladanyi (1972), Assur (1979), Fish (1983b)).

Vialov suggested that at a constant temperature the variation of strength with time be represented by

$$\sigma_f = \frac{\beta}{\ln\left[\frac{(t_f + t^*)}{B}\right]} \approx \frac{\beta}{\ln\left[\frac{t_f}{B}\right]}$$

where β and B are parameters that depend upon the type of soil and its temperature.

$t^* = B \exp(\frac{\beta}{\sigma_o})$, which can be neglected (Vialov (1959))

σ_o = instantaneous strength

t_f = time of failure

σ_f = creep failure stress, i.e. creep strength

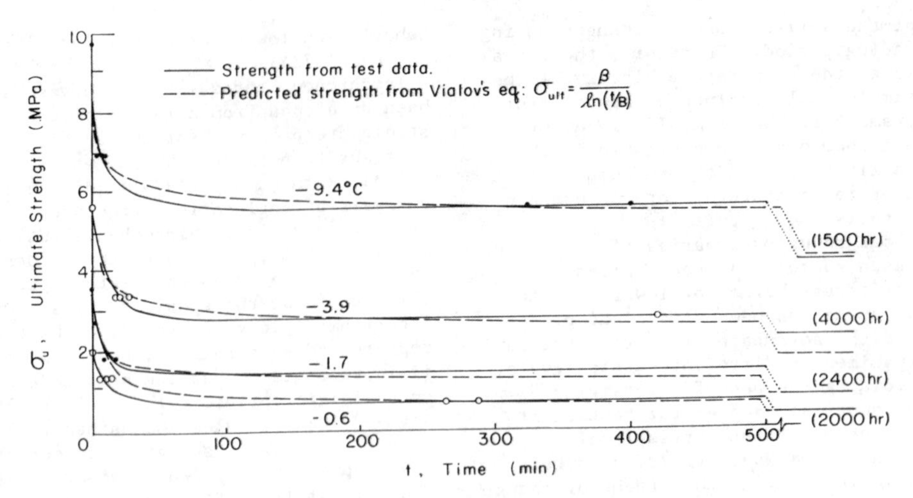

Fig. 7. Strength versus time.

This equation represents the variation of strength with time reasonably well, provided the parameters are determined by testing the given frozen soil under the stress conditions and at the temperatures that are to prevail during the period for which the prediction is made (e.g. design life of the structure). Refer to Figure 7 for a comparison of strength vs. time plots.

Ladanyi (1972) suggests that the relationship between time to failure and creep strength be expressed as

$$t_f = \frac{\epsilon_f}{\dot{\epsilon}_c(\sigma/\sigma_c)^n}$$

where t_f = time to creep failure
ϵ_f = failure strain
σ = applied stress
$\dot{\epsilon}_c$ = arbitrary creep rate
σ_c and n are temperature-dependent parameters of the material.

This expression is applicable in ice-rich soil where the apparent secondary creep dominates the soil behavior and the initial pseudoinstantaneous plastic strain can be neglected.

Expressions used to describe the entire creep curve with a single model, in general, utilize the point of inflection in the creep curve as a key parameter in determining the shape of the curve. Since this point is defined as failure, it is a question of relating the time to the inflection point (t_m) to the stresses for which the creep curves are developed. Often the failure time is related to a power function of stress (Fish (1983b), Ting (1983), Gardner et al. (1984)) in the form

$$t_m = A\,\sigma^{-n}$$

where A and n are temperature-dependent parameters for a given soil.

It is suggested by Fish (1983b) that the equation be in the form

$$t_m = t_o\left(\frac{\sigma}{\sigma_o}\right)^{-m}$$

where

t_o = the mean duration of the "settled life" of an elementary particle in position of equilibrium (Frenkel (1946))
σ_o = stress that causes failure at time t_o (see the discussion under Creep Model for additional details).

m is a dimensionless parameter that Fish (1985) relates to ice and unfrozen water content by

$$m \approx \frac{a}{\omega_i\omega_u},$$

where ω_i = ice content of the soil
ω_u = unfrozen water content
a = soil parameter

It should be mentioned that this function assumes a straight line relationship between time to failure and strength on a logarithm plot. Plots of test results by Zhu and Carbee (1984) indicate that, even though at higher stress levels and strain rates a straight line fits the data reasonably well on logarithm coordinates, the data deviate dramatically at the lower stress levels and slow strain rates. This increase in failure time for lower stresses lends support to the concept that frictional resistance of the soil particles are an important factor affecting creep strength.

Based on the concept that the ultimate long-term strength is the result of the resistance of the soil particles and that the ice contributes nothing to this ultimate strength, Vrachev et al. (1984) propose sublimating the ice from the soil down to the level of the unfrozen water content, then testing it at conventional strain or stress rates. They reason that the conventional or instantaneous strength of the sublimated, undisturbed soil would yield the ultimate long-term strength of the soil in its original frozen condition. They present data for a clay soil with a massive texture to support this conclusion. They suggest that this method reduces the testing time for determining the ultimate long-term strength and allows the investigation of the roles of the components of the frozen soil. Clearly, this method is to be applied only to soils with a massive ice texture.

Methods for determining the ultimate long-term strength of a frozen clay were investigated by Maksimjak et al. (1983) to compare the consistency of results. The tests used and the resulting strength (in percentages of the instantaneous strength) were (a) single loading constant stress creep test (14%), (b) stepped loads with constant stress during each load (20%), (c) constant stress rate (19%), (d) constant strain rate (20%), (e) relaxation test with single loading (14%), and (f) relaxation tests (19%). They concluded that for tests where the load is applied gradually, the ultimate long-term strength appeared to be greater than for tests where the load is applied quickly in a single step. It should be noted that the single loading tests (creep test and relaxation tests) gave similar and smaller values than the others.

The creep strength of frozen soil determined by triaxial compression tests can be represented by Mohr's envelopes for specific times to creep failure. Figure 8 shows the results of tests on frozen Ottawa sand for three different times. As the time to failure increases the envelope approaches a straight line with a slope of about 29°. This angle is close to that found for this unfrozen sand tested in a drained condition. The curves support the concept that the source of the ultimate long-term strength of frozen soil is the interaction between the soil particles. In effect, the loading history effects the results of the tests.

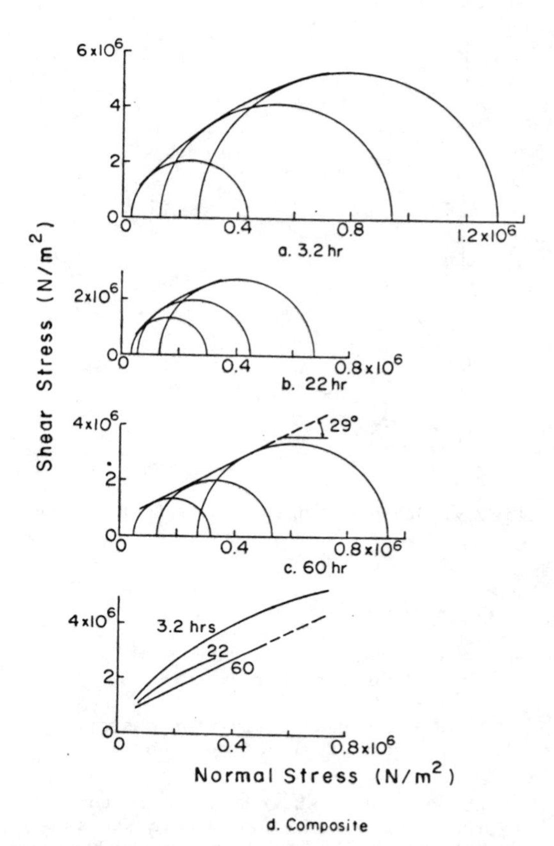

Fig. 8. Mohr's envelopes for creep failure.

Temperature Effect on Strain Rate and Strength

In general, lowering the temperature of frozen soil decreases the creep strain rate and increases the strength when other factors are constant. These effects are attributed not only to the reduction of unfrozen water by freezing but also to the gain in strength of the ice (Tsytovich (1975), Vialov et al. (1963)). Andersland and AlNouri (1970), using the concept of thermal activation, suggested that the rate of deformation of frozen soil could be represented by:

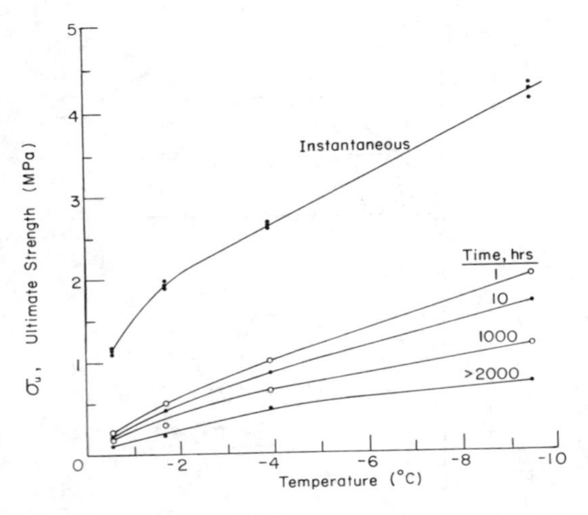

Fig. 9. Creep strength versus temperature.

$$\dot{\epsilon} = A \exp(-\frac{L}{T})$$

where $\dot{\epsilon}$ = creep strain rate
T = absolute temperature
$L = U/R$
U = apparent activation energy
R = universal gas constant.

Ladanyi (1972) showed that A in this expression can be replaced by a stress function. The proposed expression becomes

$$\dot{\epsilon} = \dot{\epsilon}_c \left(\frac{\sigma}{\sigma_{co}}\right)^n \exp\left[\frac{-L(273-T)}{273T}\right]$$

where σ = applied stress
σ_{co} = stress for $\dot{\epsilon}_c$ at 0°C (273°K)
$\dot{\epsilon}_c$ = arbitrary reference strain rate
n = creep parameter depending on temperature.

As noted earlier, the creep rate is a strong function of temperature in the expression based on the Rate Process Theory (Fish (1976)). Plots of strength vs. temperature for different periods of stress application (Figure 9) indicate that the strength of the ice within the soil has greater resistance at lower temperatures even for long periods of time. Figure 7 shows the same trend. However, if the hypothesis is accepted that the ultimate long-term strength is the strength of the unfrozen soil in a drained condition, then even at low temperatures, the ultimate strength is unchanged (Miller (1963), Sayles (1973), Andersland et al. (1978)). Of course, the time required to reach the ultimate long-

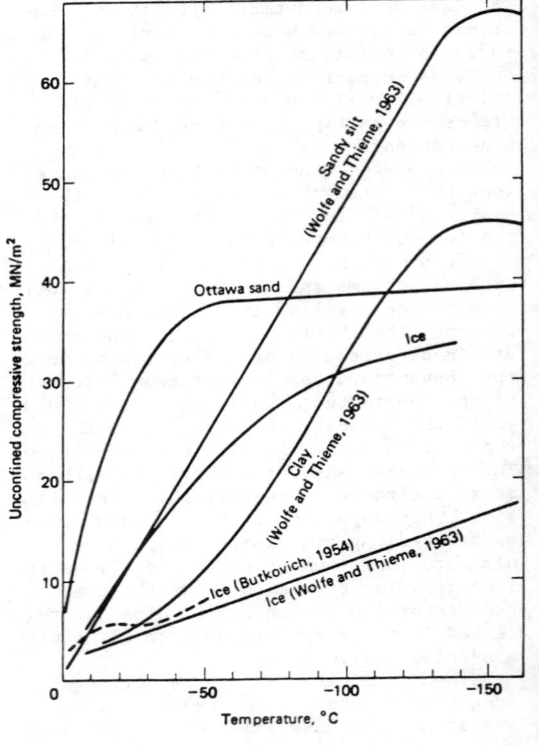

Fig. 10. Strength versus temperature.

term strength is greater at the lower temperatures. In designing of temporary artificially frozen structures, advantage is taken of the greater frozen soil strength at lower temperatures. But for permanently frozen structures, the ultimate long-term or unfrozen strength should be evaluated and used to resist the applied loads.

Constant rate of applied strain or stress tests have been performed for a few soils down to cryogenic temperatures (Wolfe and Thieme (1963), Sayles and Epaunchin (1966), Bourbonnais and Ladanyi (1985a,b). The results of some of these tests are shown in Figure 10. Differences in the shapes of the curves at the warmer temperatures may be due at least in part to the different rates of loading used by different investigators. At the lower temperatures, the failures are usually brittle and nearly independent of the rate of loading, although Bourbonnais and Ladanyi (1985a,b) reported that the ductile-brittle transition occurred for the frozen clays that they tested at temperatures as low as -110°C.

4. ICE COMPONENT OF FROZEN SOIL

It is well established that the presence of ice in soil distinguishes the mechanical properties of frozen soil from those of unfrozen soil (Tsytovich (1975), Andersland et al, (1978), Johnston et al. (1981), Phukan (1985)) when all other conditions except temperature are the same. Clearly, the amount of ice present, its crystal orientation, and its anisotropic nature all influence the strength and deformational characteristics of the frozen soil. The influence of the volume of ice in frozen soil was clearly established by the addition of sand in increasing proportion to ice (Goughnour and Andersland (1968). For low volumes of sand, the shear strength increases linearly with proportional increases in the percent of sand up to about 42%; then the uniaxial compressive strength increases rapidly with increasing sand volume. Hooke et al. (1972) conducted creep tests on ice with increasing concentrations of sand. They found that creep rates in ice with low sand concentrations were in some cases higher than clean ice. But at higher sand concentrations, the creep rate decreased exponentially with the increasing volume of sand.

Elsewhere in the literature the available strength data on frozen soils appears to be inconsistent with regard to the relationship between strength and ice content of frozen soils. Some investigators find the strength decreased with increasing ice content (Tsytovich and Sumgin (1937); Shusherina and Bobkov (1969), Kaplar (1971)) while others show an increase in strength with increasing ice content (Pekarskaya (1963)). Sayles and Carbee (1981) suggested that part of the inconsistencies may be due to the different sources of strength (i.e. the cohesive strength vs. the frictional strength of the soil). At high strain rates and low temperatures, ice cohesion can dominate the behavior, causing the initial yield that occurs at less than 1% strain to be larger than the frictional or residual strength that occurs at a much larger strain. These two types of strength can be identified by examining stress vs. strain curves.

Vialov et al. (1963) and Pekarskaya (1963) pointed out that soil texture influences the strength of frozen soil. Vialov et al. (1963) tested undisturbed soil and found the shear strength to be greater in the direction perpendicular to the ice inclusions than parallel to them. Baker and Johnston (1983) tested undisturbed frozen varved clay in uniaxial compression and found the strength parallel to the varves to be greater than that perpendicular to them. This finding was confirmed by in situ penetrometer and pressuremeter tests. Ohrai et al. (1983) tested segregated columnar ice taken from frozen soil and found the uniaxial strength perpendicular to the ice lens stronger than parallel to it. Clearly, frozen soil is an anisotropic material and this characteristic must be considered in the use of test results for design.

Saline Soils

Leonards and Andersland (1960) found that salt in the pore water of frozen soil had the expected effect of lowering the freezing point of the soil and the strengths were reduced to about the strengths for salt-free soil at higher temperatures. The salt, in effect, reduced the amount of ice in the soil and increased the amount of unfrozen water present. The salt concentrates in the unfrozen water, thus lowering the freezing point of the water even more. Sego et al. (1982) tested saline sand in uniaxial compression using salt concentrations from 0 to 100 ppt. He found for a given salt concentration the strength increased with increasing applied strain rate but at a much faster rate as the salt concentration increased. Ogata et al. (1983) confirmed the general observations of Sego et al. (1982) for both sand and a clay. Ogata et al. (1983) also found that creep tests on saline sand showed a marked increase in creep strain with increase in salt concentrations. They suggest that a relationship between unfrozen water film thickness be used to describe the decrease in compressive strength with increase in salt concentration.

Nixon and Parr (1984) conducted creep tests on a saline gravel and, in general, agreed with the findings of Ogata et al. (1983) for creep of sand. Chamberlain (1985) showed that frozen clay and sand soils with either fresh water or sea water in the pores are anisotropic by shearing test samples at different angles with the freezing plane.

He also noted that below -1°C the frozen clay was stronger than the sand with the same salinity. He suggests that the salt in the clay is concentrated in the normally larger volume of unfrozen water in the clay while salt reduces the ice content in the sand in a greater proportion, thus reducing the strength of the sand. The results of strength and creep tests on a saline clay by Nixon and Lem (1984) and Aas (1980) are in general agreement with the results of Ogata et al. (1983). Apparently much investigational work is still required to quantify the strength and creep deformation relationships to the salt concentration in saline soils.

5. CONSOLIDATION OF FROZEN SOIL

Deformations due to consolidations of frozen soil at low temperatures are usually quite small compared to deformations from creep. However, at temperatures near the freezing point, the deformation from consolidation can be as much as one third of the total settlement (Tsytovich (1975)). This deformation from the consolidation process includes: the compression of any air or vapor pockets within the pore ice; the redistribution of the initially unfrozen water to lower-stress areas; the redistribution of ice by pressure melting at the contact points at the soil grains (Vialov (1959), Brodskaia (1962), Tsytovich (1975)). Ladanyi (1975) and Kent et al. (1975) suggested that time settlement of frozen soil be separated between consolidation and creep. The data needed for using this method is lacking (Andersland et al. (1978)). Summaries of the available relative compressibility coefficients for various frozen soils for temperatures between 0 and -3.6°C are found in Tsytovich (1975) and Andersland et al. (1978). They show that finer-grain soils with clay constituents have larger coefficients than coarse soil such as sand. These results are consistent with the concept that unfrozen water has an important influence on the compressibility of frozen soil. (Tsytovich et al. (1980)) suggest that the strain due to compressibility can be represented by

$$\epsilon = A \frac{t}{B+t} (L+M\theta)\sigma^n$$

where A, B, L, M, and n are parameters determined experimentally; t is time; θ is the temperature below freezing; and σ is stress. Domaschuk et al. (1983) tested partially saturated silt in isotropic compression and found the same type of time function to be appropriate for describing their test results for a given stress and temperature.

6. INTERACTION OF FROZEN SOIL COMPONENTS

A qualitative description of the behavior of the soil grains, ice, and unfrozen water components of frozen soil has been suggested by Tsytovich (1975), Vialov et al. (1963), and others. However, by applying the principles of unfrozen soil mechanics, thermodynamics, and ice mechanics, some investigators are making progress toward understanding the detailed interactions between components as well as developing quantitative analyses of these interactions. Ladanyi (1981) suggests that the partitioning of the applied deviator stresses between the effective shear stress of the granular soil and the shear stress for the ice can be accomplished by using stress paths on Rendulic plots (1936). Utilizing the relative individual compressibilities of the soil skeleton, the pore matrix, and the soil grains (B factor by Bishop (1973)), Ladanyi (1985) demonstrated how the volumetric or hydrostatic stresses are carried by each component of the frozen soil mass. Youssef (1984) suggests that the portion of stress carried by the ice and stress carried by the soil components be evaluated by triaxial testing of both frozen and unfrozen soil at the same void ratio and comparing the resulting stress vs. strain curves.

An explanation of creep and strength of frozen soils has been related with some success to the Rate Process Theory (Fish (1976, 1980), Fish and Assur (1984), Martin et al. (1981), Ting (1981), Orth (1985, 1988)). However, this approach does not account for the effect of internal friction within the soil and ice components. For low stress levels acting for extended periods of time on granular soil, the friction becomes dominant (Miller (1963), Sayles (1973)).

It has been established that the orientation of the ice inclusions with respect to loading influence the strength of the frozen soil (Vialov et al. (1963), Radd, et al. (1978), Chamberlain and Gow (1978)). The importance of using undisturbed frozen samples that reveal the orientation of the ice inclusions is emphasized by Savigny and Morgenstern (1986) where they show that failure surfaces in the soil tend to favor the directions along the interfaces between the soil and the ice inclusions.

In considering the ice component in a frozen soil with a massive structure (i.e. soil without segregated ice), it has not been shown conclusively that the polycrystalline ice is randomly oriented in all three directions. Gow (1975) has shown that the ice crystals taken from ice lenses in a silt were columnar and C axes were randomly oriented in a plane parallel to the surface of the lens. However, in soils with a massive structure the ice could be columnar, following a tortuous path through the soil pores in the direction of heat flow, or it could be small-grained with the c-axis randomly oriented in three directions. Clearly, the two different ice structures would influence the soil behavior differently.

7. IN SITU TESTING OF FROZEN SOIL

Although the testing of frozen undisturbed soils in the laboratory permits a better evaluation of the soil properties than testing remolded soils that are frozen in the laboratory, in situ testing would permit an even better assessment of the properties. Although penetration tests and pressuremeter tests are in common use in unfrozen soil and rock mechanics, they have been used only to a limited extent in frozen soil. Pressuremeter tests have been applied to ice-rich permafrost for determining creep parameters with reasonable success for a number of years (Ladanyi and Johnston (1973), Fish (1976), Ladanyi and Huneault (1987)). This type of test seems to be directly applicable to artificially frozen ground in many construction projects where the lateral stresses are restrained by soil containing predominantly vertical ice lenses. The pressuremeter is capable of testing the frozen soil in the same direction as the anticipated loading.

Penetration tests have been used to delineate the extent of subsea permafrost and to determine some of the engineering properties of the soil. (Sellmann and Chamberlain (1979), Blouin et al. (1979)). Ladanyi (1976) performed both constant load and constant rate penetration tests in ice-rich soils. He found that, at the same equivalent applied rate of strain, the short-term cohesion determined by the penetrometer tests was similar to that determined by the pressuremeter tests. But the strength increase with strain rate in pressuremeter creep tests was much higher than that observed in the static penetration tests. This difference is thought to be due to the anisotropy of the soil.

At the present time, the published literature does not appear to contain a design and analysis of an artificial frozen structure based on the mechanical properties of frozen soil where the performance of the structure is compared with the designer's analysis. This type of in situ investigation would make an important contribution to the field of ground freezing.

8. CONCLUSIONS AND RECOMMENDATIONS

During the past decade, important advances have been made toward the better understanding of the strength and deformational behavior of frozen soils. The work in these advances can be grouped into three categories: (1) testing to increase the quantity and quality of data available, (2) modelling the frozen soil behavior through empirical expressions, and (3) applying the principles of physics to individual components of this composite material and to the interaction among the components. Although all three categories fulfill an important function, the last category holds the greatest potential for future advances.

Areas in which this review has shown a need for further investigational effort include:

1. Development of a mathematical or physical model that reliably predicts the deformational behavior of frozen soil at stress levels down to and below its ultimate long-term strength. Such a model would account for the interaction of the soil particles with each other in the resistance of the frozen soil to deformation and failure.
2. Relate the strength and deformational behavior to physical properties of the frozen soil that can be determined by the practicing engineer. These physical properties could include ice content, dry unit weight, unfrozen water content, soil composition, salinity, ice, and soil anisotropy.
3. Evaluation of the structural performance of different types of frozen ground structures from the start of construction to the end of their operational lives. Such an evaluation would include (a) a detailed description of the soils, water, and drainage condition of the unfrozen soils, (b) testing of undisturbed soil samples taken from the frozen soil structure, and (c) the periodic measurement of deformation and temperatures. A comparison between the results of the evaluation and the structural design would reveal the areas that require further research.
4. Improvement in both field and laboratory testing methods to reduce the time for determining long-term creep and strength parameters for frozen soil.
5. Further studies of the strength and deformational behavior of frozen soil in relation to the salinity, anisotropy, ice crystal orientation, and stress, deformation, and temperature history.

REFERENCES*

Aas, G. 1980. Laboratory determination of strength properties of frozen salt marine clay. Proceedings of 2nd International Symposium on Ground Freezing, Trondheim, Norway: 144-156.#

Akagawa, S. 1980. Poisson's ratio of sandy frozen soil, under long-term stress by creep tests. Proceedings of 2nd International Symposium on Ground Freezing, Trondheim, Norway: 235-268.

Alkire, B.D. and Andersland, O.B. 1973. The effect of confining pressure on the mechanical properties of sand-ice materials. Journal of Glaciology 12(66):469-481.

Andersland, O.B. 1972. Discussion on mechanical behaviour of frozen earth materials under high pressure triaxial test conditions. Geotechnique 22:136-137.

Andersland, O.B. and Akili, W. 1967. Stress effect on creep rates of a frozen clay soil. Geotechnique 17(1):27-39.

Andersland, O.B. and AlNouri, I. 1970. Time-dependent strength behavior of frozen soils. Journal of the Soil Mechanics and Foundations Division, ASCE 96(SM4):1249-1265.

Andersland, O.B, Sayles, Jr., F.H. and Ladanyi, B. 1978. Mechanical properties of frozen ground. In Geotechnical Engineering for Cold Regions (O.B. Andersland and D.M. Anderson, Eds.). McGraw-Hill, New York: 216-275.

Anderson, D.M., Pusch, R. and Penner, E. 1978. Physical and thermal properties of frozen ground. In Geotechnical Engineering for Cold Regions (O.B. Andersland and D.M. Anderson, Eds.). McGraw-Hill, New York: 37-163.

Andrade, E.M. da C. 1910. The viscous flow in metals and allied phenomena. Proceedings of Royal Society, London, Ser. A 84:1-12.

Ashby, M.F. and Duval, P. 1985. The creep of polycrystalline ice. Cold Regions Science and Technology 11:285-300.

Assur, A. 1963. Discussions. Proceedings of 1st International Conference on Permafrost, Purdue University, Lafayette, Indiana: 339-340.

Assur, A. 1979. Some promising trends in ice mechanics. Physics and Mechanics Ice Symposium, International Union of Theoretical and Applied Mechanics, Copenhagen, Denmark.

Baker, T.H.W. and Johnston, G.H. 1983. Unconfined compression tests on anisotropic frozen soils from Thompson, Manitoba, Canada. Proceedings of 4th International Conference on Permafrost, Fairbanks, Alaska: 40-43.

*See page X concerning ISGF papers

Berggren, A-L. and Furuberg, T. 1985. A new Norwegian creep model and creep equipment. Proceedings of 4th International Symposium on Ground Freezing, Sapporo, Japan, 1:181-185.

Bishop, A.W. 1973. The influence of an undrained change in stress on the pore pressure in porous media of low compressibility. Technical Note, Geotechnique 23:435-442.

Blouin, S.E., Chamberlain, E.J., Sellmann, P.V. and Garfield, D.E. 1979. Determining subsea permafrost characteristics with a cone penetrometer - Prudhoe Bay, Alaska. Cold Regions Science and Technology 1(1): 16.

Bourbonnais, J. and Ladanyi,B 1985a. The mechanical behaviour of frozen sand down to cryogenic temperatures. Proceedings of 4th International Symposium on Ground Freezing, Sapporo, Japan (1):235-244.

Bourbonnais, J. and Ladanyi, B. 1985b. The mechanical behaviour of a frozen clay down to cryogenic temperatures. In Proceedings of 4th International Symposium on Ground Freezing, Sapporo, Japan (2):237-244.

Bragg, R.A. and Andersland, O.B. 1980. Strain rate, temperature and sample size effects on compression and tensile properties of frozen sand. Proceedings of 2nd International Symposium on Ground Freezing, Trondheim, Norway, 1:34-47.#

Brodskaia, A.G. 1962. Compressibility of frozen ground. Izd-Vo Acad. Nauk SSSR, US Army Cold Regions Research and Engineering Laboratory, Translation 28.

Chamberlain, E.J. and Gow, A.J. 1978. Effect of freezing and thawing on the permeability and structure of soils. Proceedings of 1st International Symposium on Ground Freezing, Bochum, Germany: 31-44.

Chamberlain, E.J. 1985. Shear strength anisotropy in frozen saline and freshwater soils. Proceedings of 4th International Symposium on Ground Freezing, Sapporo, Japan: 2: 189-194

Chamberlain, E.J., Groves, E. and Perham, R. 1972. The mechanical behaviour of frozen earth materials under high pressure triaxial test conditions. Geotechnique 22 (3):469-483.

Chen, X. 1988. Mechanical characteristics in artificially frozen clay layers at depths under triaxial stress conditions. Preprint, 5th International Symposium on Ground Freezing, Nottingham, England.

Domaschuk, L., Man, C.S., Shields, D.H. and Yong, E. 1983. Creep behaviour of frozen saline silt under isotropic compression. Proceedings of 4th International Conference on Permafrost, Fairbanks, Alaska: 238-243.

Eckardt, H. 1979. Creep behavior of frozen soils in uniaxial compressing tests. Engineering Geology 13:185-195.

Eckardt, H. 1982. Creep tests with frozen soils under uniaxial tension and uniaxial compression. Proceedings of the 4th Canadian Permafrost Conference, March 1981, Calgary, Alberta, Canada, (Roger J.E. Brown Memorial Volume) 394-405.

Finborud, I. and Berggren, A-L. 1980. Deformation properties of frozen soils. Proceedings of 2nd International Symposium on Ground Freezing, Trondheim, Norway: 262-271,

Fish, A.M. 1976. An acoustic and pressure meter method for investigation of the rheological properties of ice. Candidate of Science thesis, Arctic and Antarctic Research Institute, Lenigrad, USSR Translated from Russian, 1978. US Army Cold Regions Research and Engineering Laboratory, Internal Report 846.

Fish, A.M. 1980. Kinetic nature of the long-term strength of frozen soils. Proceedings of 2nd International Symposium on Ground Freezing, Trondheim, Norway: 95-108.

Fish, A.M. 1982a. Comparative analysis of the U.S.S.R. construction codes and the U.S. Army Technical Manual for design of foundations on permafrost. US Army Cold Regions Research and Engineering Laboratory, CRREL Report 82-14.

Fish, A.M. 1982b. Deformation and failure of frozen soils and ice at constant and steadily increasing stresses. Proceedings of the 4th Canadian Permafrost Conference, March 1981, Calgary, Alberta, Canada (Roger J.E. Brown Memorial Volume; H.M. French, ed.). Ottawa: National Research Council of Canada: 419-428.

Fish, A.M. 1983a. Comparison of U.S.S.R. codes and U.S. Army manual for design of foundations on permafrost. Cold Regions Science and Technology 8(1):3-24.

Fish, A.M. 1983b. Thermodynamic model of creep at constant stresses and constant strain rates. US Army Cold Regions Research and Engineering Laboratory, CRREL Report 83-33.

Fish, A.M. 1985. Creep strength, strain rate, temperature and unfrozen water relationship in frozen soil. Proceedings of 4th International Symposium on Ground Freezing, Sapporo, Japan: 2:29-36.

Fish, A.M. and Assur, A. 1984. Discussion paper, Tertiary creep model for frozen sand, by J.M. Ting. ASCE Journal of Geotechnical Engineering 110(9):1373-1378.

Frenkel, J. 1946. Kinetic theory of liquids. Oxford University Press, London, England.

Gardner, A.R., Jones, R.H. and Harris, J.S. 1984. A new creep equation for frozen soils and ice. Cold Regions Science and Technology 9:271-275.

Glasstone, S., Laidler, K.J. and Eyring, H. 1941. The Theory of Rate Processes. Mc-Graw-Hill, New York.

Gonze, P., Louisberg, E. and Thimus, J.FR. 1985. Determination of rheological parameters of frozen soils by laboratory tests. Proceedings of 4th International Symposium on Ground Freezing, Sapporo, Japan: 2:195-200.

Goto, S., Minegishi, K. and Tanaka, M. 1988. Direct shear test at a frozen/unfrozen interface. Preprint, 5th International Symposium on Ground Freezing, Nottingham, England.

Goughnour, P.P. and Andersland, O.B. 1968. Mechanical properties of a sand-ice system. Journal of the Soil Mechanics and Foundations Division, ASCE 94(SM4) (Proc. Paper 6030):923-950.

Gow, A.J. 1975. Application of thin section techniques to studies of internal structure of frozen silts. US Army Cold Regions Research and Engineering Laboratory, Technical Note.

Hampton, C.N., Jones, R.H. and Gardner, A.R. 1985. Modelling the creep behaviour of frozen sands. Proceedings of 3rd National Symposium on Ground Freezing, University of Nottingham, England (R.H. Jones, ed.): 27-33.

Hawkes, I. and Mellor, M. 1972. Deformation and fracture of ice under uniaxial stress. Journal of Glaciology 11(61)-:103-131.

Haynes, F.D. and Karalius, J.A. 1977. Effect of temperature on the strength of frozen silt. US Army Cold Regions Research and Engineering Laboratory, CRREL Report 77-3.

Haynes, F.D., Karalius, J.A. and Kalafut, J. 1975. Strain rate effect on the strength of frozen silt. US Army Cold Regions Research and Engineering Laboratory, Technical Report 350, 27 pp.

Hooke, R.L., Dahlin, B.B. and Kauper, M.T. 1972. Creep of ice containing dispersed fine sand. Journal of Glaciology 11(63)-:327-336.

Hult, J.A.H. 1966. Creep in Engineering Structures: Blaisdell Publ. Co., Waltham, Mass.

Jessberger, H.L. 1977. Strength and time dependent deformation of artificially frozen soils. Proceedings of International Symposium on Frost Action in Soils, Lulea 1:157-167.

Jessberger, H.L. 1980. State-of-the-art

report. Ground freezing: Mechanical properties, processes and design. Proceedings of 2nd International Symposium on Ground Freezing, Trondheim, Norway: 1-33. #

Johnston, G.H., Ladanyi, B. Morgenstern, N.R. and Penner, E. 1981. Engineering characteristics of frozen and thawing soils. Permafrost - Engineering Design and Construction (G.H. Johnston, ed.). John Wiley and Sons, Toronto: 73-147.

Jones, S.J. and Parameswaran, V.R. 1983. Deformation behaviour of frozen sand-ice material under triaxial compression. Proceedings of 4th International Conference on Permafrost, Fairbanks, Alaska: 560-565.

Kaplar, C.W. 1971. Some strength properties of frozen soil and effect of loading rate. US Army Cold Regions Research and Engineering Laboratory, Technical Report 159.

Kent, D.D., Fredlund, D.G. and Watt, W.G. 1975. Variables controlling behavior of a partly frozen saturated soil. Presented at Conference on Soil Water Problems in Cold Regions, Calgary, Alberta, Canada May 1975.

Klein, J. 1979. The application of finite elements to creep problems in ground freezing. Proceedings of the 3rd International Conference on Numerical Methods in Geomechanics, 1:493-502.

Kuribayashi, E., Kawamura, M. and Yui, Y. 1985. Stress-strain characteristics of artificially frozen sand in uniaxially compressive tests. Proceedings of 4th International Symposium on Ground Freezing, Sapporo, Japan 2:177-182.

Ladanyi, B. 1972. An engineering theory of creep of frozen soils. Canadian Geotechnical Journal 9(1):63-80.

Ladanyi, B. 1974. Bearing capacity of frozen soils. 27th Canadian Geotechnical Conference, Edmonton, Alberta, Canada.

Ladanyi, B. 1975. Bearing capacity of strip footings in frozen soils. Canadian Geotechnical Journal 12:393-407.

Ladanyi, B. 1976. Use of the static penetration test in frozen soils. Canadian Geotechnical Journal 13:95-110.

Ladanyi, B. 1981. Shear-induced stresses in the pore ice in frozen particular materials. Proceedings of Symposium on Free Boundary Problems, Montecatini (A. Fasano and M. Primicerio, eds.), Research Notes in Mathematics, No. 78, Pitman Books Ltd. II:549-560.

Ladanyi, B. 1985. Stress transfer mechanism in frozen soils. Proceedings of 10th Canadian Congress of Applied Mechanics, London, Ontario, Canada 1: 11-23.

Ladanyi, B. and Arteau, J. 1978. Effect of specimen shape on creep response of a frozen sand. Proceedings of 1st International Symposium on Ground Freezing, Bochum, Germany 1:141-153.

Ladanyi, B. and Eckardt, H. 1983. Dilatometer testing in thick cylinders of frozen sand. Proceedings of 4th International Conference on Permafrost, Fairbanks, Alaska:677-682.

Ladanyi, B. and Huneault, P. 1987. Use of the borehole dilatometer stress-relaxation test for determining the creep properties of ice. Proceedings of 6th International Offshore Mechanics and Arctic Engineering Symposium, Houston, Texas IV:253-259.

Ladanyi, B. and Johnston, G.H. 1973. Evaluation of in situ creep properties of frozen soils with the pressuremeter. Proceedings of 2nd International Conference on Permafrost, Yakutsk, USSR: 310-318.

Ladanyi, B. and Sayles, F.H. 1979. General Report Session II: Mechanical Properties. Proceeding of First International Symposium on Ground Freezing, Bochum, Germany: 7-18.

Lade, P.V., Jessberger, H.L. and Diekmann, N. 1980. Stress-strain and volumetric behaviour of frozen soils. Proceedings of 2nd International Symposium on Ground Freezing, Trondheim, Norway, 48-64.

Leonards, G.A. and Andersland, O.B. 1960. The clay-water system and the shearing resistance of clays. American Society of Civil Engineers Research Conference on Shear Strength of Cohesive Soils, Boulder, Colorado: 743-818.

Maksimjak, R.V., Vialov, S.S. and Chapaev, A.A. 1983. Method for determining the long-term strength of frozen soils. Proceedings of 4th International Conference on Permafrost, Fairbanks, Alaska: 783-786.

Martin, R.T., Ting, J.M. and Ladd, C.C. 1981. Creep of frozen sand, Final Report, Part 1. Massachusetts Institute of Technology to U.S. Army Research Office. US Army Cold Regions Research and Engineering Laboratory, Internal Report IR 627.

Miller, R.D. 1963. Phase equilibria and soil freezing. Proceedings of 1st International Conference on Permafrost, Lafayette, Indiana: 193-197.

Mitchell, J.K., Campanella, R.G. and Singh, A. 1968. Soil creep as a rate process. Journal of the Soil Mechanics and Foundations Division, Proc. Paper 5751, ASCE 94(SM4): 231-253.

Nixon, J.F. and Lem, G. 1984. Creep and strength testing of frozen saline fine-

grained soils. Canadian Geotechnical Journal (21) 518-529.
Nixon, M.S. and Parr, G.M. 1984. The effects of temperature, stress and salinity on the creep of frozen saline soil. Journal of Energy Resources Technology 106:344-348.
Neuber, H. and Wolters, R. 1977. Mechanical behaviour of frozen soils under triaxial compression. National Research Council of Canada, Division of Building Research, Technical Translation, NRC/CNR TT-1902.
O'Connor, M.J. and Mitchell, R.J. 1978. Measuring total volumetric strains during triaxial tests on frozen soils: a new approach. Canadian Geotechnical Journal 15:47-53.
Ogata, N., Yasuda, M. and Kataoka, T. 1983. Effect of salt concentration on strength and creep behavior of artificially frozen soils. Cold Regions Science and Technology 8:139-153.
Ohrai, T., Takashi, T., Yamamoto, H. and J.Okamoto 1983. Uniaxial compressive strength of ice segregated from soil. Proceedings of 4th International Conference on Permafrost, Fairbanks, Alaska: 945-950.
Orth, W. 1985. Deformation of frozen sand and its physical interpretation. Proceedings of 4th International Symposium on Ground Freezing, Sapporo, Japan 1:245-253.
Orth, W. 1988. A creep formula for practical application based on crystal mechanics. Preprint, 5th International Symposium on Ground Freezing, Nottingham, England.
Ouvry, J.F. 1985. Results of triaxial compression tests and triaxial creep tests on an artificially frozen stiff clay. Proceedings of 4th International Symposium on Ground Freezing, Sapporo, Japan: 2: 207-212.
Parameswaran, V.R. 1980. Deformation behavior and strength of frozen sand. Canadian Geotechnical Journal 17:74-88.
Parameswaran, V.R. and Jones, S.J. 1981. Triaxial testing of frozen sand, Journal of Glaciology 27(95):147-155.
Parameswaran, V.R. 1987. Extended failure time in the creep of frozen soils. Mechanics of Materials 6:233-243.
Perkins, T.K. and Ruedrich, R.A. 1974. A study of factors influencing the mechanical properties of deep permafrost. Journal of Petroleum Technology 26:1167-1177.
Pekarskaya, N.K. 1963. Shear strength of frozen grounds and its dependence on texture. Izd-Vo Akad. Nauk, SSSR, Moscow. US Army Cold Regions Research and Engineering Laboratory, Draft Translation 115:60-77.
Phukan, A. 1985. Frozen ground engineering. Prentice-Hall, Englewood Cliffs, N.J.
Phukan, A. and Andersland, O.B. 1978. Foundations for cold regions. Geotechnical Engineering for Cold Regions (O.B. Andersland and D.M. Anderson, eds.) McGraw-Hill, New York: 276-362.
Pusch, R. 1980. Creep of frozen soil, a preliminary physical interpretation. Proceedings of 2nd International Symposium on Ground Freezing, Trondheim, Norway:190-201.
Radd, F.J. and Wolfe, L.H. 1978. Ice lens structures, compression strengths and creep behaviour of some synthetic frozen silty soils. Proceedings of 1st International Symposium on Ground Freezing, Bochum, Germany: 115-130.
Rein, R.G. 1985. Correspondence of creep data and constant strain-rate data for frozen silt. Cold Regions Science and Technology 11:187-194.
Rein, R.G. and Hathi, V.V. 1978. The effect of stress on strain at the onset of tertiary creep of frozen soils. Canadian Geotechnical Journal 15:424-426.
Rein, R.G., Hathi, V.V. and Sliepcevich, C.M. 1975. Creep of sand-ice system. Journal of Geotechnical Engineering Division, ASCE 101(GT2): 115-128.
Rendulic, I.L. 1936. Relation between void ratio and effective principal stresses for a remoulded, silty clay. 1st International Conference on Soil Mechanics, Harvard University, Cambridge, Mass. III:48-51.
Sanger, F.J. and Kaplar, C.W. 1963. Plastic deformation of frozen soils. Proceedings of 1st International Conference on Permafrost, Purdue, Ind: 305-315.
Savigny, K.W. and Morgenstern, N.R. 1986. Creep behaviour of undisturbed clay permafrost. Canadian Geotechnical Journal 23:515-527.
Sayles, F.H. 1968. Creep of frozen sands. US Army Cold Regions Research and Engineering Laboratory, Technical Report 190.
Sayles, F.H. 1973. Triaxial constant strain rate tests and triaxial creep tests on frozen Ottawa sand. Proceedings of 2nd International Permafrost Conference, Yakutsk, USSR: 384-391.
Sayles, F.H. and Carbee, D.L. 1981. Strength of frozen silt as a function of ice content and dry unit weight. Engineering Geology 18:55-66.
Sayles, F.H. and Epauchin, N.V. 1966. Rate

of strain compression tests on frozen Ottawa sand and ice. US Army Cold Regions Research and Engineering Laboratory, CRREL Technical Note.

Sayles, F.H. and Haines, D. 1974. Creep of frozen silt and clay. US Army Cold Regions Research and Engineering Laboratory, Technical Report 252.

Sego, D., Schultz, T. and Banasch, R. 1982. Strength and deformation behaviour of frozen saline sand. Proceedings of 3rd International Symposium on Ground Freezing, Hanover, N.H.: 11-18.

Sellmann, P.V. and Chamberlain, E.J. 1979. Subsea permafrost, drill core analysis, penetration tests, and permafrost depth. Proceedings of 11th Offshore Technology Conference, Houston: 1481-1493.

Shibata, T. Adachi, T., Yashima, A., Takahashi, T. and Yoshioka, I. 1985. Time-dependent and volumetric change characteristic of frozen sand under triaxial stress conditions. Proceedings of 4th International Symposium on Ground Freezing, Sapporo, Japan: 173-179.

Shusherina, E.P. and Bobkov, Yu.P. 1969. Effect of moisture content on frozen ground strength, Merzlotnye Issledovaniya 9:122-127. National Research Council of Canada Technical Translation, 1918:8-19.

Smith, L.L. and Cheatham, J.B. 1975. Plasticity of ice and sand-ice systems. Transactions of ASME Journal of Engineering for Industry: 479-484.

Takagawa, K., Nakazawa, A., Ryyokai, K. and Akagawa, S. 1979. Creep characteristic of frozen soils. Proceedings of 1st International Symposium on Ground Freezing, Bochum, West Germany 1:133-139.

Thompson, E.G. and Sayles, F.H. 1972. In situ creep analysis of room in frozen soil. Proceedings, American Society of Civil Engineers 98(SM9): 899-915.

Ting, J.M. 1981. The creep of frozen sands: Qualitative and quantitative models. Final Report, Part II. Massachusetts Institute of Technology to US Army Research Office, US Army Cold Regions Research and Engineering Laboratory.

Ting, J.M. 1983. Tertiary creep model for frozen sands. ASCE Journal of Geotechnical Engineering 109(7):932-945.

Ting, J.M. and Martin, R.T. 1979. Application of the Andrade equation to creep data for ice and frozen soil. Cold Regions Science and Technology 1:29-36.

Tsytovich, N.A. and Sumgin, M.I. 1937. Principles of Mechanics of Frozen Ground, Izdatel'stvo Akad. Nauk SSSR, SIPRE Translation 19:106-107.

Tsytovich, N.A. 1975. Mechanics of Frozen Soils. McGraw-Hill, New York.

Tsytovich, N.A., Kronik, Ya.A., Gavrilov, A.N. and Vorobyov, E.A. 1980. Mechanical properties of frozen coarse-grained soils. Proceedings of 2nd International Symposium on Ground Freezing, Trondheim, Norway: 65-74.

Vialov, S.S. and Tsytovich, N.A. 1955. Creep and long-term strength of frozen soils. Dok. Akad. Nauk 104:850-853

Vialov, S.S. 1959. Rheological properties and bearing capacity of frozen soils. USA Cold Regions Research and Engineering Laboratory, Translation 74.

Vialov, S.S. 1973. Long-term rupture of frozen soil as a thermally activated process. Proceedings of 2nd International Conference on Permafrost, Yakutsk, USSR:222-228.

Vialov, S.S. 1978. Kinetic theory of deformation of frozen soils. Proceedings of 3rd International Conference on Permafrost, Edmonton, Canada 1:751-755.

Vialov, S.S. and Shusherina, Y.P. 1964. Resistance of frozen soils to triaxial compression. USA Foreign Science and Technology Center, Technical Translation FSTC-HT-23-750-70.

Vialov, S.S., Gmoshinskii, V.G., Gorodetskii, S.E., Grigorieva, V.G., Zaretskii, Yu.K., Pekarskaia, N.K. and Shusherina, E.P. 1963. The strength and creep of frozen soils and calculations for ice-soil retaining structures. (Vialov, S.S., ed). US Army Cold Regions Research and Engineering Laboratory, Translation 76.

Vialov, S.S., Gorodetskii, S.E., Ermakov, V.F., Zatsarnaya, A.G. and Pekarskaya, N.K. 1966. Methods of determining creep, long-term strength and compressibility characteristics of frozen soils. National Research Council of Canada, Technical Translation 1364.

Vialov, S.S., Dokuchayev, V.V. and Sheynkman, D.R. 1976. Ground ice and ice-rich ground as structure foundations. US Army Cold Regions Research and Engineering Laboratory, Draft Translation 737.

Vialov, S.S., Slepak, M.E., Maximyak, R.V. and Chapayev, A.A. 1988. Frozen soil deformation and failure under different loading. Preprint, 5th International Symposium on Ground Freezing, Nottingham, England.

Vrachev, V.V., Ivaschenko, I.N. and Shusherina, E.P. 1984. A new technique for determining the static fatigue limit of frozen ground, Final Proceedings of 4th International Conference on Perma-

frost, Fairbanks, Alaska: 306-310.
Wolfe, L.H. and Thieme, J.O. 1963. Physical and thermal properties of frozen soil and ice. Society of Petroleum Engineers Journal 4(1):67-72.
Youssef, H. 1984. Indirect determination of intergranular stresses in frozen soils. PhD thesis, Ecole Polytechnique, University of Montreal, Canada.
Zaretskiy, Y.K. 1972. Rheological properties of plastic frozen soils and determination of settlement of a test plate with time. Soil Mechanics and Foundation Engineering, Moscow. Translation: Consultants Bureau, New York: 81-85.
Zaretskiy, Y.K. and Gorodetskii, S.E. 1975. Dilatancy of frozen soil and development of a strain theory of creep. Translated from Gidrotekhnicheskoe Stroitel'stvo 2:15-18.
Zaretskiy, Y.K. and Vialov, S.S. 1971. Structural mechanics of clayey soils. Soil Mechanics and Foundation Engineering, Moscow. 8(3):153-160.
Zhu, Y. and Carbee, D.L. 1983. Creep behavior of frozen silt under constant unixial stress. Proceedings of 4th International Conference on Permafrost, Fairbanks, Alaska: 1507-1512.
Zhu, Y. and Carbee, D.L. 1984 Uniaxial compressive strength of frozen silt under constant deformation rates. Cold Regions Science and Technology 9:3-15.
Zhu, Y. and Carbee, D.L. 1985. Strain rate effect on the tensile strength of frozen silt. Proceedings of 4th International Symposium on Ground Freezing, Sapporo, Japan: 153-157.
Zhu, Y. and Carbee, D.L. 1987. Tensile strength of frozen silt. US Army Cold Regions Research and Engineering Laboratory, CRREL Report 87-15.
Zhu, Y., Zhang, J. and Shen, Z. 1988. Uniaxial compressive strength of frozen medium sand under constant deformation rates. Preprint, 5th International Symposium on Ground Freezing, Nottingham, England.

Acknowledgments

The author wishes to express his appreciation to Andrew Assur, William F. Quinn, and Anatoly M. Fish for their fruitful discussions and valuable comments; and a special thanks to Donna Harp and Maria Bergstad for typing and editing.

5th International Symposium on Ground Freezing, Jones & Holden (eds)
© 1988 Balkema, Rotterdam. ISBN 90 6191 824 3

Cyclic thermal strain and crack formation in frozen soils

Orlando B. Andersland & Hassan M. Al-Moussawi
Michigan State University, Mich., USA

ABSTRACT: On cooling, frozen soil will contract if it is not constrained. Tensile stresses are generated with no observable strains. In tension, the soil will creep along with some stress relaxation. The creep deformation includes elastic, delayed elastic, and viscous flow (permanent deformation). The same sample on warming will expand, but the permanent deformation will remain. Subsequent cycles of cooling and warming allow the permanent deformation to accumulate followed by eventual rupture in tension. To study this behavior, a series of constant strain/stress relaxation tests were conducted on duplicate frozen sand samples. Other tests included measurement of the linear thermal contraction/expansion coefficients and the rate of stress increase with rate of temperature decrease on duplicate frozen sand specimens. The data suggest that tensile failure will occur when the permanent strain accumulates to about 1 percent in the frozen sand. This behavior of frozen soil in tension has implications relative to tensile strength of frozen soil in the design of complex earth structures and in the control of thermally induced cracking problems in surface soils of the cold regions.

1 INTRODUCTION

Frozen soils subjected to a decrease in temperature will contract and, if the soil is restrained, tensile stresses will develop. For a rapid temperature decrease of only a few degrees and little stress relaxation, the tensile stresses are capable of cracking weak, partially saturated frozen soils. For frozen soils in tension, cyclic temperature changes can cause an accumulation of viscous creep strain since expansion during each warming phase of the cycle will not reverse the permanent strain. Rupture and crack formation are possible when the accumulated creep strain in tension totals about one percent. This phenomenon has application to frozen surface soils in the cold regions of the world and may be applicable to certain controlled ground freezing situations.

Duplication in the laboratory of field conditions responsible for thermally induced cyclic loading requires facilities and equipment not available to most researchers. To simplify experimental work on the effect of cumulative strain on the soil behavior, a series of constant strain/stress relaxation tests were run on duplicate frozen sand samples maintained at a constant temperature. At the start of each load cycle, a preselected permanent axial deformation was imposed on the frozen soil to simulate cooling in the field. Next, stress relaxation for each cycle was observed as a function of time. At the end of the appropriate cycle, axial strain was increased to provide information on the decrease in strength and the increase in failure strain. Other tests included measurement of the linear thermal contraction/expansion coefficients for the frozen sand at several dry densities and temperatures. Limited data were also obtained on stress increase as a function of soil cooling rate. The experimental results suggest that accumulated tensile strain due to thermally induced cyclic loading is a significant factor relative to frozen soil behavior in tension.

2 THERMAL CONTRACTION OF FROZEN SAND

Thermal contraction is that change in length, or volume, of a material resulting from a decrease in temperature. The ratio of change in length ΔL to the original length L_o of the material times the change in temperature ΔT defines an average coefficient of linear thermal contraction α, thus

$$\alpha = \frac{\Delta L}{\Delta T\, L_o} = \frac{L_2 - L_1}{(T_2 - T_1)\, L_o} \quad (1)$$

Cohesion within a given material and adhesion at the interface between two materials provide the forces resisting separation during contraction. Saturated frozen sand consists of an assemblage of particles bonded together with an ice matrix. Thermal contraction involves both materials, quartz particles and the ice matrix, each with its own coefficient of thermal contraction. Experimental measurements of α will provide the composite contraction behavior with differences between specimens due to the ice and/or quartz volume fractions in the frozen specimens.

Linear coefficients of thermal contraction and expansion were determined on rectangular frozen sand beams using a 254 mm (10-inch) Whittmore gauge (a mechanical strain gauge). The sand consisted of subangular quartz particles with a uniform gradation (particle size range from 74 µm to 595 µm) and a uniformity coefficient d_{60}/d_{10} = 1.51. Steel gauge plugs, embedded in the top surface at each end of the frozen sand specimens, permitted multiple readings to be taken as the specimen was slowly (24-hour period) cooled from about -5 °C down to -25 °C and then warmed back up to -5 °C. Stress-free conditions (no base friction) were approximated along the length of the frozen beams during cooling and warming by using a series of 9 mm diameter wooden dowels to support the beams. Sublimation of the ice was minimized by covering the frozen specimens with a thin plastic covering during the cooling and warming periods.

Experimental measurements on the frozen sand specimens, summarized in Fig. 1, show that the coefficient of thermal contraction ranged from about $18 \times 10^{-6}\ ^{o}C^{-1}$ to $30 \times 10^{-6}\ ^{o}C^{-1}$ for the temperatures (-10 to -20°C) and dry densities (1.96 to 2.26 Mg/m^3) included in the study. When solid fractions are considered, these α values are in agreement with those reported for ice, about $50 \times 10^{-6}\ ^{o}C^{-1}$ at -20 °C (Butkovich,

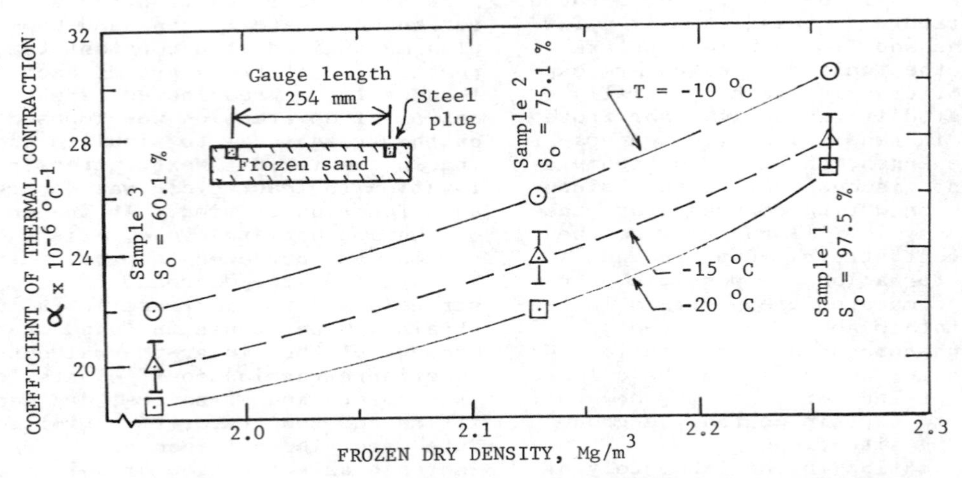

Fig. 1 Thermal contraction/expansion behavior of frozen quartz sand (data from Al-Moussawi, 1988).

1957), and about 20x10^{-6} ^{o}C for quartz at -25 ^{o}C (Butkovich, 1957). Some data scatter, indicated on the -15 ^{o}C curve, was due to limitations of the Whittmore gauge technique in measurement of dimensional changes. The change in α with dry density is related to the differences in degree of ice saturation S_o. Temperature effects on α are small with an observed change close to 4x10^{-6} ^{o}C^{-1} over the ten degree range. These values of α appeared to be suitable for use in subsequent thermal contraction calculations.

3 STRESS INCREASE WITH TEMPERATURE DECREASE

On cooling, frozen soil will contract if it is not constrained. Tensile stresses are generated with no observable strains. The thermal strain ε is given by the thermal contraction, thus

$$\varepsilon = \frac{\Delta L}{L_o} = \alpha \Delta T = \alpha (T_2 - T_1) \qquad (2)$$

where α is the coefficient of linear thermal contraction, L_o is the length at some reference temperature, and ΔL is the change in length due to a temperature change ΔT. With the soil constrained, and if the frozen soil is assumed to behave elastically, the tensile stress beneath a surface becomes

$$\sigma = \frac{E}{1 - \nu} \varepsilon = \frac{E}{1 - \nu} \alpha \Delta T \qquad (3)$$

where E is Young's modulus and ν is Poisson's ratio. A linear increase in σ with ΔT would be predicted. Stress-strain curves for frozen sand in tension (Eckardt, 1982) at -15 ^{o}C and a 10 h loading period give a maximum E close to 4.40 GPa. Using this value with ν = 0.30 gives $E/(1 - \nu) \simeq$ 6.28 GPa or σ = 170 ΔT (kPa) for equation (3). Failure would be predicted in partially saturated frozen sands with a few degrees drop in temperature. Saturated frozen sands are stronger and would require a larger decrease in temperature for failure.

Field and laboratory observations show that frozen sand is not as sensitive to thermal contraction as elastic theory would suggest. As tensile stresses increase, frozen soil in tension will deform with time in a viscous manner resulting in a reduced maximum stress. To provide information on the rate of stress increase with rate of temperature decrease, several frozen sand tensile specimens with fixed ends were subjected to rapid cooling in the laboratory. Specimens, similar to those used by Eckardt (1982), were mounted between fixed supports at a temperature close to -5 ^{o}C. The rigid frame support was designed for very minimal deflection at maximum load. A seating load of about 690 kPa was imposed on the specimen to insure no slack remained at the fixed ends. When stress relaxation was essentially complete for the seating load, specimen temperatures were lowered using dry ice (higher rates) or normal refrigeration (lowest rate, about -0.021 ^{o}C/min). Loads and temperatures were both recorded against time. A reduction in specimen diameter to 25.4 mm was required in order to obtain the -3.01 ^{o}C/min cooling rate. A cross plot of tensile stresses against temperature provided the data sum-

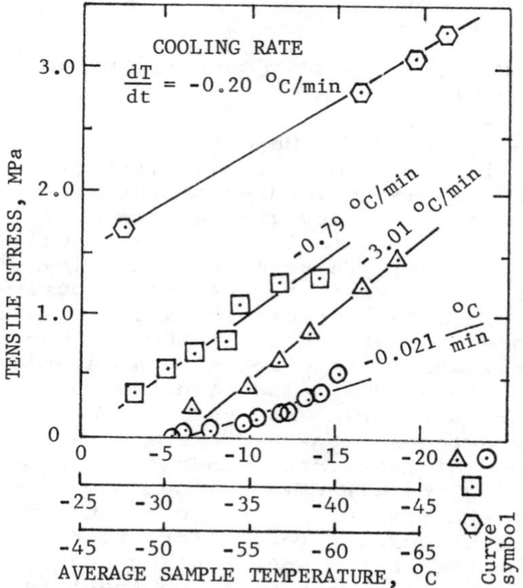

Fig. 2 Cooling rate effect on axial tensile stress in saturated frozen quartz sand (data from Al-Moussawi, 1988).

marized in Fig. 2. With very limited control over cooling rates, only data corresponding to a reasonably constant rate of temperature decrease are plotted.

A comparison between curves in Fig. 2 is difficult since different temperature ranges are represented. The frozen soil bebavior includes both an elastic and viscous response to the tensile forces generated by thermal contraction. Changes in Young's modulus with temperature and with rate of stress increase complicate the material behavior. Comparisons show that the stress increase rate for dT/dt = -0.2 °C/min at colder temperatures (-42 to -65 °C) is similar to the stress increase rate for dT/dt = -0.79 °C/min at warmer temperatures (-28 to -35 °C). The lowest cooling rate (-0.021 °C/min), corresponding to normal refrigeration, allows time for more stress relaxation and thus smaller tensile stresses. The saturated frozen sand failed in tension at close to 5.03 MPa for a rapid cooling rate (dT/dt = -0.79 °C/min). The complex material behavior demonstrated by these limited tests shows that close control of cooling rates will be required to obtain complete curves of tensile stress versus temperature.

4 FATIGUE DAMAGE ACCUMULATION AND RUPTURE

In a soil mass subjected to thermally induced cyclic loading,the process of damage leading to rupture starts with microcracks forming at initial defects in the frozen soil structure. These defects may involve ice grain boundaries, impurities in the ice matrix, and/or reduced adhesion between the ice matrix and quartz particles at specific locations. The magnitude of stress reversal may vary since it is dependent on the rate and magnitude of temperature change. Creep along the direction of maximum tensile stress will tend to expand any microcracks. Creep deformation (damage) after each cooling sequence will remain, hence the total deformation accumulates for subsequent cycles. For a frozen soil with considerable initial defects, early stages of rupture may be very short or nonexistent. Sample deformation, as defined by axial strain, does provide a measure of damage accumulation in frozen soil.

Duplication in the laboratory of field conditions responsible for thermally induced cyclic loading required equipment and facilities not available to the authors. To simplify experimental work on the effect of cumulative damage on strength, strain, and rupture, a series of constant strain/stress relaxation tests were run on duplicate frozen sand samples maintained at a temperature of -15 ± 0.2 °C. At the start of each load cycle, a given permanent axial deformation (plastic strain) was imposed on the frozen sample to simulate cooling in the field. This permanent deformation was maintained for a 24 hour period, sufficient for stress relaxation on the first cycle to be completed (Fig. 3). Note that after initial deformation, an immediate elastic stress decrease occurs which is followed by stress relaxation for each cycle shown in Fig. 3. Viscoelastic behavior is indicated with the total deformation, shown in Fig. 3, consisting of elastic, delayed or retarded elastic, and viscous flow (permanent deformation). This behavior can be modeled using a four element model consisting of a Maxwell unit and a Kelvin unit in series. Some strain hardening is indicated by cycles 2, 3, and 4. Average axial strains imposed on the samples at the start of each cycle included 0.067 percent for one cycle, 0.12 percent for two cycles, and 0.10 percent for four cycles. After one, two, and four cycles, respectively, on duplicate samples, the axial strain was carefully increased until rupture, thus providing data on the stress and strain at failure. The results are summarized in Table 1. For comparison, Eckhardt (1982) reported a failure strain in tension of 0.5 percent at -15 °C for a 100 hour loading period.

Cyclic loading of the saturated frozen sand samples illustrates several phenomenon. Deformation resulting from cyclic loading produced some strain hardening which increased the failure strain from 0.24 percent to 0.48 percent in 4 cycles (Fig. 4). Data presented by Eckhardt (1982) for uniaxial

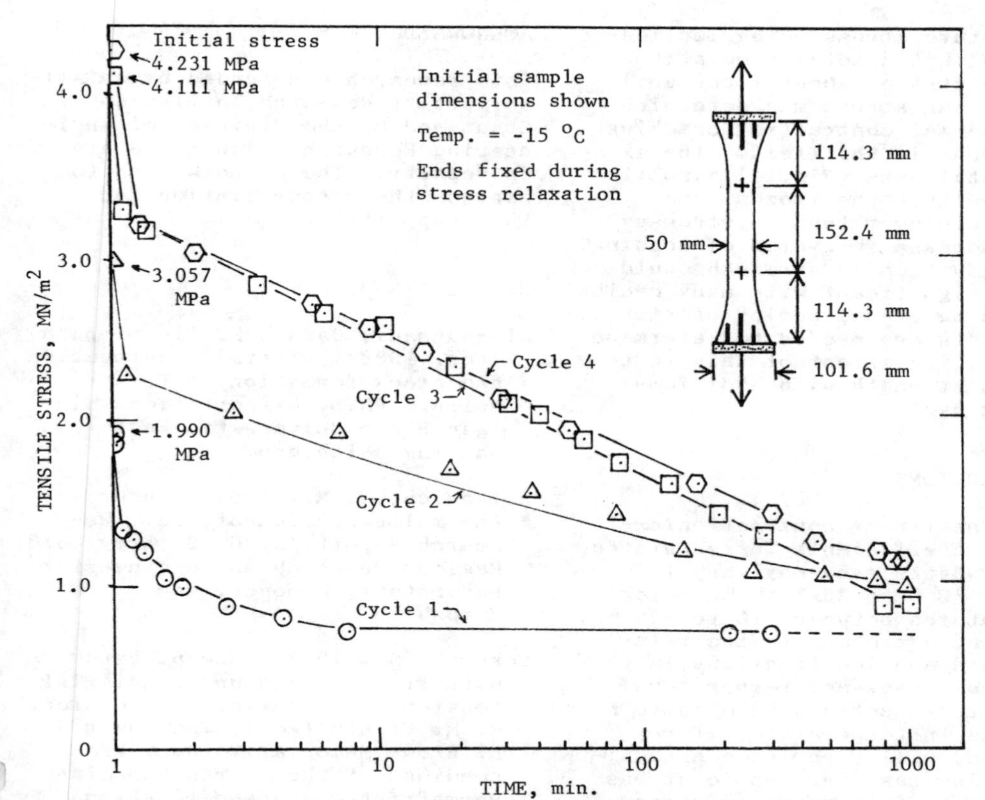

Fig. 3 Stress relaxation curves at a constant temperature (-15 °C) for frozen frozen saturated quartz sand after cyclic loading (data from Al-Moussawi, 1988).

Table 1. Results from stress relaxation tests on saturated frozen quartz sand at -15 °C.

Sample Number	Failure stress, σ_f (MPa)	Reduction in σ_f (%)	Failure strain, ε_f (%)	Increase in ε_f (%)	Cycle No.
1	722.4	0	0.24	0	1
2	709.9	1.73	0.39	60	2
3	700.8	2.99	0.48	100	4

tensile creep tests on saturated frozen sand has indicated that 0.8 percent strain may be close to an upper limit for rupture in tension. Assuming that this behavior would also be true in the field, an alternate criterion for thermal contraction cracking could be based on total strain rather than on a limit-

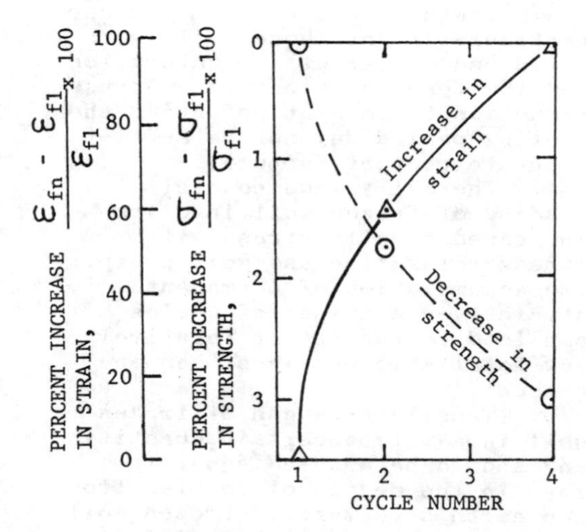

Fig. 4 Load cycle effect on failure strain and rupture stress for saturated frozen quartz sand at -15 °C.

ing rupture stress. The implication is that a total cumulative tensile strain (about 1 percent) rather than stress may determine when thermal contraction cracking will occur in the field. The experimental data (Fig. 4) also illustrate that the frozen sand failed at lower tensile stresses with increase in cycles of loading. This reduction in strength could become significant with many cycles of loading over a period of time. More tests are needed to determine the limiting extent of this reduction in strength with additional loading cycles.

5 CONCLUSIONS

1. The linear contraction/expansion coefficients for saturated frozen quartz sand may vary from 30×10^{-6} $^{\circ}C^{-1}$ to 26×10^{-6} $^{\circ}C^{-1}$ for temperatures between -10 to -20 $^{\circ}C$. Differences are due to the relative sand and ice fractions in the fozen soil mass and temperature.
2. Frozen soils subjected to a decrease in temperature, if restrained, will experience a significant increase in tensile stress. The amount of increase in stress is dependent on the cooling rate, cooling period, stress relaxation, and soil type.
3. Rupture in tension may occur, particularly for those soils with a reduced degree of ice saturation and low tensile strengths. Stress reduction by relaxation during the cooling period may not be sufficient to prevent rupture.
4. Thermally induced cyclic loading of frozen soil involves increased tensile stress with stress relaxation and soil creep. The accumulation of permanent strain over a number of cycles can lead to rupture at relatively low total strains, less than one percent.
5. Tensile strength of frozen soil is a very uncertain quantity and should be assumed equal to zero in the design of complex frozen earth structures. Frozen soil strength enhancement by reinforcement appears to be a possible future direction for thermally induced cracking problems in surface soils of the cold regions.

ACKNOWLEDGMENTS

This research was funded by an All-University Research Initiation Grant and by the Division of Engineering Research, Michigan State University. The authors wish to express their appreciation for this support.

REFERENCES

Al-Moussawi, Hassan M. (in preparation, 1988). Thermal contraction and crack formation in frozen soils. Ph.D. Dissertation, Michigan State University, East Lansing, Michigan.

Butkovich, T. R., 1957. Linear thermal expansion of ice. Research Report 40, U. S. Army Cold Regions Research and Engineering Laboratory, Hanover, New Hampshire.

Eckardt, H., 1982. Creep tests with frozen soils under uniaxial tension and uniaxial compression. H. M. French (Ed.), The Roger J. E. Brown Memorial Volume, Proceedings of the Fourth Canadian Permafrost Conference, National Research Council of Canada, Ottawa, p. 394-405.

5th International Symposium on Ground Freezing, Jones & Holden (eds)
© 1988 Balkema, Rotterdam. ISBN 90 6191 824 3

Mechanical characteristics of artificially frozen clays under triaxial stress condition

Chen Xiangsheng
Central Coal Mining Research Institute (CCMRI), Hepingli, Beijing, People's Republic of China

ABSTRACT: This paper describes mainly the mechanical characteristics of artificially frozen clays, which were taken from the depth of 200--400 m below the surface, under triaxial (Lateral stresses were the same) stress condition in the laboratory. The triaxial test results at different temperatures (-5 to -25 ℃) and at different confined pressures (1.0 to 6.0 MPa) have enabled determination of the cohesion c, the internal friction angle ϕ , triaxial creep mathematical models, and the creep parameters of these frozen clays. The paper gives the relationships of both cohesion c and the creep parameter A versus temperature; the curves of the creep parameters B and C which are independent of temperature versus water content, and the curves of the triaxial strength of these frozen clays versus time.

KEYWORDS: clay, creep, laboratory tests, mechanical properties, temperature effect

INTRODUCTION

Research on the mechanical characteristics of the frozen clay at depth now is one of the key factors in deep freeze shaft sinking. Since the clay layers are located at depth in three-dimensional stress states, research on the mechanical properties of them under triaxial stress conditions is important. This paper emphasizes the results of investigation in this field.

Temperatures in tests ranged from -5 ℃ to -25 ℃, since the mean temperature in the designed ice walls is between -7 ℃ to -15 ℃ in the Peoples Republic of China.

SAMPLES AND THEIR PREPARATION

According to the needs of the practical engineering projects, the clay samples tested were taken from mines to be constructed and under construction in three provinces in eastern China and around. Table 1 lists results of soil tests for all clays. Those with sample symbols starting with S and H were taken from soil cores, and were severely disturbed by drilling machines, since the clay layers lie deeply under the surface. Those with the sybol A were taken from a shaft as the shaft was sunk. The water contents and unit weights of samples were measured in situ, and the other soil parameters were measured in the laboratory in Beijing.

On the basis of their natural water contents and natural unit weights, all samples for triaxial tests were prepared in the laboratory according to the Regulation for Soil Test SDS 01-79, China. The diameter of the samples is 61.8mm and the heigh is 150 mm.

TEST EQUIPMENT AND METHOD

All tests were conducted on Cryo High Pressure Triaxial Test Equipment FS-1 in the intermediate cryo chamber at the freeze laboratory of CCMRI, it consists of three cryo chambers: large (25 m^2), intermediate (18 m^2) and small (14 m^2). The temperature ranges from 0 ℃ to -30 ℃ which is held to a tolerance of ± 0.4 ℃ in the test area in each chamber.

The triaxial test equipment FS-1 is shown in Fig 1. With confined pressure from 0.0 to 6.0 MPa and axial force 0.0 to 150 KN, it can carry out triaxial test with 18 levels of axial strain rates from 0.00013%/min to 1.3%/min. In order to conduct a constant stress creep test, it is equipped with a controlling system which can hold the axial stress to a tol-

Table 1. The results of soil test

Sample symbol	Depth	NWC	NUW	LL	PL	I_p	Soil composition (Grain diameter in mm)							Texture
		W	γ	W_L	W_P		>2	2-0.5	0.5-0.25	0.25-0.1	0.1-0.05	0.05-0.005	<0.005	
	M	%	N/m³	%	%		%	%	%	%	%	%	%	
S1	159	23	19915	36.8	18.8	18.0				20	18	20	42	Clay
S2	160	18	19915	39.5	22.3	17.2		3	7	10	17	22	41	Clay
S3	180	28	18350	38.5	20.4	18.1		4	6	17	14	29	30	Clay
A1	260	29	20111	77.7	38.8	38.9					8	21	71	Clay
A2	292	14	20111	43.2	24.1	19.1		1	2	13	17	31	36	Clay
A3	322	17	20307	30.4	17.5	12.9		1	6	23	26	23	21	Loam
H1	309	22	20601	59.0	30.7	28.3				1	20	47	32	Silt clay
H2	350	17	20111	31.0	18.8	12.2				2	13	52	33	Silt clay
H3	370	14	22465	29.0	16.2	12.8	5	4	3	4	5	18	61	Clay

Table 2 Cohesion c and angle φ

Sample symbol	Texture	T °C	C MPa	φ Degree
S1	Clay	-10	0.9	5.8
S2	Clay	-10	1.3	3.5
S3	Clay	-10	1.1	2.8
A1	Clay	-5	0.6	2.1
		-10	1.1	3.5
		-15	1.8	3.3
		-20	2.3	3.3
		-25	3.2	3.4
A2	Clay	-10	1.4	8.9
A3	Loam	-10	1.4	13.3
H1	Silt clay	-5	0.6	3.5
		-10	1.2	4.4
		-15	1.6	4.4
		-20	2.2	4.5
H2	Silt clay	-5	0.9	3.1
		-10	1.1	7.0
		-15	1.8	6.9
H3	Clay	-5	0.8	2.6
		-10	1.2	6.4
		-15	1.8	6.2

Table 3 Creep parameters for frozen clays

Sample symbol	Confining pressure MPa	T °C	G MPa	Creep parameters A $(MPa)^{-B}h^{-C}$	B	C	Agreement* %
S1	1.96	-10		0.124	2.17	0.38	99
S2	1.96	-10		0.115	1.87	0.31	98
S3	1.96	-10		0.193	2.12	0.38	98
A1	2.94	-5	17	0.255	1.88	0.21	94
		-10	24	0.164			97
		-15	30	5.84×10^{-2}			97
		-20	36	2.18×10^{-2}			95
		-25	40	4.85×10^{-3}			96
A2	2.94	-10		2.41×10^{-3}	3.08	0.19	98
A3	2.94	-10		3.12×10^{-4}	3.12	0.41	98
H1	3.93	-5	18	0.251	1.58	0.31	96
		-10	23	8.90×10^{-2}			97
		-15	25	4.35×10^{-2}			97
H2	3.93	-5	12	0.104	3.06	0.39	95
		-10	21	3.58×10^{-2}			98
		-15	30	1.74×10^{-2}			97
H3	4.61	-5	17	0.123	3.06	0.29	97
		-10	22	3.42×10^{-2}			97
		-15	30	4.57×10^{-3}			96

*Which is the degree approximation to the observations.

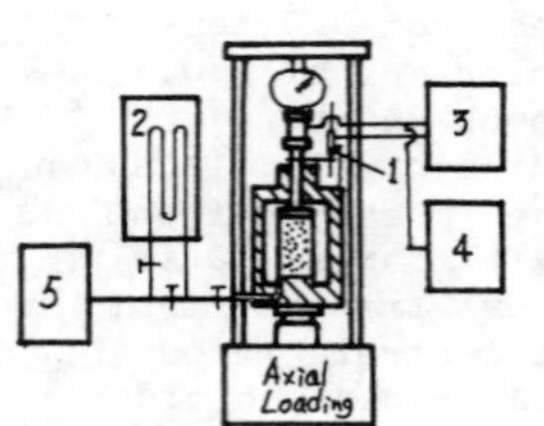

1--displace sensor
2--volume-change meassuring gauge
3--automatic recorder
4--axial stress controlling system
5--confined pressure controlling system

Fig 1 Cryo high pressure triaxial creep test equipment

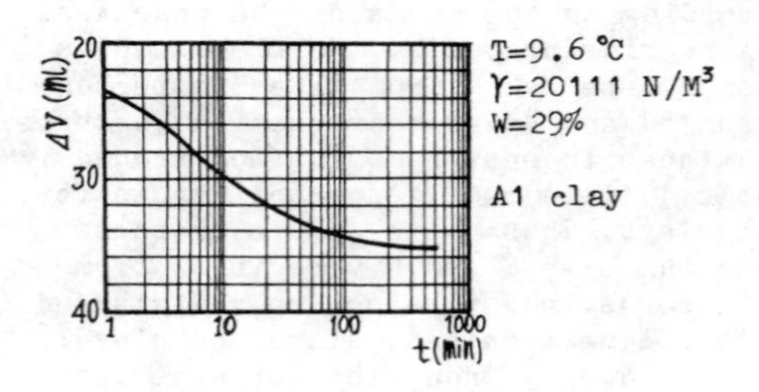

Fig 2 Consolidation curve

erance of $\pm$ 0.006N/mm^2.
Every sample for testing was kept at a selected constant temperature for at least 48 hours and was then consolidated in the triaxial cell before testing. Replicate clay samples had different confined pressures applied prior to the triaxial shear tests to determine the cohesion c and the internal friction angle ϕ. One of the consolidating pressures should conform to the calculated ground pressure at the depth at which the clay was located. For the same clay samples, triaxial creep tests were performed under the same consolidating pressure, which was equal to the calculated ground pressure, and this pressure was kept constant throughout the testing.
The consolidation time generally was about 5 hours for all samples. The consolidation degree of a sample was distinguished on the basis of the change in volume of the unfrozen liquid in the triaxial cell. Next a consolidation curve was plotted as volume change versus the logarithm of time so that the degree of consolidation can be determined (Fig 2). After consolidation of 3 to 4 hours the consolidation degree of a sample usually reached 95 per cent in our tests.

RESULTS AND DISCUSSION

1 Triaxial shear test

Fig 3 show the Mohr's envelopes of A1, H1, H2 and H3 clay samples at different temperatures, with a constant strain rate of 0.8%/min. Fig 4 shows the curves of normal strain and temperature versus deviator stress of these frozen clay samples. The cohesion c and the internal friction angle ϕ of all frozen clay samples are listed in Table 2. From these figures and Table 2 we have,

1) In the range of testing temperatures, the failure deviator stress of frozen clay increases in proportion to the absolute value of minus temperature under the same confined pressure. The failure deviator increases with temperature dropping by 0.26 -- 0.3MPa per -1 ℃.

2) At the same temperature, the failure deviator stress increases slightly with the increment in confined pressure. The small internal friction angles ϕ of the frozen clays tested ranged from 3 to 7 degrees.

3) The internal friction angles ϕ at -5 ℃ are smaller than those at other lower temperatures, however, the values are very close.

4) The deformation and failure characteristic of H1 clay under confined pressure (3.0MPa) are different from those under no confinement (uniaxial) at -20 ℃, which is shown in Fig 4. The largest deviator stress in the linear phase of $(\sigma_1 - \sigma_3)$—ε_1 curve under confinement is twice the unconfined. The failure of the former is plastic and that of the latter brittle. The failure strain of the former is twice as large as that of the latter.

5) Fig 5 shows that the cohesion c of frozen clay in the range of testing temperatures increases in proportion to the absolute value of minus temperature. The cohesion c increases with decreasing temperature by 0.08 to 0.12 MPa per -1 ℃.

2 Triaxial creep test

The triaxial creep tests of all clay samples were conducted at -10 ℃. Some auxiliary creep tests of clays were conducted at -5 ℃, -15 ℃, -20 ℃ and -25 ℃, in order to investigate the change of creep parameter A with temperature. The curves of the triaxial creep tests of A1 and H1 clays at -10 ℃ are shown in Fig 6 in solid lines. By analyzing the data of theses creep tests, when the loading coefficient $k=(\sigma_{1c}-\sigma_3)/(\sigma_1-\sigma_3)$, where $(\sigma_{1c}-\sigma_3)$ is equal to the creep deviator and $(\sigma_1-\sigma_3)$ is the failure deviator of the same clay, is smaller than 0.75, it is found that the nature of triaxial creep for the primary and the secondary phases can be expressed by the following exponential formula:

$$\gamma_i = \gamma_0 + A\tau_i^B t^C \qquad (1)$$

where $\gamma_i = \sqrt{2\sum_{i=1}^{3} e_i^2}$, $\tau_i = \sqrt{\frac{1}{2}\sum_{i=1}^{3} s_i^2}$, A is creep parameter dependent on temperature, B and C are stress and time creep parameters, $e_i = \varepsilon_i - \varepsilon_m$ and $s_i = \sigma_i - \sigma_m$, ε_m is the mean normal strain and σ_m is the mean normal stress, γ_0 is the instantaneous shear strain strength which is equal to τ_i/G and G is the instantaneous deformation modulus. Formula (1) is based on $\varepsilon = A\sigma^B t^C$ (S.S.Vyalov et al 1962 and Klein, 1978).
The shear deformation modulus G for frozen clay at the same temperature can approximately be replaced by the mean value of the secant shear deformation moduli at the different shear stress level, when k is smaller than 0.75. As shown in Fig 7 the mean value of secant shear deformation moduli of straight lines OP and OA can be taken as the approximate shear deformation modulus G.
All testing parameters in formula (1) are listed in Table 3 under the condition of constant volume and axisymmetry, that is ε_m=0, Poisson ratio μ=0.5 and then $\tau_i = (\sigma_{1c}-\sigma_3)/\sqrt{3}$, and $\gamma_i = \sqrt{3}\varepsilon_1$. The creep curves calculated by formula (1) for the same

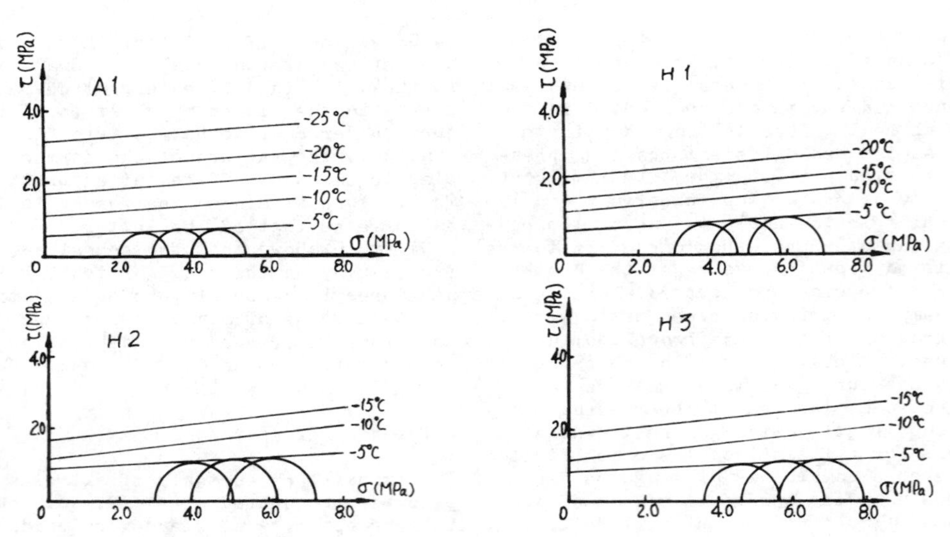

Fig 3 Mohr's envelopes of A1, H1, H2 and H3 frozen clay samples

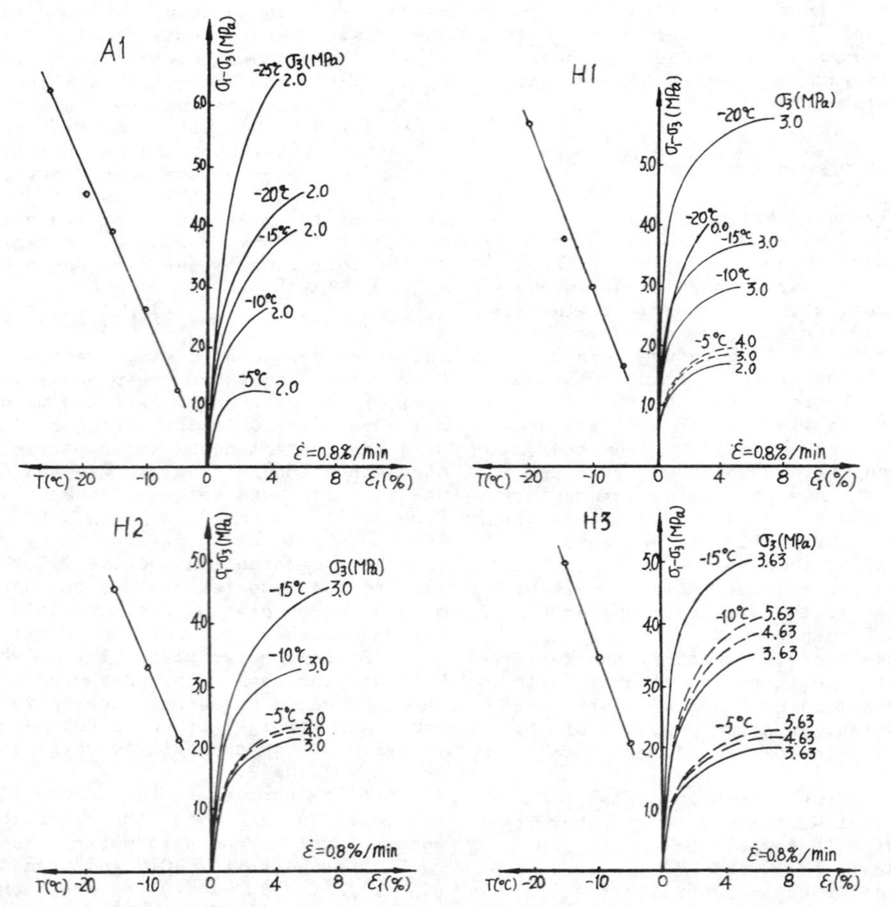

Fig 4 Curves of normal strain and temperature vs deviator stress for A1, H1, H2 and H3 frozen clay samples

stress levels as in the tests are shown in Fig 6 in dashed lines. These may be compared with the values from the tests shown in solid lines.
From Table 3, Fig 6 and the following figures certain observations are made,

1) The behaviour of the frozen clay triaxial creep in the primary and secondary phases can be described by the formula (1) when the creep loading coefficient $k<0.75$. The agreement degree between the shear strain calculated by formula (1) and that from testing is very high though the calculated is larger in the primary phase and slightly smaller in the secondary phase.

2) Fig 8a shows that the stress creep parameter B decreases with an increase in the water content in frozen clay. But the value of parameter B is almost constant when the water content is larger than 20 per cent. The value of B reaches its peak at a water content of 14 to 18 per cent, which shows that the influence of the stress strength factor on creep is greatest at this water content.

3) Fig 8b shows that the time creep parameter C increases with an increase in water content, but decreases slightly if the water content is more than 20 per cent. When the water content is about 16 to 20 per cent the parameter C reaches its peak, which shows the influence of the time factor on creep is greatest at water content of 16 to 20 per cent.

4) On the basis of the stress and time creep parameters B and C being not very sensitive to temperature and the creep parameter A being highly sensitive to temperature, the values of A at different minus temperatures can be found by performing creep tests on the same clay samples at the corresponding temperature. Fig 9 shows the curve of parameter A vs temperature for A1, H1, H2 and H3 frozen clays. It shows clearly the decrease of parameter A with decrease in temperature. The value of parameter A decreases rapidly when temperature decreases from -5 °C to -15 °C, and the decrease of the value A is reduced at temperatures below -15 °C to -20 °C. Thus creep deformation decreases rapidly with temperature decrease, but the deformation decrease tends slow at temperature below -15 °C to -20 °C. In the view of this fact it may be reasonable that the mean temperature of ice wall in the tested clay layers should be between -15 °C and -20 °C.

5) Fig 10 shows the tendency of the strength of the tested frozen clays to decrease with time. Their long-term triaxial strength is close to the half of their instantaneous triaxial strength.

CONCLUSIONS

1) In the range of testing temperatures, the failure deviator stress of frozen clay increases in proportion to the absolute value of minus temperature under the same confined pressure. The failure deviator increases with decrease in temperature by 0.26 to 0.3 MPa per -1 °C.

2) At the same temperature, the failure deviator stress increases slightly with an increase in confining pressure, which produces small internal friction angles ϕ for the frozen clays tested ranging from 3 to 7 degrees.

3) The internal friction angles ϕ of the tested clays at -5 °C are smaller than those at other lower temperatures for which the angles ϕ, are very similar. These results are in some disagreement with usual assumptions.

4) There is difference in deformation and failure characteristic of frozen clay for confined vs unconfined conditions. The largest deviator stress in the linear phase of $(\sigma_1-\sigma_3)-\varepsilon_1$ curve under confined pressure (3.0 MPa) is twice the unconfined at -20 °C. The failure of the former is plastic and that of the latter brittle. The failure strain of the former is twice that of the latter.

5) The cohesion c of frozen clay increases in proportion to the absolute value of minus temperature. The cohesion c increases with decreasing temperature by 0.08 to 0.12 MPa per -1 °C.

6) The behaviour of triaxial creep for frozen clay in the primary and secondary phases can be expressed by the formula $\gamma_i = \gamma_0 + A\tau_i^B t^C$ when the creep loading coefficient $k < 0.75$. The agreement between the shear strain calculated by formula (1) and that from testing is very high, though the calculated value is larger in the primary phase and slightly smaller in the secondary phase.

7) The value of stress creep parameter B decreases with the increase of water content in frozen clay, this decrease however lessens when the water content is larger than 20 per cent. The value of parameter B reaches its peak at the water content of 14 to 18%.

8) The time creep parameter C increases with increasing water content and slowly decreases when water content is more than 20%. The parameter C reaches its peak when the water content is about 16 to 20%.

9) From the items 7) and 8), it is found that the creep deformation of the clay tested is larger at the water content of 14 to 20% than that at other water content.

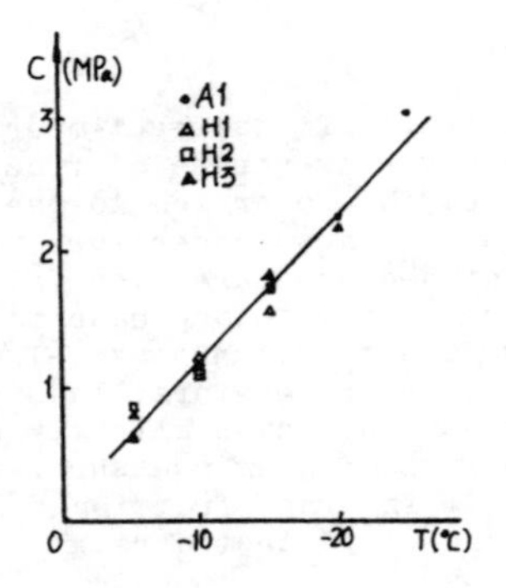

Fig 5 Cohesion c vs temperature T

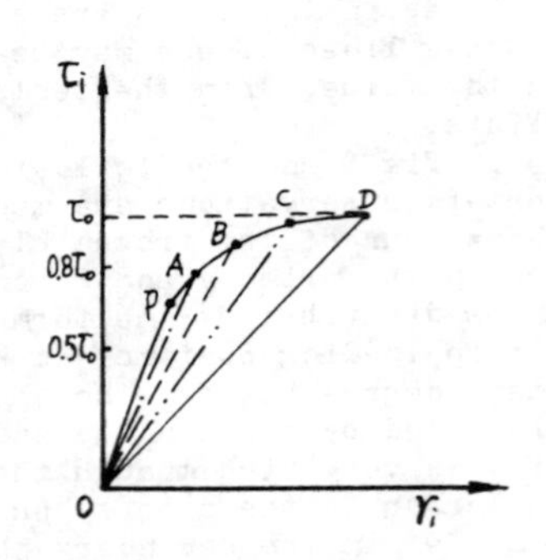

Fig 7 Instantaneous shear stress strength τ_i vs strain strength γ_i

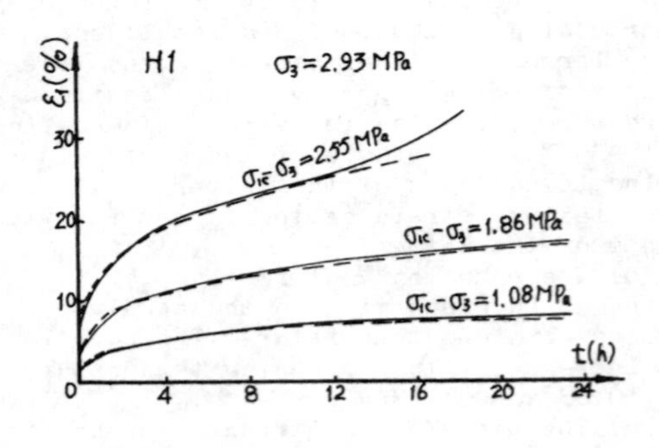

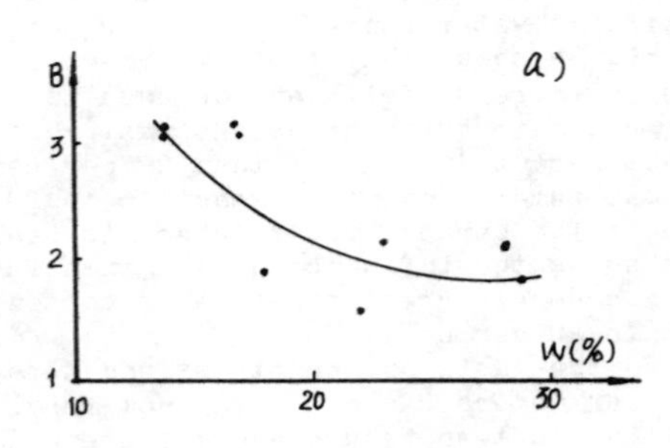

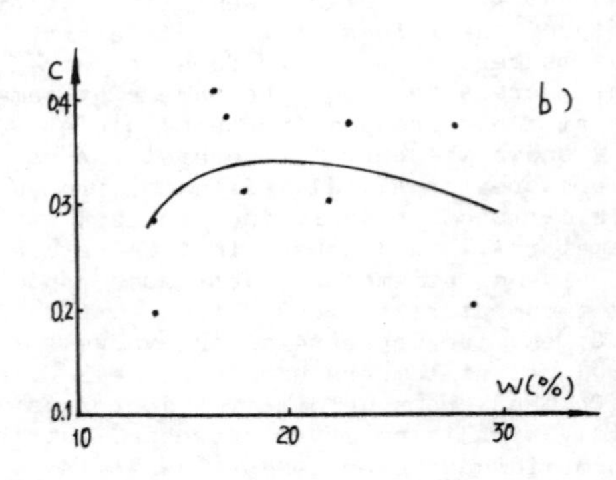

Fig 6 Creep curves of H1 and A1 clay samples at -10 ℃

Fig 8 Creep parameters B and C vs water content W

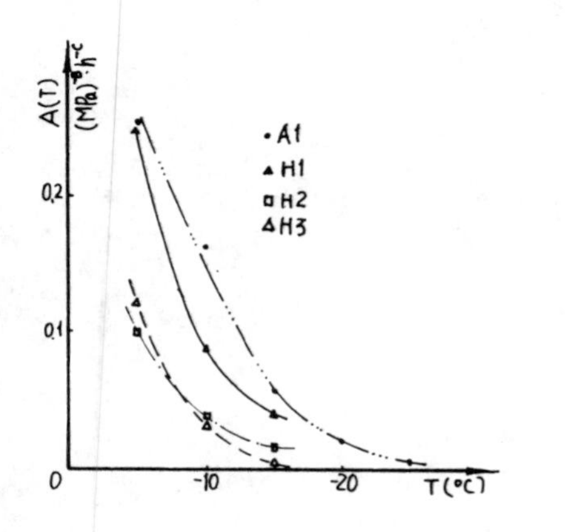

Fig 9 Creep parametr A vs temperature T

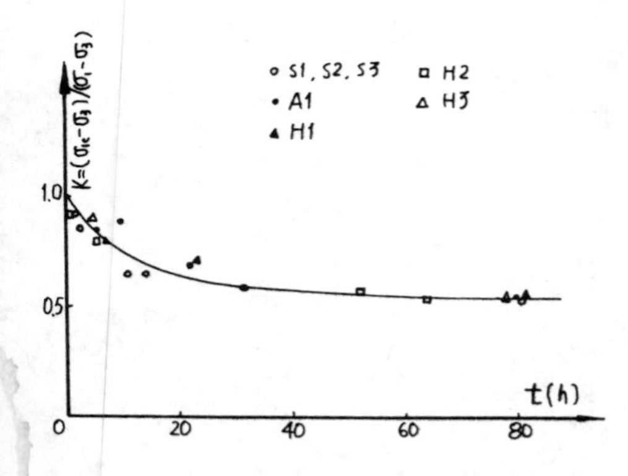

Fig 10 Creep loading coefficient k vs failure time t

10) The temperature creep parameter A decreases rapidly with decrease in temperature at 0 to -15 °C, but the decrease becomes small at the temperature below -15 °C to -20 °C. Based on this fact, it may be reasonable that the mean temperature of ice wall in the tested clay layers should be between -15 °C and -20 °C to reduce the creep deformation.

11) The failure triaxial strength decreases with time. The long-term triaxial strength for the tested clay samples is close to the half of their instantaneous triaxial strength.

ACKNOWLEDGEMENT

The author expresses his sincere appreciation to all his colleagues in the Cryo Test Laboratory of CCMRI for their help with the tests described in the paper.

REFERENCES

Chen Xiangsheng 1987. Report on tests of frozen soil of Yongcheng Mine Field in Henan Province. Report of Cryo Test Laboratory, CCMRI.

Cryo Test Laboratory of CCMRI 1986. Research on mechanical properties of frozen soil of Jining Coal Mine No.2. Report of CCMRI

Jessberger, H.L. 1981. The fundamental and application of freeze method. Lecture at Beijing Res. Inst. of Mine Construction, CCMRI

Klein, J. 1978. Nichtlineares Kriechen von künstlich gefrorenem Emschermergel. Schriftenreihe des Instituts für Grundbau, Wasser-und Verkehrswesen, Ruhr-Universität Bochum, Serie G,Heft 2

Liang Huisheng, Xu Bingzhuong, Chen Xiangsheng, etc 1987. Cryo high pressure creep test. Coal Science and Technology, Beijing: Coal Industry Press, 1:13-15

Ministry of Water Conservancy and Power 1980. The regulation for soil test SDS 01-79. Vol. 1:1-94. Beijing: Press of Water Conservancy and Power

Vyalov, S.S. et al 1962. The strength and creep of frozen soils and calculations for ice-soil retaining structures. Transl. 76, CRREL, 1965.

5th International Symposium on Ground Freezing, Jones & Holden (eds)
© 1988 Balkema, Rotterdam. ISBN 90 6191 824 3

Direct shear test at a frozen/unfrozen interface

S.Goto & K.Minegishi
Tokyo Gas Company, Tokyo, Japan

M.Tanaka
Kajima Institute of Construction Technology, Tokyo, Japan

ABSTRACT: The estimation of the frost heave of an LNG in-ground storage tank as a freezing front advances in the ground, requires an understanding of the shear strength of soil at a frozen/unfrozen interface.
Concerned that a weak shear plane may occur at a frozen/unfrozen interface, we conducted direct shear tests at the interface (as 0°C line) and in unfrozen soil.
The results show that the shear strength at the frozen/unfrozen interface is greater than in completely unfrozen soil.
The increase of shear strength is caused by a water migration from the unfrozen to the frozen side during test.
The weakest zone of surrounding ground for in-ground tanks lies in the unfrozen ground free of the influence of suction force near the freezing front. Therefore, the shearing strength of completely unfrozen soil at the initial water content should be chosen for the stability analysis.

INTRODUCTION

The three types of in-ground LNG storage tanks are classified by the form of the freezing of the surrounding ground, as shown in Fig. 1. These tanks of (a) and (b) types are lifted by the frost heaving force.
The design of these tanks requires knowledge of the mechanical properties of the surrounding soil for anticipated loading conditions. (Fig. 2)
Vertical heaving forces may be resisted by the surrounding ground structural strength, and can result in two potential failure modes: (1) localized edge failure in unfrozen soil close to ground surface, (2) shear failure at or near the frozen/unfrozen interface.
Based on the result of the preliminary model test of pulling up the model tank as shown in photo 1, we were concerned that the failure surface could occur at or near the frozen/unfrozen interface.
To clarify the shearing strength at the interface, we developed a direct shear test for three kinds of soils.

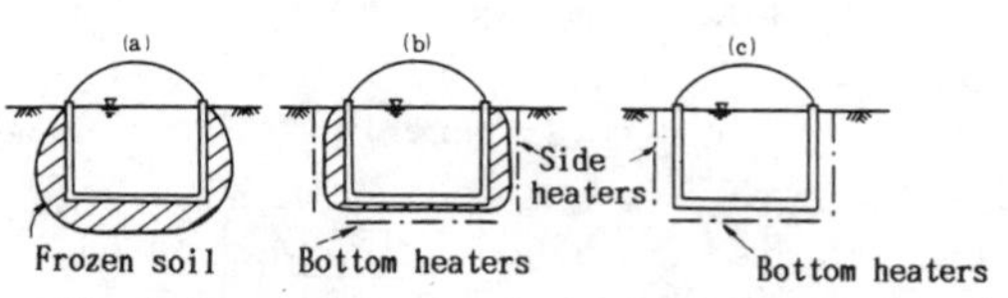

Fig–1 Types of storage tanks, by form of freezing of surrounding ground.

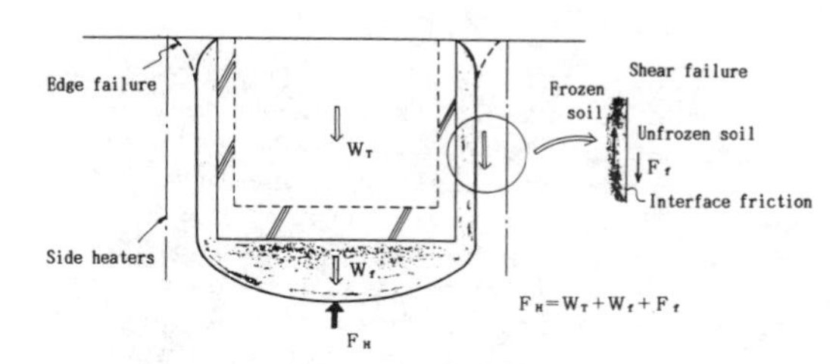

Fig–2 In-ground failure modes

SAMPLE PREPARATION

Three kinds of undisturbed samples were used for the experiments, volcanic ash soil (Kanto loam), Mud stone (Dotan) and diluvial sand.
Kanto loam was taken as blocks of 20*20*30cm at a depth of about 2m and Dotan in the same size of blocks at a depth of about 5m from the ground surface in Yokohama hill, respectively.
Diluvial sand was taken as cylinders of

Photo–1 Test of pulling up model freezing tank

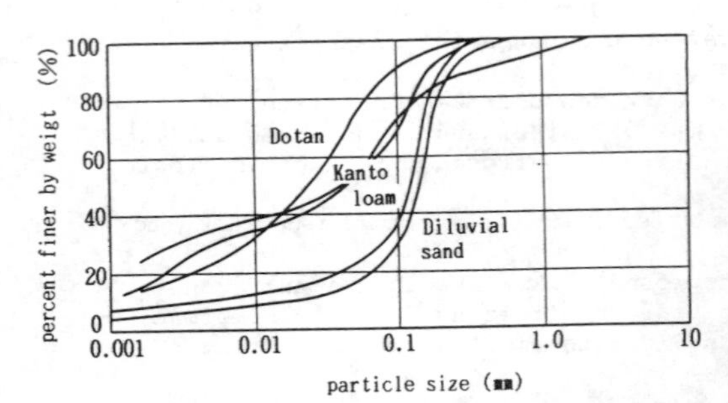

Fig–3 Particle size accumulation curve

Table–1 Soil properties

Item / Sample	Specific gravity	Water content (%)	Grain size (%) sand	silt	ciay	L.L (%)	P.L (%)
Kanto loam	2.70~2.74	72~126	47	17	36	106	65
Diluvial sand	2.66~2.71	21~29	77	15	8	—	—
Dotan	2.59	32~55	13	73	13	54	39

20cm in diameter and 40cm in height at a depth of about 35m from the ground surface at the Tokyo Gas Sodegaura Works. The particle size distribution curve and physical properties of the respective samples are shown in Fig. 3 and Table 1. For testing, cylindrical specimens of 15cm diameter and 5cm height were cut out of the blocks with a trimming-ring container.

DIRECT SHEAR TEST APPARATUS

The direct shear test apparatus was designed to permit inducing shear failure at a controlled plane within the partially frozen freeze-front zone.
The apparatus, shown schematically in Fig. 4 and photo 2, is composed of three components:
1) Vertical and horizontal load system,

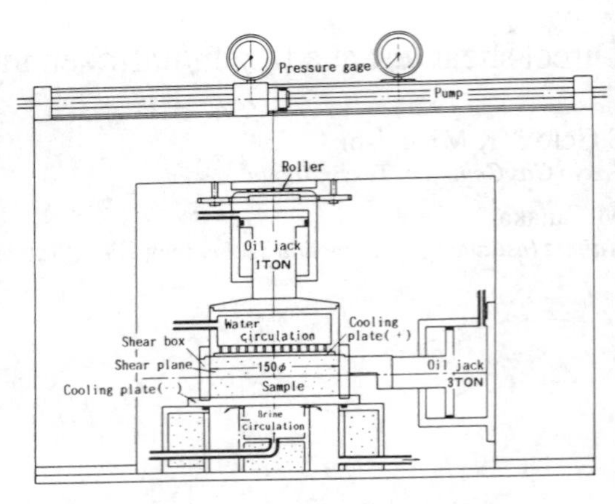

Fig–4 Schematic of direct shear test apparatus

Photo–2 Direct shear test machine

2) Shear box
3) Thermal system.
Together, this apparatus is capable of shearing a cylindrical specimen of 15cm in diameter and 5cm in height under vertical loads up to 2 tonne and horizontal loads of 3 tonne, while maintaining the desired thermal conditions.
The shear box halves are separated by an adjustable spacer to allow a shear plane gap to accommodate testing of a variety of maximum soil particle size.
Upper and lower cold plates with circulating water and brine respectively are transmittal and reaction platens for vertical loads.
The water temperature is constant at +10°C ±0.1°C, and the brine temperature can be varied from -30°C to +30°C.
The shear test apparatus is situated in a cold room at +10°C ±1°C.

TEST PROCEDURES

The common laboratory undrained rapid

shear test was adopted as the method of testing. The unfrozen sample was placed in the shear apparatus and then frozen from the bottom up with brine circulation in the lower cooling plate until the frozen/unfrozen interface coincided with the preferred shear plane.
Temperature monitoring of the sample allows direct shear testing when the desired thermal condition exists at the shear plane. The sample was then subjected to an overburden pressure and sheared at a constant rate of displacement. Horizontal load, vertical load, horizontal deformation, vertical deformation and temperature were monitored during the shear test. Upon completion of the test, the frozen and unfrozen side were removed for moisture content determination.
Test conditions are shown in Table 2.

TEST RESULTS

Temperature

Typical thermal data during preparation (freezing) are shown in Fig. 5 for diluvial sand. From this figure, we can see that the shear plane temperature is 0°C.
The temperature is measured by copper-constantan thermocouple (stranded wire 0.6mm in diameter) in the center and at both surfaces of the specimen.

Table–2 Test condition

Sample	Freezing	Vertical Load ×kN	Number
Kanto loam	unfrozen	2～5	23
	partially frozen	2～5	23
Diluvial sand	unfrozen	1～5	10
	partially frozen	2～5	17
Dotan	unfrozen	1～5	23
	partially frozen	2～5	24
			Total 120

The thermocouple in the center is inserted in a hole made by a needle from the top surface of specimen to the shear plane, at 2.5cm depth.
The thermocouple at the shear plane is used to determine an exact testing time.

Water migration

Fig. 6 shows the water migration during uniaxial freezing from the bottom.
The water migration is expressed as (w-wi) where wi is initial water content and w is after the test.
The results indicate that the water migration of clayey soil with high water contents is larger than that of sandy soil with low water contents.
The water migration should be caused by the suction force corresponding to the temperature gradient at the freezing front.

Shear strength

A comparison of the load-displacement results for partially frozen and completely unfrozen sandy soils is shown in Fig.7. For the sandy soil, the peak load of the frozen/unfrozen interface was slightly larger than that of unfrozen state. However for the clayey soil, the former was much larger than the latter.
The residual strength approximates that of completely unfrozen material for the sandy soil.
The comparison of the shear strength at the frozen/unfrozen interface with that of completely unfrozen soil is shown Figs. 8-10. It is seen from the tests that the shear strength of a clayey soil with a frozen/unfrozen interface is greater than that of completely unfrozen state, but there is little difference between the values for the sandy soil.
Table 3 shows the angle of shearing resistance and cohesion of average shear strength from Figs. 8-10. It is seen that the reason for larger shear apparent cohesion. The larger apparent cohesion

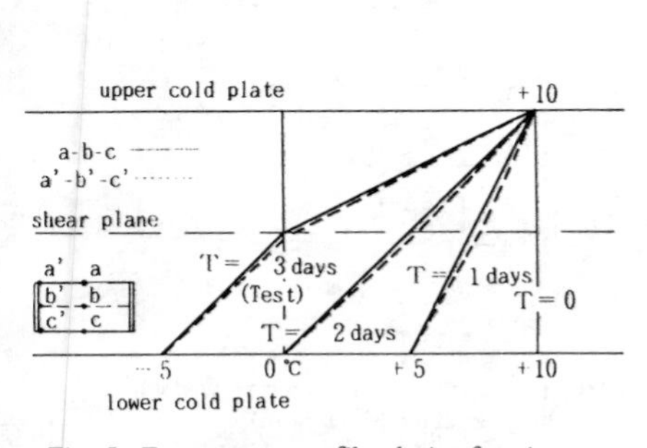

Fig–5 Temperature profiles during freezing

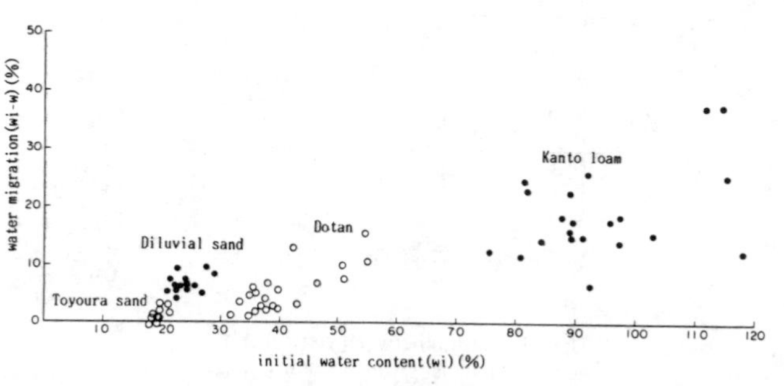

Fig–6 Water migration after freezing

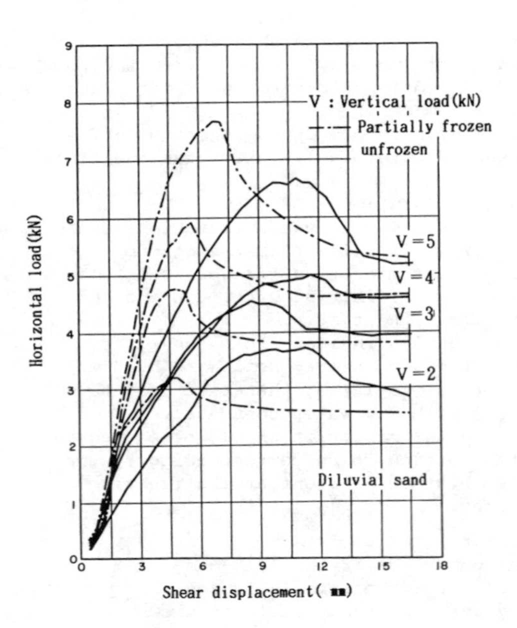

Fig–7 Typical load-displacement results

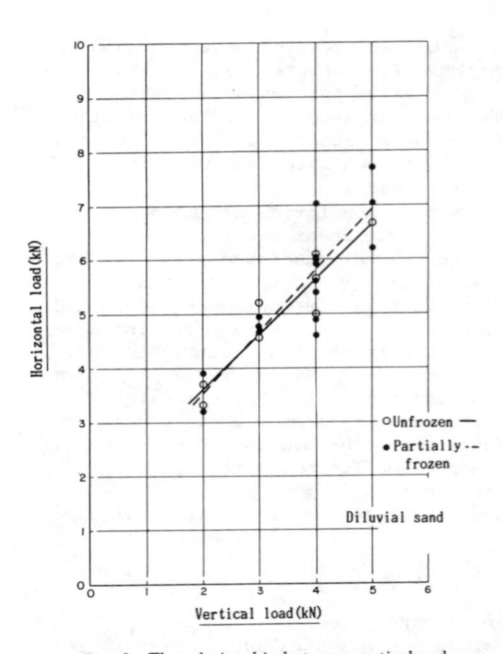

Fig–9 The relationship between vertical and horizontal loads (Diluvial sand)

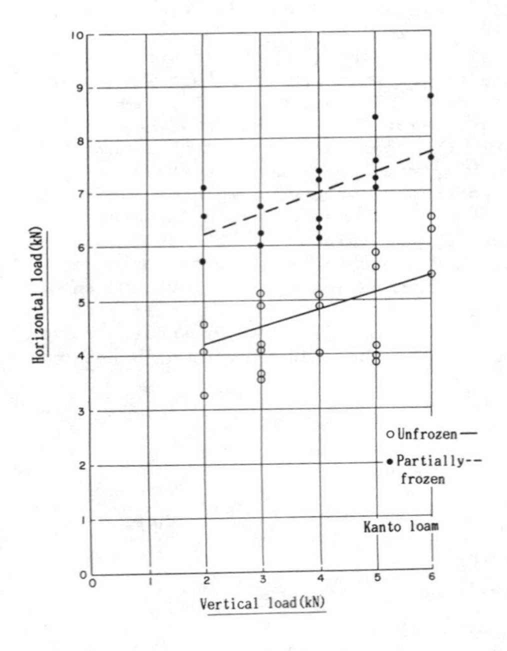

Fig–8 The relationship between vertical and horizontal loads (Kanto loam)

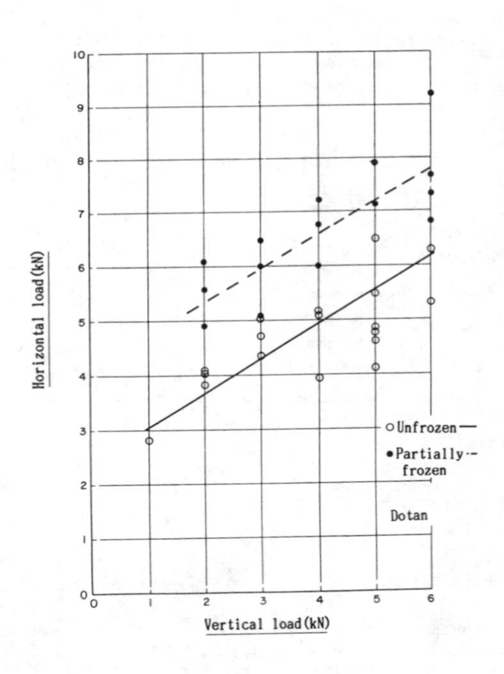

Fig–10 The relationship between vertical and horizontal loads (Dotan)

Table–3 Shear strength parameters

Sample		Angle of shearing resistance ϕ	Cohesion c(MPa)
Kanto loam	unfrozen	18.4	0.20
	partially frozen	21.8	0.31
Diluvial sand	unfrozen	45.4	0.09
	partially frozen	48.1	0.08
Dotan	unfrozen	31.0	0.14
	partially frozen	33.0	0.23

may be caused by the decrease in the water content in the unfrozen side. Therefore, the unfrozen soil after testing may be considered as soil overconsolidated by the process of freezing.
Typical shear test results are presented in Fig. 11 for Kanto loam. The relationship between the shear strength and the water content is approximately linear for completely unfrozen and partially frozen soil, therefore implying that the unfrozen strength after testing is appropriately equal to the strength at freezing front of the same water content.

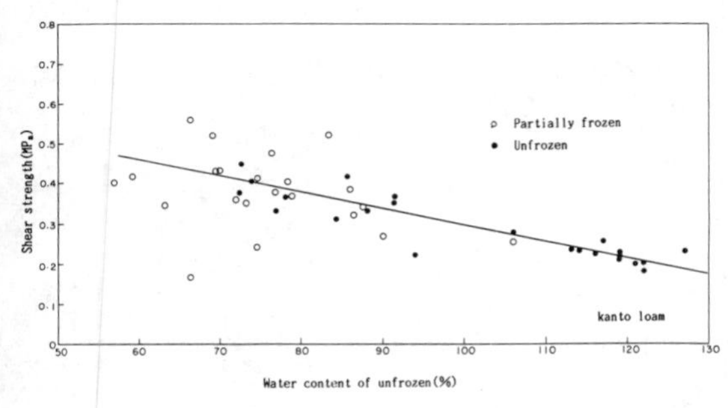

Fig–11 The relationship between water content and shear strength

CONCLUSION

From the direct shear test the following conclusion can be drawn;
1. The shear strength of a frozen/unfrozen interface is greater than that of a completely unfrozen soil with initial high water contents for clayey soil, but there is little difference between them for sandy soil. The reason may be a water migration from the unfrozen to the frozen side in partially frozen soil during freezing.
2. The weakest zone of surrounding ground for in-ground tanks lies in the unfrozen ground free of the influence of suction force near the freezing front. Therefore, the shearing strength of completely unfrozen soil at the initial water content should be chosen for the stability analysis.

ACKNOWLEDGEMENT

Regarding the execution of the direct shear tests, we are grateful to the Tokyo Gas Company, Kajima Corporation and other persons concerned, for their support and advice.

REFERENCES

Roggensack, W.D. and Morgenstern, N.R. 1978, "Direct Shear Test on Natural Fine-grained Permafrost Soils", Proceedings, Third International Conference on Permafrost, Edomonton, Canada.
Brain J.A.S. and Larry, J.M. 1983, "Large-scale Direct Shear Test System for Testing Partially Frozen Soils, "Fourth International Conference on Permafrost, Fairbanks, Alaska, USA.
Japan Gas Association, 1979, "Recommended Practice for LNG Inground Storage".

5th International Symposium on Ground Freezing, Jones & Holden (eds)
© 1988 Balkema, Rotterdam. ISBN 90 6191 824 3

Mechanical properties of circular slabs of frozen sand and clay in bending tests

H.Izuta, H.Yamamoto & T.Ohrai
Seiken Co. Ltd, Kawarayamachi Minami-ku, Osaka, Japan

ABSTRACT: The model test of a circular slab made of frozen soil have been performed for designing the adequate thickness of an frozen soil wall to be formed when a shield tunnel is driven. A series of experiments was carried out under a simple confining condition. Yield tensile stresses were found to be twice those determined by a pure bending test of beam. These experiments may also suggest that the frozen clay exhibits noticeable shear deformation with an increase in thickness. In addition, another kind of experiment was conducted under conditions closer to the conditions at the actual site as to the range of loading and confining conditions. The experiment showed that the frozen soil slabs adfrozen to the region outside the open diameter behaved similarly to a clamped slab. It also showed that the fracture pressures were two to three times those measured by above mentioned simple condition tests.

1 INTRODUCTION

The excavation of a cofferdam for a shield machine starting often requires the building of ground retaining walls. In recent years, the maximum diameter of the shield machine increased to more than 10 meters and the depth of excavation reached to more than 30 meters from the ground surface. As a result, the number of opportunities has increased in the forming of a frozen circular slab for securing a highly reliable wall. Two ways are considered to determine the mechanical properties of the slab. One is monitoring at the site and the analysis of the data obtained; and the other is the conducting of a model test. Various ground strata exist and stress distribution is complex at the site, and any fracture dose not occur. Consequently, model tests are useful in understanding the basic and ultimate mechanical characteristics of the frozen circular slab.

The model tests of the structural frozen soils have been conducted by a large number of investigators about a frozen circular cylinder relevant to the construction of a mining shaft (Vialov,1962; Orth and Meissner, 1985). On the other hand, reports on model tests about a frozen circular slab are not available to the best of the authors knowledge. Thus three kinds of experiments have been conducted: the one is under a simple confining condition and the other two are under conditions closer to those at the actual site.

2 EXPERIMENT

2.1 Tested Soils and Specimens

Tested soils are Toyoura-standard sand and Fujinomori-blue clay named sand and clay in this paper respectively. The grain size distributions are shown in Fig. 1. The sand contains no fine-grained particles less than 0.1 mm in diameter.

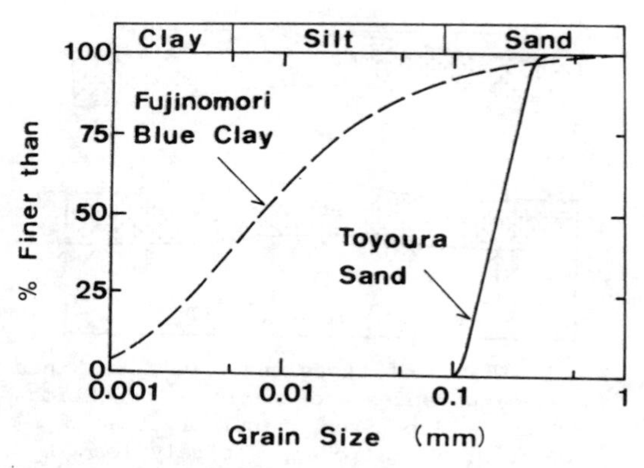

Fig. 1 Grain size distribution curves

Firstly, an adequate amount of water was added to the air dried soil to make it into a slurry. The saturated soil was placed in a cylindrical mould and rapidly frozen unidirectionally from the bottom. The circular specimen was obtained by planing both the top and bottom surfaces of the frozen soil to ensure the roughness of the surfaces less than 0.5 mm. Specimens were saturated, and the physical properties of the specimens are given in Table 1.

Table 1 Index properties of specimens

		Frozen sand	Frozen clay
Liquid limit	%	-	53.6
Plastic limit	%	-	20.7
Specific gravity	-	2.64	2.61
Water content	%	24.5±1.6	54.2±9.5
Unit weight	g/cm³	1.91±0.08	1.60±0.08

2.2 Experimental Method

Specimens were subjected to three kinds of loading ranges and confining conditions as shown schematically in Fig. 2. Mechanical conditions of a circular slab in Test-S in which the test was conducted at the supporting state were simplest stress distribution among the three kinds of tests. That is, a load is applied uniformly only to a portion of the circular slab with the supported diameter. This test was intended both to investigate the fundamental bending deformation behavior of a frozen soil in the shape of a circular slab and to find out the yield tensile strength of the slab. The conditions of the other two kinds of tests were closer to those at the actual site. In both tests, called Test-N and Test-A, a load was applied entirely throughout the one side of the slab, with confining conditions of the no-adfrozen state and adfrozen state on another side, respectively.

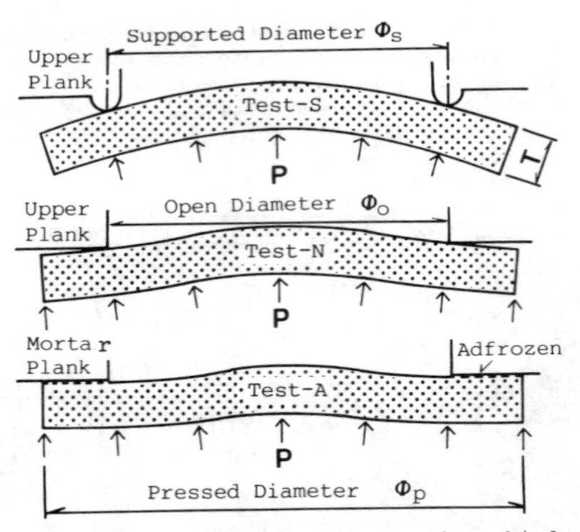

Fig.2 Sketch of tests under various kinds of loading ranges and confining conditions
Test-S: Supported, partially loaded.
Test-N: No-adfrozen, entirely loaded.
Test-A: Adfrozen, entirely loaded.

The experimental apparatus for Test-S is illustrated in Fig. 3. The upper surface of a circular-shaped slab specimen (1) is supported by the upper plank (2) of which the support diameter is 20 cm. The bottom surface of the specimen contacts with the brine (5) supplied from the cylinder (8) through the rubber membrane (4) 1 mm thick within the supported diameter.

The pressing down of the piston (7) of a cylinder for supplying brine at constant speed enables an increase in amount of brine in the sliding tube (6) at constant speed, and so, this allows the speed of increase in deflection of the volume of the slab at a constant rate. The deformation rate of the slab is expressed by the flow of the brine pressed out from the brine supplying cylinder.

A change in brine pressure P resistive to the deflection of the slab is measured by the pressure transducer. The maximum allowable pressure of this apparatus is 4 MPa. The apparatus also includes four differential transducers (9) and strain gauges (11) frozen to the surface of the slab.

Tests have been carried out with this experimental apparatus installed in an insulated box in the cold room, where the

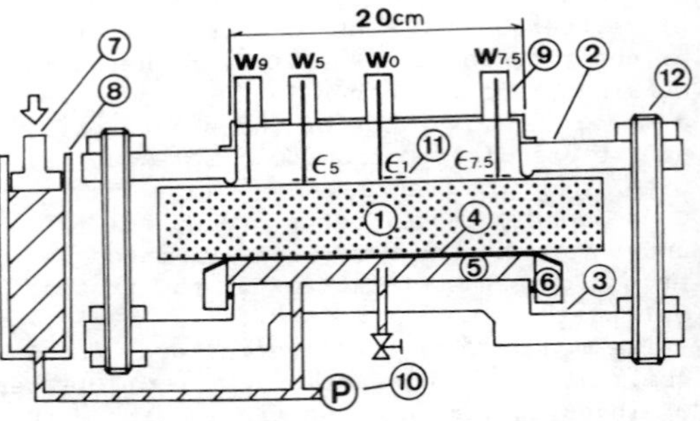

1 Frozen circular slab 2 Upper plank 3 Lower plank 4 Rubber membrane 5 Brine 6 Sliding tube 7 Piston 8 Brine supply cylinder 9 Differential transducers 10 Pressure transducer 11 Strain gauges 12 Support

Fig. 3 Experimental apparatus

air temperature was maintained constant at -10 C.

The apparatus used for Test-N and Test-A was the same as the one for Test-S except for the upper plank and the loading range. In case of Test-N, a specimen contacts the plain iron plank with a circular cavity 10 cm in diameter; and in case of Test-A a specimen is frozen to the mortar plank contacted with the iron plate. In both cases, a load was applied to the entire area of the bottom surface of the specimen.

3 RESULTS

3.1 Test-S (supported, partially loaded)

Figure 4 shows changes with time in displacement of center, w_0, strain ϵ_1 at a point of 1 cm from center on the top surface and brine pressure P, which provides typical changes. In all cases, each test indicates a gradual increase at the initial stage, and then a linear increase with the lapse of time. The gradual increase was caused by a seating effect between the top surface of the specimen and the upper plank.

The frozen sand, when the deformation rate, v, was large, exhibited elastic failure, and instantaneous rupture was formed by the formation of cracks throughout the specimen. However, when v was small, the frozen sand exhibited plastic failure, with gradual formation of many vague cracks. The terminologies of "elastic" and "plastic" failure were used to indicate whether the plot of pressure versus time was respectively linear and linear plus nonlinear.

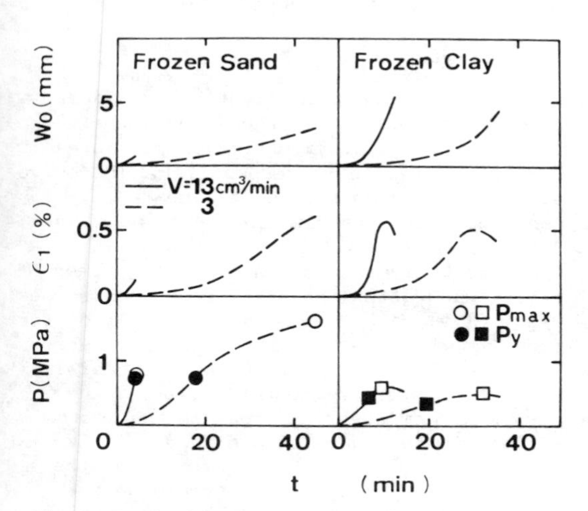

Fig.4 Typical changes with time t in displacement w_0 of center, strain ϵ_1 at a point 1 cm from center and brine pressure P (T=3 cm, $\Phi_s = \Phi_p$=20 cm)

Over the range of v studied here, the failures of the frozen clay were all plastic ones. In the plastic failure, the time of departure from the linear portion between P and t nearly coincided with the one between ϵ_1 and t. The yield pressure P_y was defined as the pressure at the time.

Figure 5 sketches cracks in a frozen circular slab after failure when v is large. The cracks of both frozen sand and clay specimens extend in a radial direction from the center. In the former case, five to eight cracks extend toward the fringe of the slab, and throughout the bottom surface. These cracks formed instantaneously, when the pressure was the maximum pressure P_{max}. On the other hand, in the latter case the cracks were three to four thick ones with several thin ones around them. Furthermore, these cracks in the frozen clay existed within the support diameter, and did not reach the bottom of the slab.

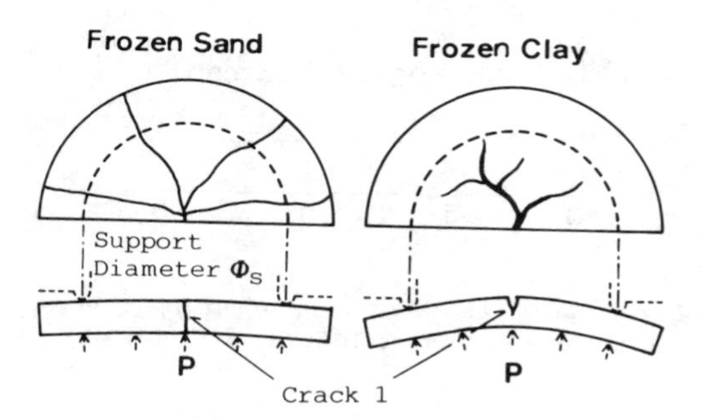

Fig.5 Sketches of cracks in a circular slab after failure in case of Test-S (T=3 cm, $\Phi_s = \Phi_p$=20 cm)

The yield pressure P_y and the maximum pressure P_{max} versus the deformation rate v are presented in Fig. 6, where P_y of both frozen sand and clay gradually decrease with decreasing v. Meanwhile, P_{max} of frozen clay also shows a similar monotonous decrease to P_y. However, P_{max} of frozen sand shows a rapid increase with decreasing v in a range of v less than 5 cm^3/min, accompanying a great increase in deflection.

Whether or not shear deformation occurred was studied by measuring the strain distribution on the top surface of the slab in the radial direction. The normalized strain ϵ' at P_{max} versus radius r is presented in Fig. 7 for frozen sand and clay specimens of various thicknesses. The value of ϵ' was obtained from dividing the measured strain at each radius by the strain

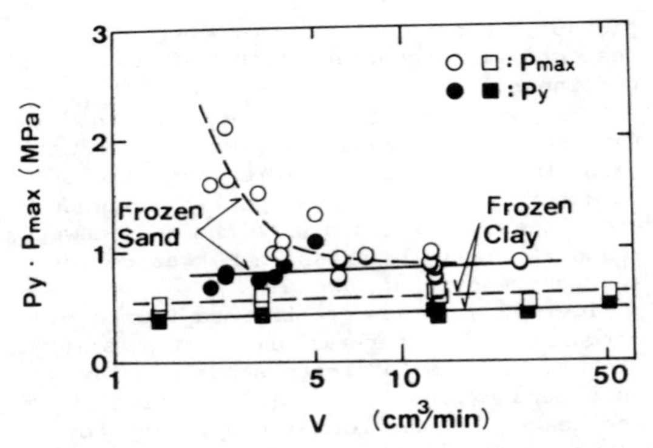

Fig.6 Yield pressure P_y and maximum pressure P_{max} versus deformation rate v in case of Test-S (T=3 cm, $\Phi_s = \Phi_p$=20 cm)

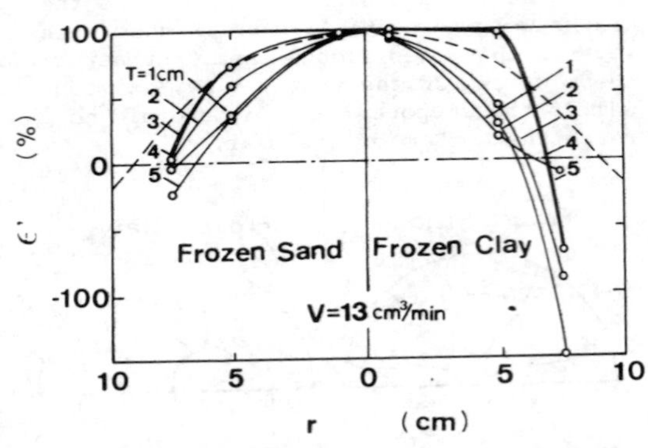

Fig.7 Normalized strains ϵ' at P_{max} versus radius r in case of Test-S ($\Phi_s = \Phi_p$=20 cm)

at the center. The ϵ'-r curves of frozen sands are gentle over the range of the specimen. The curves of frozen clay, thinner than 3 cm, are also gentle; however the curves, thicker than 4 cm, change sharply at a point between 5 and 7.5 cm in radius. Similar changes near a point 6 cm in radius are also observed, as to the relationship between the deflection on the top surface and the radius. Rapid changes in surface strain and deflection may suggest that the frozen slab exhibits conspicuous shear deformation near P_{max} with increasing thickness.

3.2 Test-N (no-adfrozen, entirely loaded)

Typical curves of brine pressure P versus time are shown by solid lines in Fig. 8 for Test-N. In case of frozen sand, instantaneous release in P as shown in Fig. 8 and strain on the top surface occur twice, usually before the maximum pressure P_{max} is

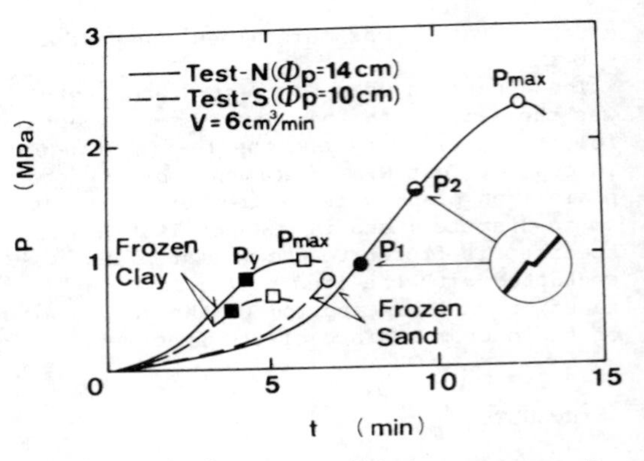

Fig. 8 Typical curves of brine pressure P versus time in case of Test-N and Test-S (T=1.5 cm, Φ_o=10 cm)

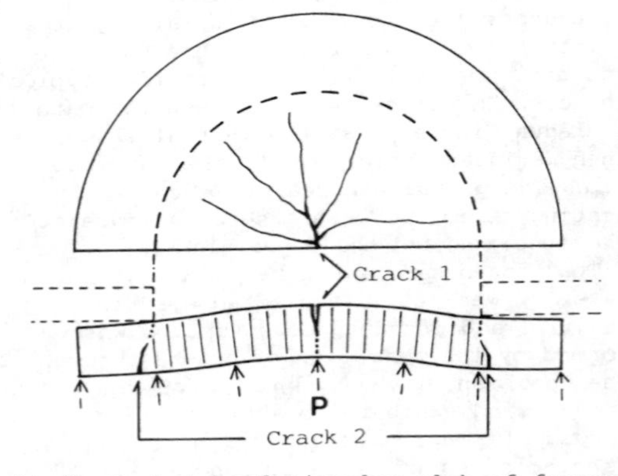

Fig.9 A sketch of circular slab of frozen sand after P_{max} in case of Test-N (T=1.5 cm, Φ_o=10 cm, Φ_p=14 cm)

reached. These changes might be considered to have been caused by the formation of different types of cracks. Such pressures were named P_1 and P_2 respectively. The curve of frozen clay consisted of linear and nonlinear portions similar to these of Test-S as shown in Fig. 8 by a dotted line.

Figure 9 gives a sketch of the specimen of frozen sand after the brine pressure reaches P_{max}, of which the P-t curves largely differ from those of Test-S. Two types of cracks are observed: type 1 cracks extend radially from the center on the top surface ; the type 2 circumferential crack on the bottom surface in the shape of a circle approximately 12 cm in diameter. From the observation of a change in strain on the surfaces and cracks formed, it was confirmed that the type 1 cracks broke out at the time of appearance of P_1, followed by the starting of formation of the type 2 cracks.

Dotted lines in Fig. 8 show the P-t curves of Test-S, conducted under the same experimental conditions as those of Test-N except the upper plank and loading range. A comparison of the P-t curves of Test-N to those of Test-S shows that the P_{max} of frozen sand and clay on Test-N was three times and one and half times ones of Test-S, respectively. A significant difference in P_{max} observed in case of frozen sand is considered to come about from the manner in which cracks were formed as follows: in case of Test S, even when a slight crack starts forming, it brings about the entire rupture of the specimen. However, in case of Test-N, as mentioned above, type 1 cracks did not extend to the fringe of the specimen and the pressing upward of the central portion of the specimen, which is hatched in Fig. 9, was prevented by an wedge effect. These account for the reason why P_{max} of Test-N is by far the larger than that one of Test-S.

3.3 Test-A (adfrozen, entirely loaded)

A change in pressure with time and the manner in which cracks were formed were similar between Test-A and Test-N.

Plots of the yield pressure P_y ($P_1 = P_y$ in the case of frozen sand) and the maximum pressure P_{max} are given in Fig. 10 against the adfrozen distance f, which is the distance in a radial direction. Both P_y and P_{max} increase with increasing f in a range of f between 0 and 2 cm, but beyond 2 cm they no longer increase. Under the perfectly clamped condition in a range over 10 cm in open diameter, P_y can be calculated by substitution of a yield tensile strength obtained from Test-S in an equation based on the elastic theory under pure bending (Yuasa, 1964). The calculated values of P_y in frozen sand and frozen clay, which are 1.6 MPa and 0.8 MPa, respectively and are shown by a dotted lines in Fig. 10, agree almost with the measured values, in the range of f over 2 cm. In addition, in such a range of f any separation of the specimen from the mortar plank was not found. These results suggest that the confining condition of Test A in a range of f over 2 cm against the open diameter of 10 cm is similar to the clamped condition.

4 DISCUSSION

4.1 Shape effect on yield tensile strength

The yield tensile strength σ_y of a frozen circular slab was obtained by substituting the yield pressure P_y of Test-S into the following equation on the basis of the elastic theory under pure bending obtained by Timoshenko and Goodier (1951).

$$\sigma_y = \frac{3(3+\nu)}{8} \cdot \left(\frac{a}{T}\right)^2 \cdot P_y$$

where a is the supported radius, T is the thickness of the circular slab and ν is the Poisson's ratio, which is assumed to be 0.3.

The σ_y for various T are presented in Fig. 11. In frozen sand, the dependency of σ_y on T was not found, but σ_y of frozen clay gradually decreased with increasing T over 3 cm. This decrease in σ_y can be interpreted as due to the shear deformation presented in Fig. 7. Therefore it was found that the shape effect concerned with the thickness

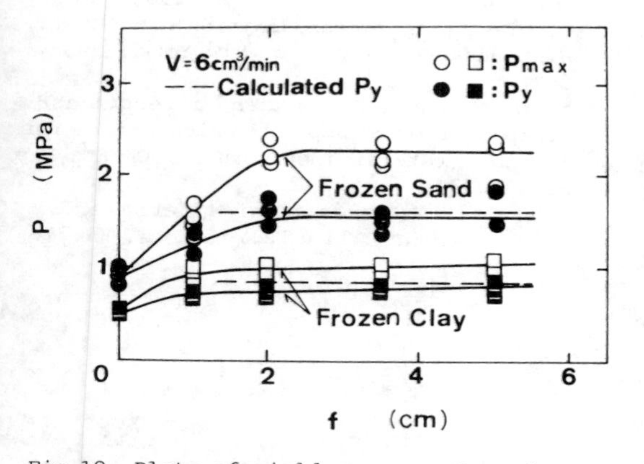

Fig.10 Plots of yield pressure P_y and maximum pressure P_{max} versus adfrozen distance f in case of Test-A (T=1.5 cm, Φ_o=10 cm)

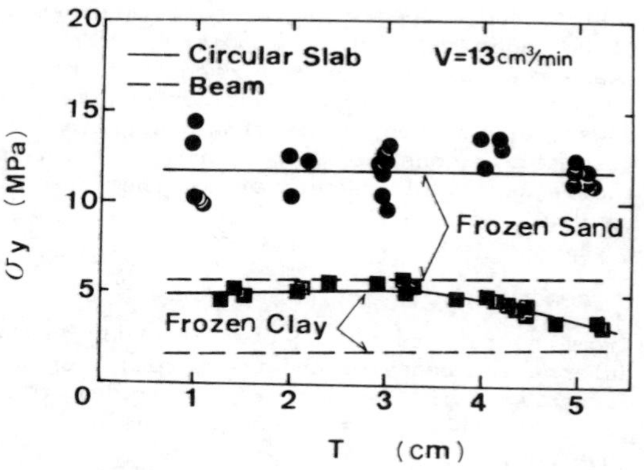

Fig.11 Yield tensile strength σ_y of frozen circular slab versus thickness T in case of Test-S ($\Phi_s = \Phi_p$=20 cm)

existed in the case of frozen clay.

Another consideration is the effect of the difference in shapes of frozen soils, that is whether the shape is a circular slab or a rectangular beam which is commonly used to investigate the tensile strength. Values of σ_y for the rectangular beam of the same frozen sand and clay obtained by Izuta et al.(1988) are presented in Fig. 11 by dotted lines. A comparison between the values of σ_y showed that the σ_y of circular slab without noticeable shear deformation were approximately twice the σ_y of the beam. It is interesting to note that the ratio of the σ_y of the slab to the σ_y of the beam agrees with the ratio for polycrystalline ice reported by Gold (1977).

4.2 Effect of loading range and confining condition on yield pressure

As already mentioned in the argument concerning the result of Test-N, an increase in loaded area from the central portion of the specimen within the supported diameter to the entire surface of the specimen resulted in an increase in yield pressure. This increase in P_y seemed to be result from an increase in resistance against the deflection of the specimen, and agreed with the solution of an equation on the basis of the elastic theory.

The value of adfreeze strength between the frozen soil and the mortar plate at -10 C has been found to be 1.5 to 2.0 MPa by the authors. A change in the confining condition from the no-adfrozen state to the adfrozen state had been expected to make P_y larger. However, the obtained result suggested that when the diameter of the specimen exceeded a certain value, P_y of Test-A was hardly different from that of Test-N, and agreed with the calculated value of P_y under the clamped condition. From these results, it was confirmed that the frozen circular slab without noticeable shear deformation larger than a certain diameter is considered to be almost clamped, independently of whether or not the frozen soil was adfrozen.

5 CONCLUSIONS

From the results of the present study on the deformation behavior and the strength of frozen soil in the shape of a circular slab, the following tentative conclusions were obtained:

1) From the results obtained by supported and partially loaded tests (Test-S), the yield pressure P_y of both frozen sand and clay gradually decreased with the decreasing deformation rate v. The maximum pressure P_{max} of frozen clay also exhibited the similar decreasing tendency, but P_{max} of frozen sand did not; that is , its P_{max} showed a rapid increase with decreasing v in a range of v less than a certain value. The measurement of the surface strain and deflection distributions suggested that the shear deformation became noticeable with the increasing thickness T of the slab in the case of frozen clay.

2) The yield tensile strengths σ_y of the frozen circular slabs without noticeable shear deformation were found to merely depend on T, and also found to be approximately twice the values of rectangular beams.

3) P_{max} of entirely loaded tests (Test-N and A) were considerably larger than these of Test-S, especially in case of frozen sand the magnification of P_{max} of Test-N to that of Test-S was found to be two to three.

4) The confining condition of the no-adfrozen tests (Test-N) and adfrozen tests (Test-A) can be considered to be a clamped condition, when the diameter of a slab is larger than a certain value.

REFERENCES *

Gold, L. W. 1977. Engineering properties of fresh-water ice. J. Glaciology. 19. 81. : 197-212.

Izuta, H., Yamamoto, H. & Ohrai. T. 1988. Deformation behavior and mechanical properties of frozen soils in a bending test. J. Japanese Society of Snow and Ice. 50. 1.

Orth, W. & Meissner, H. 1985. Experimental and numerical investigations. Proceedings of Fourth International Symposium on Ground Freezing. 2.: 259-262.

Timoshenko, S. & Goodier, J.N. 1951. Theory of elasticity (second edition). McGRAW-HILL Book Company. Inc. New York.: 343-352.

Vialov, S.S. et al. 1962. The strength and creep of frozen soil and calculations for ice-soil retaining structures. CRREL. Rep. AD-484093.

Yuasa, K. 1964. Strength of Materials. 2. (in Japanese).: 110-121. Korona. Tokyo.

*See page X concerning ISGF papers

5th International Symposium on Ground Freezing, Jones & Holden (eds)
© 1988 Balkema, Rotterdam. ISBN 90 6191 824 3

On the regularities of the change of shear strength of soils with thawing and dynamic loadings

V.D.Karlov & S.V.Arefyev
Leningrad Institute of Constructional Engineers, Leningrad, USSR

ABSTRACT: The complex investigations of the strength properties of the thawing sandy soils were carried out under static (up to 0.3 MPa) and dynamic (up to 1500 mm/s^2) loadings. The experiments were made using artificially prepared samples of frozen sand with different granulometric structure on a specially constructed installation. The installation has permitted us to determine the shear strength limits directly on the boundary of thawing and beyond it. The samples were 113mm in diameter and 50mm high.

The experiments made with thawing soils have shown a considerable reduction of shear strength in these soils (with vibration acceleration of 0.5g, 1.0g and 1.5g under constant static loading) and its linear dependence on vibration acceleration.

The influence of the vibration frequency is complex. The maximum decrease of shear strength occurs with the vibration frequencies of 20-25 cycles per second. The degree of strength decrease of thawing soils depends on their granulometric structure and look like second order curves with homogeneous character (with different values of loading). Thawing sandy soils with the fraction size from 0.5 to 1.0mm have maximum values of shear strength.

1 INTRODUCTION

The problems of the bearing capacity decrease of sandy soils with thawing and vibrodynamic loadings in recent years have become important in connection with the intensive industrial development of Northern regions and artificial soil freezing. This problem appears to be still more important in the light of the fact that there are no sufficiently serious investigations dedicated to the change of the strength properties of thawing soils under the action of vibrodynamic loadings.

The experiments for determining the shear strength limit of thawing soils with different types of loading were carried out in two stages.

The first stage of investigations was the main one, and its aim was to find out the dependence of the shear strength limit of thawing soils upon the value of the vertical static pressure, parameters of vibro-dynamic influences and granulometric structure of the investigated sandy soils in laboratory conditions. The second stage was the in situ testing without vibro-dynamic influence upon the soil. The purpose of this was the more detailed study of the process of the strength properties decrease of sandy soils under thawing conditions with consideration of the time elapsed since the beginning of thawing.

2 LABORATORY TESTS

The laboratory tests were carried out on a special installation, intended for determining the strength properties of thawing soils, which was developed and constructed in the laboratory of frost studies at the Department of Foundation Engineering and Soil Mechanics in Leningrad Institute of Constructional Engineers. The most characteristic parts of the installation are: the hydraulic system to cause some vertical static pressure on a sample; the ability to provide the plane-parallel thawing of a frozen ground sample in the non heat-conducting oedometer by warm water circulation at a fixed temperature; the laboratory blade paddle impeller intended for the direct determination of the shear strength limit of the thawing soil (τ dyn.thaw lim.) which enables a shear to be applied directly on the boundary of thawing, i.e. in the mostly weakened zone. A special device makes it possible to mount the apparatus on the vibrating test-bench. The parameters of the vibro-dynamic influences were recorded by means of seismo-sensing element mounted on the apparatus to determine the strength properties of thawing soils.

In the first series of the laboratory experiments the influence of the value of the vertical static pressure on the strength properties of melting and thawing soils, both under static loading and under dynamic effect, was studies (Figure 1).

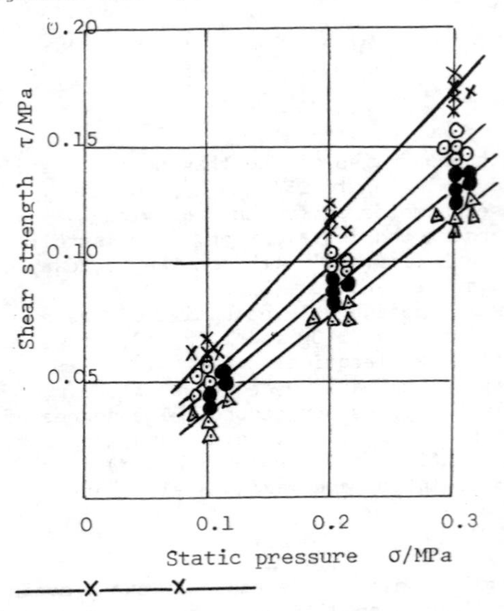

—x—x— melting soil with static pressure

—⊙—⊙— thawing soil with static pressure

—●—●— melting soil with static pressure and dynamic

—△—△— thawing soil with static pressure and dynamic

Figure 1 The dependence of the shear strength limit on the vertical pressure

For the experiments, a median-grained sandy soil was used taken from the deposits of the river Tosno with the following granulometric structure of natural composition:
fractions larger than 1mm - 16%; 0.5mm - 27%; 0.25mm - 52%; 0.1mm - 3%; and the rest - 2%. The samples were prepared from entirely water-saturated sandy soils with the unit weight of 1.6 t/m^3.

The experiments were carried out with three different values of static pressure 0.1, 0.2 and 0.3 MPa. The dynamic loading was applied with the vibration frequency of 50 cycles per second and with the vibration acceleration of 0.5g. All the experiments were repeated six times. The graph (Figure 1) shows the dependence of
1) the shear strength limit of the melting sandy soil on the vertical static pressure;
2) the shear strength limit of the thawing sandy soil on the vertical static pressure;
3) the relation 1 with consideration of dynamic loading
4) the relation 2 with consideration of dynamic loading.
The average of strength as to the relation 1 was 18.3; 20.5. 27.0% for the 2nd, 3rd and 4th relations respectively. As one can see from these data the degree of strength decrease reaches sufficiently great values, which should be taken into account without fail in designing buildings and constructions on the thawing soils subjected to dynamic loading. Besides, the greater the value of the vertical static pressure, the greater the increase of the degree of the strength decrease. This tells especially on the samples under the combined effect of thawing and dynamic loading.

The second series of the laboratory experiments aimed to find out the dependence of shear strength limit for the melting and thawing of sandy soils (with consideration of dynamic loading and without it) upon the granulometric structure of the investigated soils.

For this purpose the same soil, as in the first series of experiments, was screened through several sieves, then the soil fractions which were left on the sieves with the mesh-sizes of 0.25; 0.5 and 1.0mm were selected. The density and moisture of the investigated samples, as well as the parameters of the dynamic influences, were taken the same as in the first series of experiments. The value of the vertical static pressure was constant and equal to 0.1 MPa. The results of the experiments are shown in the graph (Figure 2). The symbols are given in Figure 1. The average decrease of strength as to the relation 1 was 19.1;22.4;28.0%. The maximum value of the shear strength limit was found for the soils with the fraction sizes from 0.5 up to 1.0mm. The relative degree of the decrease of τ dyn, thaw lim. for the soil subjected both to the combined and to the separate effects of thawing and dynamic loading, increases with the reduction of the fraction size of the investigated soil.

Some series of experiments were also performed with varying parameters of the dynamic effects (with frequencies from 20 to 100 cycles per second and vibration acceleration from 0.5 to 1.5g). Vertical static

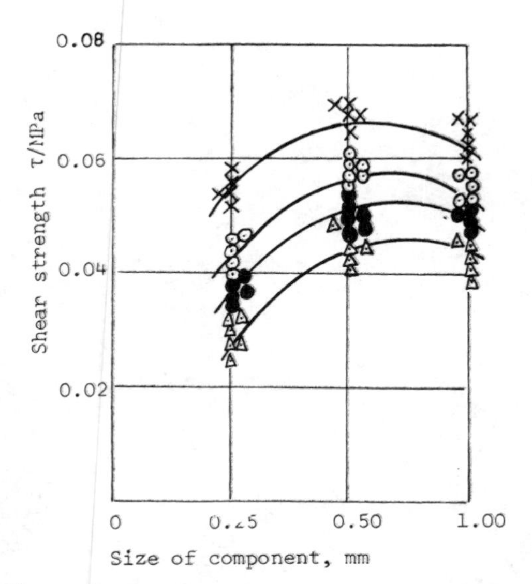

Figure 2 The dependence of the shear strength limit of sandy soils on their granulometric structure with static pressure of 0.1 MPa, vibration acceleration of 0.5g and frequency of 50 cycles per second.

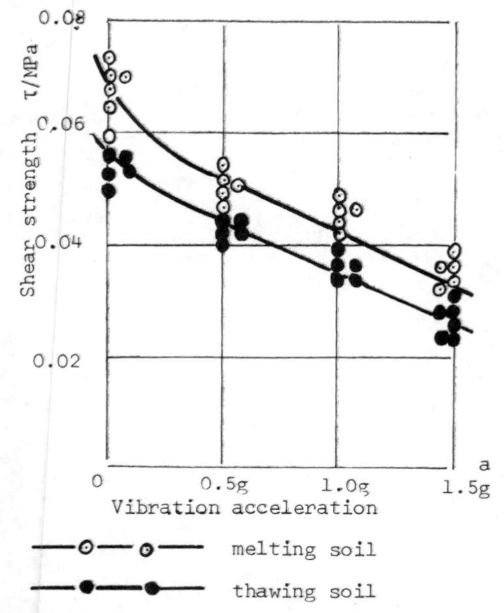

Figure 3 The dependence of the shear strength limit of sandy soils on the vibration acceleration with static pressure of 0.1 MPa

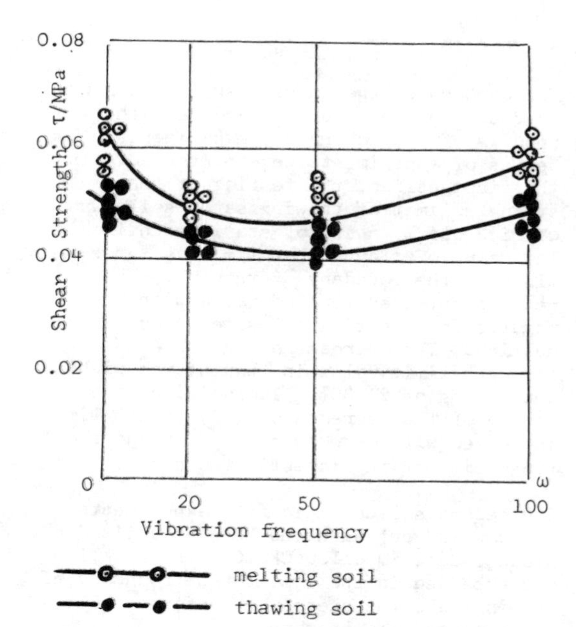

Figure 4 The dependence of the shear strength limit of sandy soils on the vibration frequency with static pressure of 0.1 MPa.

pressure was constant - 0.1 MPa, and all other characteristics were the same as in the first series of experiments. It was ascertained that the decrease of τ dyn. thaw lim. with the increase of vibration acceleration is linear (Figure 3).

The study of the influence of vibration frequency (Figure 4) has shown that the maximum decrease of the shear strength limit is at the frequencies of 20-25 cycles per second. The average decrease of strength with the combined effect of thawing and dynamic loading is about 30%, in both cases.

Drawing a general conclusion from the results of all the laboratory experiments, we shall note that in all cases the decrease of the shear strength limit for sandy soils with thawing only was about 20%, and with the combined effect of thawing and dynamic loading, it was about 30%. These are significant values , and it is not permissible to ignore them in the process of designing foundations for buildings and other structures and constructions under such conditions. Some other interesting quotient relations were obtained for shear strength limits: τ dyn thaw lim. on one side, and vertical static pressure, granulometric structure of sandy soils and the parameters of dynamic actions, on the other side.

3 IN SITU TESTS

The purpose of the in situ experiments was to check and to make more accurate the results of the laboratory experiments. Two series of experiments were performed. In the first series, the testing of the shear strength limit of thawing sandy soils was carried out by means of static soundings. The value of τ dyn. thaw lim. was determined at the boundary of thawing. The results obtained were compared with the results from testing the same soils during melting. The decrease of the shear strength limit value in thawing was within the limits of 25-30%. In addition, the degree of the decrease of τ dyn. thaw lim. increased with decreasing density and the finer size of the investigated sands.

The second series of in situ experiments was carried out with the help of a field blade paddle impeller CK-10. The impeller was embedded in the soil to the boundary of thawing, and the shear was performed. It was repeated at the same depth in 6,12,18 and 24 hours after the first measurement. When investigating the flour sandy soils it was found that the character of the recovery curve remained the same with the general decrease of the value, in the limits of 40%.

CONCLUSION

The main conclusion of all the experiments is:
the decrease of the shear strength limit of sandy soils subjected to thawing (and more so with dynamic effects) may be very substantial and this must be taken into account when calculating the stability of footings, slopes and retaining walls.

5th International Symposium on Ground Freezing, Jones & Holden (eds)
© 1988 Balkema, Rotterdam. ISBN 90 6191 824 3

Acoustic properties of frozen near-shore sediments, southern Beaufort Sea, Arctic Canada

P.J.Kurfurst & S.E.Pullan
Geological Survey of Canada, Ottawa, Ontario, Canada

ABSTRACT: Uphole seismic surveys, carried out in offshore boreholes up to 32 m deep, yielded compressional wave velocities of frozen marine sediments between 2200 and 3800 m/s. Acoustic wave velocities measured in the laboratory on core samples recovered from the boreholes tested in the field produced compressional wave velocities between 2400 and 3700 m/s and shear wave velocities ranging from 800 to 1900 m/s. Sea ice samples were also tested, producing compressional wave velocities between 3300 and 3600 m/s and shear wave velocities between 1650 and 1900 m/s.

Standard engineering property tests were carried out on undisturbed core samples. Using the values of bulk density and laboratory acoustic wave velocities, acoustic elastic constants were calculated. Young's modulus (E), Poisson's ratio (ν), shear modulus (G) and bulk modulus (K) were then compared for various soils (clays, silts and sands) and for sea ice. These data demonstrate that acoustic wave velocities and dynamic moduli are a function of the ice content of the soil, while their dependence on soil type is only minor. Acoustic wave measurements from the field and from the laboratory are in good agreement, thus confirming reliability of both techniques used and their suitability for determination of acoustic elastic constants.

INTRODUCTION

Data on the physical and mechanical properties of frozen seabottom sediments are vital to the engineering design of structures such as submarine or subseabed pipelines and marine terminals. The near-shore areas represent a transition zone between the onshore and offshore environments, a zone where only limited data on soil properties are available.

Twenty boreholes were drilled and tested during the springs of 1986 and 1987 along two onshore-offshore transects off Richards Island in the southern Beaufort Sea (Figure 1). This was part of a detailed geotechnical investigation of the near-shore sediments carried out to provide information on physical and engineering properties of seabottom sediments and to correlate and compare them with their geophysical properties. Besides geotechnical measurements such as shear vane tests and cone penetration tests, acoustic wave velocities of the sediments were measured. The details of field investigations are described by Kurfurst (1986, 1988).

This paper describes the results of

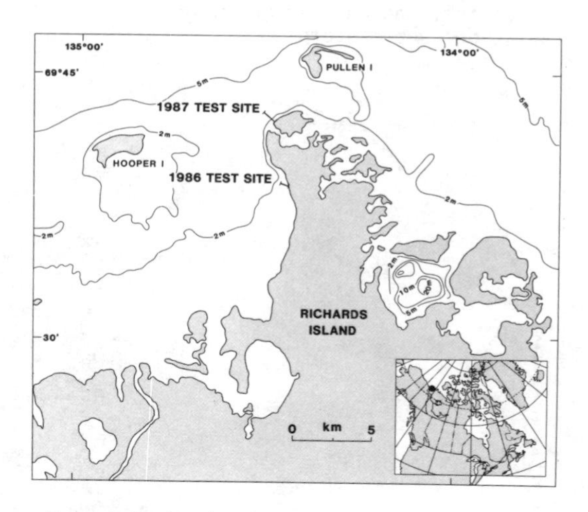

Figure 1. Site location plan

field seismic surveys and laboratory acoustic velocity measurements on representative frozen samples of marine sediments, summarizes the calculated acoustic elastic constants and discusses their dependence on ground ice content.

TESTING PROCEDURES AND EQUIPMENT

Uphole seismic velocity measurements were made in the PVC-cased boreholes drilled through the sea ice and into seabottom. A seismocap detonated beneath the bottom of the ice alongside the borehole was the signal source. The receiver used was a Mark Products P-44 hydrophone hanging freely in the antifreeze-filled borehole. The receiver was moved upward in 1 m intervals and the shots were recorded using a Nimbus 1210F engineering seismograph and digital tape recorder. The data were transferred to floppy-disk storage and first arrival times were obtained using computer-based routines. After correction for shot offset and depth, the distance-travel time data were plotted and interpreted. Typical plots from boreholes 86-6 and 86-7 are shown on Figures 2 through 5. The first plots (Figures 2 and 4) show shot-hydrophone distances (after correction for shot position) and the observed travel times with the interpreted interval velocities shown as straight line segments. The second plots (Figures 3 and 5) show these interval velocities plotted against depth of hydrophone in the borehole. Timing errors are on the order of ±0.05 ms and hydrophone position errors are in the order of ±5 cm. A conservative approach has been used to interpret interval velocities, and only large scale variations were interpreted.

Acoustic wave velocity measurements were made in the laboratory on core samples from the boreholes where uphole seismic measurements were carried out in the field. An OYO Sonic viewer 5217A was used to measure the compressional and shear wave velocities independently. Two sets of transducers with different frequencies were used for the measurements: 33 kHz and 100 kHz shear wave transducers, and 63 kHz and 200 kHz compressional wave transducers. Pulse travel time was read directly from the display screen and the pulse trace was simultaneously recorded and printed. The details of the technique used are described by Kurfurst (1977).

Compressional (V_p) and shear (V_s) wave velocities were calculated, the density (ρ) of samples was determined in the laboratory and acoustic elastic constants [Young's modulus (E), Poisson's ratio (ν), shear modulus/modulus of rigidity (G) and bulk modulus (K)] were calculated using formulas recommended by the American Society for Testing and Materials (ASTM, 1984).

FIELD AND LABORATORY RESULTS

Boreholes 86-6 and 86-7 were chosen as representative examples of lithological, ground ice and acoustic velocity profiles. Detailed core logging identified several distinct lithologies, shown on Figures 3

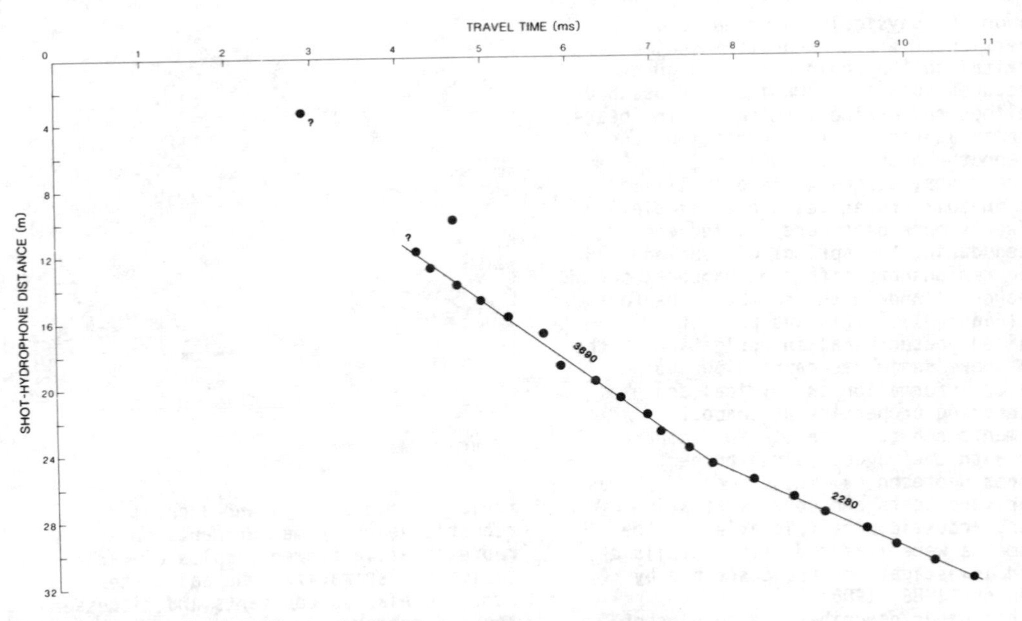

Figure 2. Compressional velocity vs. shot-hydrophone distance - borehole 86-6

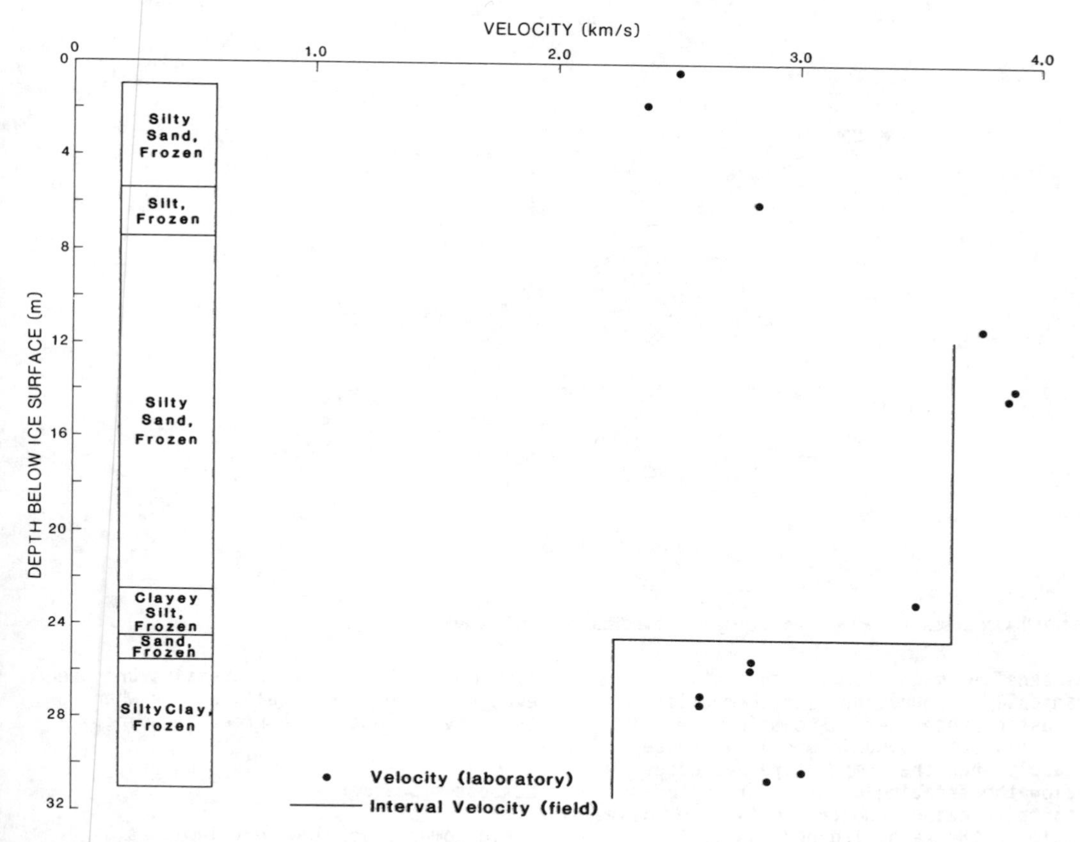

Figure 3. Compressional velocity vs. hydrophone depth-borehole 86-6

and 5. Laminated marine silts and reworked sands of Holocene age are underlain by Wisconsinan fluvial sands and marine silts and clays. Ice contents of all sediments, calculated as a percentage of the total volume, ranged from 22% to 40%; the bulk density was between 1.77 and 2.02 g/cm^3 with a mean of 1.87 g/cm^3.

Results of the vane shear tests and cone penetrometer tests indicate that, although most of the sediments have relatively low shear strengths (less than 20 kPa), a zone of slightly overconsolidated sediments with shear strengths up to 50 kPa is present between 6.5 m and 10 m below seabottom.

Uphole seismic velocities (V_p) of the sediments have been calculated as described above. Most interval velocities, as shown on Figures 2 and 4, ranged from 2230 to 3725 m/s, although an anomalously low velocity of 1130 m/s was recorded for sediments around 6 m depth in borehole 86-7. Detailed study of all results failed to explain this aberration.

Acoustic wave velocities (V_p and V_s), measured on 11 samples from borehole 86-6 and 11 samples from borehole 86-7, ranged from 2360 to 3990 m/s and from 780 to 1870 m/s respectively. Relationship between acoustic wave velocities and total ice content is shown on Figure 6. All acoustic measurements, the corresponding uphole seismic measurements and calculated acoustic elastic constants (E, ν, G and K) are summarized in Table 1.

All laboratory measurements were made at temperatures approximating those measured in the field, i.e. -6°C for samples from borehole 86-6 and -4°C for samples from borehole 86-7.

DISCUSSION

The marine sediments tested had undergone freezing for an extended period of time. Although the freezing process has little or no effect on such physical properties

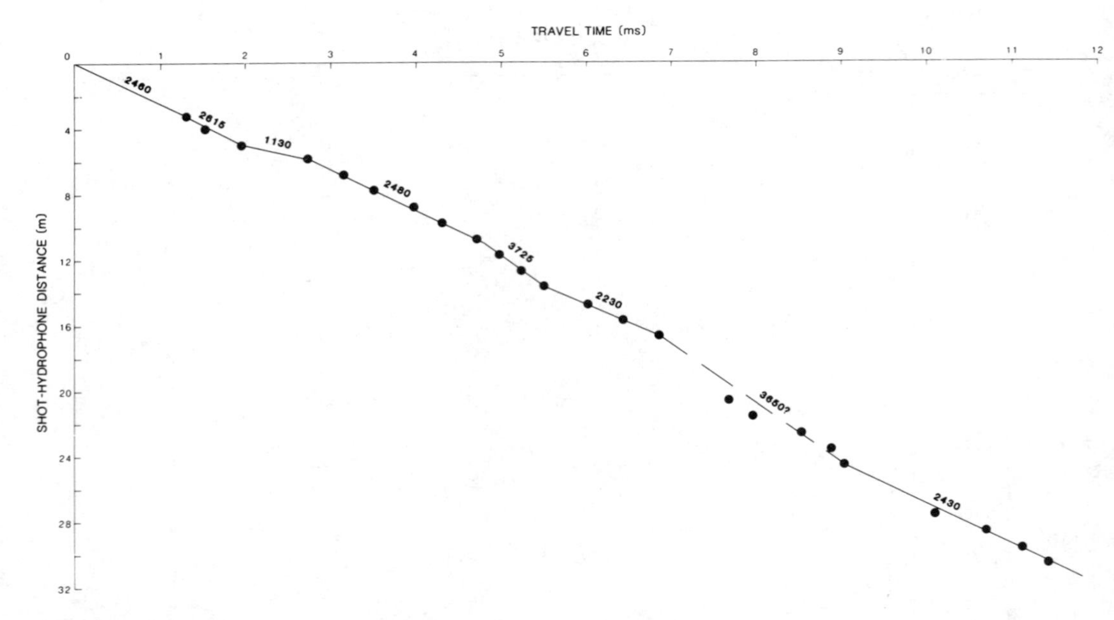

Figure 4. Compressional velocity vs. shot-hydrophone distance -borehole 86-7

as density, magnetism or radioactivity, it radically changes the elastic moduli and acoustic properties. Acoustic wave velocities and dynamic moduli increase sharply when the temperature decreases below the freezing point. Pore water starts freezing, and the surface-to-volume ratio in the remaining pore space increases, thus accelerating the freezing of the remaining pore water. However, a portion of the pore water remains unfrozen even at temperatures below 0°C or when salinity depresses the freezing point.

Seabottom Sediments

Field compressional wave velocities, determined by the seismic uphole measurements, agree well with

Table 1. Borehole 86-6 and 86-7 data

Sample depth (cm)	Lithology	Density (g/cm^3)	V_p field (m/s)	V_p lab (m/s)	V_s lab (m/s)	Young's Modulus E(GPa)	Poissons's ratio ν	Shear Modulus G(GPa)	Bulk Modulus K(GPa)
Borehole 86-6									
15-19	silty	1.95	--	2553	1241	8.08	.345	3.00	8.71
165-175	sand	1.88	--	2358	--	--	--	--	--
597-610	silt	1.17		2851	970	3.16	.435	1.10	8.04
1060-1078		1.99		3872	1875	18.71	.337	19.14	7.00
1383-1399	silty	1.96	3690	3918	1544	13.16	.408	23.86	4.67
1404-1422	sand	2.02		3901	1754	17.07	.373	22.45	6.21
2262-2310		1.82		3529	1119	6.58	.444	19.63	2.28
2536-2554		1.80		2843	958	4.74	.436	12.35	1.65
2562-2580	silty	1.79	2280	2843	883	4.04	.447	12.61	1.40
2688-2706	clay	1.85		2615	948	4.74	.424	10.43	1.66
2709-2727		1.90		2645	850	3.96	.442	11.46	1.47
Borehole 86-7									
2400-2418	clayey			3957	1764	15.16	.376	5.51	20.37
2502-2520	silt		3660	3991	1673	14.12	.376	5.07	22.07
2540-2558		1.86		3370	1264	8.43	.418	2.97	17.16
2596-2614		1.80		2885	1000	5.15	.432	1.80	12.58
2621-2619		1.85		2828	879	4.13	.447	1.43	12.89
2640-2658	silty	1.84		2662	933	4.58	.430	1.60	10.90
2660-2678	clay	1.86	2430	2631	1028	5.54	.410	1.97	10.25
2687-2705		1.81		2616	918	4.36	.430	1.52	10.35
2839-2857		1.86		2432	778	3.25	.443	1.13	9.50
3052-3070		1.76		2660	810	3.35	.449	1.15	10.91
3078-3096		--		2647	882	--	.446	--	--

Note: Samples from borehole 86-6 were tested at -6°C, samples from borehole 86-7 at -4°C

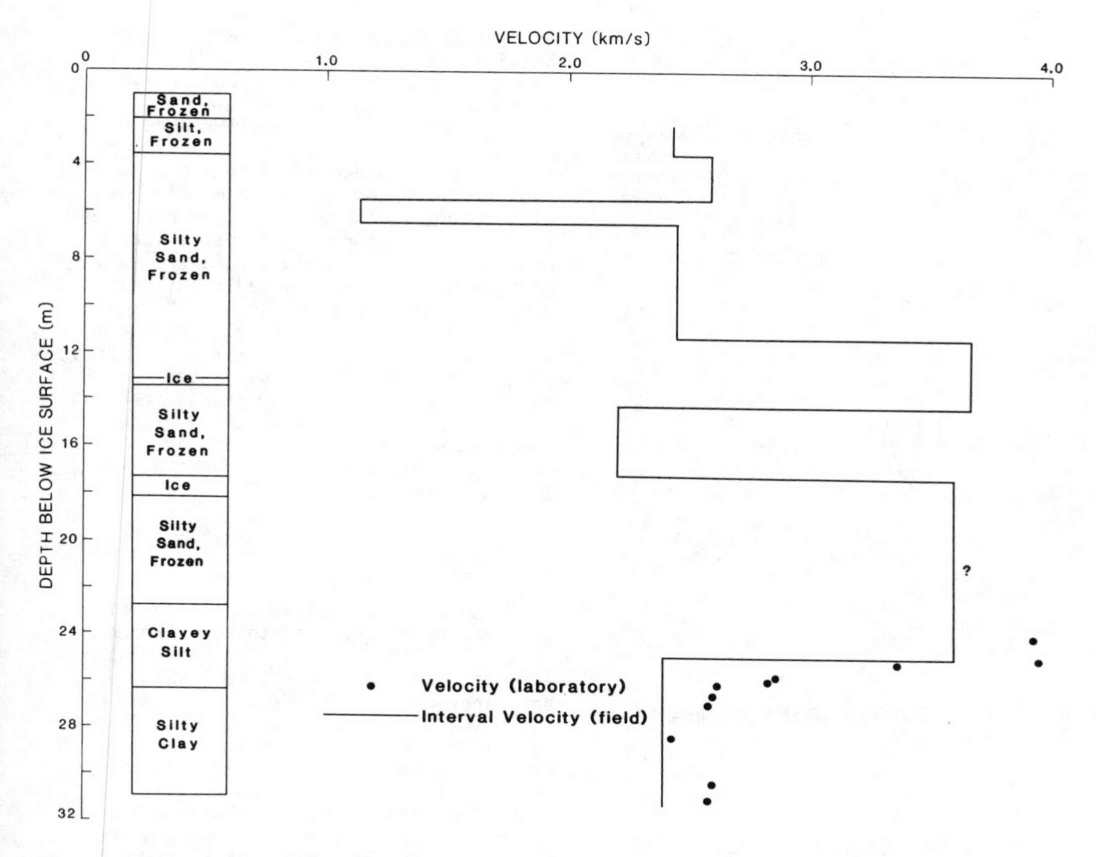

Figure 5. Compressional velocity vs. hydrophone depth -borehole 86-7

compressional wave velocities measured in the laboratory on core samples from the corresponding depths. In general, fine-grained, poorly ice-bonded silty clays display lower compressional wave velocities (2280-2430 m/s) than coarse, well ice-bonded silts and sands (3660-3690 m/s), as illustrated on Figures 3 and 5 and Table 1. The highest velocities were measured in horizons containing ice lenses or massive ice.

Laboratory measurements of both compressional and shear wave velocities show similar behaviour. The compressional and shear wave velocities of fine-grained silty clays range from 2432 m/s to 2885 m/s and 778 m/s to 1000 m/s respectively, while values for silts and sands vary between 3370 m/s and 3991 m/s and between 1120 m/s and 1875 m/s respectively. These values correspond well with values of interval compressional wave velocities measured by the seismic uphole surveys. In several cases, the velocities measured in the laboratory were 10 to 15% higher than the velocities measured in the field. This discrepancy was caused by the higher ice content in the tested specimens, as core samples with highly visible ice bonding or core samples including ice lenses or solid ice were selected preferentially for testing.

Review of the field and laboratory results showed that neither the uphole seismic survey in the field nor the acoustic measurements in the laboratory detected any change in acoustic wave velocities when measured in the horizons of overconsolidated sediments or on overconsolidated samples.

Acoustic elastic constants showed behaviour similar to that of acoustic wave velocities. Values of Young's modules, shear modulus and bulk modulus for fine-grained silts and silty clays were much lower than those for coarse sands and silty sand. However, values of Poisson's ratio were generally constant for all soil types.

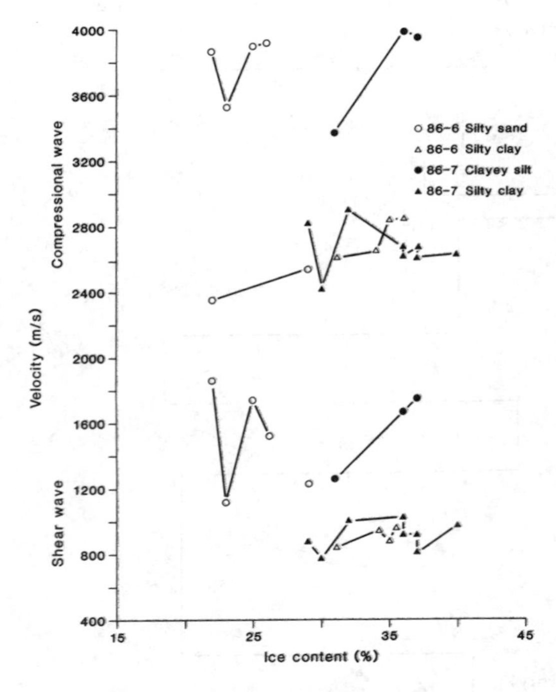

Figure 6. Relationship between acoustic velocities and ice content

Sea ice

Laboratory measurements of acoustic wave velocities have been also carried out in eight samples of artifical sea ice. The samples were cored from a sheet of ice grown in the laboratory, using saline water concentration (11 ppt) and temperatures (-20°C and -35°C) approximating those encountered in the Beaufort Sea.

Compressional and shear wave velocities averaged 3300 m/s and 1700 m/s respectively at -20°C and increased to 3560 m/s and 1820 m/s respectively when the temperature decreased to -35°C. Similar increases of approximately 10% with decreased temperature were recorded for values of Young's modulus, shear modulus and bulk modulus; however, values of Poisson's ratio remained relatively constant throughout the tests.

The increased value of acoustic wave velocities and acoustic elastic constants of sea ice samples with decreased temperature is due to additional freezing of free saline pore water still present in the sample at higher temperatures. The results of the tests carried out at -20°C and -35°C are summarized in Table 2.

CONCLUSIONS

The results of the field uphole seismic and laboratory acoustic tests of frozen marine sediments and sea ice show that compressional and shear wave velocities are dependent on temperature and thus are functions of the ice content and unfrozen pore water in the specimens. The velocities increase proportionally to the increase in ice content caused by decreased temperature. Although coarse-grained sands and well ice-bonded silts display higher velocities than fine-grained clays, acoustic velocities in general are more dependent on the ice content than on the soil type. The higher ice content in the coarse sediments causes a marked increase in the velocities.

Table 2. Sea ice samples data

Sample	Temperature (°C)	V_p (m/s)	V_s (m/s)	Young's Modulus E(Gpa)	Poisson's ratio ν	Shear Modulus G(Gpa)	Bulk Modulus K(GPa)
TS 1	-35	3559	1822	7.90	0.322	2.99	7.42
TS 2	-35	3605	1879	8.35	0.313	3.18	7.46
TS 3	-35	3517	1759	7.43	0.333	2.78	7.42
S 1	-20	3379	1727	7.10	0.323	2.68	6.70
S 2	-20	3306	1690	6.80	0.323	2.57	6.41
S 3	-20	3288	1681	6.73	0.323	2.54	6.34
S 4	-20	3297	1685	6.76	0.323	2.56	6.38
S 5	-20	3297	1657	6.58	0.331	2.47	6.49

Note: All samples had pore water salinity 11.5 ppt and density 0.9 g/cm^3

Acoustic elastic constants (Young's modulus, shear modulus and bulk modulus) also increase with increased ice content and decreased temperature; however Poisson's ratio does not seem dependent on these changes and remains relatively constant.

Sea ice samples display similar behaviour to the samples of frozen marine sediments. When temperature is decreased from -20°C to -35°C, acoustic wave velocities and acoustic elastic constants increase in proportion to the temperature change. This behaviour confirms that with salinity of 11 ppt, a portion of pore water is unfrozen even at temperatures as low as -20°C.

Detailed comparison of the field and laboratory data shows very good agreement between acoustic wave velocities measured in the field (seismic uphole) and in the laboratory (acoustics). This technique, combined with the temperature and salinity measurements, can be used to estimate the ground ice content and monitor its changes, thus reducing the need for costly drilling. However, neither the field nor the laboratory measurements of acoustic wave velocities can detect thin overconsolidated horizons documented in the field.

ACKNOWLEDGMENTS

The authors wish to thank Messrs. J. Bisson and M. Nixon for their assistance during the field and laboratory tests and Messrs. S. Dallimore, J.A. Heginbottom and Dr. J. Hunter for their comments and suggestions.

REFERENCES

American Society for Testing and Materials, 1984. Standard test method for laboratory determination of pulse velocities and ultrasonic elastic constants of rock; in Annual book of Standards, v. 04.08, No. D2845-83, Philadelphia.

Kurfurst, P.J., 1977. Acoustic wave velocity apparatus; in Geological Survey of Canada, Paper 77-1A.

Kurfurst, P.J., 1986. Geotechnical investigations of the near-shore zone, North Head, Richards Islands, N.W.T.; in Geological Survey of Canada, Open File No. 1376.

Kurfurst, P.J., 1988. Geotechnical investigations off northern Richards Island, N.W.T. - 1987; in Geological Survey of Canada, Open File No. 1707.

5th International Symposium on Ground Freezing, Jones & Holden (eds)
 ISBN 90 6191 824 3

A creep formula for practical application based on crystal mechanics

Wolfgang Orth
Wibel & Leinenkugel, Karlsruhe, FR Germany

ABSTRACT: Creep of frozen sand under compressive stress has been investigated in both stress and strain-rate controlled uniaxial and triaxial compression tests. It was found that creep curves obtained under different uniaxial stresses and temperatures can be represented by a unique mastercurve when the time axis is transformed linearly. Consequently, it is sufficient to calculate one representative point of the stress and temperature dependent creep curve. This is done by using a formula derived from consideration of the micromechanical processes occurring in pore ice. In this paper, the theoretical background of this formula is briefly described and its practical application is demonstrated in the case of frozen sand. Although the theoretical background is based on consideration of pore ice, this formula is also applicable to other types of frozen soil.

KEY WORDS: Creep, deformations, laboratory test, mathematical models, mechanical properties, sands, stress - strain relations, structural analysis, temperature effects, theoretical.

1 INTRODUCTION

The mechanical behaviour of frozen sand under compressive stress has been investigated by means of uniaxial and triaxial creep tests, constant strain-rate tests and relaxation tests. The entire testing program and its results, as well as the theoretical interpretation based on crystal mechanics, is given by Orth (1986). In this paper, a creep formula for creep under compressive stress is derived from consideration of thermally-activated processes. Its application for practical creep problems is also presented.

The uniaxial and triaxial laboratory tests were run in the Institute for Soil Mechanics and Rock Mechanics of Karlsruhe University on a specially-designed testing apparatus which is automatically controlled by a micro-computer connected to a servo-hydraulic system. The water-saturated frozen-medium sand samples, - 10 cm in diameter and 10 cm high - were densely packed (dry unit weight: 17.2 kN/m^3) and tested at temperatures from -2°C to -20°C. The end platens were lubricated. Preparation of the specimen by a well-defined procedure and use of the automatic test control led to results of high accuracy and extremely low scattering. All details of the specimen preparation and the testing apparatus are given by Orth (1985).

2 TEST RESULTS

Figure 1a shows an example of a set of creep curves from tests with different uniaxial stresses at -10°C; the corresponding strain-rate versus time curves are shown in Figure 1b. From Figure 1b it can clearly be seen that the strain rate decreases initially and after some time increases again. The so-called secondary creep with constant creep rate was not found in any of the tests that reached accelerating creep (the tests with lower stress levels had to be finished before creep accelerated again because of lack of time). The time when creep rate begins to increase is called t_m and the corresponding strain is ε_m (Fig. 2).

An important result of the creep tests at various stresses and temperatures is that all the creep curves have nearly the same shape when they are normalized by a linear transformation of the time axis

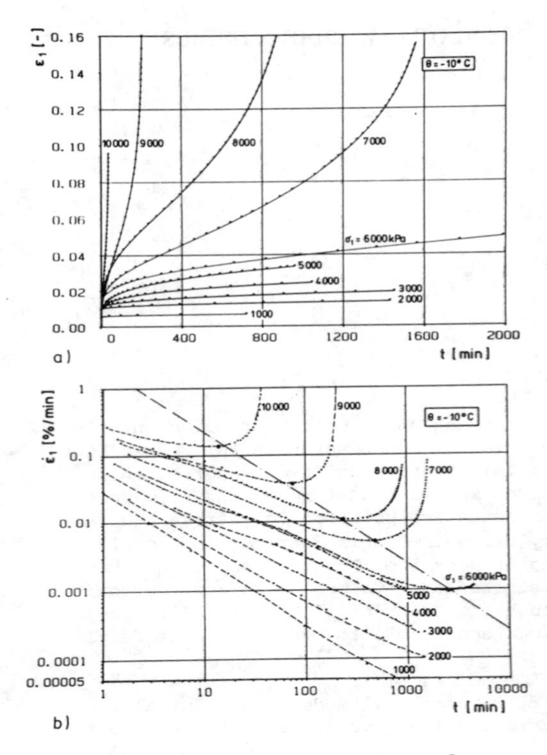

Fig. 1 Uniaxial creep tests at -10°C: a) strain ε_1 and b) strain rate $\dot{\varepsilon}_1$ versus time

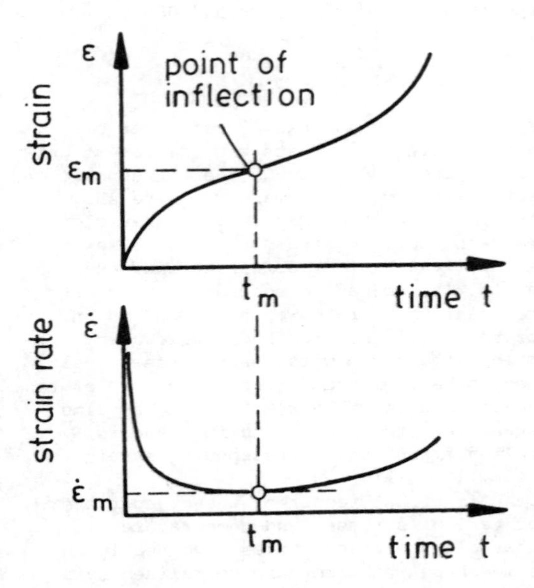

Fig. 2 Definition of t_m, ε_m and $\dot{\varepsilon}_m$

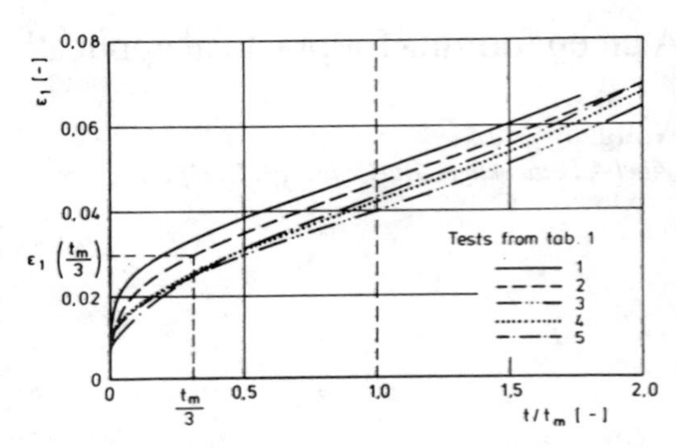

Fig. 3 Creep curves obtained at different stress and temperature, normalized time axis

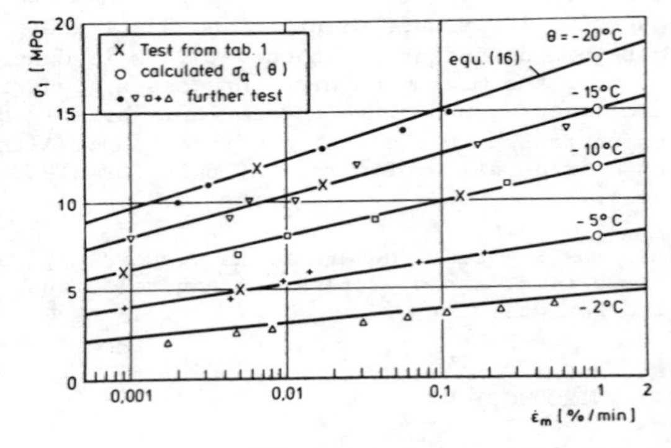

Fig. 4 Uniaxial stress σ_1 versus $\dot{\varepsilon}_m$ from creep tests

(Fig. 3). It can be seen that the scatter of strain is mainly an offset caused by displacement of the end platen during loading at the beginning of the tests. This is confirmed by tests that yielded constant strain ε_m even under partly different test conditions (e.g. variations in height-to-diameter-ratio of specimen and tests without lubrication of the end platens) (Ladanyi 1972, Klein 1978, Ting 1983). Consequently, the different creep curves can be represented by a unique mastercurve when the time axis is transformed linearly. The dependence on temperature and stress can be described by one characteristic point of the curve.

It is convenient to use the minimum creep rate $\dot{\varepsilon}_m$ that can easily be found from the strain-rate versus curves (Fig. 1b).

The empirically-found dependence of $\dot{\varepsilon}_m$ on uniaxial stress σ_1 and temperature θ is given in Figure 4. It shows that the logarithm of $\dot{\varepsilon}_m$ is proportional to the stress σ_1. Although this relation can also be fitted to a power law, the fit is slightly worse and only the logarithmic law can be supported by theoretical considerations.

An extended study of micromechanical processes occurring in crystalline bodies under stress finally led to formula (1) (Orth 1986):

$$\dot{\varepsilon}_m(\sigma,T)=\dot{\varepsilon}_\alpha \exp\left[\left(\frac{K_1}{T}+\ln\dot{\varepsilon}_\alpha\right)\left(\frac{\sigma}{\sigma_\alpha(T)}-1\right)\right] \quad (1)$$

$\dot{\varepsilon}_m$: minimum strain rate during creep (Fig. 2)

σ: the applied deviatoric stress, in uniaxial stress states proportional to uniaxial stress σ_1

T: absolute temperature [K]

$\dot{\varepsilon}_\alpha$: reference strain rate, here $\dot{\varepsilon}_\alpha = 1$ %/min

$\sigma_\alpha(T)$: that stress under which at temperature T there is $\dot{\varepsilon}_m = \dot{\varepsilon}_{\alpha(T)}$

K_1: a characteristic temperature [K] that was found to be constant if $T < 268.4$ K ($\hat{=}$ -5^oC)

From the similarity of the creep curves and the strain-rate curves, another important result can be derived, which was also found in research on pure ice by Mellor & Cole (1982):

$$\dot{\varepsilon}_m \cdot t_m = C = \text{constant} \quad (2)$$

and

$$t_m = C/\dot{\varepsilon}_m \quad (3)$$

Knowing the minimum creep rate $\dot{\varepsilon}_m$, the time t_m (when the approximately constant strain ε_m is reached) can be calculated and vice versa. The time needed to obtain any partition of ε_m can also be determined from the mastercurve.

3 PHYSICAL BACKGROUND OF THE FORMULA (1)

In this section, the physical background of the formula interrelating uniaxial stress (σ) (or, as can be shown, deviatoric stress in general), temperature (T) and minimum creep rate ($\dot{\varepsilon}_m$) is briefly given.

The stress/strain/temperature behaviour of frozen soil shows similarities with that of other crystalline materials, such as ceramics and metals.

However, it should be noticed that differences arise from the special structure of the ice due to the dipole water molecules.

Deformation of crystal is related to displacements of single molecules. These occur preferably at crystal defects, where the crystal lattice is less stable. Thermal oscillations support molecule displacement. Therefore, crystals become softer with increasing temperature. Molecule displacements can also be caused by the movements of dislocations (Frost & Ashby 1982). The rate of their motion and, thus, the deformation rate of the crystals can be calculated from the available thermal and stress-induced mechanical energy (Prandtl 1928). As already noted, the stress/strain-rate relation found in this study (see Fig. 4) can be fitted to a logarithmic law and - still with a good agreement - to a power law. This indicates that two different deformation mechanisms might be active in the stress, strain-rate and temperature range used in the tests:

Thermally-activated dislocation glide occurs when the dislocation movement is limited by discrete obstacles (dispersoids, forest dislocations, precipitates) (Frost & Ashby 1982). This process is usually named "thermally-activated process" or "rate process", but it should be stated that it is only one type of thermally-activated process interrelating stress, strain rate and temperature. It leads to a linear relation between the hyperbolic sine of the deviatoric stress and the strain rate:

$$\sinh\sigma \sim \dot{\varepsilon}\ .$$

This is in good agreement with the test results given in Figure 4 because $2\cdot\sinh\sigma \simeq \exp\sigma$ if $\sigma \geq 1$.

Diffusional flow occurs when the diffusive flux of matter through and around the surfaces of crystal grains is directed by a deviatoric stress (Frost & Ashby 1982). Further, it can be caused by dislocation climb. With these deformation mechanisms, a power law interrelates stress and strain rate. The power law approximates the test results almost as well as the hyperbolic sine formula.

An extensive study of the test results (Orth 1986) yielded several indications for the thermally-activated dislocation glide limited by discrete obstacles, this being the dominant deformation process in the creep tests with frozen sand. The micromechanic activation volume was found to be temperature dependent(Fig. 5a). The obtained micromechanical parameters (activation energy, activation volume - Fig.5) have the same order of magnitude as in many other crystalline materials (Frost & Ashby 1982). This is not the case of diffusional flow is assumed.

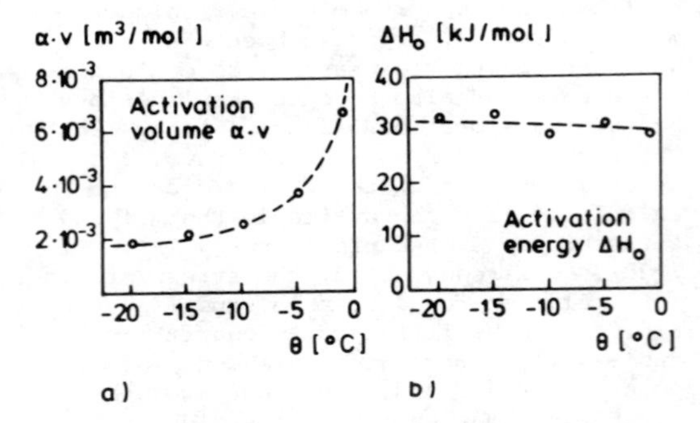

Fig. 5 a) Activation volume and
b) activation energy, calculated from creep tests

The relaxation tests of this study, however, indicate that diffusinal flow becomes dominant at very low stress and strain rates. For practical purposes, the latter are irrelevant as the in-situ stress is usually not low enough.

4 DETERMINATION OF $\dot{\varepsilon}_m$ (θ,σ)

This section gives an example of how to calculate the time t_m when accelerating creep in a frozen soil under a given stress and temperature begins, or when a certain strain is reached. This can be done by using Eq. 1:

$$\dot{\varepsilon}_m(\sigma,T)=\dot{\varepsilon}_\alpha \exp\left[\left(\frac{K_1}{T}+\ln\dot{\varepsilon}_\alpha\right)\left(\frac{\sigma}{\sigma_\alpha(T)}-1\right)\right] \quad (1)$$

First, the parameter K_1 and the function $\sigma_\alpha(T)$ have to be determined. The function $\sigma_\alpha(T)$ represents the dependence of the micromechanic activation volume on temperature (Fig. 5a). At present, it is not understood theoretically, so $\sigma_\alpha(T)$ has to be fitted to test results. This can be done easily, as will be seen later.

At least two creep tests performed at the same temperature are needed to determine the value of K_1. The applied stress should differ sufficiently for the values of $\dot{\varepsilon}_m$ to differ by at least one order of magnitude, because the logarithm of $\dot{\varepsilon}_m$ is used in the calculation.

Transforming Eq.(1) yields

$$\sigma = \sigma_\alpha(T)\left(\frac{\ln(\dot{\varepsilon}_m/\dot{\varepsilon}_\alpha)}{\frac{K_1}{T}+\ln\dot{\varepsilon}_\alpha}+1\right) \quad (4)$$

and

$$\frac{d\,\sigma}{d\,\ln(\dot{\varepsilon}_m/\dot{\varepsilon}_\alpha)} = \frac{\sigma_\alpha(T)}{\frac{K_1}{T}+\ln\dot{\varepsilon}_\alpha} = A(T) \quad (5)$$

The function A(T) is the slope of a line connecting two points in the diagram where the creep stress σ is plotted versus the logarithm of the measured minimum strain rate $\dot{\varepsilon}_m$ (Fig. 4).

Knowing A, K_1 can be calculated by transforming Eq.(1):

$$\ln(\dot{\varepsilon}_m/\dot{\varepsilon}_\alpha) = \frac{\frac{K_1}{T}+\ln\dot{\varepsilon}_\alpha}{\sigma_\alpha}\cdot\sigma - \frac{K_1}{T} - \ln\dot{\varepsilon}_\alpha \quad (6)$$

$$\ln\dot{\varepsilon}_m - \ln\dot{\varepsilon}_\alpha = \frac{\sigma}{A(T)} - \frac{K_1}{T} - \ln\dot{\varepsilon}_\alpha \quad (6a)$$

$$K_1 = T\left(\frac{\sigma}{A(T)} - \ln\dot{\varepsilon}_m\right) \quad (7)$$

Using σ, A(T) and $\dot{\varepsilon}_m$ from one of the tests, we obtain K_1.

Now to determine $\sigma_\alpha(T)$ we need just one creep test at each of two or three different temperatures. As these tests in general will yield a minimum creep rate $\dot{\varepsilon}_m$ different from $\dot{\varepsilon}_\alpha$, we have to calculate $\sigma_\alpha(T)$. This is done again via the slope A(T) in the σ-ln $\dot{\varepsilon}_m$ diagram for each temperature (see Eq. 7).

$$A(T) = \frac{\sigma}{\frac{K_1}{T}+\ln\dot{\varepsilon}_m} \quad . \quad (8)$$

Knowing K_1 and inserting σ and $\dot{\varepsilon}_m$ from a test at any temperature T, we obtain A(T).

A further transformation gives

$$\sigma = A(T)\left(\frac{K_1}{T}+\ln\dot{\varepsilon}_m\right)$$

and with $\dot{\varepsilon}_\alpha$ instead of $\dot{\varepsilon}_m$

$$\sigma_\alpha(T) = A(T)\left(\frac{K_1}{T}+\ln\dot{\varepsilon}_\alpha\right) \quad .$$

The measured values of $\sigma_\alpha(T)$ are now fitted to an appropriate function, and then $\dot{\varepsilon}_m(\sigma,T)$ can be determined by using Eq. (1). Using Eq. 3, $t_m(\sigma,T)$ may also be determined.

Example: To determine the parameters of Eq. (1) we use five uniaxial creep tests:

Table 1. Uniaxial creep test results.

No.	Test cond. θ [°C]	σ_1 [kPa]	Test results $\dot{\varepsilon}_m$ [%/min]	t_m [min]	C (Eq.2)
1	-5	5000	$4.963 \cdot 10^{-3}$	552.7	0.0274
2	-10	6000	$9.0183 \cdot 10^{-4}$	2534.4	0.0229
3	-10	10000	0.13194	16.1	0.0212
4	-15	11000	$1.752 \cdot 10^{-2}$	123.5	0.0216
5	-20	12000	$6.351 \cdot 10^{-3}$	356.5	0.0268

The average value of C is 0.024.

We chose the reference strain rate $\dot{\varepsilon}_\alpha$ = 1 %/min. From tests number 2 and 3 at T = 263.4 K (≙ -10°C) we calculate by Eq. (5)

$$A(263.4\ K) = \frac{10000 - 6000}{\ln 0.13194 - \ln 9.0183 \cdot 10^{-4}} = \tag{11}$$

$$= 802.3\ kPa \tag{11}$$

and by Eq. (7) from test no. 2

$$K_1 = 263.4\ (\frac{6000}{802.3} - \ln 9.0183 \cdot 10^{-4}) =$$

$$= 3817\ K \tag{12}$$

Now to determine the function $\sigma_\alpha(t)$, A(T) from test no. 1 at T = 268.4 K (≙ -5°C) is first evaluated by Eq. (8):

$$A(268.4\ K) = \frac{5000}{\frac{3817}{268.4} + \ln 4.963 \cdot 10^{-3}} =$$

$$= 561\ kPa \tag{13}$$

and then σ_α from Eq. (10):

$$\sigma_\alpha(268.4\ K) = 561\,(\frac{3817}{268.4} + 0) = 7977\ kPa \tag{14}$$

Doing this with the other temperatures, we obtain the values given in table 2:

Table 2. Calculated values of A(T) and $\sigma_\alpha(T)$ for four different temperatures.

θ[°C]	T [K]	A [kPa]	σ_α[kPa]
-5	268.4	561	7977
-10	263.4	802	11622
-15	258.4	1025	15153
-20	253.4	1199	18065

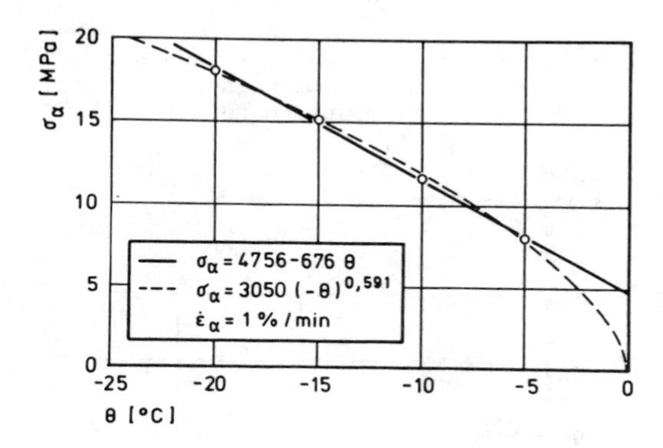

Fig. 6 σ_α versus θ

In Figure 6, σ_α is plotted versus temperature θ in °C. Two different functions obtained from a least square fit are given and plotted in Figure 6. The power law (dotted line) is preferred, as this gives the correct value at θ = 0°C (where the uniaxial strength of frozen sand vanishes).

Now, substituting T by θ we finally obtain

$$\dot{\varepsilon}_m(\sigma,\theta) = 1\ \exp\left(\frac{3817}{\theta+273.4} + 0\right)\left(\frac{\sigma}{3050(-\theta)^{0.591}} - 1\right) \tag{15}$$

or reversed

$$\sigma = 3050(-\theta)^{0.591}\ \left(1 + \frac{\ln(\dot{\varepsilon}_m/1)}{\frac{3817}{\theta+273.4}}\right) \tag{16}$$

In Figure 7, function (16) and the test results listed in Table 1 as well as additional tests are plotted.

Now using Eq. (2) and Eq. (15), t_m can also be determined, since the average value of C is known:

$$t_m = 0.024/\dot{\varepsilon}_m\ [min]$$

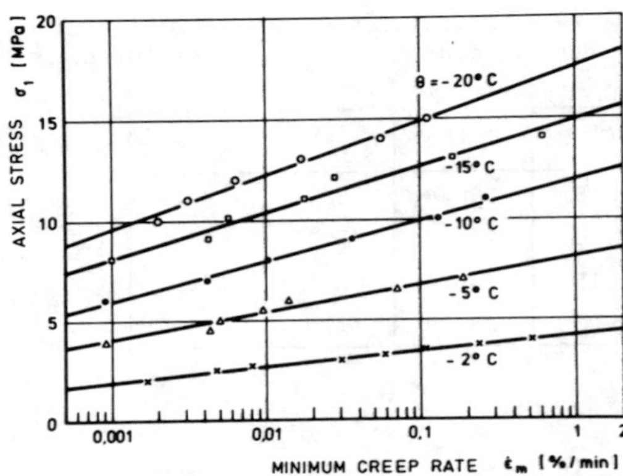

Fig. 7 Uniaxial stress σ_1 versus $\dot{\varepsilon}_m$. The figure shows both tests and calculations using Eq. (16).

The calculated values of t_m and the test results are given in Figure 8.

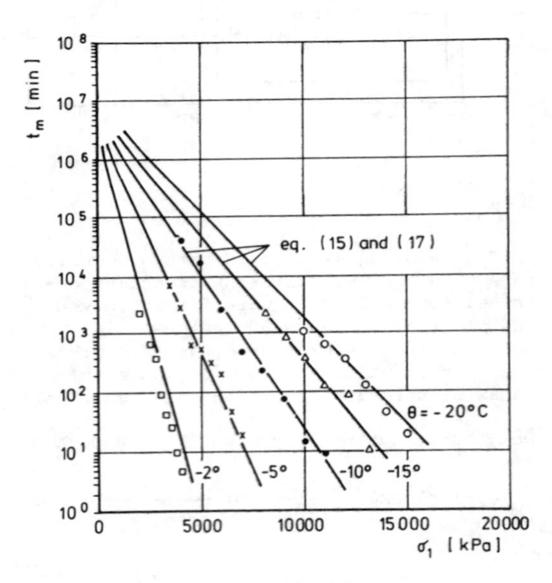

Fig. 8 t_m versus σ_1 at different temperatures θ: calculation and test results

It can be seen that there is a good agreement between test results and calculations obtained from only five tests. However, it should be considered that artificially made specimens of exceptional low scatter were used in the study. For practical purposes, more tests should be carried out to detect failed tests. A convenient way of doing this is to plot all test results as in Figure 4. In this way, any irregular results are found quickly.

5 CALCULATING DEFORMATIONS

Usually, for practical purposes, the maximum stress under which deformations do not exceed a given limit within a certain time at a given temperature has to be determined. From the normalized creep curves (Fig. 3), the partition of t_m can be found, dependent on any desired deformation. For example, if we take $\varepsilon_1 = 0.03$, then we see this deformation is reached after $t_m/3$. If the frozen body is designed so that t_m is three times its lifetime, then its deformations reach $\varepsilon_1 = 0.03$ just within lifetime.

Now $\dot{\varepsilon}_m$ can be calculated from t_m by Eq. (2) or (17), respectively, and then by Eq. (16) the allowed stress can be calculated at a given temperature.

6 TRIAXIAL STRESS STATES

The influence of triaxial stress states on the behaviour of frozen soil depends on the grain skeleton (grain size, grain shape, type of mineral, density) as well as on the degree of water saturation. However, for many practical applications of soil freezing, triaxial stress can be neglected for two reasons:

Firstly, significant grain friction in frozen soils is mobilized at comparatively high strain (more than 10 % axial strain). This is in most cases unacceptable.

Secondly, the triaxial stress 5 or 10 metres below ground surface is too low to cause considerable frictional strength. Calculations showed that the frictional strength 10 m below the face with an angle of friction of $\phi = 35^o$ is only about 10 % of the uniaxial stress that can be applied at -10^oC and $t_m = 6$ month. Of course, friction can become important at high triaxial stress states (e.g. in deep frozen shafts) if there is coarse soil. For many practical problems, however, it is sufficient to use uniaxial tests and to design using deviatoric stress only.

7 CONCLUSION

From the presented test results and calculations the following conclusions can be drawn:

- Creep curves from uniaxial creep tests on frozen soil at various stresses and temperatures can be presented with a unique mastercurve when the time axis is transformed linearly. So the dependence on temperature and stress can be described by one characteristic point of the creep curve.

- A formula derived from consideration of micromechanical processes properly describes the temperature and stress dependence of creep under uniaxial stress. As this formula is based on obstacle-limited dislocation glide,that micromechanical process can be assumed to be dominant in the tests carried out. The obtained micromechanical parameters (activation energy, activation volume) have the same order of magnitude as in many other crystalline materials, which supports that assumption.

- Only five creep tests are needed to determine the parameters of the given creep formula. By means of the creep mastercurve, the time it takes to reach a given strain under any given temperature and stress can be calculated.

- Although the creep formula is made for uniaxial creep tests, it can also be used in many practical cases with triaxial stress states even in coarse soils, because in most practical applications triaxial stress and allowed strain are too low to cause a significantfrictional effect from triaxial stress.

8 ACKNOWLEDGEMENT

The experimental work for this study was supported by the Deutsche Forschungsgemeinschaft (DFG), which is gratefully acknowledged.

9 REFERENCES*

Frost, H.J. & M.F. Ashby 1982. Deformation Mechanism Maps. The Plasticity and Creep of Metals and Ceramics. Pergamon Press, Oxford.

Klein, J. 1978. Nichtlineares Kriechen von künstlich gefrorenem Emschermergel. Schriftreihe Inst. f. Grundbau, Wasserwesen u. Verkehrswesen, Univ. Bochum, Heft 2.

Ladanyi, B. 1972. An Engineering Theory of Creep of Frozen Soils. Can.Geotech. Journal, 9, 63-80.

Mellor, M. & D.J. Cole 1982. Deformation and Failure of Ice under Constant Stress of Constant Strain Rate. Cold Regions Science and Technology, 5, p. 201-219.

Orth, W. 1985. Deformation Behaviour of frozen Sand and its physical Interpretation. Proc. of the Fourth Int. Symp. on Ground Freezing, Sapporo, Japan, Aug. 1985, 245-253

Orth, W. 1986. Gefrorener Sand als Werkstoff - Elementversuche und Materialmodell. Veröff. Inst. f. Bodenmech. u. Felsmech., Univ. Karlsruhe, Heft 100.

Prandtl, L. 1928. Ein Gedankenmodell zur kinetischen Theorie fester Körper. Zeitschrift angew. Mathematik und Mechanik (ZAMM), Bd. 8, S. 85-106.

Ting, J. 1983. Tertiary Creep Model of frozen Sands. ASCE, Journ. of Geotech. Eng., Vol. 109, No. 7, 932-945

*See page X concerning ISGF papers

5th International Symposium on Ground Freezing, Jones & Holden (eds)

ISBN 90 6191 824 3

Laboratory determination of pore pressure during thawing of three different types of soil

C.G.Rydén
Swedish National Road Administration, Borlänge, Sweden

K.Axelsson
Department of Civil Engineering, Luleå University of Technology, Luleå, Sweden

ABSTRACT: Thawing tests have been conducted on frozen samples of remoulded illite clay, undisturbed iron sulphide-rich silty clay, and inorganic silt. All soil specimens have been water-saturated and isotropically frozen without external water supply. During the thawing tests in a specially designed oedometer, temperature distribution, pore pressure, development, and compression of the sample were continuously measured. The measured excess pore pressure were compared with analytical results according to the Morgenstern -Nixon model. It was found that the impact of the rate of the consolidation coefficient is significant. Since the correct choice of this is uncertain the measured pore pressure was utilized for the calculation of the consolidation coefficient by the use of the theory. These obtained values of the consolidation coefficient agreed well with those calculated from the degree of consolidation but showed an unexpected correlation with the rate of thawing.
KEYWORDS: Clays, consolidation, pore pressures, silts, thawing, thermal properties

Introduction

The reduction of bearing capacity on the road network is an annual problem in arctic and sub-arctic regions that calls for better methods of prediction. For example 20 000 kms of Swedish roads, representing about 20% of the Swedish road network, are annually subjected to load restrictions, which usually last for about 40 days. It is desirable to be able to predict the loss of bearing capacity as a function of ambient temperature and geotechnical properties of the soil in question.

The reduction of bearing capacity is assumed to be caused by excess pore pressure in thawing soil. This is caused by release of excess pore water as pore ice and ice lenses within the soil melts.

Theory

A complete solution to the one-dimensional thaw-consolidation problem is proposed by Morgenstern and Nixon, (1971). It is based on the heat conduction equation and the classical theory of consolidation accord- to Terzaghi. The thawed depth is described by

$$X = \alpha \cdot \sqrt{t} \tag{1}$$

and by introducing the thaw consolidation ratio R

$$R = \frac{\alpha}{2\sqrt{c_v}} \tag{2}$$

the complete solution will have the form

$$u(x,t) = \frac{q_o}{\frac{e^{-R^2}}{\sqrt{\pi R}} + \operatorname{erf}(R)} \operatorname{erf}\left(\frac{x}{2\sqrt{c_v t}}\right) + \frac{v_b x}{\frac{1}{2R^2} + 1} \tag{3}$$

where u denotes excess pore pressure and erf denotes the error function.

The surface temperature is assumed to be a step function, increasing from the original value T_0 to T_s at the time t=0, and remaining constant thereafter. Pore water liberated at the thaw line will dissipate upwards, and drain off from the surface.

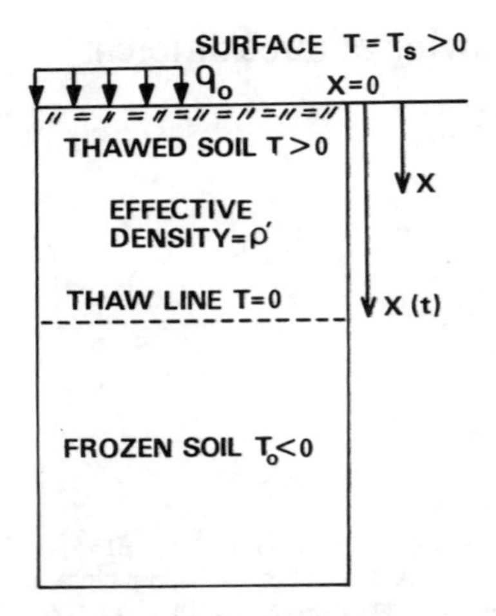

Fig. 1 One-dimensional thawing and consolidation

Eq. (3) is arrived at under the assumption that the residual stress σ_0^1 in the thawing soil equals zero. Values of σ_0^1 higher than zero will give a lower value of the apparent excess pore pressure during thawing. Nixon and Morgenstern (1973) describes the residual stress as being dependent on soil type and stress history of the soil in question.

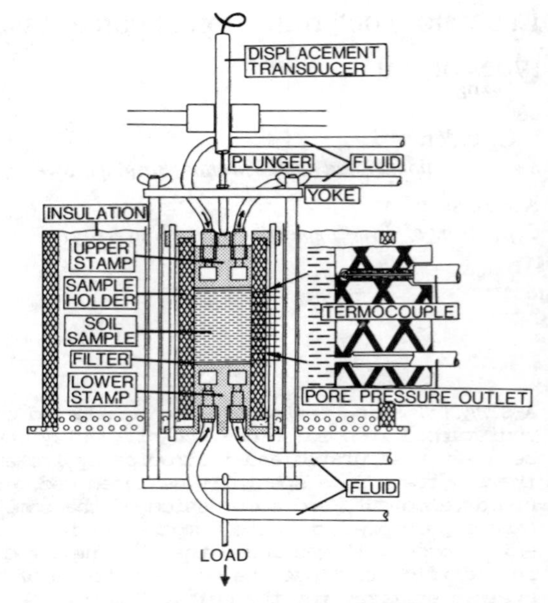

Fig. 2 Overall design of the CBT oedometer. The soil sample is 50 mm high and 46 mm in diameter. The sample holder has 11 thermocouples and 11 pore pressure outlets, each mounted in one vertical row respectively

Testing equipment

In order to test the validity and applicability of this solution, a laboratory device, the CBT oedometer (Constant Boundary Temperature oedometer), has been developed, see Fig. 2. In this device, one-dimensional thawing tests upon frozen soil samples can be performed by subjecting them to the same boundary conditions as assumed in the analytical solution.

During thawing tests in the CBT oedometer, temperature distribution, pore pressure, and compression of the sample is measured continuously.

Thawing tests have been conducted on frozen samples of remoulded illite clay, undisturbed iron sulphide-rich silty clay, and remoulded inorganic silt. All soil specimens have been water-saturated and isotropically frozen without external water supply. Due to freezing, alteration of the microstructure of the soil has occurred which has led to an internal redistribution of pore water and formation of ice lenses.

At this stage of the experimental work, it was considered reasonable to put most of the attention on the excess pore pressure at the thaw line (x=X), because of its assumed significance for eg. the stability of slopes. From eqs. (1) and (3), excess pore pressure at the thaw line will be

$$u(X,t) = \frac{q_0 \operatorname{erf}(R)}{\frac{e^{-R^2}}{(\sqrt{\pi}R) + \operatorname{erf}(R)}} + \frac{\gamma_b \cdot X}{\frac{1}{2R^2} + 1} \quad (4)$$

When comparing experimental results with theoretically calculated data, the small size of the soil sample makes it reasonable to neglect the influence of buoyant forces. Hence, eq. (4) can be reduced to

$$u(X,t) = \frac{q_0 \operatorname{erf}(R)}{\frac{e^{-R^2}}{\sqrt{\pi}\,R} + \operatorname{erf}(R)} \quad (5)$$

Test results

The surface temperatures have not been found to equal a step-function perfectly, see Fig 3. The course of thawing have however for all tests agreed very well with eq. (1), see Fig. 4. The actual value of the rate of thawing α has been difficult to predict.

The excess pore pressure at the thaw line is evaluated from the pore pressure recording. The peak value is taken as the pore pressure at the thaw line. The normalized pore pressure U* can then be calculated as

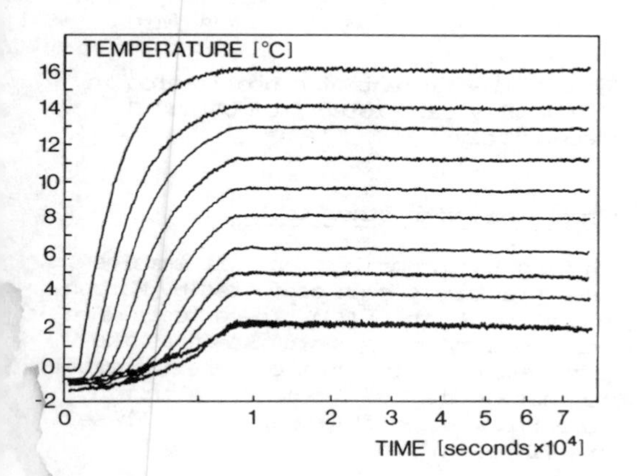

Fig. 3 Temperature at 11 fixed positions during thawing versus time in square-root time-scale. The uppermost curve represents surface temperature

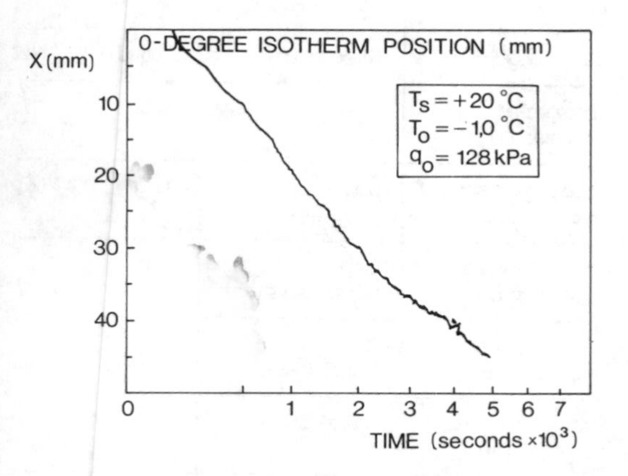

Fig. 4 Thaw front graph in square-root time scale for illite clay

$$U^* = \frac{U(peak)}{q_0} \tag{6}$$

A number of thawing tests have been performed on each of the three soil types. For none of the soils, the normalized pore pressure U* at the thaw line has been found to increase significantly, as the rate of thawing α increases.

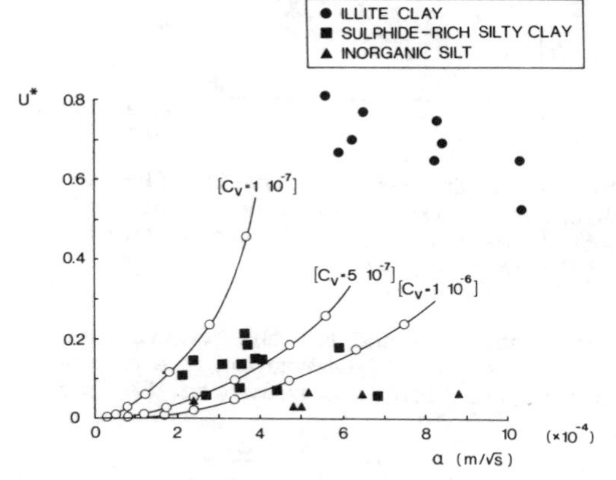

Fig. 5 Measured values of U* as a function of rate of thawing α

The theoretical values of U* according to eq. (5) are plotted as curves in Fig. 5, representing different values of the coefficient of consolidation.

The obtained values of U* can be utilized for calculation of the corresponding c_v, by use of eqs (2) and (5). The calculated values of the coefficient of consolidation show an unexpected correlation with the rate of thawing α, see Fig. 6.

The values of coefficient of consolidation plotted in Fig. 6 for "thawed" samples are evaluated from compression tests performed on samples identical to the ones used for thawing tests, that were allowed to thaw entirely before compression.

The degree of consolidation D during thawing has been possible to evaluate only from thawing tests performed on illite clay. The degree of consolidation during thawing is proposed to be defined as the settlement of the thawed soil at time t divided by the settlement of the thawed soil at time = ∞ assuming that thawing has stopped at time = t ($X(t) = X(\infty)$).

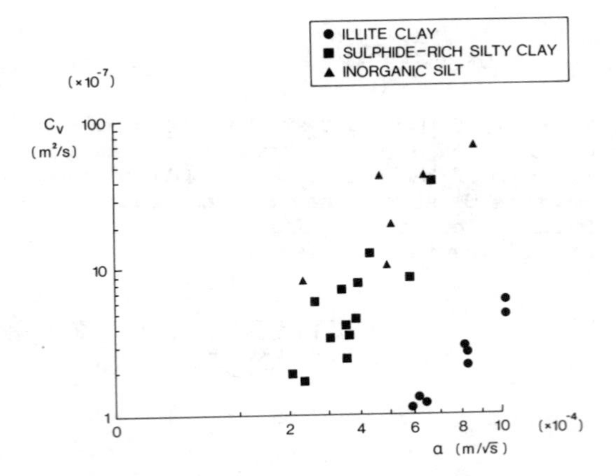

Fig. 6 Coefficient of consolidation c_v calculated from measured values of U*, plotted as a function of rate of thawing α (illite clay)

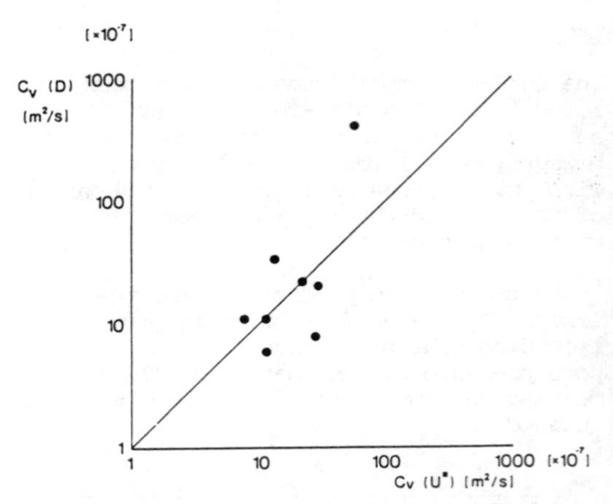

Fig. 7 Coefficient of consolidation in illite clay calculated from U* and D respectively

From eq. (3), and for the boundary conditions in the CBT-oedometer, the degree of consolidation is obtained on the form

$$D = 1 - \frac{\mathrm{erf}(R) - \frac{e^{-R^2}-1}{\sqrt{\pi}\,R}}{\frac{e^{-R^2}}{\sqrt{\pi R}} + \mathrm{erf}(R)} \qquad (7)$$

The degree of consolidation D depends primarily on the the thaw consolidation ratio R. This means that the degree of consolidation according to eq. (7) is constant as long as thawing proceeds according to eq. (1). Low values of R yield a degree of consolidation D close to unity. High values of R represent a case where substantial excess pore pressure remains in the soil after that thawing is completed and considerable post-thawing settlement may occur.

The evaluated values of the degree of consolidation D show, like the normalized peak pore pressure U*, little dependence on the rate of thawing α. Hence, values of the coefficient of consolidation, calculated by use of known values of D and eq. (7), also show a similar and unexpected correlation with the rate of of thawing α. The two methods of calculating the value of the coefficient of consolidation, c_v(U*) and c_v(D) respectively, are compared in Fig. 7. It can thus be concluded that measured pore pressure and measured course of settlement are consistent with each other.

Comparison with other tests

Thawing tests upon frozen soil samples have previously been conducted by Morgenstern and Smith (1973). These tests were performed by an apparatus known as the permode. Its main functions are identical to that of the CBT Oedometer, although some important differences exist. These are listed in Table 1.

Table 1. Comparison between Permode and CBT Oedometer

	Permode	CBT
Heating	Electric	liquid
Temperature measuring	?	side
pore pressure measuring	base	side
compression measuring	top displacement	top displacement
Pore water drainage	top (and base)	top and base
sample size φ x h [mm]	64 x 30	46 x 50

Also slight differences in test procedure exist. Morgenstern and Smith have estimated the value of the coefficient of consolidation from the course of compression after thawing is finished. This allows the coefficient of consolidation to be determined for every individual soil sample. On the other hand, it could be questioned whether the value of the coefficient of consolidation in completely thawed soil is equal to that of a soil in which thawing proceeds.

Conclusions

By use of measured values of rate of thawing α and the well-known simplified Stefan's solution, see Lunardini (1981), the heat conductivity of the thawed soil can be calculated. Such calculations have in some cases yielded results within very wide ranges, see Rydén (1985). As is shown by Nixon and McRoberts (1973), the simplification made in Stefan's solution only have limited influence of calculated values of α, compared to more sophisticated solutions of the thawing problem. The authors believe that difficulties in making a proper estimation of the unfrozen water content in the frozen soil prior to thawing makes prediction of rate of thawing almost impossible.

Although the true relationship between surface temperature and rate of thawing has not yet been determined, there is still a point in studying the relationship between rate of thawing, pore pressure development during thawing, and type of soil.

Pore pressure measurements indicate less impact of the rate of thawing on pore pressure at the thaw line, than what is given by eq. (5). If the maximum pore pressure in a thawing soil depends on the type of soil only, the solution of the thermal problem will be just a matter of how much time will be needed before the entire frozen soil is thawed.

The degree of consolidation during thawing evaluated from the measured course of compression of the soil samples have in all cases been found to be a constant fraction of the thawed depth. The computed values of the degree of consolidation have also been consistent with measured pore pressure values, and seemingly not depending on the rate of thawing.

If, after all, the theoretical solution proposed by Morgenstern and Nixon is assumed to be applicable to the thawing tests performed in the CBT oedometer, this implies one or both of the following:

* The coefficient of consolidation is dependent on the rate of thawing. Higher rate of thawing gives higher values of the coefficient of consolidation
* The residual stress in a thawing soil is dependent on the rate of thawing, where higher rate of thawing gives higher values of the residual stress

Neither of the two statements above is easy to believe.

Acknowledgement

The CBT oedometer was developed as a main part of the research project "Bearing capacity and settlements in thawing soil", financed by the Swedish Council for Building Research, contract No. 790162-5.

References

Axelsson, K. and Rydén, C.G. Experimental Determination of Pore Pressure Development in a Thawing Soil, 9th International Conference on Port and Ocean Engineering under Arctic Conditions - POAC87, Fairbanks, Aug. 17-21, 1987

Carslaw, H.S. and Jaeger, J.C. 1985. Conduction of Heat in Solids. 2nd ed., Oxford University Press, Oxford

Lunardini, V.J. 1981. Heat Transfer in Cold Climates. Van Nostrand, New York

McRoberts, E.C. and Nixon, J.F. A study of some factors affecting the thawing of frozen soils. Canadian Geotechnical Journal, 10 (1973) 3, 439-452

Morgenstern, N.R. and Nixon, J.F. 1971. One-dimensional consolidation of thawing soils. Canadian Geotechnical Journal, 8, 558-564

Morgenstern, N.R. and Smith, L.B. Thaw-consolidation tests on remoulded clays. Canadian Geotechnical Journal, 10 (1973), 25-40

Nixon, J.F. and Morgenstern, N.R. The residual stress in thawing soils. Canadian Geotechnical Journal, 10 (1973), 571-580

Rydén, C.G. 1985. Pore pressure in thawing soil. Proc. Fourth Int. Sympt. on Ground Freezing, Sapporo, ed. S. Kinosita & M. Fukuda, Balkema, Rotterdam, 223-226

Rydén, C.G. 1986. Pore pressure in thawing soil - theory and laboratory determination. Licentiate thesis 1986 - 14L, University of Luleå, Luleå.

5th International Symposium on Ground Freezing, Jones & Holden (eds)
 ISBN 90 6191 824 3

Soil freezing and thaw consolidation results for a major project in Helsinki

I.T.Vähäaho
Geotechnical Department of the City of Helsinki, Finland

ABSTRACT: Settlements caused by thawing and extra loads can be determined from the results of water content and normal oedometer tests. The outflow of water from frozen and thawed clay has been observed to decrease the water content by one third. Shear strength decreases during the thawing process but later on begins to increase until greater than the original. Increase in the remoulded shear strength is even larger and even the drop under thawing was not observed with remoulded shear strength. This work contains the results of tests on both frozen and thawed clays with water content from 40 to 120 % of the weight of solids. Tests were made in laboratory and in the field.
KEYWORDS: clays, deformation, insitu testing, laboratory testing, shear strength, thawing.

1 INTRODUCTION

This work was started towards the end of 1979. At first it was regarded as a normal but difficult geotechnical planning project. The purpose was to build a tunnel under a major railway with ground freezing technique. The plan was presented at the Third International Symposium in Hanover (Vähäaho & Eronen 1982). Afterwards the working limits in the railway area were changed by the authorities and therefore it became possible to use conventional methods, too. However, no contract competition for the project has yet been arranged.

So many new and interesting soil conditions were found that full scale ground freezing test was considered necessary. The insitu test was made with brine at - 24 °C and the laboratory tests at - 10 °C temperature. The full scale test was made with two freezing tubes imitating the actual plan and with instrumentation to follow the property changes in the ground.

This work contains the results of tests on both frozen and thawed sedimentational origin Litorina-clays with water content from 40 to 120 % of the weight of solids. The mineral composition of Finnish clays (grain size < 0,001 mm) is quartz (< 10%), feldspar (< 10 %), some amphibole and clay minerals (Soveri 1956). The most common clay minerals are illite, vermiculite, chlorite and some mixed-layer minerals of these. Because of the atomic structure of these clay minerals, water cannot penetrate between the unit layers. The conductivity of these clays is about 400 μs and they shrink strongly during drying.

The main test results (Vähäaho 1987) are briefly presented in the following.

2 THAW SETTLEMENT

Results of thaw settlement without extra

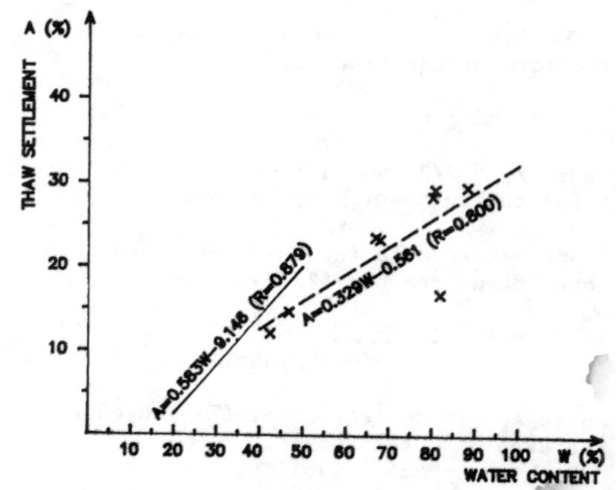

Fig. 1 Diagrams showing the relationship between the original water content of clay and the thaw settlement from China and Finland.

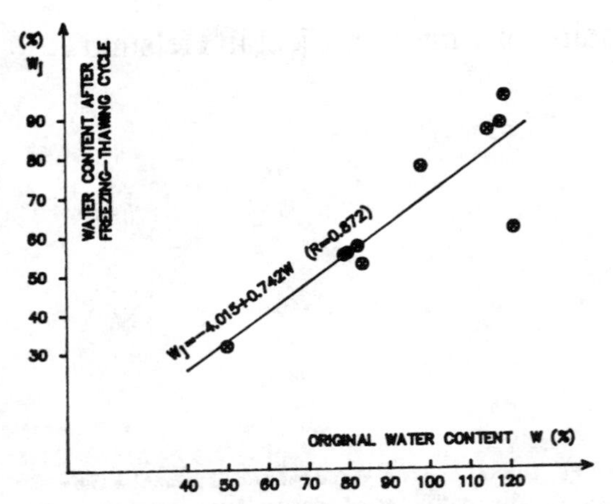

Fig. 2 Diagram showing the change of water content caused by freezing-thawing process.

load are shown in fig. 1. The left-hand line was calculated from the results of Tong Changjiang & Chen Enyvan (1985) from China. The right-hand line presents the results in Helsinki. The author suggests that thaw settlements could be calculated with the following equation:

$$A = \frac{w}{k_1} \quad (1)$$

where $k_1 \cong 3$, A = thaw settlement (%) and w = original water content (%).

To check these results, additional test from different areas of Helsinki were made. The clay water content was measured both before freezing and after the freezing-thawing cycle (fig. 2).

The line in fig. 2 can be presented in the form of equation (2)

$$w_j = k_w \cdot w \quad (2)$$

where $k_w \cong 2/3$ and w_j = water content after freezing-thawing cycle (%).

Another formula for thaw settlement can be written from eqn (2).

$$A_k = \left(1 - \frac{100 + kw \,.\, w \,.\, 2,65}{100 + w \cdot 2,65}\right) \cdot 100 \quad (3)$$

where A_k = thaw settlement (%) caused by the drop of water amount.

Both in normal consolidation tests and in freeze-thaw tests the decrease in water amount was observed to be about 4/5 of the settlement.

Thus, eqn. (3) can be rewritten as:

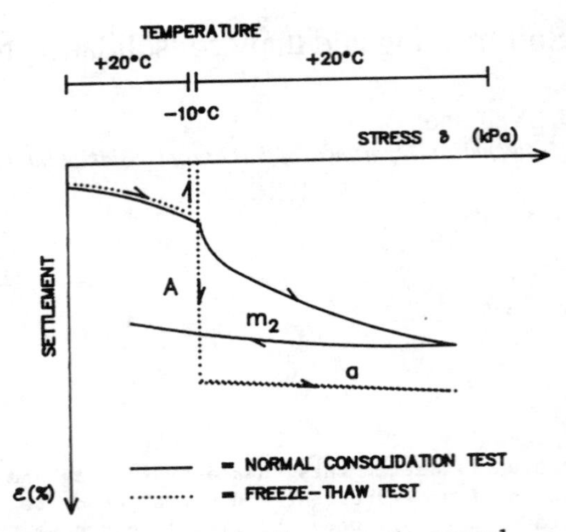

Fig. 3 The shapes of curves in normal consolidation and freeze-thaw tests with clays.

$$A = 5/4 \cdot \left(1 - \frac{100 + 2/3 \,.\, w \,.\, 2,65}{100 + w \,.\, 2,65}\right) \cdot 100 \quad (4)$$

3 SETTLEMENT AFTER THAWING CAUSED BY EXTRA LOADS

Settlement caused by extra loads is relatively small after thawing, more or less like in overconsolidated condition (fig. 3).

The author has determined that the modulus of elasticity after thawing (a) is about 1/2 of the rebound modulus (m_2) in normal consolidation test (see eqn. 5)

$$a = \frac{m_2}{k_4} \quad (5)$$

where k_4 = 2...4.

When no freeze tests are made, a value of 4 for k_4 is recommended.

4 ADDITIONAL DEFORMATION IN THE SECOND AND THIRD FREEZING-THAWING CYCLES

Thaw settlements after the second or third freezing cycle are much smaller than after first freezing (fig. 4).

Eqn. (6) is suggested to estimate additional deformations for two or three times frozen clays.

$$\Delta\mathcal{E}_2 = \frac{w}{k_2} \quad \text{and} \quad \Delta\mathcal{E}_3 = \frac{w}{k_3} \quad (6)$$

where $k_2 \cong 20$ and $k_3 \cong 40$.

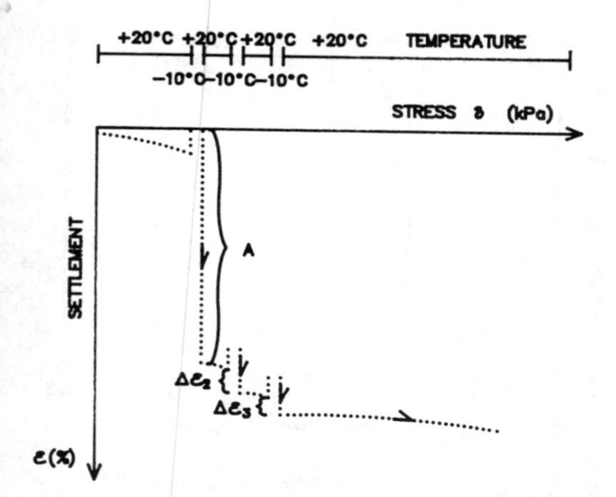

Fig. 4 Typical curve for three times frozen clay.

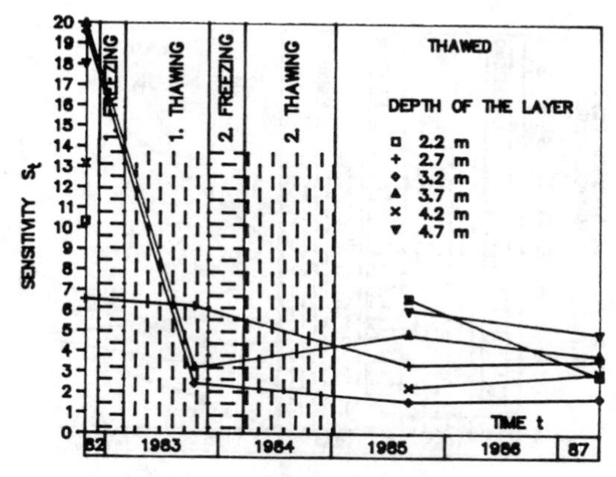

Fig. 6 Effect of freeze-thaw cycles on clay sensitivity at several depths in the full scale test area.

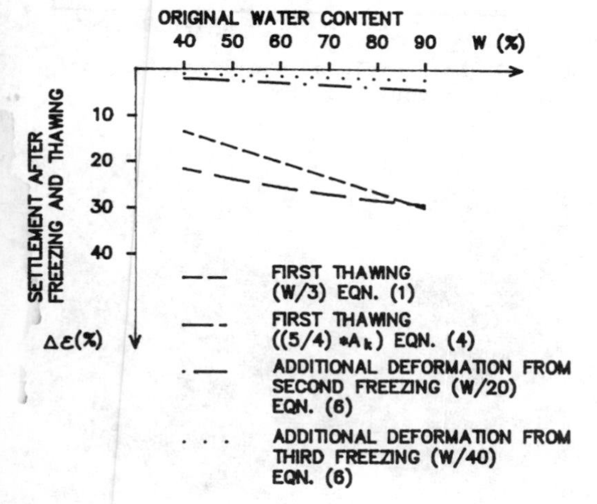

Fig. 5 Comparison between eqn. 1, 4 and 6.

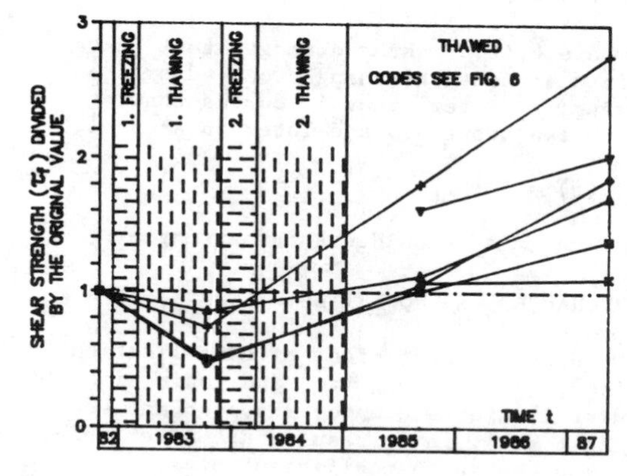

Fig. 7 Development of shear strength as shown by vane tests.

5 COMPARISON BETWEEN DIFFERENT THAW SETTLEMENT EQUATIONS

Eqn. (1) and (4) are both formulae for thaw settlement, although they are obtained from separate test areas. The two thaw settlement equations can be compared in fig. 5.

Eqn. (1) and (4) give exactly the same value for thaw settlement when the water content is about 90 %. With a smaller water content, eqn. (4) gives a slightly bigger settlement. However, the author considers eqn. (1) better because it is developed from direct observations. Formulae to estimate second or third thawing are also presented in fig. 5.

6 SHEAR STRENGTH

The original shear strength of clay in the full scale test area is 8 to 15 kPa and the water content 45 to 95 %.

The development of the shear strength was observed for almost five years. Vane tests were made even between first and second freezing when the lower ground was still frozen. Changes in the shear strength can be seen in fig. 6-8, fig. 6 showing the development in sensitivity, fig. 7-8 in shear strength and in remoulded shear strength, divided by the original values.

The author suggests that the following type of formula could be taken up for consideration:

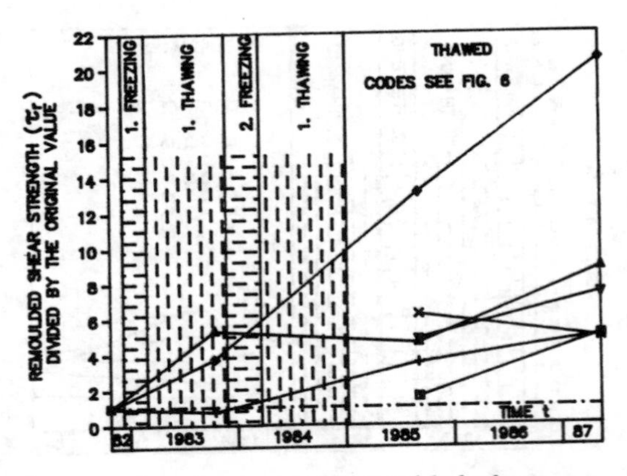

Fig. 8 Development of remoulded shear strength as shown by vane tests.

$$\tau_f(t) = r_t \cdot \tau_f \qquad (7)$$

where $\tau_f(t)$ = shear strength that takes into account the changes caused by freezing, r_t = term that is during thawing process about 0,5 and later on about 1,5.

$$\tau_r(t) = 3 \cdot r_t \cdot \tau_r \qquad (8)$$

where τ_r = remoulded shear strength.

7 CHANGES IN PROPERTIES

Changes in coefficient of consolidation (c_v), liquid limit (w_L) and grain size distribution are under research and the first preliminary results are given here.

The change in coefficient of consolidation is considerable as can be seen in fig. 9. The minimum value of c_v becomes about triple and even the stress state where the minimum is situated increases.

The decrease in the liquid limit (w_L) is observed to be about 30 % and even the change in appearance is easy to notice by the eye. One example is given in fig. 10 where we find out that the change is from clay to silty material.

When we burn clay we get bricks, but when we freeze clay we get also a new material and the author calls this frozen and thawed clay "frost crust". "Frost crust" is still clay if you consider it by grain size distribution, but the water content, liquid limit, consolidation properties, strength, sensitivity and appearance are more like those of a silty material.

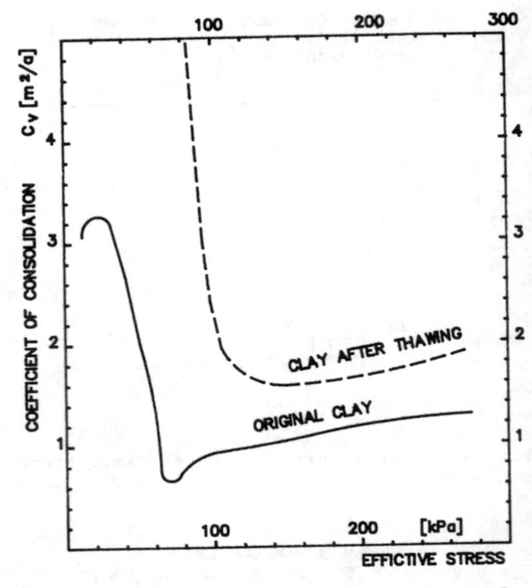

Fig. 9 An example of the change in coefficient of consolidation.

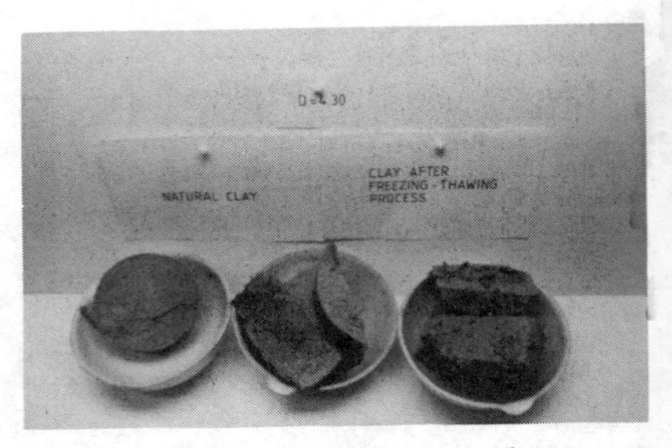

Fig. 10 Clay samples from same layer, where the left-hand one is in original state and the two others after two freezing-thawing cycles.

8 CONCLUSION

Both natural frost and artificial freezing changes the properties of clay, and these changes are not known sufficiently. Some new formulae presented in this paper are meant to be taken up for discussion. The author considers the trends of those equations important while the exactness of terms needs much more research.

As an example of how to use this "new" knowledge is fig. 11, which shows what happened to a road that was built a couple of years ago about 0,5 m below the natural

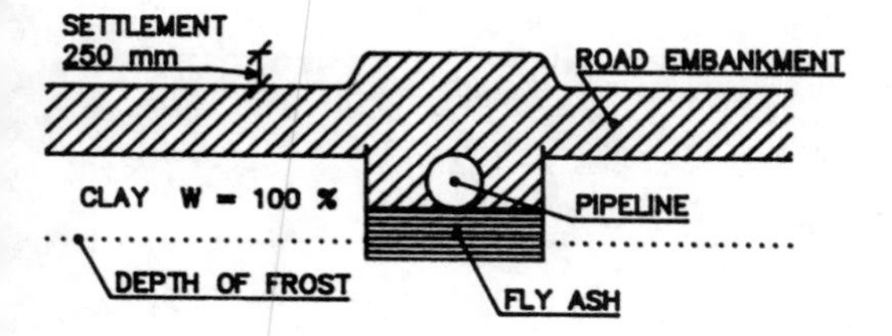

Fig. 11 A road embankment in Helsinki with thaw settlement failure.

ground. When the road was also kept clean of snow, frost could penetrate about 0,7 m deeper into the clay layer than ever before. So a thawing settlement of 250 mm was observed. From eqn. (1) the value is 233 mm (= 100/3 · 700/100).

This settlement is easy to see and feel when driving because pipelines are founded in a better manner on fly ash bed and this bed has also isolated clay under pipe line from frost. Now the moral of the previous example is this. Because we had the coldest winter for a hundred years in Finland, the clay under the road was stabilized so that no extra improvement was needed for the clay, just a levelling of the road surface.

All presented information indicates that freezing changes the structure of clay permanently and that the change in soil type is from clay to silty material. Furthermore, the origin of dry crust has at least to the author got a new natural explanation and it is the frost. Similarly, some landslides after a very cold winter have got at least one good explanation which is the decrease in shear strength during the thawing after cold periods.

REFERENCES*

Tong C. and Chen. E. 1985. Thaw consolidation behaviour of seasonally frozen soils. Proc. Fourth Int. Symp. on Ground Freezing, Sapporo, ed. S. Kinosita & M. Fukuda, Balkema, Rotterdam, 159-163

Soveri, U. 1956, On the mineralogical composition of argillaceous sediments of Finland. Ann. Acad. Scient, Fennicaea, Ser A, III, 48.

Vähäaho, I.T. and Eronen, T. 1982. Construction of a tunnel under a major rail way with the aid of temporary bridges and V-shaped icewalls. Proc. Third International Symposium on Ground Freezing, Hanover, New Hampshire, USA. 423-428.

Vähäaho, I.T. 1987, Jäädytysmenetelmän käyttö (= The Use of Ground Freezing), Geoteknisen osaston tiedote 44, Helsinki (only in Finnish).

*See page X concerning ISGF papers

5th International Symposium on Ground Freezing, Jones & Holden (eds)
© 1988 Balkema, Rotterdam. ISBN 90 6191 824 3

Uniaxial compressive strength of frozen medium sand under constant deformation rates

Zhu Yuanlin, Zhang Jiayi & Shen Zhongyan
Lanzhou Institute of Glaciology & Geocryology, Lanzhou, People's Republic of China

ABSTRACT: Uniaxial compressive strength tests were conducted on remolded, saturated Lanzhou medium sand under various constant cross-head speeds and temperatures ranging from -2 to -15°C. The specimens were very densely compacted, with an average dry density of 1.80 g/cm3. It is found from the tests that a ductile-brittle transition of failure mode for the sand occurs at a strain rate of about $3x10^{-4}s^{-1}$. The peak strength (σm) of the sand increases with increasing strain rate ($\dot{\varepsilon}$) in a power-law when $\varepsilon<3x10^{-4}s^{-1}$. It also increases with decreasing temperature in a power function within the range of temperature tested. Test results show that the failure strain remains almost the same within the range of strain rate from $3x10^{-4}s^{-1}$ to $10^{-6}s^{-1}$, while it decreases when both $\dot{\varepsilon}>3x10^{-4}s^{-1}$ and $\dot{\varepsilon}<10^{-6}s^{-1}$. The elastic (linear) modulus of the sand was found to be almost constant when both $\dot{\varepsilon}>3x10^{-4}s^{-1}$ and $\dot{\varepsilon}<10^{-6}s^{-1}$, whereas it increases with increasing strain rate in a power-law within the range of strain rate from $3x10^{-4}s^{-1}$ to $10^{-6}s^{-1}$. It is also increases exponentially with decreasing temperature.

KEYWORDS: compressive strength, failure, sand, strain rate, stress-strain behaviour, temperature effects.

INTRODUCTION

Since sandy soils are frequently encountered in soil foundation and used as pads or fill in geotechnical engineering in cold regions, it is, therefore, necessary to investigate the stress-strain behaviour of frozen sand so as to make the design of foundation more reasonable. As early as 1930, Tsytovich carried out uniaxial compressive tests on frozen sand under various loading rates and temperatures and reported that the compressive strength increased with increasing loading rate and decreasing temperature (Tsytovich, 1975). Since then, various researchers (Vialov, 1959; Chamberlain et al. 1972; Sayles, 1974; Bragg et al. 1980; Parameswaran, 1980; Ladanyi, 1981; Baker et al. 1981; Parameswaran and Jones, 1981; Ting et al. 1983; Jessberger et al. 1985; Kuribayashi et al. 1985 and others) have studied the time-and temperature-dependent strength behaviour of frozen sandy soils under various conditions. However, only a few compressive tests were conducted on frozen sand in China (Wu et al. 1983; Ma et al. 1983).

To provide a better understanding of the time-and temperature-dependent strength behaviour of frozen sand in China, uniaxial compression tests were intensively performed on remolded, partially saturated, Lanzhou medium sand under various cross-head speeds ranging from 6 to 0.0048 mm/min and temperatures from -2 to -15°C. The test specimens were densely compacted having an average dry unit weight of 1.80 g/cm3 with a water content of 14%, corresponding to a degree of saturation of 93%.

EXPERIMENT

Test Material

The test material used in this study was a remolded medium sand taken from the flood land of Yellow River near the Lanzhou Highway Bridge. The gradation curve of the sand is shown in Fig.1. Its specific gravity is 2.65.

Preparation of Specimens

The specimens were prepared by compacting the air-dried sand to a desired dry unit weight of 1.80g/cm^3 in a cylindric copper mold. After the specimens were saturated with no vacuum, they were quickly frozen in a cold room. After freezing, the specimens were taken out from the molds, and then the ends carefully trimmed flat and parallel on a lathe in a cold room. The prepared specimens had the nominal dimensions of 61.8 mm in diameter and 150 mm long.

Test Method

The uniaxial unconfined compressive tests were conducted in a cold room on a screw-driven universal testing machine equipped with a temperature-controlled chamber. Five head speeds of 6, 0.9, 0.08, 0.0096 and 0.0048 mm/min were employed in this investigation. All specimens were tempered at the appropriate test temperature (with fluctuation less than $\pm 0.1^{\circ}C$) for at least 12 hours before testing. The axial force and deformation produced during the compression of each specimen were measured with a load cell and a DCDT. After testing, the specimens were photographed, and the bulk densities and water contents were determined.

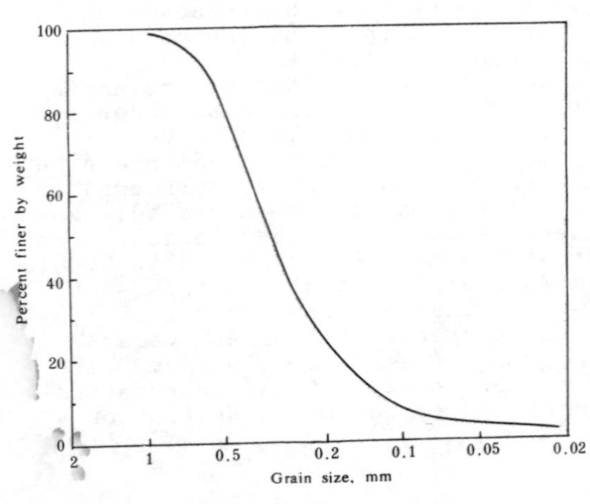

Fig.1. Gradation curve for the material tested

TEST RESULTS

Asial stress-strain curves for typical specimens tested at various strain rates and temperatures are shown in Fig.2. It is seen from the figure that the stress-strain behaviour of frozen sand is strongly strain-rate-(except for relatively high and low strain rates) and temperature-dependent. Based on the peak of the curves, the peak strength σ_m, failure strain ε_f and the time to failure t_m of the specimens were determined, and these are summarized in Table 1. According to initial yielding, at which the stress-strain curves start to perceptibly deflect from their initial near linear portion (corresponding to a strain of about 1%), the elastic (or quasielastic) modulus E_o of the samples have been determined, and these are also given in Table 1.

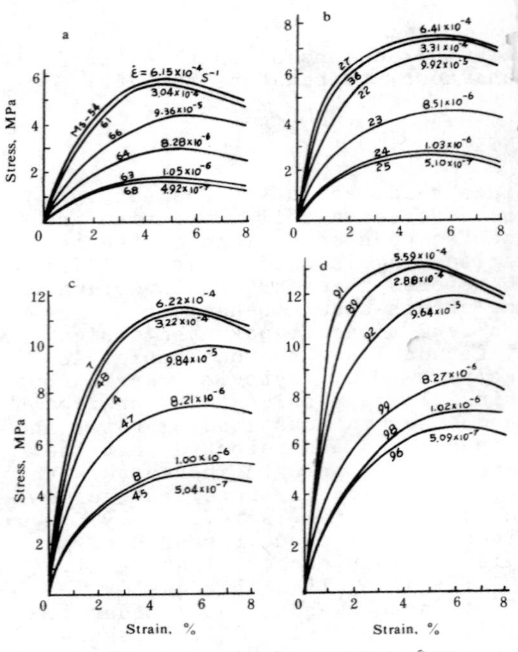

Fig.2. Stress-strain curves for various strain rates at temperatures of: (a) -2°C, (b) -5°C, (c) -10°C and (d) -15°C

Table 1. Summary of test results and conditions

Sample number	Applied strain rate $\dot{\varepsilon}$, s^{-1}	Peak strength σm, MPa	Failure strain ε_f, $x10^{-2}$	Time to Failure t_m, min	Elastic modulus E_o, MPa
		Temp. = -2°C			
MS-53	$6.14x10^{-4}$	5.55	4.91	1.33	230
MS-54	$6.15x10^{-4}$	5.73	5.17	1.40	233
MS-62	$6.12x10^{-4}$	5.72	4.99	1.36	226
MS-57	$3.22x10^{-4}$	5.49	4.54	2.35	230
MS-61	$3.04x10^{-4}$	5.41	5.03	2.76	203
MS-55	$9.59x10^{-5}$	3.95	5.38	9.35	166
MS-65	$9.37x10^{-5}$	4.36	5.16	9.18	191
MS-66	$9.36x10^{-5}$	4.27	5.49	9.78	158
MS-56	$8.67x10^{-6}$	2.51	5.10	98.0	115
MS-64	$8.28x10^{-6}$	2.87	5.12	103	121
MS-52	$1.05x10^{-6}$	1.69	5.31	840	66
MS-63	$1.05x10^{-6}$	1.76	4.59	730	71
MS-58	$5.04x10^{-7}$	1.62	3.39	1120	78
MS-67	$5.05x10^{-7}$	1.54	3.79	1250	67
MS-68	$4.92x10^{-7}$	1.63	3.84	1300	65
		Temp. = -5°C			
MS-19	$6.35x10^{-4}$	7.19	5.52	1.45	405
MS-36	$6.41x10^{-4}$	7.27	4.54	1.18	440
MS-26	$3.19x10^{-4}$	7.64	5.16	2.70	426
MS-27	$3.31x10^{-4}$	7.62	5.96	3.01	402
MS-21	$9.88x10^{-5}$	6.36	6.26	10.6	322
MS-22	$9.92x10^{-5}$	6.58	5.66	9.5	336
MS-23	$8.51x10^{-6}$	4.41	5.82	114	200
MS-24	$1.03x10^{-6}$	2.80	4.89	786	132
MS-25	$5.10x10^{-7}$	2.58	3.92	1276	130
MS-37	$5.15x10^{-7}$	2.59	4.14	1340	127
		Temp. = -10°C			
MS-3	$6.18x10^{-4}$	11.38	4.93	1.33	608
MS-7	$6.22x10^{-4}$	11.45	5.27	1.41	588
MS-14	$3.00x10^{-4}$	11.23	4.77	2.65	595
MS-48	$3.22x10^{-4}$	11.44	5.03	2.60	626
MS-1	$9.80x10^{-5}$	10.00	6.47	11.0	491
MS-4	$9.84x10^{-5}$	9.82	6.21	10.5	472
MS-2	$8.29x10^{-6}$	6.96	6.46	130	309

Table 1 (continued)

MS-47	$8.21x10^{-6}$	7.29	6.21	126	337
MS-5	$1.03x10^{-6}$	5.20	7.03	1140	198
MS-8	$1.00x10^{-6}$	5.29	6.70	1120	210
MS-45	$5.04x10^{-7}$	4.86	4.93	1630	208
MS-46	$5.13x10^{-7}$	4.70	5.29	1720	189
		Temp. = -15°C			
MS-90	$5.49x10^{-4}$	12.43	4.28	1.30	830
MS-91	$5.59x10^{-4}$	13.28	4.53	1.35	870
MS-89	$2.88x10^{-4}$	13.22	4.85	2.81	823
MS-94	$2.98x10^{-4}$	12.75	5.02	2.81	884
MS-100	$3.01x10^{-4}$	12.71	4.99	2.76	787
MS-88	$9.48x10^{-5}$	11.63	6.20	10.9	662
MS-92	$9.64x10^{-5}$	11.95	6.19	10.7	698
MS-93	$8.65x10^{-6}$	9.23	7.21	139	471
MS-99	$8.27x10^{-6}$	8.47	6.85	138	420
MS-97	$1.01x10^{-6}$	7.25	7.32	1210	289
MS-98	$1.02x10^{-6}$	7.29	7.24	1180	297
MS-95	$5.14x10^{-7}$	6.50	7.24	2350	286
MS-96	$5.09x10^{-7}$	6.74	6.11	2000	290

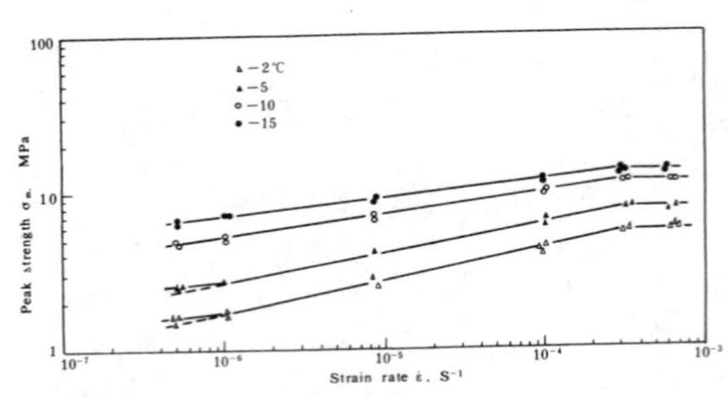

Fig.3, Peak strength as a function of strain rate

DISCUSSION

Peak Compressive Strength

Strain-rate-and temperature-dependence of peak strength

The peak compressive strength (σ_m) as a function of strain rate ($\dot{\varepsilon}$) for various temperatures is plotted in Fig.3 on a log-log scale. It is seen from the figure that the peak strength increases with increasing strain rate in a power law when $\varepsilon < 3x10^{-4}s^{-1}$, while it remains almost constant when $\dot{\varepsilon} > 3x10^{-4} s^{-1}$, indicating that a ductile-brittle (D-B) transition of failure mode occurs at a strain rate of about $3x10^{-4}s^{-1}$ for the frozen medium sand with high density. According to Fig.3, the peak strength as a function of strain rate for the sand can be described by

$$\sigma_m = \sigma_o(\dot{\varepsilon}/\dot{\varepsilon}_o)^m, \text{ when } \dot{\varepsilon}<\dot{\varepsilon}_o$$
$$\sigma_m = \sigma_o, \qquad \text{when } \dot{\varepsilon}\geqq\dot{\varepsilon}_o \qquad (1)$$

where $\dot{\varepsilon}_o=3x10^{-4}s^{-1}$ is the critical strain rate of D-B transition, σ_o is the limiting shortterm strength of the sand tested for $\dot{\varepsilon}=\dot{\varepsilon}_o$, and m is the exponent of strain rate. The values of σ_o and m obtained by linear regression analysis of the data at various temperatures are shown in Table 2. The values of m observed in the present study are higher than those reported by Sayles (1974) and Parameswaran (1980) for Ottawa sand over a similar temperature range. They reported that the value of m is nearly independent of temperature. It is, however, clearly dependent upon temepratureaccording to the present study (see Table 2). It is seen from Table 2 that the value of σ_o is also strongly temperature-dependent. They can be expressed in terms of temperature

Table 2. Values of σ_o and m in eq (1) and n in eq(5)

θ, °C	σ_o, MPa	m	n
-2	5.30	0.191	0.199
-5	7.71	0.171	0.180
-10	11.42	0.136	0.134
-15	13.00	0.105	0.100

by

$$\sigma_o = 3.8(\theta/\theta_o)^{0.466} \quad (2)$$

and

$$m = 0.204 + 0.0066\theta \quad (3)$$

where θ_o=-1°C is the reference temperature and θ is the negative temperature of the samples in °C. The value of the exponent, 0.466, in eq (2) is close to that found by earlier authors (Tsytovich, 1975; Parameswaran, 1980).

Substituting eqs (2) and (3) into (1), produces the constitutive equations for the strength of the Lanzhou as follows:

$$\sigma_m=3.8\left(\frac{\theta}{\theta_o}\right)^{0.466}\left(\frac{\dot{\varepsilon}}{\dot{\varepsilon}_o}\right)^{0.204+0.0066\theta}, \quad (\dot{\varepsilon}<\dot{\varepsilon}_o)$$

$$\sigma_m=\sigma_o\left(\frac{\partial}{\theta}\right)^{0.466} \quad (\dot{\varepsilon}\geq\dot{\varepsilon}_o) \quad (4)$$

where $\dot{\varepsilon}_o=3\times10^{-4}s^{-1}$, θ_o=-1°C, $\dot{\varepsilon}$ is in s^{-1} and σ_m is in MPa.

Note that there may be another transition in the mode of deformation, indicated by the change in slope of the log σ_m vs log $\dot{\varepsilon}$ curves for -2 and -5°C, at a strain rate of about $10^{-6}s^{-1}$. A similar phenomenon was observed earlier by Zhu and Carbee (1983) from creep tests on Fairbanks silt.

Time-dependence of strength

The peak strength as a function of time to failure for various temperatures is plotted in Fig.4 on a log. log scale. As expected, a D-B transition of the failure mode can be seen in this figure, correspongging to a time to failure of about 3 min. Therefore, the peak strength a function of time to failure for

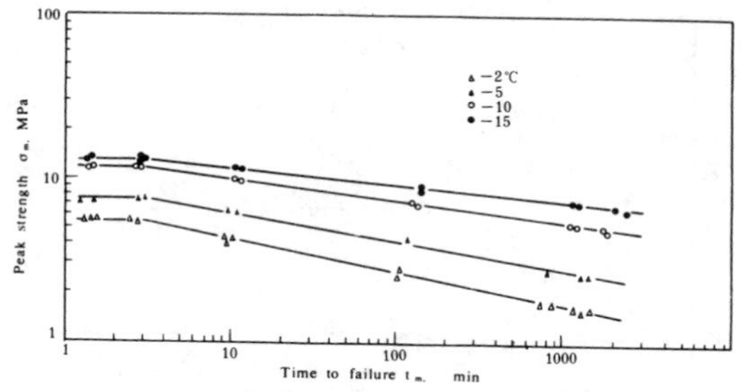

Figure.4, Peak strength as a function of time to failure

$$\sigma_m = \sigma_o(t_m/t_o)^{-n}, \quad \text{when } t_m>t_o$$

$$\sigma_m = \sigma_o, \quad \text{when } t_m\leq t_o \quad (5)$$

where t_o=3 min is the reference time corresponding to the D-B transition of failure mode for the sand, t_m is in minutes and the exponent n is temperature-dependent. Its variation with temperature is also shown in Table 2 and can be evaluated by

$$n = 0.215 + 0.0077\theta \quad (6)$$

As expected, the value of n is close to that of m. For a first approximation, they can be taken as the same. The value of n was found to be of 0.07 by Wu et al. (1983) for Huinan sand at -15°C.

Substituting eqs (2) and (6) into (5), produces the strength relaxation equations for the Lanzhou sand as follows:

$$\sigma_m=3.8\left(\frac{\theta}{\theta_o}\right)^{0.466}\left(\frac{t_m}{t_o}\right)^{-(0.215+000770)}, \quad (t_m>t_o)$$

$$\sigma_m = \sigma_o = 3.8(\text{——})^{0.466}, \quad (t_m\leq t_o) \quad (7)$$

where t_o=3 min, t_m is in m minutes and σ_m is in MPa.

Failure Strain

Fig.5 illustrates the variation of the failure strain with strain rate for various temperatures. An abrupt change of failure strain was observed at a strain rate of $3\times10^{-4}s^{-1}$, also indicating the D-B transition in failure mode, which consists with the change in slope of log σ_m vs

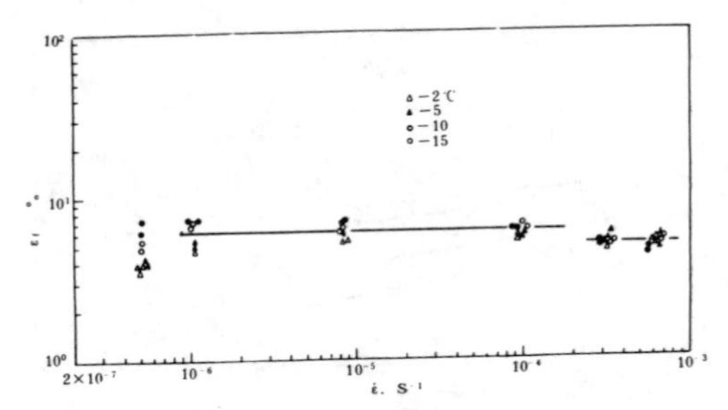

Fig.5, Variation of failure strain with strain rate

log $\dot{\varepsilon}$ curves, shown in Fig.3. A similar result was observed earlier by Bragg and Andersland (1980) and Baker et al. (1981) for Ottawa sand and Kuribayashi et al. (1985) for Kiso River sand. Note that another reduction in the failure strain can be seen from Fig.5, for the temperatures higher than -10°C at a strain rate of $5x10^{-7}s^{-1}$, indicating another transition in the deformation mode.

The failure strain for the sand was found to be relatively insensitive to temperature for $\dot{\varepsilon}>10^{-6}s^{-1}$, whereas it decreased significantly with increasing temperature for $\dot{\varepsilon}<10^{-6}s^{-1}$, as has also been observed by Zhu and Carbee (1984) from uniaxial compression tests on silt. It has average values of about 5% and 6% for the sand failed in brittle and ductile modes, respectively.

Elastic Modulus

Strain-rate-and temperature-dependence of elastic modulus

The elastic modulus, E_o, of the medium sand is shown in Fig.6 as a function of strain rate, for various temperatures, on a log-log scale. It is interesting to note that the curves in this figure are parallel, and that two identical break points (change in slope) are seen on each curve at the strain rates of $3x10^{-4}$ s^{-1} and $10^{-6}s^{-1}$, respectively. The elastic modulus increases with increasing strain rate in a power law when $10^{-6}s^{-1}<\dot{\varepsilon}<3x10^{-4}s^{-1}$, while it remains almost the same when $\dot{\varepsilon}\geq3x10^{-4}s^{-1}$ and $\dot{\varepsilon}\leq10^{-6}s^{-1}$. Therefore, the variation of elastic modulus with strain rate can be described by

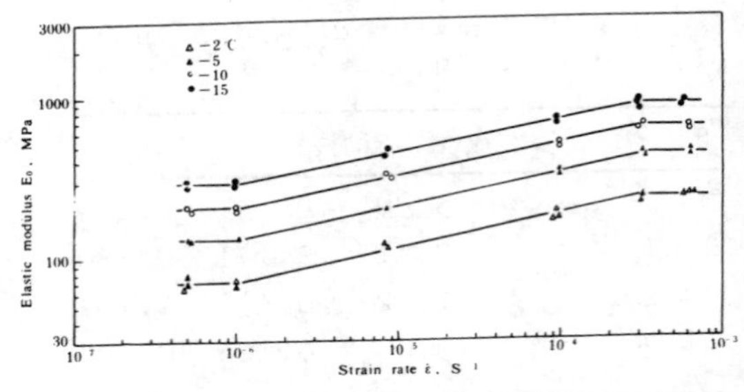

Fig.6, Elastic modulus as a function of strain rate

$$E_o=E_s\left(\frac{\dot{\varepsilon}}{\dot{\varepsilon}_o}\right)^r, \text{ when } 10^{-6}s^{-1}<\dot{\varepsilon}<3x10^{-4}s^{-1}$$

$$E_o=E_s \text{ , when } \dot{\varepsilon}\geq3x10^{-4}s^{-1} \quad (8)$$

$$E_o=E_l \text{ , when } \dot{\varepsilon}\leq10^{-6}s^{-1}$$

where E_s is called the limiting short term elastic modulus corresponding to the critical strain rate of D-B transition, i.e., $\dot{\varepsilon}_o=3x10^{-4}s^{-1}$, E_l is called the limiting long-term elastic modulus corresponding to a strain rate of $10^{-6}s^{-1}$, and r is the temperature-independent exponent, having an average value of 0.191 from linear regression analysis of the data.

The averaged values of E_s and E_l from Fig.6 for the various temper tures are given in Table 3 can be related to temperature by power-law equations:

$$E_s = 134\left(\frac{\theta}{\theta_o}\right)^{0.686} \quad (9)$$

$$E_l = 44\left(\frac{\theta}{\theta_o}\right)^{0.684} \quad (10)$$

where E_s and E_l are in MPa.

Note that the exponents of temperature in the above two equations are almost the same. Thus, we have the simple relation of

$$E_l = 0.33E_s \quad (11)$$

i.e., the limiting long-term elastic modulus is about one third of the limiting short-term elastic modulus for the frozen sand at the same temperature. It is also interesting to note that the value of exponent in eqs (9) and (10) are very close to the value (0.651) observed by Zhu

Table 3. Values of E_s and E_l in eq (8)

θ, °C	E_s, MPa	E_l, MPa
-2	220	70
-5	410	132
-10	610	210
-15	850	290

and Carbee (1987) for silt, indicating that it may be independent of the type of soil.

Substituting eqs(9) and (10) and the value of r into (8), then we have the expressions relating elastic modulus with temperature and strain rate as follows:

$$E_o=134\left(\frac{\theta}{\theta_o}\right)^{0.686}\left(\frac{\dot{\varepsilon}}{\dot{\varepsilon}_o}\right)^{0.191}, \quad (10^{-6}s^{-1}<\dot{\varepsilon}<3\times10^{-4}s^{-1})$$

$$E_o=E_s=134\left(\frac{\theta}{\theta_o}\right)^{0.686}, \quad (\dot{\varepsilon}\geq3\times10^{-4}s^{-1}) \qquad (12)$$

$$E_o=E_l=44\left(\frac{\theta}{\theta_o}\right)^{0.684}\doteq0.33E_s, \quad (\dot{\varepsilon}\leq10^{-6}s^{-1})$$

where $\dot{\varepsilon}_o=3\times10^{-4}s^{-1}$, $\theta_o=-1°C$ and E_o is in MPa.

Time-dependence of elastic modulus

The elastic modulus, as a function of time to failure for different temperatures, is plotted in Fig.7 showing a family of parallel curves with two break points for each cur at the time to failure of about 3 and 1000 min, which is similar to Fig.6. Based on Fig.7, the natu of the relaxation of elastic modulus for the sand can be well described by

$$E_o=134\left(\frac{\theta}{\theta_o}\right)^{0.686}\left(\frac{t_m}{t_o}\right)^{-0.187}, \quad (3\ \text{min}<t_m<1000\ \text{min})$$

$$E_o=E_s=134\left(\frac{\theta}{\theta_o}\right)^{0.686}, \quad (t_m\leq3\ \text{min}) \qquad (13)$$

$$E_o=E_l=44\left(\frac{\theta}{\theta_o}\right)^{0.684}\doteq0.33E_s \quad (t_m\geq1000\ \text{min})$$

where t_o=3 min, t_m is in minutes and E_o is in MPa.

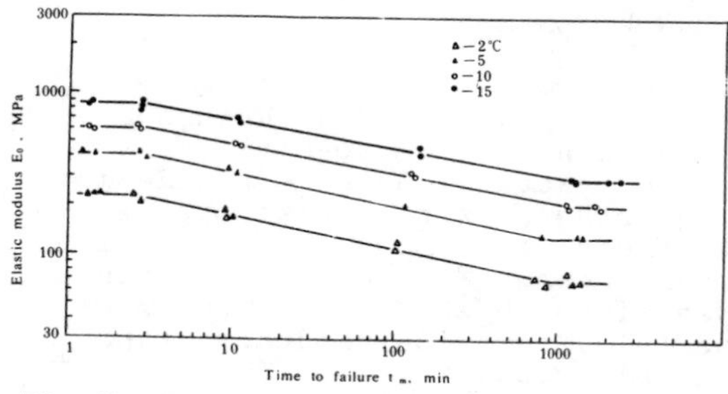

Fig.7, Elastic modulus as a function of time to failure

CONCLUSIONS

It can be concluded from this investigation that:

1. A clear break was observed on log σ_m vs log $\dot{\varepsilon}$ and log εf vs log $\dot{\varepsilon}$ curves at a strain rate of 3×10^{-4} s^{-1}, indicating the ductile-brittle transition of failure mode occured at this strain rate (corresponding to the time to failure of about 3 min) for the medium sand with high density within the testing temperature range.
2. Both strain rate (or time) and temperature are important factors influencing the peak strength of frozen sand. They can be related to each other by a power-law equation. The exponent m of $\sigma_m\alpha\dot{\varepsilon}^m$ is not a constant, but varies with temperature.
3. The failure strain of medium sand is found not to be sensitive to either strain rate or temperature for a particular failure mode. It has an average value of 5% and 6% for brittle and ductile failure of the sand with high unit weight.
4. The elastic modulus (E_o) of the frozen sand is sensitive to temperature and strain rate. But, as for the effect of strain rate, there exist two limits, i.e., the limit of short-term elastic modulus and the limit of long-term elastic modulus, which occur at the strain rates of $3\times10^{-4}s^{-1}$ and $10^{-6}s^{-1}$, respectively. The variation of E_o with temperature and strain rate (or time) can be well described by a power-law equation. The exponents in the form of $E_o\alpha\dot{\varepsilon}^r\theta^s$ are found to be constants, having a value of r= 0.191 and s=0.686, respectively. for the sand.

ACKNOWLEDGEMENTS

This study was funded through a program numbered Geo-85865 by the Chinese Natural Science Foundation. The authors express their thanks to their colleagues, Prof. Wu Ziwang, Messrs Wang Maohai and Pen Wanwei and Miss Miao Linafor some useful discussions and assistance in this study.

REFERENCES *

Baker, T.H.W., S.J. Jones and V.R. Parameswaran, 1981. Contined and unconfined compression tests on frozen sands. Proceedings of 4th Canadian Permafrost Conference, Calgary, Alberta, pp.387-393.

Bragg, R.A. and D.B. Andersland, 1980. Strain rate, temperature and sample size effects on compression and tensile properties of frozen sand. 2nd Int. Symp. on Ground Freezing, Trondheim, Preprints, pp.34-47. #

Chamberlain, E., C. Grores and R. Perham, 1972. The mechanical behaviour of frozen earth materials under high pressure triaxial test conditions. Geotechnique,22, No. 3, pp.469-483.

Jessberger, H.L., W. Ebel and P. Jordan, 1985. Temperature dependent strength and creep behaviour of frozen saline sand. Proc. of 4th Int. Symp. on Ground Freezing, Sapporo, preprint (unpublished).

Kuribayashi, E., M. Kawamura and Y. Yui, 1985. Strees-strain characteristics of an artificially frozen sand in uniaxially compressive tests. Proc. of 4th Int. Symp. on Ground Freezing, Sapporo, Vol.2, pp.177-182.

Ladanyi,B., 1981. Mechanical behaviour of frozen soils. Proc. of Int. Symp. on the Mechanical Behaviour of Structured Media, Ottawa, pp.205-245.

Ma Shimin and Peng Wanwei, 1983. Failure behaviour of frozen soils in uniaxial compression. Proc. of 2nd Chinese Permafrost Conference, Lanzhou, pp.281-283.

Parameswaran, V.R., 1980. Deformation behaviour and strength of frozen sand. Canadian Geotechnical Journal, Vol.17, No.1, pp.74-88.

Parameswaran, V.R. and S.J. Jones, 1981. Triaxial testing of frozen sand. Journal of Glaciology, Vol. 27, No.95, pp.147-156.

Sayles, F.H., 1974. Triaxial constant strain rate tests and triasial creep tests on frozen Ottawa sand. U.S. Army CRREL Technical Report 253.

Ting, J.M., R.T. Martin and C.C. Ladd, 1983. Mechanisms of strength for frozen sand. Journal of Geotechnical Engineering, 109(10), pp.1286-1302.

Tsytovich,N.A., 1975. The mechanics of frozen ground. McGraw-Hill, 426p.

Vialov,S.S., 1959. Rheological properties and bearing capacity of frozen soils. USA Snow, Ice and Permafrost Research Establishment, Translation 74. AD48156.

Wu Ziwang, Zhang Jiayi and Zhu Yuanlin, 1983. Strength and failure behaviour of frozen soils. Proc. of 2nd Chinese Permafrost Conference, Lanzhou, pp.275-280.

Zhu Yuanlin and D.L. Carbee, 1983. Creep behaviour of frozen silt under constant uniaxial stress. Proc. of 4th Int. Conf. on Permafrost, Fairbanks, Alaska, pp. 1507-1512.

Zhu Yuanlin and D.L. Carbee, 1984. Uniaxial compressive strength of frozen silt under constant deformation rates. Cold Regions Science and Technology, 9(1984), pp. 3-15.

Zhu Yuanlin and D.L. Carbee, 1987. Creep and strength behaviour of frozen silt in uniaxial compression. U.S. Army CRREL Report 87-10.

*See page X concerning ISGF papers

3. Engineering design

5th International Symposium on Ground Freezing, Jones & Holden (eds)
© 1988 Balkema, Rotterdam. ISBN 90 6191 824 3

State of the art: Engineering design of shafts

J.Klein
Bergbau-Forschung GmbH, Essen, FR Germany

ABSTRACT: The development of freezing methods in non-competent water-bearing strata initially took place simultaneously with the development of shaft lining techniques which, at that time, were exclusively based on cast-iron tubbings. Even today, the choice of the permanent shaft lining controls the dimensioning of the frozen body. If for example concrete is chosen as lining material, the cylindrically frozen body undergoes full stress during shaft sinking work only for a short period. In case of "sliding" shaft design, as standard in the Federal Republic of Germany, this cylindrical frozen body needs to secure the excavation against water and ground pressure over a considerable period of time. The different geotechnical aspects and thermophysical boundaries are demonstrated for 10 freeze shafts in coal mining. A short outline on the actual state of development of freezing techniques in shaft sinking in other mining countries is given.

1 INTRODUCTION

In water-bearing strata, cast-iron was used as lining material in freeze shafts from the very beginning. The tubbings invented in England (HOFFMANN 1967 <1>) underwent further development in Germany from the beginning of this century and became bolted watertight support elements with an inside rib pattern. The ring-wise bolted tubbings were assembled in short lengths or preferably fitted to the next higher tubbing element immediately after the corresponding length of shaft had been sunk into the frozen body. With respect to the strength of the frozen cylinder it is obvious that the latter need to withstand all water and strata induced stresses from outside over only a fairly short period. The tubbing lining took over its role as resistant support almost immediately, and gave stress relaxation to the frozen body. This needs to be considered, in particular when comparing various dimensioning formulae for freeze shafts. For the conditions prevailing during the early years of freezing technology the well known DOMKE-formula (DOMKE 1915 <2>) in which the material is considered as cohesive and frictionless,was sufficient. Present-day freeze shafts with immediately placed concrete lining can be assessed in a similar way, however, the situation looks much different when the dimensioning of cylindrical frozen bodies intend for a longer useful life.

2 FROM TUBBING TO"SLIDING" LINING

The collapse of the freeze shafts FRANZ-HANIEL 2 in 1925 and AUGUSTE VICTORIA 3 in 1927 (SCHLATTMANN 1930 <3>) led to initial intensive research into the structural and geotechnical correlations. On the structural side the lamellar-graphite cast-iron was closely investigated, theoretical knowledge of stability from civil engineering was applied to shaft construction, and the influence of the back-fill concrete was given more consideration. On the geotechnical side, intensive solid investigations of the strength of frozen soils and also of thermophysical analyses with respect to freezing methods were made. As happens frequently in civil engineering, the reasons for the damage to both shafts were found neither in the freezing process nor in the cast-iron lining, but in the workmanship.

In German coal mining, in particular in the Ruhr district, more than 300 shafts totalling a length of 30 km had been sunk with tubbing lining by that time. More than one hundred tubbing shafts totalling a length of 12 km are still in service in German collieries; 15 of them are older than 100 years, and the oldest dates back to 1872.

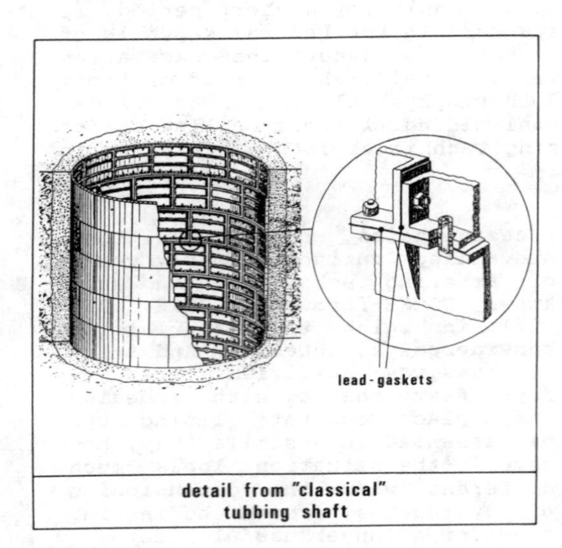

detail from "classical" tubbing shaft

Fig.1

While in England and other European countries shaft lining material made its way quite early into shaft construction technology (AULD 1987 <4>), cast-iron tubbings were used in German freeze shafts until the end of the 1950's. In all cases, tubbing-supported freeze shafts exhibited satisfactory watertightness and considerable stability even when subjected to stresses far beyond the calculated values.

The tubbing method lost much of its importance in Germany, when high priority was given to mining operations near the shaft, and the trend went towards asphalt-jointed "sliding" shaft linings not bonded to the strata. 20 "sliding" shafts have been built during the last 30 years. Separated from the surrounding strata by an asphalt mass the load conditions for the inner shaft lining are clearly defined.

Whereas in England the overburden, although water bearing was stable, in Germany always water-bearing quick-sand formations had to be worked through. The same applied to the mining industries of the BeNeLux-countries, and still applies today to Polish coal and ore mining districts and, to a certain extent, to present-day exploration areas in China.

In Polish copper ore mining lamellar graphite tubbing still ranks first among the shaft lining materials. Nodular-graphite iron lining was used first in English potash mining, in the BOULBY shaft (CLEASBY et al 1975 <5>). In Canada, too, shafts were successfully lined with the material until recently. The superior properties of this material as well as the modern vacuum mould technology give reason to predict a renaissance of this kind of support.

Incidentally, the Channel tunnel, between England and France has been designed using this material in the cross-passages between the running and service tunnels.

Modified cross-sectional design with the same mass, can result in extremely cost-effective solutions for any stress configuration of the shaft in question, as shown by calculations made within BERGBAU-FORSCHUNG GMBH. In particular attractive composite lining designs can be arrived at by combining nodular-graphite cast-iron and concrete. Such a composite structure was actually placed in SOPHIA-JACOBA No. 8 shaft (KLEIN/RIEß/RITTER 1987 <6>) in the form of an inner-liner. For a lining bonded to the strata, cast-iron liners being used as a concrete formwork is quite an imaginable solution. In particular for the concrete wall thicknesses preferred in England (BELL 1986 <7>) - in some cases more than one meter of wall thickness - this seems to be an interesting and cost-effective alternative. The fact that the lining and its placement have a substantial influence on the formation of the cylindrical frozen body justifies a review on the development of lining technology for shafts in coal mining being given.

3 ENGINEERING DATA OF 10 FREEZE SHAFTS IN GERMAN COAL MINING

In general shaft sinking development in the Federal Republic of Germany recorded for mining a steady regression over the past 30 years. The only transient upwards trend, after the oil crisis, went downwards again quite soon, and an increased need for new main shafts is not within sight. The shafts sunk in competent non-aquiferous strata for finished internal diameters up to 8.10 meters are lined with concrete, as practised for the RADBOD No. 6 shaft just under construction. The 4.20 m high concrete ring lining elements with 30 cm joints ensure sufficient flexibility of the shaft structure. Among the shaft sinking projects those which require freezing up to 600 m are particularly prominent. I quote here the shafts VOERDE, SOPHIA-JACOBA No. 8, and the RHEINBERG shaft just being started. The latter is the 21st "sliding" shaft sunk in West German coal mining.

It is obvious that the complex structure needing the placement of a concrete block lining up from the foundation level, and subsequently the build-up of the internal lining similar to a chimney structure is correspondingly expensive. "Sliding" shafts including freezing of strata are approximately three times as expensive as a conventional shaft. However they exhibit the following advantages:

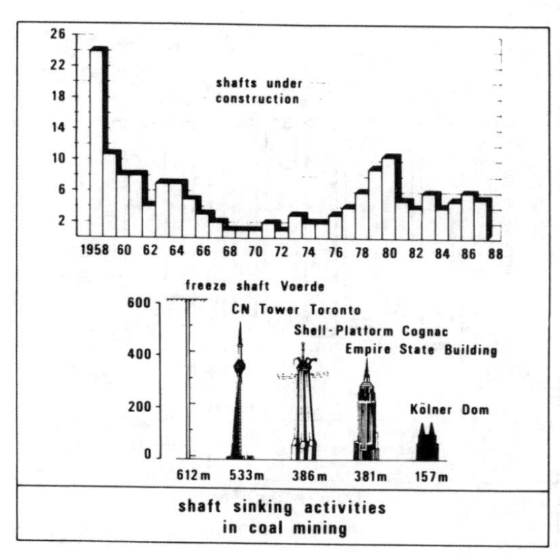

Fig.2

- mining operations within the shaft pillar are possible
- the external welded steel membrane guarantees absolute water-tightness.

Indeed in the cases of several "sliding" shafts (LÜTGENDORF 1986 <8>) working directly underneath resulted only in damages to the strata-bonded concrete lining of the shaft, and not in damage to the "sliding" portion of the lining. In addition, the advantages for ventilation technology of an absolutely dry shaft should not be underestimated.

Down to about 400 m the economic advantages of "sliding" outer steel-membrane/reinforced-concrete lining for shafts are unrivalled by those of any other lining system. For greater depths composite structures with concrete/steel or concrete/nodular cast iron systems become necessary to obtain an efficient total wall thickness of the inner shaft lining.

In the following let us take a closer look to the freeze shafts sunk in West German coal mining (a total of 10 in this decade). 8 of these shafts have been completed already. The frozen por-

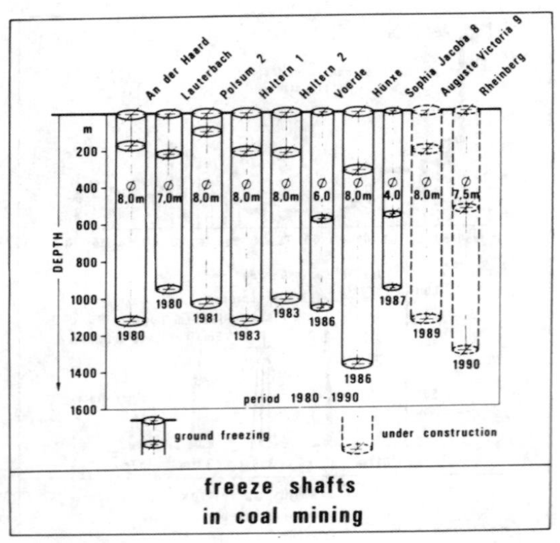

Fig.3

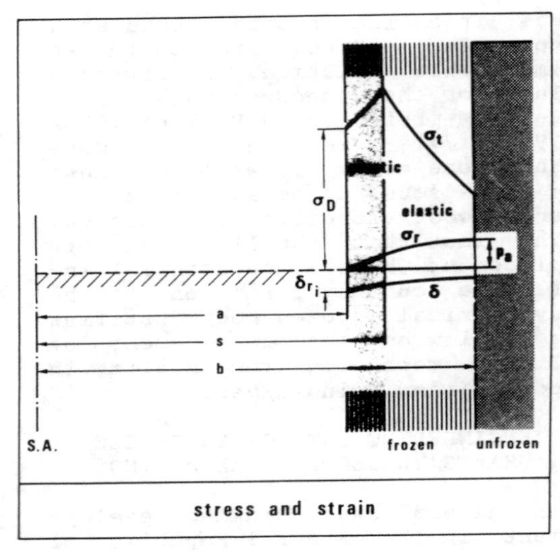

Fig.4

tion of AUGUSTE VICTORIA No. 9 shaft has just been completed, and for RHEINBERG shaft the freezing installations and the cooling systems are in working order. This year the hauling of the first kibble will start up the biggest freeze shaft to date near the river Rhine. With a freezing circle diameter of 22 m this shaft is one of the largest "sliding" shafts ever sunk in Germany. Also in this case, the external lining set during shaft sinking is intended to be made from shaped concrete blocks arranged over some length in several rows. This is the distinctive feature to other, even deeper freeze shafts sunk in Belgium, Canada or England since the frozen body for "sliding" systems needs to be maintained over a long period. For this reason comprehensive work was done in the field of freeze shaft design, in particular over the past few years. The significance of this research work may be revealed less by formulae than by discussion of the geotechnical aspects and thermophysical boundaries.

3.1 Geotechnical aspects

To quote the example of a 500 m deep freeze shaft (KLEIN 1982 <9>) average strata conditions require already 8 m of frozen wall thickness. By the vertical freezing tube arrangement this frozen body is cylindrical and, necessarily, generally overdimensioned in its upper part. The thickness of the external lining is determined by the soil-physical parameters of the frozen subsoil characteristics.

The internal lining resting on a foundation placed in competent and watertight zones undergoes stresses induced by the asphalt mass. The asphalt mass exerts fluid-pressure effects - according to density - on the internal lining. The respective value was set to be of 1,3 g/cm³ for the majority of shafts. In all cases the reinforced concrete is surrounded by tight-welded steel shells so that the water-tightness of the system is assured.

When looking at the frozen cylinder for the elastoplastic calculations this body is subdivided into two sections. The radial pressure is supposed to be induced from outside by water and strata pressure onto the elastic thick-walled cylinder of the frozen body. The formation of an internal plastic zone is deduced by classic strata mechanics. As to the material properties, this zone must be described by an adequate flow rule.In this case, substantial differences are found among various dimensioning formulae known from literature (AULD 1985 <10>). On the internal rim of the frozen body the uniaxial compressive strength of the frozen body is active. By the effect of a support resistance such as the one of a wall made from shaped concrete blocks - this means an additional confining pressure - the stress pattern can be influenced.

When setting immediately the permanent lining - either tubbings or concrete - only the short-term strength of the frozen soil is of importance. Cohesion determines essentially the strength of the material, and a friction effect can hardly develop. In case of long-time maintenance and shaped concrete block lining (for the RHEINBERG shaft the scheduled period was of more than 1 year), of course, the friction angle gains increasing importance, i.e. cohesion decreases with time while particle-to-particle friction within the structure of strata is activated. For the determination of the soil-physical characteristics, triaxial compression tests are necessary. In this case the description of the shear strength according to TRESCA is not used but rather the known yield criterion as per MOHR-COULOMB. In many cases the long maintenance of the frozen body requires also time-dependent statements on the stress and strain history. For this purpose, creep tests are run. The variety of parameters to be catered for doesn't make design easier, whether by empirical descriptions or by rheological models.

The essential questions are centered on the assessment of radial convergence of the un-lined shaft lengths (WÖLFFER 1985 <11>) and the "aging" of the frozen material. The latter, of course, is again a function of the kind of external lining, that is whether a soft lining made from shaped concrete blocks or a harder concrete panel lining is chosen. For daily shaft sinking work in-situ measurements and comparative assessment against the calculated data are important. Most calculations assume constant external pressure and constant support resistance. However, rheological descriptions are also known (BORM 1986 <12>, GILL and LANDANYI 1987 <13>). Already one different friction angle - all other parameter unchanged - proves to have considerable importance on the size of the plastic zone. We may state generally that the necessary thickness of the frozen cylindrical body is inversely proportional to the friction angle and the support resistance. Unfortunately an outside and an inside frozen body relative to the freezing tubes is spoken of, and this is actually not justified by facts but results from elasto-mechanical idealization. For the frozen body it is not the position of the freezing tubes but the interface between water and ice which is of importance - here the external pressure is actually acting.

The approximate equation of DOMKE can easily be extended in that the friction effects are integrated into the dimensioning equation. As assumed for said equation the circle of transition from the plastic to the elastic zone is defined by the geometric average of radii s = SQR (a*b). Accordingly, the elastic portion of the frozen cylinder is always larger than the plastic one, and this results in a safety margin for the stability. For the plastic zone the MOHR-COULOMB yield criterion (KLEIN 1985 <14>) may be adopted. Water and soil pressure p as well as the uniaxial compression strength K serve as input parameters. It is obvious that the necessary freeze wall thickness decreases with an increasing friction angle. In case of the representative friction angle $\Phi = 30°$ as applicable for example for sand formations, the

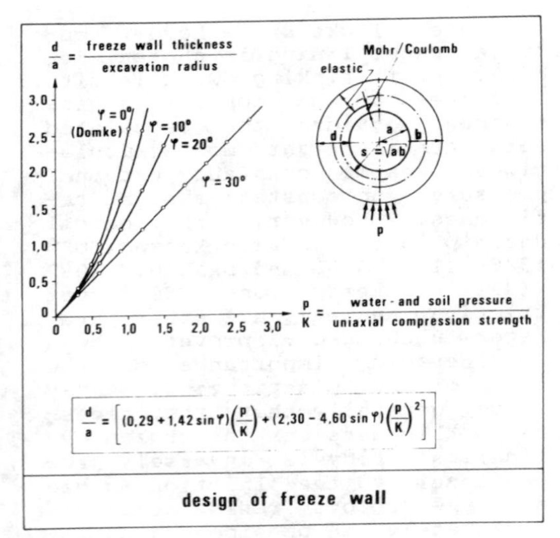

Fig.5

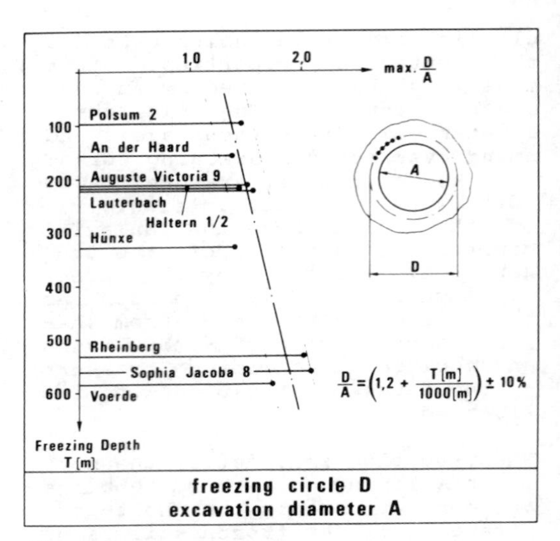

Fig.6

equation becomes linear reducing to d/a = p/K which is easy to remember. For freeze shafts with immediately effective support by tubbings or concrete, this is a simple formula for approximate computation.

3.2 Thermophysical boundaries

In the following the freezing technological data of 10 freeze shafts in West German coal mining are compiled for statistics. These data clearly show the relation freezing circle D/excavation diameter A being a function of depths. This logically means increasing frozen cylinder thickness as a function of increasing depth. The average relation is D/A = 1.2 + T (m)/1000 (m). All projects carried out, i.e. the deepest shaft with 581 m of frozen length (VOERDE) as well as the shallowest shaft with 99 m of frozen length (POLSUM No. 2) are situated within this range of ±10 %.

The comparison of the ground freezing installations is similar interesting. The performance of the refrigeration plants is seen to reach up to 15 GJ/h. The pipe capacity per 1 m of installed freezing pipe averages 600 kJ/h which corresponds to approx. 143 kcal/h. This value is still almost twice as high as the theoretically calculated requirements (RIES 1981 <15>): this means that a sufficient percentage of thermal losses is compensated by such capacity. On the other hand, the variation range of data was stated to be of ± 20 % (when comparing the projects carried out). Obviously, the necessary rating depends essentially on the stratigraphic features, the soil-physical data, and in particular on the water content of the frozen strata.

When comparing the various freeze-pipe arrangements the picture looks more uniform again. Three of the German companies specialising in freeze shaft technology prefer mostly a spacing ranging between 1.2 m and 1.3 m so that for all of the 10 freeze shafts an arithmetic average of 1.28 m was calculated. A fourth specialist company prefers larger freeze-pipe spacing. Accordingly, the shafts VOERDE (HEGEMANN and JESSBERGER 1985 <16>) and RHEINBERG go beyond this 10 % band-width. Certainly also, the use of high precision drilling merits consideration. The total of the freeze-pipe installations recorded for

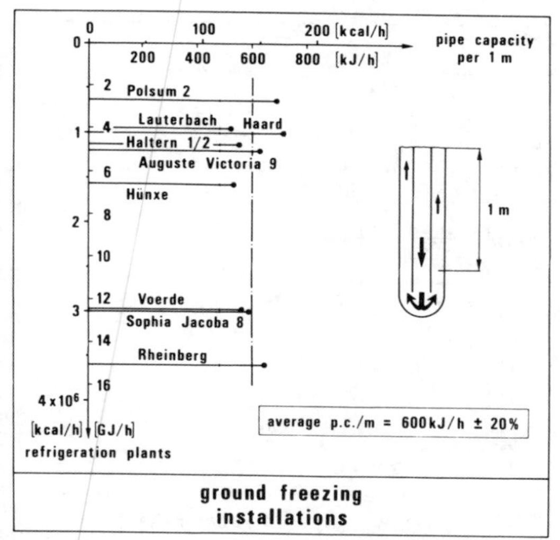

Fig.7

three projects carried out totals more than 20 km of length. With these considerable freezing depths the material choice for the casings and the connections of the pipes are of most importance. R & D carried out clearly has shown the un-suitability of API long-socket connections so that Ω-type connectors were preferred in cases of deeper freeze shafts.

4 FREEZE SHAFT PROJECTS IN OTHER COUNTRIES

Not only Germany but also in other mining countries remarkable results in the field of freeze shafts were obtained. In Britain the SELBY-project within which 10 freeze shafts were sunk (WILD and FORREST 1981 <17>) should be mentioned. For geological reasons the total length of freeze-pipe installations is relatively limited. The 1.39 m average freeze-pipe spacing is larger than that adopted on projects carried out in Germany. In Britain particular experience is on hand in the field of sectionwise freezing, a method only rarely applied in Germany. At present, the expert's interest is centered on the ASFORDBY-project we may see on the technical visit within this symposium.

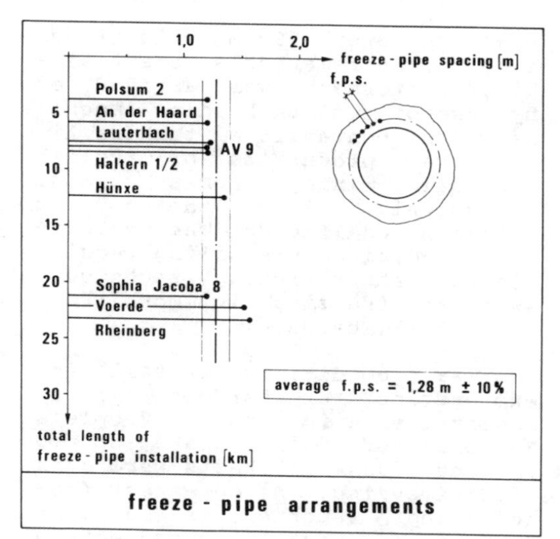

Fig.8

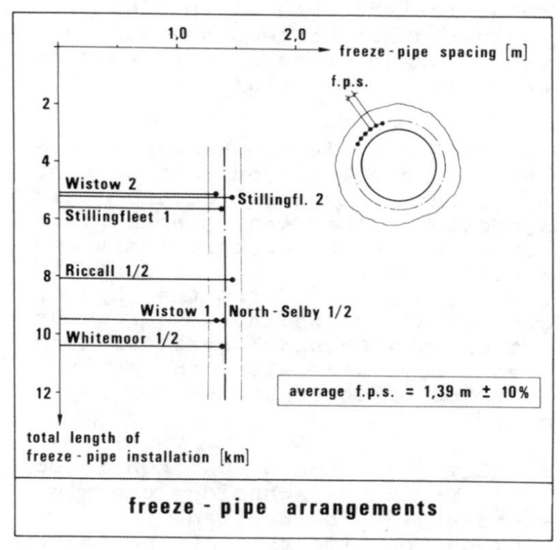

Fig.9

In Polish mining (GARUS 1985 <18>) the shaft sinking performances in copper mining deserve particular mention. In the past 25 years, 22.7 km of shafts where sunk in the Legnica-Glogow copper region, half of this total length by freezing. The shafts exhibit useful cross-sections of 6.0 m to 7.5 m and reach depths of up to 1 200m. Recently, the work on the 28th shaft - on SIEROSZOWICE mine - was completed. The shaft is of 7.5 m useful diameter and was sunk over 580 m with freezing. This remarkable shaft sinking performance was arrived at by use of modern heading machinery and conventional tubbing lining.The production of Polish hardcoal mining comes almost exclusively from the Upper Silesian coalfields. The geological conditions prevailing require special shaft sinking techniques such as freezing in more than 90 % of all cases.

Similar conditions prevail for exploration of hardcoal mining industries in the People's Republic of China where up to present 39 km of shafts were sunk with freezing. The greatest frozen length reported is 415 m at 8.0 m of useful cross-section (ZHANG 1987 <19>). If, for example, the freezing technological data for the PAN JI project are evaluated it becomes obvious that the average freeze-pipe spacing actually was relatively small.

Further details on Chinese freeze shaft technology you may be hear by Chinese authors during this symposium. The stratigraphic features which in certain areas are similar to those of the European coal deposits - from the hydrogeological viewpoint - are the reason why Chinese shaft sinking technology will also in the future be of interest for our regions.

Finally it should be mentioned that freezing technology can also successfully be applied in shaft sinking outside the mining industries. The underground workings of the LEP (Large Electron Positron Storage Ring) may be discussed here as an example. For the research organisation CERN (Organisation Europeenne pour la Recherche Nucleaire) in Geneva, Switzerland, a European consortium (WANNIGER 1987 <20>) headed a tunnel of 3.8 m useful cross-section which described a circle of 8.6 km diameter. The average depth of the underground working sites was approx. 100 m and implied particular requirements with respect to the heading methods used.The access to the un-

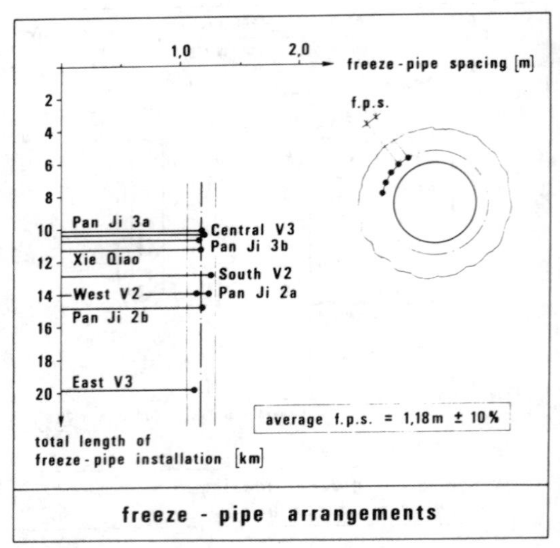

Fig.10

derground sites was assured by a total of 17 shafts on 8 individual shaft sinking sites. Also within this project freezing technology was successfully applied in 5 cases, and turned out to be a reliable shaft sinking technology in water-bearing and non-competent strata.

5 CONCLUSIONS

Frozen-soil cylinders for protecting shaft sinking activities for mining shafts are an internationally recognized construction technique. The engineering problems implied in this method are extremely complex and range from thermal/physical correlations to civil engineering considerations. Even the preliminary drilling-technological methods require, besides theoretical considerations, a high degree of precision. In all cases the final shaft lining system chosen has a decisive influence on the design. Formerly this lining used to be as cast-iron tubbing system. Nowadays either concrete or "sliding" shaft lining is chosen according to the mining-technological requirements implied. The necessary water-tightness of the shaft lining, the characteristics of the overburden to work through, and the infrastructure of the site are important parameters as well.

The manifold calculation methods for the stress-strain analysis as well as for the thermo-physical conditions were the subjects of extensive discussions throughout the time of their development. The present state of freeze shaft technology and the successes recorded are mainly due to the know-how of specialised companies and to the creativity of many experts from various disciplines.

ACKNOWLEDGEMENTS

The author wishes to express his thanks to STEINKOHLENBERGBAUVEREIN as well as to the German companies for special mining technological tasks for the many technical data relative to the projects carried out and for the possibility to use these data for this paper.

REFERENCES*

1. HOFFMANN,D. (1967) :"175 Jahre gußeiserner Schachtausbau." Glückauf 103,S.481-489

2. DOMKE,O. (1915) :"Über die Beanspruchung der Frostmauer beim Schachtabteufen nach dem Gefrierverfahren." Glückauf 51,S.1129-1135

3. SCHLATTMANN (1930) :"Bericht des Ausschusses zur Verhütung des Zusammenbruches von Tübbingschächten." Abhandlung 78 des Grubensicherheitsamtes im Preußischen Ministerium für Handel und Gewerbe.

4. AULD F.A. (1987) :"A decade of deep shaft concrete lining." Concrete, Volume 21,Number 2,pp.4-8

5. CLEASBY,J.V et al. (1975): "Shaft sinking at Boulby mine, Cleveland Potash, Ltd.." Institution of Mining and Metallurgy, Number 1,pp.A7-A28

6. KLEIN,J.;RIEß,H.G.;RITTER,H. (1987) :"Abteufen und Ausbauen des Gefrierschachtes Sophia Jacoba 8." Glückauf Heft 22

7. BELL,M.J. (1986) :"Modern mine shaft design in the UK." Colliery Guardian, September edition, pp. 423-424

8. LÜTGENDORF,H.O. (1986): "Rückblick auf 30 Jahre gleitenden Schachtausbau." Glückauf 122, S.1101-1262

*See page X concerning ISGF papers

9. KLEIN,J. (1982) :"Present state of freeze shaft design in mining." Symposium on Strata Mechanics, 5-7. April, Newcastle upon Tyne, UK; see Elsevier, Developments in Geotechnical Engineering Vol.32,pp. 147-153

10. AULD,F.A. (1985) :"Freeze wall strength and stability design problems in deep shaft sinking - is current theory realistic?" 4.International Symposium on Ground Freezing, 5-7. August, Sapporo,Japan,pp.343-349

11. WÖLFER,K.H. (1985) :"Einfluß der Teufsohle und des tragfähigen Außenausbaus auf Stoßschiebungen in Gefrierschächten." Glückauf-Forschungshefte 46, S.150-158

12. BORM,G.(1985):"Wechselwirkung von Gebirgskriechen und Gebirgsdruckzunahme am Schachtausbau." Felsbau 3, Nr.3, S.153-158

13. GILL,D.E.;LADANYI,B. (1987): "Time-dependent ground response curves for tunnel lining design." 6.International Congress on Rock Mechanics, Montréal, Canada; see Balkema, Volume 2, pp. 917-921

14. KLEIN,J. (editor) (1985): "Handbuch des Gefrierschachtbaus im Bergbau."Glückauf-Betriebsbücher 31,Glückauf, Essen

15. RIES,A.(1981):"Geschichtliche und technische Entwicklung des Gefrierverfahrens im Schachtbau." Schacht- und Tunnelbaukolloquium, 12. Juni, Düsseldorf

16. HEGEMANN,J; JESSBERGER,H.L. (1985) : "Deep frozen shaft with gliding liner system." 4. International Symposium on Ground Freezing, 5.-7.August, Sapporo, Japan, pp.357-373

17. WILD,W.M.; FORREST,W. (1981): "The application of the freezing process to ten shafts and two drifts at the Selby project." The Mining Engineer, June ,pp. 895-904

18. GARUS, B. (1985) :"The underground construction in 25 years of the Mine Construction Company in the Legnica-Glogow copper region." International Symposium, 22. November, Lubin, Poland.

19. ZHANG, Y. (1987) :"Die neue Entwicklung der Sonderschachtbautechnik für tiefe Schächte in China." 13. World Mining Congress, Stockholm, Schweden; see Balkema S.737-745

20. WANNINGER,R. (1987): "Untertagebauten des LEP für die Forschungsorganisation CERN in Genf." Technischer Bericht, Ausgabe März, Philipp Holzmann AG, Frankfurt/Main

5th International Symposium on Ground Freezing, Jones & Holden (eds)
© 1988 Balkema, Rotterdam. ISBN 90 6191 824 3

State of the art: Tunnelling using artificially frozen ground

J.S.Harris
Consultant, Nottingham, UK
(Formerly: Ground Freezing Division, British Drilling and Freezing Co. Ltd, Nottingham, UK)

ABSTRACT. The flexibility of the ground freezing method allows the practitioner to use his capabilities to meet the needs of a wide range of local conditions. This paper brings together achievements in many facets of the practice of frozen ground tunnelling, collectively representing the State of the Art, and which demonstrate the versatility and reliability of the method. The main advances in tunnel freezing works relate to the control of heave by careful use of intermittent freezing, minimizing thaw settlement and dealing with groundwater movement; but the advances in, and everyday use of, FEM computing techniques to produce reliable thermal and structural designs are particularly significant.

1 INTRODUCTION

The only previous State of the Art paper relating to frozen ground tunnelling that I have located is that presented by Jones (1980) to the second ISGF. Since then the amount of published matter in our discipline, including that generated by our own Symposia, has increased dramatically. Whilst this is a welcome development, particularly when compared with the scarcity of authentic material less than 20 years ago, the task of identifying and digesting all references to frozen tunnelling is a large one. Jones (1980) represents the obvious starting point for this update, but some earlier contributions, if they appear relevant, have not been entirely disregarded; equally there may be recent contributions which, because they have not come to my notice, have not been considered. It is inevitable that any review of a specific subject will overlap with related topics : where this happens my comments are brief.

Examples of frozen tunnels which have been described in the literature are listed in table 1.

The many elements of a ground freezing design are inter-related and include the purpose(s) for which the icewall is required, the possible mechanism(s) of failure, the capacity of AGF to counter adverse side effects (and conversely the degree of distress generated by AGF that can be tolerated nearby), the properties of the soils/rocks to be frozen, the refrigeration system, the method of construction, and structural design. It is probably true of all practitioners that improvement of detail in one aspect leads to a review of other matters, and this focusses attention on some facet which, until then, was considered to be adequately dealt with and understood. Thermal and structural design methods based on computing techniques, which can easily handle the many variables, are a feature of most recent contributions to the literature.

2 PURPOSE

Ground freezing is chosen to create a stable and/or impermeable boundary - the "ice-wall" - around an intended excavation until the permanent structure has been constructed, thus providing a safe and dry working environment with minimal (adverse) effects on adjacent property. There is a growing recognition that inclusion of the ground freezing method at design stage offers time and cost advantages to the end result; nevertheless there are many occasions during non-frozen construction when instability has halted progress, and the tunnel or other excavation has been recovered by freezing, often after large expenditure on other methods.

Table 1. Major tunnel freezings reported since 1980.

Location	Dia	Cover	Length(s)	H/V/I	B/LN	Completed	Reported
Milchbuck	14.4	8	12x34/45	H	B	1979	eg Mettier 1985
Gascoigne	6	180	2x105	V	B	1980	Wild & Forrest 1981
Runcorn	3	15	2x13	H	B&LN	1980	Harris & Norie 1982
Oslo			26	H	B	1980	Josang 1981
Antwerp	1.7	6	210+400	H	B	1981	Gonze et al 1985
Brussels		3		H	B	1982	Gonze et al 1985
Du Toitskloof	12.7	10/42	5x32	H&V	B	1982	Harvey 1983
Mol	3.5	220	25	H	B	1982	Funcken et al 1983
Iver	2.8	30	52	V	LN	1984	Hieatt & Draper 1985
Nunobiki	11	70	50	H	B	1984	Murayama 1985
Tokyo	9.7	37		R	B	1985	Murayama et al 1988
Keihin	9.7	15	2	H	B	1985	Numazawa et al 1988
Stonehouse	2	10	10	V	LN	1986	Harris 1987
Agri Sauro	4	150	24	H	LN	1986	Restelli et al 1988
Zurich						1986	Jessberger 1987
Vienna a	7	1.6	65	H	B	1987	Deix & Braun 1988
Vienna b	6.5	3	2x35	H	LN	1987	Martak 1988

H=horizontal, V=vertical, I=inclined (freeze-tubes),
B=brine, LN=liquid nitrogen

3 MECHANISM OF FAILURE

3.1 It is generally recognised that frozen ground behaves as an elasto-plastic material, although it has been shown that clays behave in a brittle manner at temperatures colder than -110°C (Bourbonnais & Ladanyi 1985).

Plastic deformation,which is temperature and time related, is more significant for cohesive soils than sands; creep may be the limiting parameter in a frozen soil tunnel design.

Workers at many institutions have reported studies of laboratory preparation of frozen specimens, constant strain-rate tests to determine compressive and tensile strength parameters, and constant stress tests at chosen fractions of the ultimate compressive strength to determine the creep parameters, each at a few controlled sub freezing-point temperatures. Many of these comply with the recommendations for laboratory testing of artificially frozen ground prepared by the IOC Working Party (Sayles et al 1986). Collectively the results from these various studies constitute the basis for a useful databank.

In parallel with the laboratory work the creep law itself has been studied, and several researchers have produced refinements to the basic equation in their attempts to eliminate discrepancies between experimental and predicted time-deformation curves (Fish 1983,1985; Klein 1978; Hampton et al 1985). Although each author achieves a good fit for the soil type he has tested, particularly when applied to a single inflection point characteristic curve, a universal relationship has yet to be found. However, based upon published results for several soil types, Hampton (1986) concluded that a modified form of the creep equation proposed by Klein (1978), based on a single inflection point characteristic curve, yields reasonable agreement over the timescales experienced in engineering practice, and is simple to use.

3.2 Although **groundwater flow** is a hazard to ground freezing, steps can be taken to limit its effect.

A velocity of 2m/day groundwater flow is universally recognised as the maximum above which ground freezing carried out with brine warmer than about -20°C will be unduly slow or impractical. Several measures have been taken to alleviate this situation. Colder refrigerating temperatures, coupled with a denser pattern of freeze-tubes on the upstream side of the area being protected, have been used (eg Einck & Weiler 1982). Calcium chloride brine can be chilled to -40°C before increasing viscosity has a significant effect on pumping resistance; colder temperatures dictate the use of alternative secondary refrigerants or the circulation of the primary refrigerant itself through the freeze-tube.

In severe cases it is necessary to reduce the velocity of flow, the most practical way being to inject cement or chemicals (eg Murayama et al 1988) to part fill the voids, and thereby reduce the transmissivity by orders of magnitude. Alternatively (or as well) it has been possible to introduce an insulating blanket. The latter method is much favoured in Japan where several tunnels have been successfully driven at shallow depth below watercourses (eg Ohrai et al 1985).

Heated barriers, as utilised in England in the 1960's, are also valid together with, or alternatively to, insulating blankets to limit ice growth or minimise ice erosion; Miyosha et al (1978) report use of this method beneath the bed of a river.

In an extreme case in Zurich, Gysi & Mader (1987) report the construction of a physical barrier to water flow, in addition to an insulating cover, prior to frozen ground tunnelling. In this case sheet pile walls were driven through the Limmat riverbed gravels on each side of the intended tunnel where there were no surface obstructions, supplemented by jet grouting below the foundations of the riverbank wall. The gravel thus entrapped was then vibrocompacted to reduce its permeability by more than an order of magnitude. The rate of groundwater flow was thereby reduced to an acceptable level.

By comparison, phased isolation of sections of the area to be excavated - a railway station in Duisberg - was achieved by temporarily sealing 1.4m gaps between 5.4m sections of a diaphragm wall by ground freezing; when the ice thawed on completion the full original groundwater flow was re-established, thus minimising undesirable side effects on the local environment (Weiler & Vagt, 1980).

Kessuru et al (1987) have developed an analytical solution which they have verified against a full-scale project where the groundwater flow rate was 4m/d.

3.3 Soils with **low natural moisture contents** exhibit only small improvement in compressive strength on freezing; they may also remain porous. Controlled irrigation during the primary freeze period will increase the moisture content, and therefore improve the frozen strength and permeability characteristics to the desired level. Maishman (1988), Gonze et al (1985) and Murayama (1985) report such activity in Chicago, Brussels and Nunobiki respectively.

4 DISTRESS

The main undesirable side effects of ground refrigeration on neighbouring property and utilities arise from the formation and decay of ice lenses, creep deformation of the exposed face of the excavation, and the placement of in situ concrete.

4.1 **Ground heave** on freezing, and settlement on thawing, is usually confined to the area within the plan boundary of the icewall. The likelihood and scale of such movements must therefore be related to their effect on existing or future structures within the zone of influence.

The best documented example of control of a soil/groundwater regime prone to heave is Milchbuck (eg Mettier 1985). Of 11 sections of tunnel the first had no buildings above its line; a maximum heave of 105mm was recorded during a period of continuous refrigeration whilst creating and maintaining the icewall. Subsequent sections were to pass beneath tall buildings which could not tolerate movement of this order.

By applying intermittent refrigeration during the maintenance stage the boundary of the icewall was effectively stabilised while preserving the low temperature of the core of the ice-wall. The total heave was thereby reduced to 5mm or less for 8 of the sections, and heave damage to the structures avoided.

A similar procedure has been applied in Vienna according to Deix & Braun (1987).

4.2 Thaw settlement has attracted less attention than heave, it having generally been expected that settlement would amount to some 50-90% of the heave dimension, with ground level reverting approximately to its start level. But the two phenomena are independent, as was also evident at Milchbuck (Grob 1984).

Where cyclic maintenance of refrigeration was practiced, the scale of settlement remained of similar magnitude to that for continuous refrigeration, ie 30-80mm; expressed as a percentage of heave it would be up to 3000% ! Clearly successful control of heave must be matched by control of thaw settlement. Grouting was undertaken at Milchbuck during the thawing stage to fill any voids resulting from the freeze-thaw cycle, and minimise thaw settlements at the surface. Theoretical predictive methods have been offered, eg by Blanchard & Fremond (1985) and Meissner (1985,1988a,b).

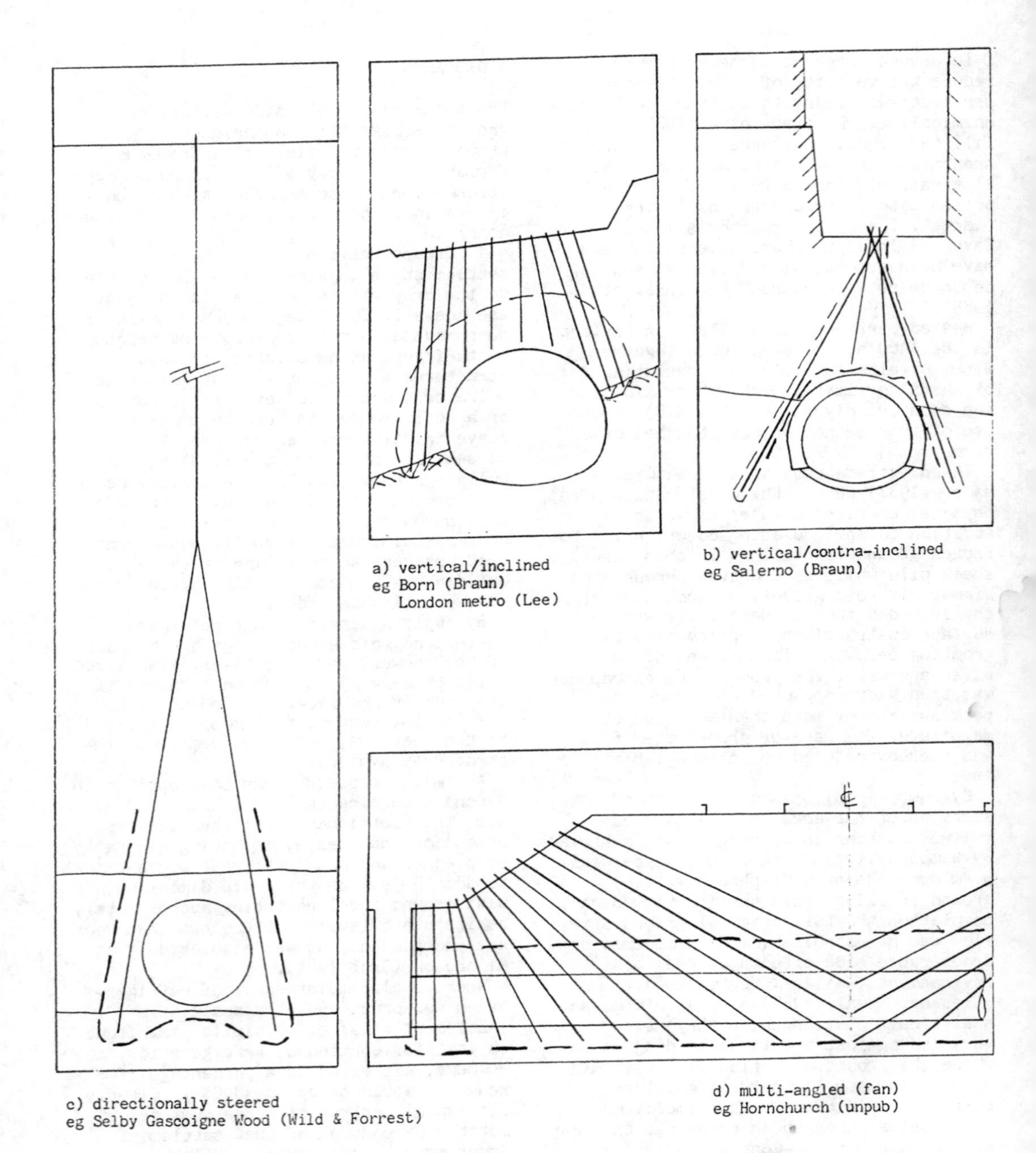

Figure 1. Freeze-tube geometries : surface installation

4.3 **Creep** is temperature and time related for each soil type. In many instances the period of unsupported exposure is sufficiently short, and the temperature depression below freezing point sufficiently large, that the scale of creep relaxation is not significant. If a soil is susceptible to creep the design should embody a creep analysis as discussed earlier. For very shallow tunnels it may also be necessary to consider whether creep relaxation into the excavation (if allowed) would also lead to settlement at the surface.

4.4 **Concrete placement.** The volume of concrete pours is usually sufficient that the heat of hydration released prevents advance of the zero isotherm into the concrete until after the initial set/chemical reaction is complete. Heated aggregates and hot mixing water and/or rapid setting additives may be used with small volume pours to ensure that this sequence occurs. On the other hand, with very large pours, it may be necessary to fix insulation to the exposed icewall before the concrete is poured, to limit the thawing effect on the integrity of the icewall.

5 SOIL PROPERTIES: are reviewed elsewhere in these Proceedings.

6 REFRIGERATION SYSTEM

6.1 A typical system is made up of three distinct parts : the source of refrigerative energy, the distribution system to effect removal of heat from the ground, and the monitoring system to control the whole.

Mechanical refrigeration is based on commercial machines of reciprocating or screw type with single or multistage compression, charged with freon or ammonia primary refrigerant. Most practitioners use calcium chloride as secondary refrigerant.

Alternatively the refrigerative energy is obtained from the evaporation of liquid nitrogen (LN) during a single pass of the freeze-tubes. To limit energy losses via the exhaust gas, use has been made of series connections - a "liquid" freeze-tube followed by one or more "gas" freeze-tubes - if the design could accomodate ice columns of different diameters. Experimental work on the relative sizes of frozen ground column generated by liquid and gas freeze-tubes has been reported by Owstrowski (1985).

To ensure that the nitrogen reaches the freeze-tubes in liquid form a well insulated delivery line is essential : vacuum-insulated lines are expensive but extremely efficient and compact; in at least one case the liquid feed line has been installed within the exhaust gas line to achieve this objective.

Recent examples where two refrigeration systems were used on one project have been reported by Harris & Norie (1982); Gallavresi (1980); Maishman (1988). In the first case the two systems were independent, the brine system being used for the major part of the works and LN for the minor part which was "unforeseen" at the outset. In the other cases work was started using LN while the mechanical plant was being assembled and commissioned; in the latter cases the transition from cryogenic to normal refrigeration temperatures was carefully and safely controlled.

6.2 The disposition of freeze-tubes is fundamental to successful application of the method; a wide selection of the many patterns which have been used to provide protection for tunnels is illustrated in figure 1.

It is sometimes unnecessary to freeze the shallowest stratum. Measures that have been taken to restrict refrigeration of these zones include insulation, locally enlarged sections of freeze-tube, twin inner tubes, and closer spacing of freeze-tubes; the efficiency of some of these methods is debatable and remains unassessed/unpublished. Freeze-tubes installed parallel with the axis of the excavation (ie horizontal for a tunnel) normally avoid the problem.

To encourage good flow characteristics of LN, and uniform ice-wall growth along the length of horizontal (or horizontally fanned) freeze-tubes, several practitioners have utilised perforated inner pipes and half bore weirs. Practical reports by Rebhan (1982) and Maishman (1988) indicate that the latter technique is particularly effective in achieving these objectives.

6.3 The essence of economical formation of an ice-wall of uniform thickness is accurate placement of the freeze-tubes.

For vertical drilling deeper than 20m or so, or horizontal drilling penetrating more than 12m or so, it is usual to survey the course of particular/all holes with one of the accurate in-hole instruments now marketed, eg single/multi-shot gyro, strain gauged elbows and photographic recording of the displacements of three equally spaced "O" rings (fotobor).

Corrective measures will be needed if severe deviations are detected eg the use

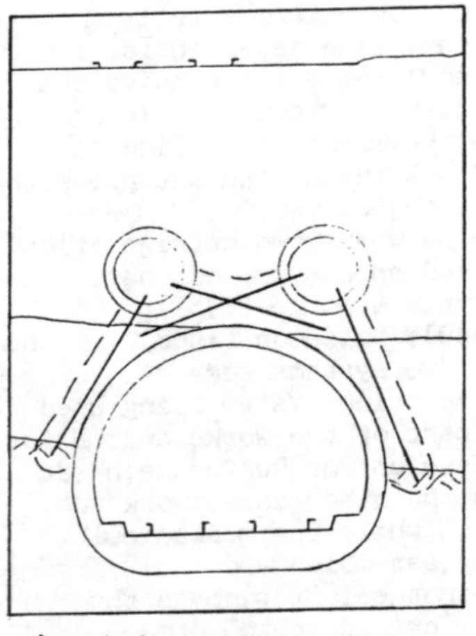

e) arched canopy
eg Essen

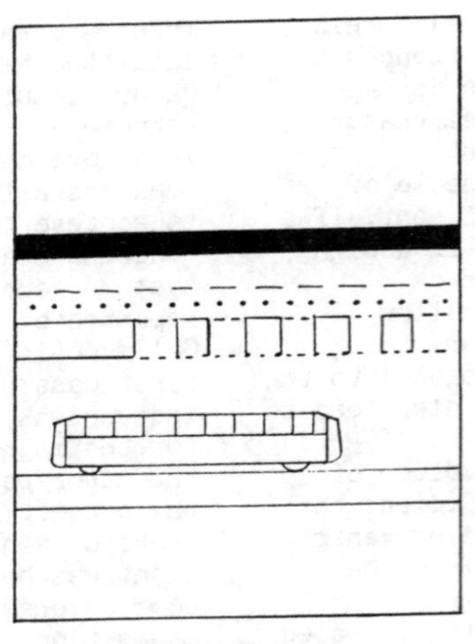

f) contiguous beam
eg Brussels (Gonze et al)
Vienna (Deix & Braun)

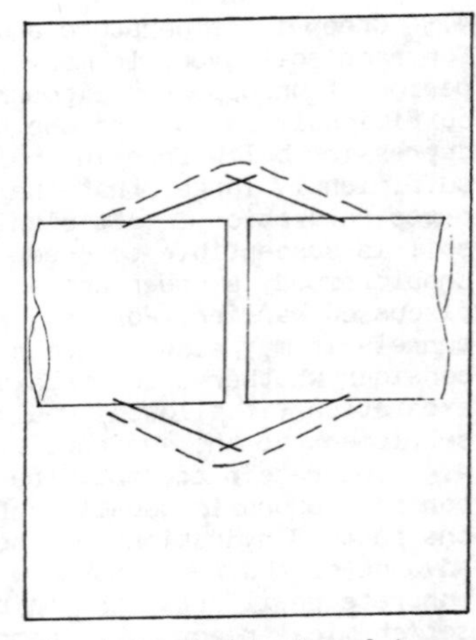

g) sub-level link, tunnel to tunnel or to shaft
eg Keihin (Numazawa et al)

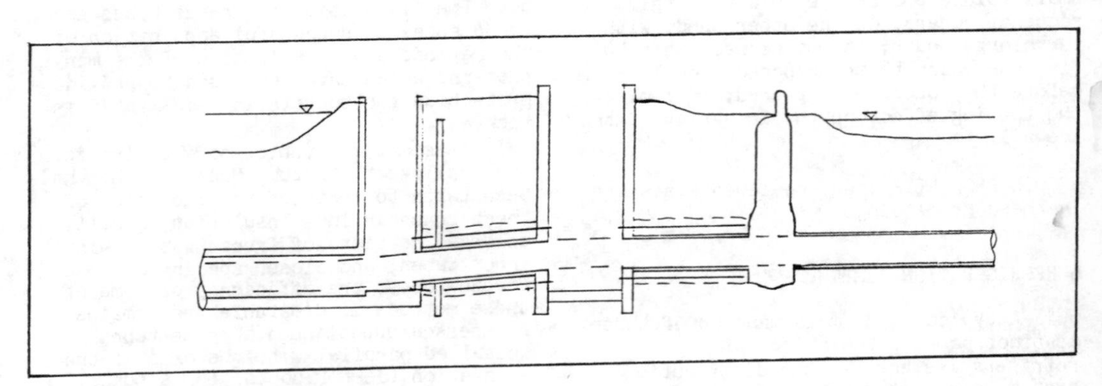

h) sub-level link between two shafts
eg Runcorn (Harris & Norie)

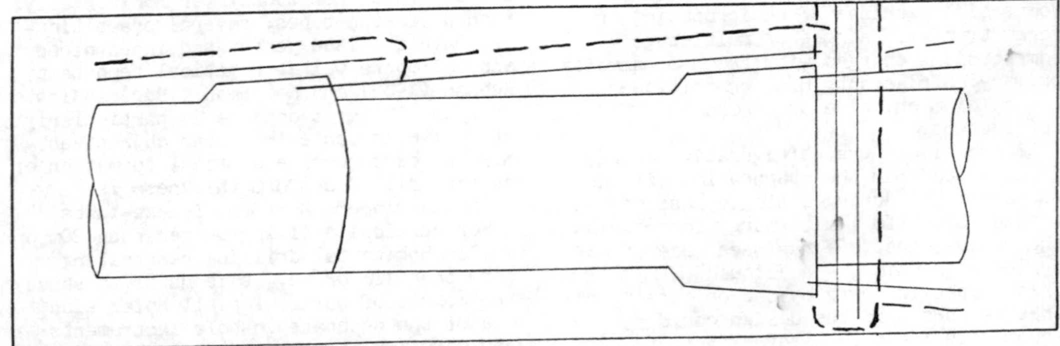

j) horizontal stages without/with vertical bulkheads
eg Milchbuck (Mettier)
Agri Sauro (Restelli et al)
Du Toits Kloof (Cockcroft et al)

Figure 2. Freeze-tube geometries : underground installation

of directional drilling techniques to recover the correct course, or the provision of additional freeze-tubes to fill any wide gaps between adjacent tubes. Oilfield type downhole motor and bent-sub tools, which are often employed for deep shaft freezings in this situation, have been adopted to intentionally steer freeze-hole drilling at a steep declination to pass either side of the path to be followed by a 1 in 4 drift (fig 1, example c) (Wild & Forrest 1981).

6.4 Whereas for vertical freezeholes it is usual to predrill a hole into which the freeze-tube can be placed, for horizontal freeze-tubes it is becoming common practice to utilise the freeze-tube as its own drillpipe - the so-called overburden drilling method. This may require special attention at the start point of each hole to prevent loss of water or soil fines from the strata during drilling (Harris & Norie 1982; Deix & Braun 1987).

7 METHOD OF CONSTRUCTION

The practice of permitting a controlled amount of creep relaxation prior to installing the lining, thus limiting the stress imposed on the lining, is now practiced in tunnelling as well as shaft sinking. In Europe this is one element of the NATM, and is reported by many authors.

Relaxation is arrested by applying a reinforced gunite or shotcrete primary lining during each stage of excavation. The structural inner lining can then be installed when the full section has been excavated.

8 STRUCTURAL DESIGN

The ice-wall must have sufficient integrity to resist all stresses, and survive all thermal loads, likely to be imposed on it during its working life.

Any design must first identify the parameters which characterise the possible failure mechanisms appropriate to the particular case, and then establish their values.

The science of specimen preparation and laboratory testing has developed rapidly,in parallel with the advances in computing that we have witnessed, as reported elsewhere in this symposium.

Nearly all recent case descriptions indicate use of the finite element method (FEM) to analyse both the thermal and the stress patterns. This numerical technique can easily model complex excavation geometries and loading conditions, and analyse the behaviour of an ice-wall with time.

Since frozen soil strength and creep parameters vary with the sub freezing-point temperature, a prerequisite to a stress/strain analysis is a thermal analysis.

The formation of frozen ground is time dependent, and as FE cannot handle changes in physical properties or the position of elements, it is necessary to perform a series of analyses using data obtained from one analysis as the input for the next. This iterative process is continued for a predetermined time or until the change is small.

When the temperature distribution is known, the elements can be accorded property values appropriate to their position with respect to temperature.

The non-linear creep problem is handled on an incremental basis, ie a series of linear problems each occurring within its own time interval.

Principal among recent contributions are:

1. Gonze 1983. Theory and examples are given of thermal analyses for single and twin circles of freeze-tubes, and stress/creep of a rectangular section drift.
2. Jessberger 1987 & 1988. A comparison between analytical and numerical solutions, embodying earlier work and contributions by Klein and others, is given in a reference chapter on AGF. Use of FEM for a contract in Austria is described in these proceedings; the effect of varying the on/off cycles during intentional intermittent freezing was studied. In this way the size of the frozen body, and amount of heave, were controlled.
3. Meissner 1985 & 1988(b). These complementary papers present the results of parametric studies by means of normalised relationships for various tunnel diameters and loads, including comparisons with the results of model tests on a single soil type (medium sand).

REFERENCES*

BLANCHARD,D. & FREMOND,M. 1985. Soils frost heaving and thaw settlement. Proc ISGF85 Japan, 209-216, Balkema.

BOURBONNAIS,J. & LADANYI,B. 1985. The mechanical behaviour of a frozen clay down to cryogenic temperatures. Proc ISGF85 Japan, 2, 237-244.

*See page X concerning ISGF papers

BRAUN, B. 1985. German and Swiss experiences with ground freezing. Tunnels & Tunnelling Dec 1985, 47 - 50.

COCKCROFT,T.N., DENNE,R.J., & HARVEY,S.J. 1982. Construction of a section of the Du Toitskloof tunnel by use of AGF. Proc Tunnelling82 England, 105-116.

DEIX, F. & BRAUN, B. 1987. Vienna subway construction - use of NATM in combination with ground freezing. Proc ISGF88, England, 1.

EINCK, H. B. & WEILER, A. 1982. Experiences and investigations using gap freezing to control groundwater flow. Proc ISGF82 CRREL USA, 193-203.

FISH, A.M. 1983. Thermodynamic model of creep at constant stresses and constant strain rates. CRREL Report 83-33.

FISH, A.M. 1985. Creep strength, strain rate, temperature and unfrozen water relationship in frozen soil. Proc ISGF85 Japan, 2,29-36.

FUNCKEN, R. GONZE, P. VRANCKEN, P. MANFROY, P & NEERDAEL, B. 1983. Construction of an experimental laboratory in deep clay formation. Proc Eurotunnel83 Switzerland, 9, 79-86.

GALLAVRESI, F. 1980. Ground freezing: application of the mixed method brine-liquid nitrogen. Proc ISGF80 Norway, I, 928-939.#

GONZE,P. 1983. Verification of the stability of excavations in artificially frozen ground. Proc Int Symp on Field Msts in geomechanics,Switzerland, 1021-1031.

GONZE, P., LEJEUNE, M., THIMUS, J.F. & MONJOIE, A. 1985. Sand ground freezing for the construction of a subway station in Brussels. Proc ISGF85 Japan, 277-283, Balkema.

GROB, H. 1984. Ground freezing as tunnelling support. Proc Beijing Intl Colloquium. Advanced tunnelling technology & subsurface use, 4, 4, 265-279.

GYSI, H. & MADER, P. 1987. Ground freezing technique : its use in extreme hydrological conditions. Proc ISSMFE87 Dublin,I,165-168.

HAMPTON, C.N., JONES, R.H. & GARDNER, A.R. 1985. Modelling the creep behaviour of frozen sands. Proc BGFS Nottingham England, 27-33.

HAMPTON,C.N. 1986. Strength and creep testing for AGF. Ph D Thesis, 221pp, University of Nottingham, October 1986.

HARRIS, J.S. 1987. Ice walls contain the bad ground problem. Construct Jnl, Feb 1987, 32-35.

HARRIS, J.S. & NORIE, E.H. 1982. Construction of two short tunnels using AGF. Proc ISGF82 CRREL USA, 383-388.

HARVEY, S.J. 1983. Ground freezing successfully applied to the construction of the Du Toitskloof tunnel. Proc BGFS83 Nottingham England, 51-58.

HARVEY, S.J. & MARTIN,C.J. 1988. Construction of the Asfordby mineshafts through the Bunter Sandstone by use of ground freezing. Proc ISGF88 England, 1.

HIEATT, M.J. & DRAPER, A.R. 1985. The Three Valleys tunnel in the reality of a rolling freeze. Proc BGFS85 Nottingham England, 45-52.

JESSBERGER,H.L. 1987. Artificial freezing of the ground for construction purposes. In FG Bell (Ed) Ground engineers reference book, Butterworth.

JESSBERGER, H.L., JAGOW, R. & JORDAN, P. 1988. Thermal design of a frozen soil structure for stabilisation of the soil on top of two parallel metro tunnels. Proc ISGF88, 1.

JONES, J.S. 1980. Engineering practice in AGF - the State of the Art. Proc ISGF80 Norway, 1, 837-856. #

JONES, R.H. 1982. Ground surface movements associated with artificial freezing. Proc ISGF82 CRREL USA, 1, 295-304.

JOSANG, T. 1980. Ground freezing techniques in Oslo city centre. Preprints ISGF80, 969 - 979. #

KESSURU, Z. DUSZA, L. WIDDER, A. & MATE, K. 1987. Simultaneous modelling of rock freezing and water seepage and its practical applications. Intl J of Mine water, Hungary, 6, 1-32.

KLEIN, J. 1978. Nichtlineares kriechen von kunstlich gefrorenem emscher- mergel. 2, G, 123pp, Bochum.

LAKE, L.M. & NORIE, E.H. 1982. Application of horizontal ground freezing in tunnel construction - two case records. Proc Tunnelling82, England, 283-290.

LEE, I. 1969. The Victoria Line escalator tunnel at Tottenham Hale Station. Proc ICE, 7250 - 5.

MAISHMAN, D. 1988. A short tunnel in Seattle frozen using liquid nitrogen. Proc ISGF88, England, 2.

MAISHMAN, D. & POWERS, J.P. 1982. Ground freezing in tunnels - three unusual applications. Proc ISGF82 CRREL USA, 397-410.

MARTAK, L.V., 1988. Ground freezing in non-saturated soil conditions using liquid nitrogen(LN). Proc ISGF88, England, 1.

MEISSNER, H. 1985. Bearing behaviour of frost shells in the construction of tunnels. Proc ISGF85, Japan, 2, 37-45.

MEISSNER, H. 1988a. Ground movements by tunnel driving in the protection of frozen shells. Proc 6th Int Conf on Num Meth in Geomech. Austria.

MEISSNER, H. 1988b. Tunnel displacements under freezing and thawing conditions. Proc ISGF88, England, 1.

METTIER, K. 1985. Ground freezing for the construction of the Milchbuck road tunnel in Zurich, Switzerland. Proc ISGF85, Japan, 2, 263-270.
MIYOSHI, M., KIRIYAMA,S., & TSUHAMOTO, T. 1978. Large scale freezing work for subway construction in Japan. Proc ISGF78,255-268.
MURAYAMA, S., MONITANI., S, & MATSUMOTO, Y. 1985. Application of freezing method to construction of tunnel through weathered granite ground. Proc ISGF85 Japan, 2, 253 - 258.
MURAYAMA, S. KUNIEDA, T. SATO, T. MIYAMATO,T. & GOTO, K. 1988. Ground freezing for the construction of a drain pump chamber in gravel between the twin tunnels in Kyoto. Proc ISGF88 England, 1.
NUMAZAMA, S., TANAKA, M. 1988. Application of freezing method to undersea connection of large diameter shield tunnel. Proc ISGF88, England, 1.
OHRAI, T., ISHIKAWA, I. & KUSHIDA, Y. 1985. Actual results of ground freezing in Japan.Proc ISGF85 Japan, 2, 289-294.
ORTH, W. 1988. Two practical applications of soil freezing by liquid nitrogen. Proc ISGF88 England, 1.
ORTH, W. & MEISSNER, H. 1985. Experimental and numerical investigations for frozen tunnel shells. Proc ISGF85 Japan, 2, 259-262.
OWSTROWSKI, W. J. 1985. Industrial tests on application of liquid nitrogen for soil freezing. Proc ISGF85 Japan, 1, 265-276.
REBHAN, D. 1982. New freeze pipe systems for nitrogen freezing. Proc ISGF82, CRREL USA,429-435.
RESTELLI, A.B. & TONOLI, G. 1988. Ground freezing solves tunnelling problem at Agri Sauro, Potenza, Italy. Proc ISGF88 England, 1.
ROTH, B. 1981. Tunnelling methods for the Essen underground Part 2.Tunnel 2, 106-114.
SAYLES, F.H., BAKER, T.H.W., GALLAVRESI, F., JESSBERGER, H.L., KINOSITA, S., SADOVSKI, A.V., SEGO, D., & VYALOV, S.S. 1986. Classification and laboratory testing of artificially frozen ground. ASCE Jnl of Cold Regions Engineering, I, 1, 22-48.
WEILER, A. & VAGT, J. 1980. The Duisberg method of metro construction, a successful application of the gap freezing method. Proc ISGF80, Norway, I, 916-927.
WILD, W.M. & FORREST, W. 1981. The application of the freezing process to ten shafts and two drifts at the Selby project. Proc IME - The Mining Engineer, June 1981, 895-904.

5th International Symposium on Ground Freezing, Jones & Holden (eds)
© 1988 Balkema, Rotterdam. ISBN 90 6191 824 3

Design and installation of deep shaft linings in ground temporarily stabilized by freezing – Part 1: Shaft lining deformation characteristics

F.A.Auld
Cementation Mining Ltd, Doncaster, UK

ABSTRACT: This paper is the first part of a two-part treatise which quantifies deformation characteristics for both shaft linings and freeze walls and studies their compatibility. The shaft sinking and lining background to the study is first outlined. Time related deformation versus outer support pressure relationships for cast in situ concrete shaft linings are developed. Deformation versus outer support pressure characteristics, which are not time dependent, have also been determined for other shaft linings. These cover precast concrete solid blocks (with and without sqeeze packs), precast concrete bolted segments, fabricated steel tubbing and spheroidal graphite (S.G.) cast iron tubbing. The paper provides a convenient reference document containing deformation characteristics for a full range of shaft lining types in the context of current structural design codes of practice. It sets the scene for the second paper which quantifies freeze wall deformation characteristics and studies shaft lining and freeze wall deformation compatibility.

1 INTRODUCTION

The stresses and strains in, the deformation of, and the pressures on deep shaft linings are directly related to the prevailing strata conditions. Strong, competent (self standing when excavated) rocks impose minimal ground pressures on a shaft lining. This is predominently the case in the United Kingdom (Black and Auld 1985) where normally shaft linings are designed to withstand hydrostatic pressures without any additional ground pressure loading component.

In other areas of the world, such as Germany and China, less competent strata conditions occur and deep shaft linings must be capable of withstanding the resulting ground pressures and associated deformations.

Where unstable ground conditions are encountered, a stabilization process is necessary to permit practicable and safe excavation. Ground freezing provides both strength enhancement for the unstable ground and the ability to prevent water ingress into the excavation. In Germany freezing to a depth of 600m for ground stabilization purposes has been achieved in shaft sinking (Klein 1982). Depths of up to 425m have been frozen in China (Yu, Wang and Zhong 1985) but, in some areas of China, soft alluvial strata can be found to a depth of 770m (Qiu and Wang 1982).

Although, in the final installed state, shaft linings only have to be capable of withstanding the loads from the unfrozen ground, during sinking and lining it is the relative deformations of the freeze wall and the lining which dictate whether the latter can be installed successfully or not. Since the deformation of an unsupported freeze wall at depth is normally much greater than the permissible deformation of the lining it is very important to ensure that when the two act together, their resulting deformations are compatible. Quantifying accurately the deformation characteristics of both the lining and the freeze wall is therefore of particular importance.

How linings and freeze walls act together also depends upon the method of excavation and lining installation. Two basic methods exist (Auld 1985):

1. Sink and line permanently as the excavation proceeds downwards (normal United Kingdom practice - see Figures 1(a) to (e)).

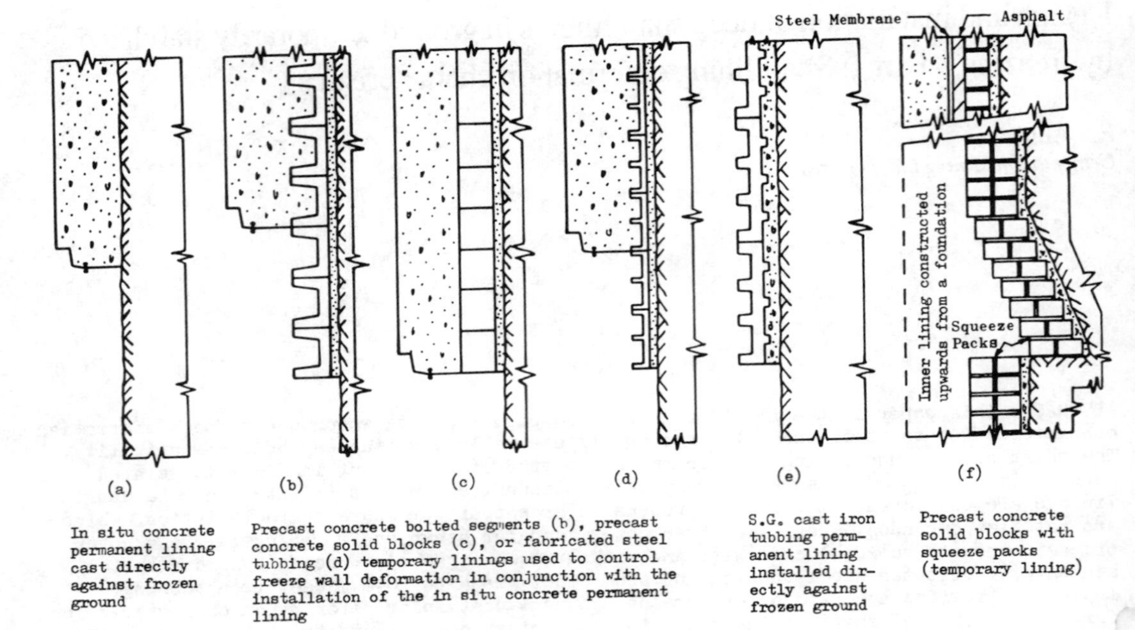

Fig.1 Temporary and permanent shaft lining systems

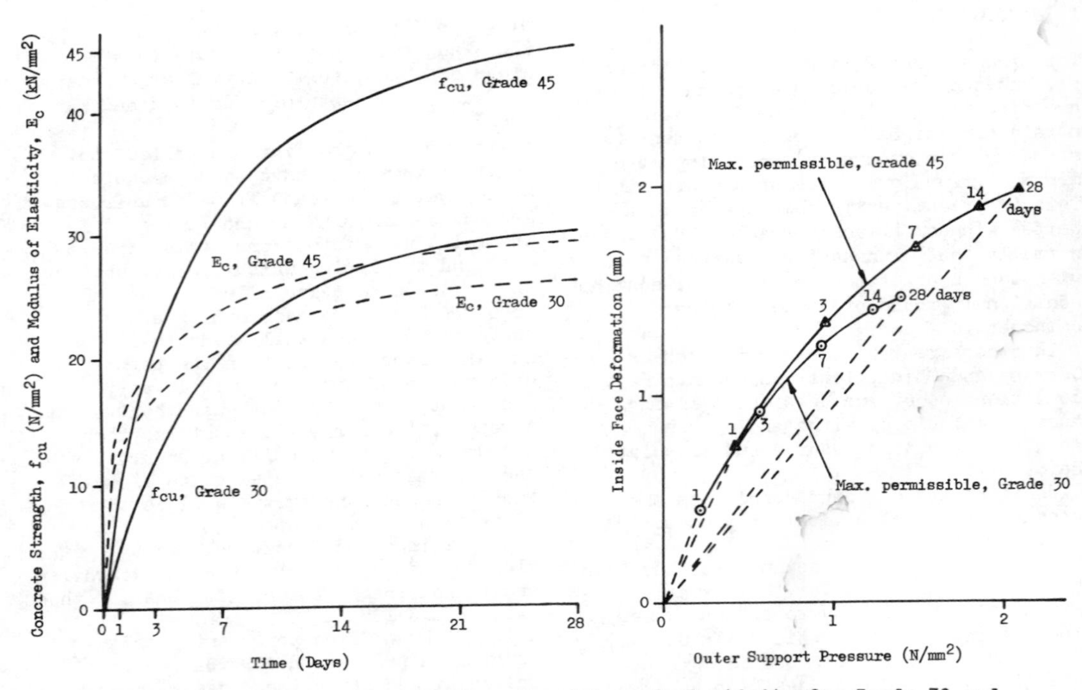

Fig.2 Time related values of strength and modulus of elasticity for Grade 30 and Grade 45 concrete (left). Time related curves of deformation versus support pressure for 750mm thick, cast in situ, Grade 30 and Grade 45 concrete linings in an 8m dia. shaft (right)

2. Sink and line temporarily and then construct the permanent lining upwards from a foundation at rockhead (normal German and Chinese practice - see Figure 1(f)).
The German standard lining design (Klein 1982) consists of a temporary lining in precast concrete solid blocks with squeeze packs (see Figures 1(f) & 4) to accommodate the freeze wall movement. A foundation is then constructed at rockhead, and the inner permanent lining of concrete, with its outer membrane of steel, is installed upwards from the foundation. Between the inner and outer linings an asphalt separating layer is introduced to accommodate ground movements resulting from mineral extraction in the shaft pillar (unworked zone surrounding a shaft in UK mining).

The Chinese employ a similar construction method (Qiu and Wang 1982). Their temporary linings however consist of in situ reinforced concrete cast directly against the freeze wall or against either limestone or precast concrete blocks which are provided as additional support and thermal insulation to the freeze wall at depth. The inner (permanent) reinforced concrete lining is slipformed upwards from a foundation, situated at rockhead, with a polyethylene ("sliding") membrane separating the temporary and permanent linings.

Currently in the United Kingdom all the new coal mine shafts have been or are being lined with in situ concrete in a single pass downwards as sinking proceeds (Auld 1987). Concrete is particularly suitable for installation in United Kingdom ground conditions, where the ability to mould itself to the excavated rock face, and the fact that relatively small increments of the lining can be constructed systematically in conjunction with the excavation process, are of paramount importance. Strengthwise, concrete shaft linings can withstand adequately the design loadings applicable in United Kingdom mining practice, and concrete also satisfies the necessary requirements for impermeability and durability.

Sinking and lining a shaft permanently downwards in a single pass has distinct advantages:

1. The shaft is fully secured with the permanent lining as sinking progresses.

2. For the same shaft sinking conditions, construction time is reduced through using the single pass system instead of temporary support first and then an inner lining constructed upwards from the foundation at rockhead (in some instances of the single pass system, short lengths of temporary support can be used ahead of the lining to facilitate its installation in high pressure zones).

3. The problem of differential shrinkage and thermal contraction at the interface between the inner slip formed concrete lining and an outer temporary concrete lining (Chinese method - Qiu and Wang 1982), which requires the polyethylene ("sliding") membrane to prevent cracking at the rear of the inner lining, is avoided.

4. Circumferential cracking, resulting from vertical ground movement due to differential thawing (Chinese method), is controlled by having construction joints at regular intervals (normally 6m construction lengths in the United Kingdom).

To use the single pass shaft sinking and lining system is therefore most beneficial, and it is on this basis that the requirements for the successful design and installation of deep shaft linings, in ground temporarily stabilized by freezing, are considered.

The design of freeze walls for deep shaft sinking has been reviewed by Auld (1985). Numerous freeze wall strength and deformation formulae were appraised and their use put into context with construction procedures. However, the main conclusion was that most of the design formulae were unrealistic and inapplicable when considered in relation to construction methods. It is also the author's opinion that previously freeze walls have largely been studied in isolation and sufficient attention has not been paid to the inner support. The latter, being permanent works, is the predominant factor for design, and the freeze wall temporary works design should be tailored to achieve satisfactory installation of the permanent works and not vice versa.

A suggestion was made in the previous paper (Auld 1985) that an improved method of design may be possible if a concept is used whereby graphs of radial closure versus support pressure are developed for both the inner lining and the freeze wall. Compatibility between the two can then be checked, making due allowance for the time interval before the lining is installed. This concept has been followed up, and the results of the study are presented. The factors evaluated in the study were as follows:

1. Time related deformation versus outer support pressure for cast in situ concrete

shaft linings and deformation versus outer support pressure for other lining systems of precast concrete solid blocks (with or without squeeze packs), precast concrete bolted segments, fabricated steel tubbing, and spheroidal graphite (S.G.) cast iron tubbing (Paper Part 1. Shaft lining deformation characteristics).
To carry out this evaluation in the case of cast in-situ concrete, a knowledge of the increase in concrete strength with time and the increase in modulus of elasticity, E_c, with time were required.

2. Elastic and time related (creep) deformation versus inner support pressure for the freeze wall (Paper Part 2. Shaft lining and freeze wall deformation compatibility). In this evaluation, determination of the external ground pressure and reference to the freeze wall thickness, the average temperature through the freeze wall, and the deformation characteristics of the chosen temporary lining system were necessary.

3. Modification of the freeze wall deformation characteristics determined previously in Item (2) to allow for the period of time before any lining is installed (Paper Part 2. Shaft lining and freeze wall deformation compatibility). Rate of advance and distance of the lining from the sump are important factors.

2 TIME RELATED DEFORMATION VERSUS OUTER SUPPORT PRESSURE FOR CAST IN SITU CONCRETE SHAFT LININGS

The stresses and deformations for a long, thick cylinder, determined by an elastic analysis, were derived originally by Lamé and Clapeyron (1833). The following formulae are quoted directly from Auld (1979):

$$\sigma_{tlmax} = 2p_{ol}r_{ol}^2/(r_{ol}^2-r_{il}^2) \quad (1)$$

$$\delta_{il} = 2p_{ol}r_{il}r_{ol}^2/E(r_{ol}^2-r_{il}^2) \quad (2)$$

where:

σ_{tlmax} = maximum tangential stress in the lining (at the inside face)

δ_{il} = inside face radial deformation of the lining

p_{ol} = uniformly distributed external pressure on the lining

E = modulus of elasticity (E_c, E_s, E_{ci} are the moduli of elasticity for concrete, steel and S.G. cast iron respectively)

r_{il} = internal radius of the lining

r_{ol} = external radius of the lining

Equation (2) divided by Equation (1) gives:

$$\delta_{il} = \sigma_{tlmax}\, r_{il}/E \quad (3)$$

Equation (1) relates, in simple terms, the applied external pressure to the maximum tangential stress in the lining. For a concrete lining, by equating the maximum tangential stress to the maximum permissible stress for the concrete:

$$\sigma_{tlmax} = 0.67\, f_{cu}/(1.4 \times 1.5) \quad (4)$$

where f_{cu} is the characteristic strength of the concrete, the maximum permissible external presure, p_{olmax}, on the lining can be deduced. As defined in BS 8110: Part 1: 1985 for the structural use of concrete, the characteristic strength is the value of the cube strength below which not more than 5% of the test results fall. The multiplication factor of 0.67 relates characteristic strength, which is based on 150mm cubes tested at any specific time, to actual strength in situ, and the values 1.4 and 1.5 are partial safety factors for load and material strength respectively.

For any known shaft lining geometry, as an example consider a shaft diameter of 8m and a lining thickness of 750mm, giving r_{ol} = 4.75m, Equations (1) and (4) combine to form a very simple relationship:

$$p_{olmax} = 0.67f_{cu}(4.75^2-4^2)/(1.4x1.5x2x4.75^2)$$
$$= 0.0464f_{cu} \quad (5)$$

Similarly the maximum permissible deformation, δ_{ilmax}, of the lining occurring at the maximum permissible concrete stress, and hence at the maximum permissible external pressure, can be deduced from Equations (3) and (4):

$$\delta_{ilmax} = 0.67f_{cu}x4000/(1.4x1.5E_c)$$
$$= 1276.19\, f_{cu}/E_c \quad (6)$$

The maximum permissible external pressure and the deformation associated with it are now quantified in terms of design characteristic strength and modulus of elasticity. However, for concrete cast in situ, both of these parameters vary with time from initial placing. Therefore it is necessary to know the relationships for concrete strength, f_{cu}, versus time and modulus of elasticity, E_c, versus time (see Figure 2, left diagram). The latter has been prepared from the former using the recommendations of BS 8110: Part 2: 1985:

$$E_{c,T} = E_{c,28}(0.4+0.6f_{cu,T}/f_{cu,28}) \qquad (7)$$

where T is greater than or equal to 3 days.

From Figure 2 (left diagram) values for concrete strength, f_{cu}, and modulus of elasticity, E_c, can be abstracted for any given time and fed into Equations (5) and (6) to give the maximum permissible support pressure, and equivalent maximum permissible deformation at that time. For 750mm thick, cast in situ, Grade 30 and Grade 45 concrete linings in an 8m diameter shaft, curves of time related deformation versus outer support pressure are plotted in Figure 2 (right diagram).

3 DEFORMATION VERSUS OUTER SUPPORT PRESSURE FOR OTHER SHAFT LININGS

Various shaft linings have been reviewed in addition to in situ concrete. Precast concrete solid blocks (with or without squeeze packs), precast concrete bolted segments, fabricated steel tubbing and S.G. cast iron tubbing were considered. The basic difference between in situ concrete and the other materials is that the latter are not time dependent strengthwise and therefore have a unique design strength value.

3.1 Precast concrete solid blocks without squeeze packs

The theory is the same as for in situ concrete but with a unique characteristic strength value, f_{cu}, at 28 days. Assuming precast Grade 50 concrete, the characteristic strength based on 150mm cubes tested at 28 days is 50 N/mm^2 and the corresponding modulus of elasticity, E_c, is 30 kN/mm^2 (BS 8110: Part 2: 1985). From Equation (4), the value of σ_{tlmax} is 15.95 N/mm^2. This data has been used in conjunction with Equations (1) and (3) to produce Figure 3 (left diagram), which gives values of maximum permissible support pressure and maximum permissible deformation for 300, 400 and 600mm thick, precast Grade 50 concrete solid block linings in a 9.5m diameter shaft. The results show a straight line relationship between deformation and support pressure for precast concrete blocks, as opposed to the time dependent curves for concrete cast in situ.

3.2 Precast concrete bolted segments, fabricated steel tubbing and S.G. cast iron tubbing

Straight line relationships between deformation and support pressure can also be determined for precast concrete bolted segments, fabricated steel tubbing and S.G. cast iron tubbing. However, it is first necessary to deduce an equivalent membrane thickness (E.m.t.) for each shaped segment by dividing the total area of the cross-section by its height. The thick cylinder design approach, using Equations (1), (2) and (3), can then be applied.

Validity of the thick cylinder design method when applied to segmental linings, in conjunction with the equivalent membrane concept, can be accepted on the following basis:

1. The surrounding ground maintains the lining in a "locked-in" condition.
2. Any localized high stresses in the segment cross-section would lead to yielding and hence a more uniform stress distribution throughout the section.
3. High stresses produced from bending moments set up by the segment vertical side flange centre of bearing (i.e. centre of tangential hoop thrust) not being co-incident with the cross-section centre of area would be alleviated causing the centre of hoop thrust to adjust rearwards into line with the cross-section centre of area.
4. For large diameter shafts, with relatively thin linings, the difference in hoop stress between the inner edge and outer edge is small. Therefore the difference from these stresses when calculated by the equivalent membrane concept, as opposed to a more accurate method allowing for the shape of the segment cross-section, such as that used by Link, Lütgendorf and Stoss (1976), is also small. The deformation formula (Equation (3)), in addition, can therefore be applied with a reasonable degree of accuracy.

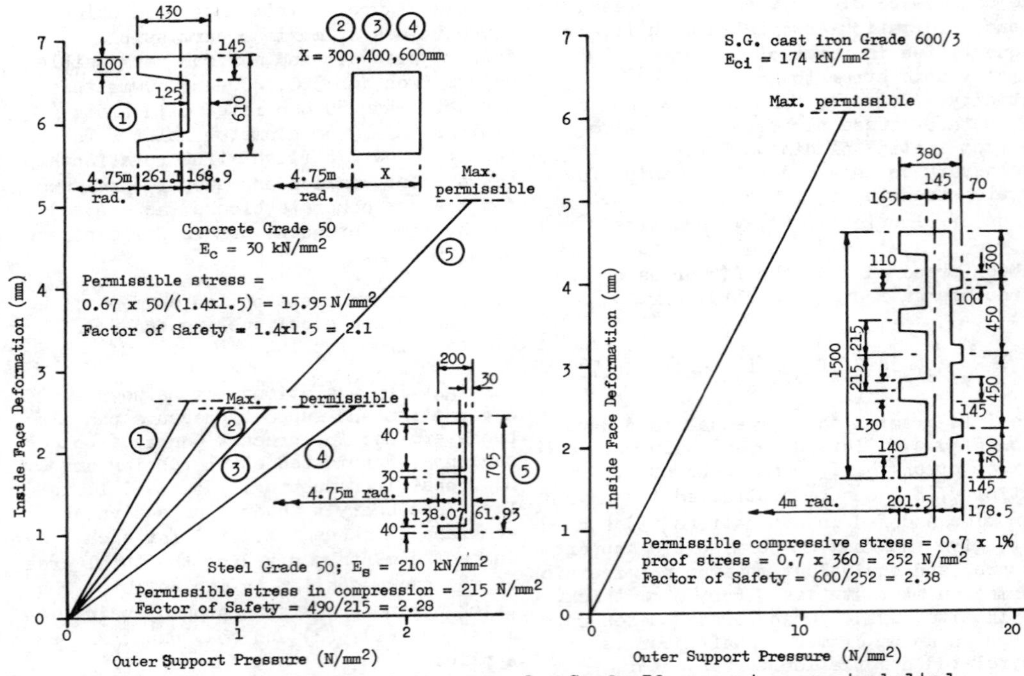

Fig.3 Deformation versus support pressure for Grade 50 precast concrete bolted segment ①, Grade 50 precast concrete block ②, ③, ④ and Grade 50 fabricated steel tubbing ⑤ linings in a 9.5m dia. shaft (left). Deformation versus support pressure for a Grade 600/3 S.G. cast iron tubbing lining in an 8m dia. shaft (right)

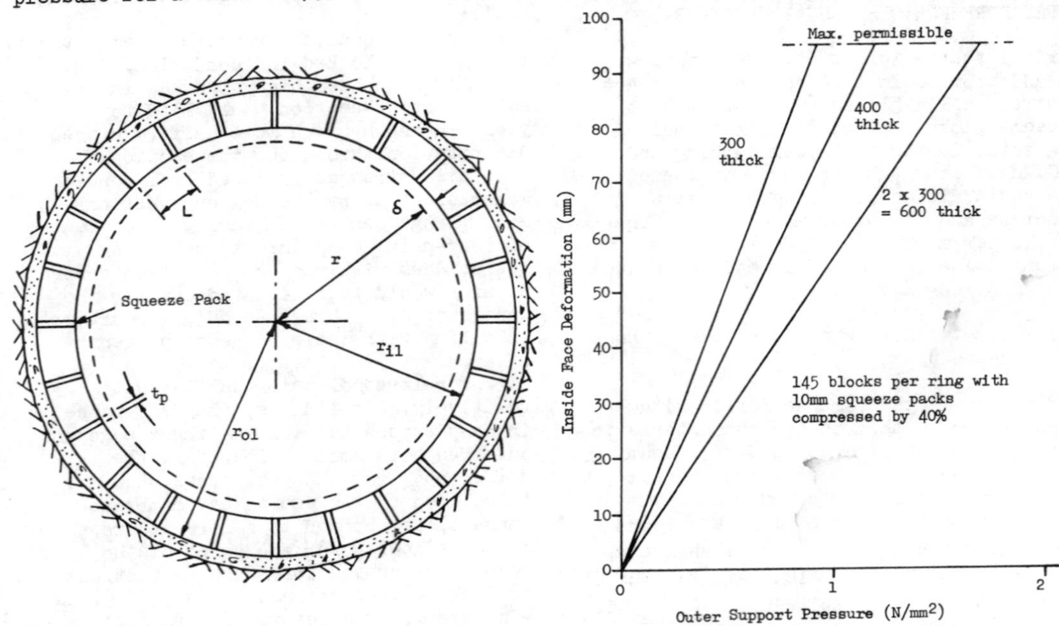

Fig.4 Schematic plan of shaft lining with precast solid blocks and squeeze packs (left). Deformation versus support pressure for Grade 50 pre-cast concrete blocks with squeeze packs in a 9.5m dia. shaft (right)

Section properties, radii, maximum permissible stresses and moduli of elasticity for precast Grade 50 concrete bolted segment and fabricated Grade 50 steel tubbing linings in a 9.5m diameter shaft and for a Grade 600/3 S.G. cast iron tubbing lining in an 8m diameter shaft are contained in Table 1. The deformation versus support pressure relationships for these linings are shown graphically in Figure 3.

3.3 Precast concrete solid blocks with squeeze packs (see Figure 4, left diagram)

The total inward deformation δ, consists of the radial movement due to compression of the squeeze packs, δ_1, plus the radial deformation from compression of the blocks, δ_2.

For a final shaft radius of r and an inward deformation of the lining equal to δ, then initial $r_{i1} = r + \delta$, and the circumference at r_{i1} is $2\pi(r + \delta)$. Assuming the squeeze packs are allowed to deform by X% of their original thickness, t_p, at the permissible stress for the concrete of $0.67\ f_{cu}/(1.4 \times 1.5)$, the new circumference becomes:

$$2\pi(r+\delta) - n \times X \times t_p = 2\pi(r+\delta_2) \qquad (7)$$

where:

n = number of blocks around the circumference

Simplifying Equation (7) gives:

$$\delta = \frac{n \times X \times t_p}{2\pi} + \delta_2 \qquad (8)$$

The radial deformation resulting from the compression of the blocks in the circumferential direction, δ_2, is found from Equations (3) and (4), which are the standard shaft lining deformation formulae.

The number of blocks, n, around the circumference can be determined from:

$$n = \frac{2\pi(r_{i1} + \delta)}{L + t_p} \qquad (9)$$

Substituting the value from Equation (9) into Equation (8) gives:

$$\delta = \frac{Xt_p(r_{i1} + \delta)}{L + t_p} + \delta_2 \qquad (10)$$

from which:

$$\delta = \frac{Xt_p\, r_{i1} + \delta_2(L + t_p)}{L + t_p(1-X)} \qquad (11)$$

For a 9.5m diameter shaft, a Grade 50 concrete block lining thickness of 300mm with 10mm thick squeeze packs compressed to a practicable construction amount of 40% of their original thickness, and limiting the inner face width of the blocks, L (see Figure 4, left diagram), to 200mm, then:

$$\delta = \frac{0.4 \times 10 \times 4750 + 2.53(200+10)}{200 + 10(1-0.4)} = 95\text{mm}$$

where the value of δ_2 is 2.53mm, (see Figure 3). From Equation (9) the value of n is 145 and the external pressure $p_{olmax} = 0.92\ \text{N/mm}^2$ (see Figure 3). The result is plotted as a straight line in Figure 4 (right diagram).

For a double thickness of 300mm solid blocks (i.e. 2 x 300 = 600mm for practicable in-shaft man-handling purposes), the same deformation of 95mm occurs but with a value of $p_{olmax} = 1.69\ \text{N/mm}^2$ (see Figure 3). This result is also plotted in Figure 4 (right diagram) together with that for a 400mm thick block lining.

4 COMMENTS

Deformation versus support pressure relationships have been evaluated for a full range of shaft lining types. The theory is also presented in a form which can be applied to linings in shafts of any diameter. Although not a paper directly concerned with the subject of artificial ground freezing, it provides a convenient reference document containing deformation characteristics for a full range of shaft lining types in the context of current structural design codes of practice. The author is unaware of any other such paper where the deformation characteristics of shaft linings can be readily observed on a relative basis. Suitability of the linings for installation in ground stabilized by freezing can be judged once the freeze wall deformation characteristics are known and Part 2 of the paper goes on to study these aspects.

Table 1. Section properties, radii, maximum permissible stresses and moduli of elasticity for precast Grade 50 concrete bolted segment (Type 1) and fabricated Grade 50 steel tubbing (Type 5) linings in a 9.5m diameter shaft and for a Grade 600/3 S.G. cast iron tubbing (S.G.T.) lining in an 8m diameter shaft

Lining	Section height, h (mm)	Section area, A (mm^2)	E.m.t.[1] A/h (mm)	r_{ol} (mm)	r_{il} = r_{ol}-E.m.t. (mm)	σ_{tlmax} (N/mm^2)	E (kN/mm^2)
Type 1	610	150,975	247.5	5180	4932.5	15.95	E_c = 30
Type 5	705	39.850	56.525	4950	4893.475	215[2]	E_s =210[3]
S.G.T.	1500	331,950	221.3	4380	4158.7	252[4]	E_{ci}=174[4]

(1) E.m.t. = Equivalent membrane thickness
(2) BS 449: Part 2: 1969 (Tensile strength 490 N/mm^2. BS 4360: 1979)
(3) BCSA CONSTRADO Structural Steelwork Handbook. 1984
(4) British Cast Iron Research Association. Engineering data on nodular cast irons. 1974

REFERENCES*

Auld, F.A. 1979. Design of concrete shaft linings. Proc. Instn Civ. Engrs. Part 2. 67. Sept; 817-832.

Auld, F.A. 1985. Freeze wall strength and stability design problems in deep shaft sinking. Is current theory realistic? Proc. of The Fourth International Symposium on Ground Freezing, Sapporo, Japan. 5-7 August: 343-349.

Auld, F.A. 1987. A decade of deep shaft concrete lining. Concrete, Journal of The Concrete Society. 21. No.2. February: 4-8.

BCSA CONSTRADO. 1984. Structural steelwork handbook. The British Constructional Steelwork Association Limited and Constructional Steel Research and Development Organisation, London.

Black, J.C. and Auld, F.A. 1985. Current and future UK practice for the permanent support of shaft excavations. International Journal of Mining Engineering. 3. No.1. March: 33-48.

British Cast Iron Research Association. 1974. Engineering data on nodular cast irons.

British Standards Institution. BS 449: Part 2: 1969. Specification for the use of structural steel in building.

British Standards Institution. BS 4360: 1979. Specification for weldable structural steels.

British Standards Institution. BS 8110: Part 1: 1985. Structural use of concrete. Part 1. Code of practice for design and construction.

British Standards Institution. BS 8110: Part 2: 1985. Structural use of concrete. Part 2. Code of practice for special circumstances.

Klein, J. 1982. Present state of freeze shaft design in mining. Proc. of the Symposium on Strata Mechanics, The University of Newcastle upon Tyne. 5-7 April: 147-153.

Lamé and Clapeyron. 1833. Mémoire sur l'équilibre intérieur des corps solides homogenès. Mém. divers savans. 4.

Link, H., Lütgendorf, H.O. and Stoss, K. 1976. Richtlinien zur Berechnung von Schachtauskleidungen in nicht Standfestem Gebirge (Instructions for design of shaft linings in unstable ground). Verlag Glückauf GmbH, Essen.

Qiu, S. and Wang, T. 1982. A study of sinking deep shafts using artificial freezing, design of shaft linings and method of preventing seepage. Proc. of The Third International Symposium on Ground Freezing, Hanover, New Hampshire, U.S.A. 22-24 June: 363-366.

Yu, X., Wang, Z. and Zhong, F. 1985. Thirty years of freezing shaft sinking in China. Coal Science and Technology (China). No.7: 2-6.

*See page X concerning ISGF papers

5th International Symposium on Ground Freezing, Jones & Holden (eds)
© 1988 Balkema, Rotterdam. ISBN 90 6191 824 3

Design and installation of deep shaft linings in ground temporarily stabilized by freezing – Part 2: Shaft lining and freeze wall deformation compatibility

F.A.Auld
Cementation Mining Ltd, Doncaster, UK

ABSTRACT: This paper is the second part of a two-part treatise which quantifies the deformation characteristics of both shaft linings and freeze walls and then studies their compatibility. The problem of quantifying freeze wall deformation is first put into context with shaft lining design. It is then reviewed in relation to elastic deformation during excavation and subsequent time dependent creep deformation. Distance of the lining from the sump and excavation rate of advance are appraised in relation to imposed deformation on the lining. Control of excavation to avoid damaging the freeze tubes is also considered. Formulae are quoted to facilitate the calculation of freeze wall elastic and creep deformation and an example is included. Shaft lining and freeze wall deformation compatibility is studied and advice follows on how to ensure compatibility and prevent damage to the freeze tubes. Finally comments are included on freeze wall design and where future study is required.

1 INTRODUCTION

In Part 1 of the paper, deformation versus outer support pressure relationships were developed for the various shaft lining types with an upper limit governed by the maximum permissible stress for the material. It was also stated in Part 1 that since the deformation of an unsupported freeze wall at depth is normally much greater than the permissible deformation of the lining it is very important to ensure, when the two act together, their resulting deformations are compatible. This relates to the method of excavation and lining installation (see Part 1), but the important factor is control of the freeze wall deformation by regulating the opening up of the excavation ahead of the previously lined section and, if necessary, providing short lengths of temporary lining in high pressure zones prior to installing the permanent system. On this basis, freeze wall deformation over the lining installation period is the dominant criteria for determining freeze wall thickness, and not freeze wall strength which is automatically catered for by controlling the deformation in conjunction with the lining to ensure that the maximum permissible stress in the lining is not exceeded. This is the concept which is concentrated on in the following sections of the paper.

2 SHAFT LINING DESIGN

Part 1 of the paper also emphasizes the fact that it is the successful installation of the shaft lining, in order to withstand the permanently imposed loads and those occurring during the construction period, which is the predominant factor for design and the freeze wall temporary works design should be tailored accordingly. The starting point is therefore the design of the shaft lining and this must be known before the freeze wall design can be put into context with it.

Shaft lining design is outwith the scope of this paper but, using the design guidelines of Link, Lütgendorf and Stoss (1976) for shaft linings through unstable ground, it can be shown that a 750mm thick, cast in situ, Grade 45 concrete lining in an 8m diameter shaft is satisfactory to a depth of 200m.

3 FREEZE WALL DEFORMATION IN RELATION TO DISTANCE OF THE LINING FROM THE SUMP, EXCAVATION RATE OF ADVANCE, AND FREEZE TUBE DEFORMATION

The two major factors which govern how

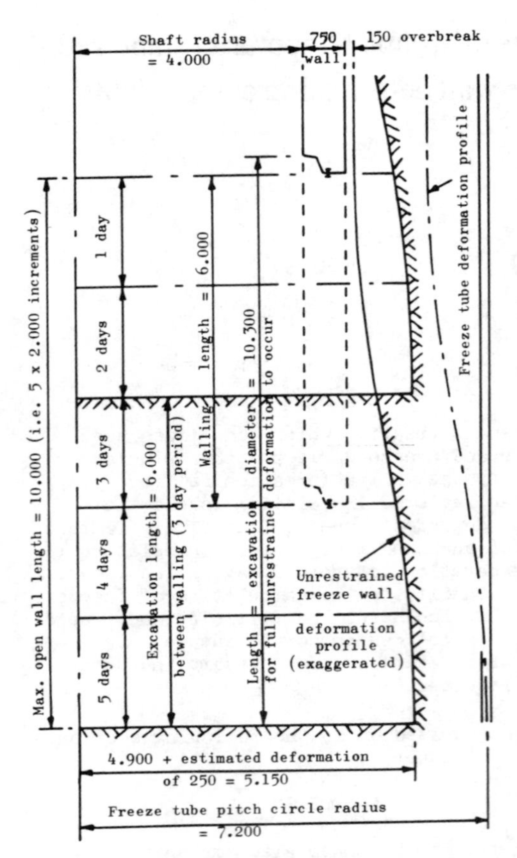

Fig.1 Excavation and lining sequence, unrestrained freeze wall deformation profiles and freeze tube deformation profile for an 8m diameter shaft with a 750mm thick concrete lining

much of the freeze wall deformation is imposed on a shaft lining are (see Figure 1):

1. distance of the lining from the sump;
2. excavation rate of advance.

In the case of the former, this distance determines the extent of the unrestrained elastic and time related unrestrained (creep) deformation which occurs prior to the lining being installed. Only subsequent deformation is imposed on the lining. Excavation rate of advance sets the time interval for the evaluation of both the unrestrained and restrained creep deformation.

By varying either of the two above major factors, different freeze wall deformations will be imposed on a shaft wall. It is therefore the control of these factors which dictates whether a lining can be installed successfully or not. However, an additional control measure which also has to be considered is the freeze tube deformation. It is very important to ensure that the freeze tubes are not allowed to rupture during the excavation process and both the distance of the lining from the sump and the excavation rate of advance must be regulated accordingly.

Figure 1 illustrates a typical situation, the shaft internal diameter adopted being 8m. For a specified set of excavation and lining rates, in this case:

1. target rate of advance 10m per 5 day working week, sink and line;
2. maximum unlined freeze wall length 10m;
3. lining length 6m,

the problem can be studied in detail. Roesner et al. (1982) have drawn conclusions regarding the distribution of stresses around a circular shaft excavation during sinking. One of their findings was that the displacement and thrust on the lining decreases rapidly as the unsupported excavation height increases in excess of one shaft diameter. This is in accordance with the well known phenomenon in underground excavation work, where the instantaneous inward movement at a particular point is only fully mobilized when the excavation face has moved some distance beyond that point. The philosophy can be applied to the case chosen (see Figure 1) to determine the extents of unrestrained and restrained freeze wall deformation which occur.

Figure 2 indicates the percentages of freeze wall deformation carried by the shaft lining (Figure 1) in relation to time from installation and distance of the lining from the sump. The percentages are estimates obtained by proportion over the length of the unrestrained freeze wall deformation profile. Symmetrical curvature is assumed in opposite directions from the centre point to each end, where full restraint against rotation is provided by the excavation floor at the bottom and the last section of lining at the top. At installation the lining carries none of the unrestrained elastic or 5-day creep displacement, of which 30% has already occurred. After one day, the lining provides a restraint equivalent to its one day strength but further excavation means 70% - 38% = 32% of the restrained elastic and 6-day creep displacement has been imposed on the lining in that time. Two days after installation, the lining has

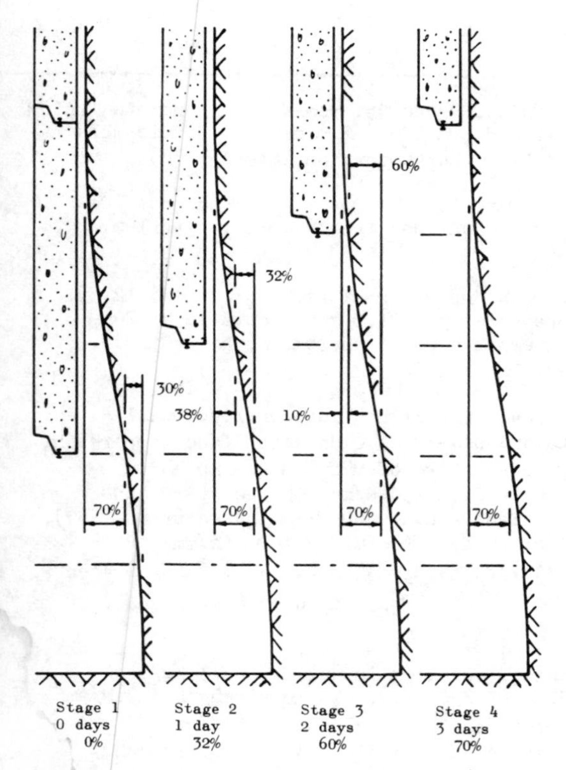

Fig.2 Percentages of freeze wall deformation carried by shaft lining subsequent to installation (values apply to parameters given in Figure 1)

been subjected to 70% - 10% = 60% of the restrained (2-day lining strength) elastic and 7-day creep movement. Finally, after 3 days, the total displacement applied to the lining becomes 70% of the restrained (3-day lining strength) elastic and 8-day creep movement. No further restrained elastic displacement occurs, only restrained creep over subsequent periods of time.

Having determined the percentages of freeze wall deformation which are imposed on the lining, the actual values need to be quantified fully.

4 QUANTIFICATION OF FREEZE WALL DEFORMATION

The first step towards quantifying freeze wall deformation is to define the stress versus strain behaviour for the frozen material. Once the stress/strain relationship is established, it can then be used in conjunction with the standard equilibrium equation (Timoshenko 1976) for the analysis of thick cylinders. The resulting differential equation is solved to produce a formula for freeze wall deformation. This approach can be applied to find the instantaneous or short term elastic deformation in the freeze wall. It can also be adopted to determine the long term creep deformation of freeze walls constructed in cohesive, frictionless soils (clays) and, in some specific cases, for those exhibiting friction (sandy soils). The total deformation to be expected is the summation of both the short term elastic and the long term creep inward movement. Appropriate formulae are quoted in the following sections.

4.1 Elastic deformation versus inner support pressure for the freeze wall

The general formula for the deformation, δ, of a thick cylinder (Timoshenko 1976), applied to the freeze wall, is:

$$\delta = (1-\nu_f)\,(r_o^2 p_o - r_i^2 p_i) r / E_f\,(r_o^2 - r_i^2) + (1+\nu_f)\, r_i^2 r_o^2 (p_o - p_i) / E_f\,(r_o^2 - r_i^2) r \qquad (1)$$

where:

p_o = uniformly distributed external pressure

p_i = uniformly distributed inner support pressure

ν_f = Poisson's ratio for the frozen material

E_f = modulus of elasticity for the frozen material

r = radius to the point in the freeze wall at which the deformation is being determined

r_i = internal radius

r_o = external radius

From Equation (1) the elastic deformation at the inside face of the freeze wall can be deduced by replacing r with r_i and simplifying:

$$\delta_i = \frac{2p_o r_i r_o^2 - p_i r_i [r_i^2(1-\nu_f) + r_o^2(1+\nu_f)]}{E_f\,(r_o^2 - r_i^2)} \qquad (2)$$

Table 1. Freeze wall deformation example design data

1. Shaft lining design. 750mm thick, cast in situ, Grade 45, reinforced concrete lining in an 8m diameter shaft. Maximum permissible depth 200m at a maximum permissible pressure of 2.6 N/mm^2 (1.3 x hydrostatic - Link, Lütgendorf and Stoss 1976).

2. Specified shaft sinking parameters (see Figure 1). Target rate of advance 10m per 5 day working week, sink and line. Maximum unlined freeze wall length 10m. Lining length 6m.

3. Freeze wall parameters. Internal radius, r_i = 5m, external radii, r_o = 10, 12.5, 15m, thicknesses = 5, 7.5, 10m. Ground pressure, p_o = 1.3 x hydrostatic at 200m depth = 2.6 N/mm^2 as for shaft lining. Internal support pressures, p_i = 0, 0.5, 1.0, 1.5, 2.0, 2.5 N/mm^2.

4. Frozen ground characteristics (assumed average temperature across freeze wall = -10°C for freeze tube brine inflow temperature of -20°C, excavation face temperature of approximately - 5°C and outer boundary temperature of 0°C - Hegemann and Jessberger 1985). Elastic properties: Clay - E_f = 130 N/mm^2 (Morgenstern and Wittebolle 1985), ν_f = 0.3 (Ladanyi 1982). Sand - E_f = 400 N/mm^2 (Jessberger 1980), ν_f = 0.3 (Ladanyi 1982). Creep properties: Clay - A = 8.7 x 10^{-5} $(N/mm^2)^{-B}h^{-C}$, B = 4.25, C = 0.12 (Klein 1980, Tertiary clay). Sand - A* = 8.2 x 10^{-3} $(N/mm^2)^{-B}h^{-C}$, B = 2.25, C = 0.24, ϕ = 22°, ULS = 1.7 N/mm^2 (Klein 1985, Walsumer Meeressand). Time interval, t = 192 hours (8 days).

5. Freeze tube limiting deformation. VAM casing, 127mm (5 in.) O.D., 22.3 kg/m (15 lb/ft), Grade K55, wall thickness 7.52mm (0.296 in.). Yield strength 379 N/mm^2 (55,000 lb/in.^{2}) minimum. Permissible stress, f_{FT} = 0.7 x 379 = 265 N/mm^2 (38,500 lb/in.^{2}). Joints assumed to have strength equivalent to tube strength. E_s = 210,000 N/mm^2. Maximum permissible deformation (10.3m height shown in Figure 1),

$$\delta_{FTmax} = \frac{265 \times 10.3^2 \times 1000^2 \times 2}{6 \times 210{,}000 \times 127} = 350\text{mm from Figure 3.}$$

Table 2. Freeze wall deformation (mm) versus inner support pressure for the 5m thick freeze wall

	Internal support pressure, p_i (N/mm^2)	0	0.5	1.0	1.5	2.0	2.5
Clay	Elastic deformation, δ_i	266.66	228.84	191.02	153.20	115.38	77.56
	8-day creep deformation, δ_{ic}	207.72	83.81	26.38	5.37	0.41	0.00
	70% δ_i	186.66	160.19	133.71	107.24	80.77	54.29
	70% δ_{ic}	145.40	58.67	18.47	3.76	0.29	0.00
	Σ 70% δ	332.06	218.86	152.18	111.00	81.06	54.29
Sand	Elastic deformation, δ_i	86.68	74.39	62.10	49.80	37.51	25.21
	8-day creep deformation, δ_{ic}	252.51	168.88	101.65	50.23	15.74	0.46
	70% δ_i	60.68	52.07	43.47	34.86	26.26	17.65
	70% δ_{ic}	176.76	118.22	71.16	35.16	11.02	0.32
	Σ 70% δ	237.44	170.29	114.63	70.02	37.28	17.97

This is the deformation of the freeze wall restrained by the inner support pressure, p_i. For the unrestrained deformation, substituting $p_i = 0$ into Equation (2) gives:

$$\delta_{i(p_i=0)} = 2\, p_o\, r_i r_o^2 / E_f (r_o^2 - r_i^2) \qquad (3)$$

which is similar to Equation (2), applied to the shaft lining in Part 1 of the paper.

4.2 Time related (creep) deformation versus inner support pressure for the freeze wall

An appropriate form of stress/strain relationship which is representative of the time and temperature dependent creep behaviour of cohesive, frictionless soils (clays) is given by Jessberger (1980) and Klein (1980, 1985):

Uniaxial primary creep strain, $\varepsilon = A \sigma^B t^C$ (4)

where:

A = temperature dependent viscosity parameter in $(N/mm^2)^{-B} h^{-C}$ which is a constant for the particular temperature under consideration

B,C = dimensionless material dependent creep parameters

σ = constant uniaxial creep stress

t = time in hours (h)

Using this stress/strain relationship for the analysis of thick cylinders produces the deformation formula quoted by Klein (1980, 1985):

$$\delta_{ic} = \left(\frac{\sqrt{3}}{2}\right)^{B+1} r_i \left[\frac{(p_o - p_i)\,\frac{2}{B}}{1 - \left(\frac{r_o}{r_i}\right)^{-\frac{2}{B}}}\right]^B A t^C \qquad (5)$$

where δ_{ic} is the creep deformation at the inside face of the freeze wall.

Klein (1981, 1985) has developed the above approach further to include the behaviour of soils containing different friction angles (sandy soils). The temperature dependent viscosity parameter A modifies to A* when dependent additionally on the angle of friction. Closed form solutions, based on the analysis of thick cylinders, are given by Klein (1981, 1985) for the cases where $\phi > 0$ and B = 2, 1 and ∞. This set of standard results encompasses the full range of friction angle soil conditions and permits a judgement to be made on soils which do not specifically fit the standard form.

The creep deformation at the inside face of the freeze wall when the friction angle $\phi > 0$ and B = 2 is:

$$\delta_{ic} = \left(\frac{\sqrt{3}}{2}\right)^{B+1} r_i \left[\frac{2\lambda(1-\mu)}{\frac{r_i}{r_o} - \mu}\right] A^* t^C \qquad (6)$$

where:

$$\mu = \left[(p_o + \lambda)/(p_i + \lambda)\right]^{\frac{1}{2}} \qquad (7)$$

and:

$$1/\lambda = \tan\phi / c \qquad (8)$$

The friction angle, ϕ, is constant with time at the particular temperature under consideration and c is the cohesion which is both temperature and time dependent. Alternatively λ may be found from the uniaxial long term strength (ULS):

$$\lambda = \text{ULS}\,(1 - \sin\phi)/2\sin\phi \qquad (9)$$

The ULS is also temperature and time dependent.

5 FREEZE WALL DEFORMATION EXAMPLE

5.1 Data

The design data for the freeze wall deformation example is contained in Table 1.

5.2 Calculations

Typical results from the calculation exercise are contained in Table 2. Elastic deformations were evaluated using Equation (2) and creep displacements were determined by Equation (5) for the clay and Equation (6) for the sand. The eight-day creep and 70% imposed deformation on the lining parameters, defined earlier for the set of excavation and lining rates related to Figure 1, were adopted.

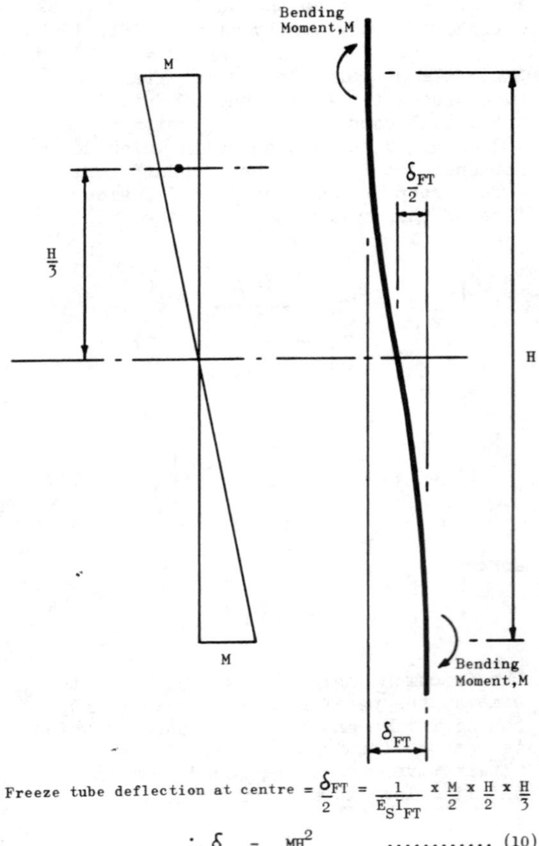

Freeze tube deflection at centre $= \frac{\delta_{FT}}{2} = \frac{1}{E_S I_{FT}} \times \frac{M}{2} \times \frac{H}{2} \times \frac{H}{3}$

$$\therefore \delta_{FT} = \frac{MH^2}{6E_S I_{FT}} \qquad \text{............ (10)}$$

Stress in freeze tube, $f_{FT} = \frac{My}{I_{FT}}$ .... (11)

From Equations (10) and (11):

$$\delta_{FT} = \frac{f_{FT}H^2}{6E_S y} \qquad \text{............ (12)}$$

Fig.3 Freeze tube deformation

Values from Table 2 for the 5m thick freeze wall, together with corresponding results for the two other freeze walls considered of 7.5m and 10m thickness, have been plotted in Figure 4 to produce deformation versus support pressure curves.

6 COMMENTS

Observation of Figure 4 and reference to the right hand diagram of Figure 2 in Part 1 of the paper indicate that the calculated freeze wall deformations are in excess of the permissible deformation for a 750mm thick, cast in situ, Grade 45

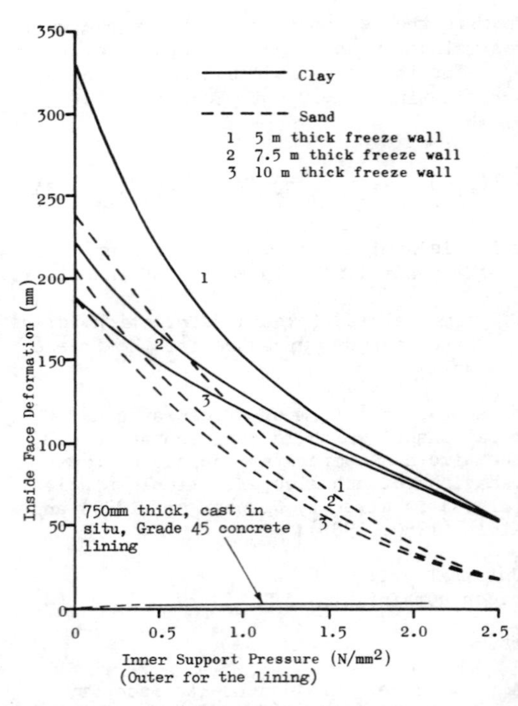

Fig.4 Freeze wall deformation versus support pressure with 70% of the restrained (3-day lining strength) elastic and 8-day (excavation) creep movement applied to the concrete lining

concrete lining and therefore the two are not compatible. As the lining has been designed adequately to withstand ground pressures in the long term, it would not be economical to over design for the short term temporary works condition. Therefore what are the options to minimize freeze wall deformation from being imposed on the shaft lining? The following courses of action are possible.

1. Increase freeze wall design thickness. Recourse to Figure 4 again, indicates that a substantial increase in wall thickness would be needed to reduce the deformation to a level which is compatible with the permissible lining deformation and this would not be economical in terms of refrigeration production.

2. During sinking, open up the excavation ahead of the previously cast lining to a length equal to the excavation diameter plus the next wall length. This increased unsupported sump depth would be 16m for the parameters indicated in Figure 1. By doing this the time interval before the next wall length is installed becomes 8 days, instead

Table 3: Freeze wall (5m thick) deformation in mm for the increased unsupported sump depth of 16m

		Internal support pressure, p_i (N/mm^2)	0	0.5	1.0	1.5	2.0	2.5
Clay	1	9-day creep deformation (1 day after lining installed)	210.68	85.00	26.76	5.44	0.41	0.00
	2	11-day creep deformation (3 days after lining installed)	215.81	87.07	27.41	5.58	0.42	0.00
	3	36-day creep deformation (28 days after lining installed)	248.81	100.38	31.60	6.43	0.49	0.00
	4	373-day creep deformation (1 year after lining installed)	329.40	132.90	41.84	8.51	0.65	0.00
	5	8-day creep deformation (prior to lining installation)	207.72	83.81	26.38	5.37	0.41	0.00
		Difference 5-1	2.96	1.19	0.38	0.07	0.00	0.00
		Difference 5-2	8.09	3.26	1.03	0.21	0.01	0.00
		Difference 5-3	41.09	16.57	5.22	1.06	0.08	0.00
		Difference 5-4	121.68	49.09	15.46	3.14	0.40	0.00
Sand	1	9-day creep deformation (1 day after lining installed)	259.76	173.72	104.59	51.67	16.21	0.47
	2	11-day creep deformation (3 days after lining installed)	272.57	182.29	109.75	54.22	17.01	0.49
	3	36-day creep deformation (28 days after lining installed)	362.29	242.29	145.87	72.07	22.61	0.65
	4	373-day creep deformation (1 year after lining installed)	634.98	424.65	255.67	126.31	39.63	1.14
	5	8-day creep deformation (prior to lining installation)	252.51	168.88	101.65	50.23	15.74	0.46
		Difference 5-1	7.25	4.84	2.94	1.44	0.46	0.01
		Difference 5-2	20.06	13.41	8.10	3.99	1.27	0.03
		Difference 5-3	109.78	73.41	44.22	21.84	6.87	0.19
		Difference 5-4	382.47	255.77	154.02	76.08	23.89	0.68

of 5 (see Figure 1), and all of the elastic deformation has taken place. It can be seen from Table 2 that the elastic deformation does not reduce rapidly with increasing internal support pressure whereas the creep effect does. Therefore allowing the elastic deformation to occur fully before installing the lining appears to be a major factor in reducing the effect of freeze wall deformation on the lining.

In Table 3 creep deformation versus support pressure for a 5m thick freeze wall has been calculated over various time periods for the increased unsupported sump depth. The calculations have been specifically related to the time from the lining installation (8 days prior plus 1, 3, 28 days and 1 year). The differences between the eight day values and the extended time periods give the amount of deformation which would be imposed on the shaft lining. Figure 5 contains the curves derived from Table 3.

It can be seen from Figure 5 that the deformation imposed on the lining by the freeze wall is considerably reduced. For the clay, the lining and freeze wall deformations are now almost compatible. The solid lines representing deformation versus inner support pressure for the clay are the relationships for 1, 3, 28 days and 1 year after the lining has been installed. Correspondingly the numbers 1, 3, 7, 14, 21 and 28 on the lining curve represent its developed deformation/outer support pressure characteristics at the particular time in days. Only at the early age of 1 day does the freeze wall create a deformation on the lining slightly in excess of the permissible, whereas at any time subsequent to one day the lining has developed the capability to withstand the imposed deformation within its permissible limits. In the case of the

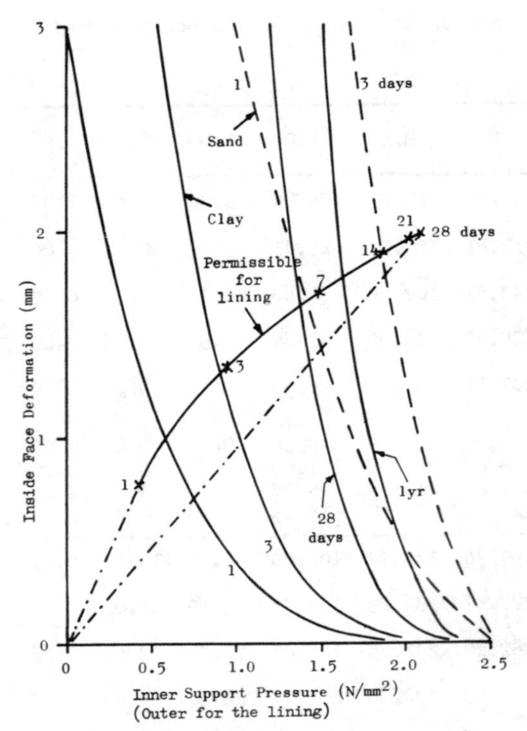

Fig.5 Compatibility of shaft lining (8m diameter, 750mm thick, cast in situ, Grade 45 concrete) and freeze wall (5m thick) deformation for increased unsupported sump depth of 16m

sand, the freeze wall imposed deformation on the lining would be in excess of the latter's permissible and therefore the two are still not compatible.

3. The next step is to find the means of further reducing the sand freeze wall deformation to ensure compatibility with the lining. However, first a check on freeze tube deformation is necessary for both the clay and the sand. Increasing the unsupported sump depth to 16m means that 100% of the unrestrained elastic and 8 day creep deformation will have occurred prior to the lining installation. For the clay, adding together the values from Table 2 gives a total unrestrained deformation of 474.38mm for the 5m thick freeze wall. Values of 316.25 and 267.76mm were also obtained for the freeze wall thicknesses of 7.5 and 10m respectively. Similarly, the corresponding values for the sand were 339.19 (Table 2), 292.03 and 267.12mm. From Table 1, Item 5 it can be seen that the maximum permissible deformation for the freeze tubes is 350mm. Only in the case of the 5m freeze wall in clay is the maximum permissible freeze tube deformation exceeded. This shows that, although the clay freeze wall deformation shown in Figure 5 appears to be compatible with the lining, the unrestrained deformation allowed to occur before its installation would be excessive for the freeze tubes. In this case, the unsupported sump depth must be shortened back to 13m (again by proportion) to control the freeze tube deformation to 74% of 474.38m which equals 350mm (see Figure 2) with 26% being applied to the lining. This obviously negates any chance of the clay freeze wall in Figure 5 being compatible with the shaft lining.

4. Additional temporary support for the excavated freeze wall face prior to installation of the concrete cast in situ lining now comes under consideration. A review of Figure 3 in Part 1 of the paper gives various options for additional temporary support. It can be seen that a fabricated steel tubbing could easily accommodate both the clay and sand freeze wall deformations in Figure 5 at the depth of shaft under consideration (200m). At shallower depths the various precast concrete types 1 to 4 would suffice. For greater pressures at deep levels it could be necessary to resort to the S.G. cast iron tubbing for permanent support which may at the same time accommodate the freeze wall deformation. The subject of temporary and permanent lining combinations and their compatibility with freeze wall deformation is a subject for future study.

The question to be answered still is what are the implications of the above observations on freeze wall design, in particular the current normal approach of designing for strength first and then checking for deformation. Reference to Klein's (1980a) modified Domke (1915) formula to determine freeze wall thickness based on strength:

$$\frac{r_o-r_i}{r_i} = 0.29\left(\frac{p_o-p_i}{ULS}\right) + 2.30\left(\frac{p_o-p_i}{ULS}\right)^2 \quad (13)$$

which is applicable to clay soils, and requires further modification to accommodate the angle of internal friction for sandy soils, gives the answer. Once the lining has been installed and achieved its 28-day design strength, the difference between p_o and p_i becomes minimal negating the need for any freeze wall thickness, r_o-r_i. This confirms the statements made at the beginning of both parts of this paper that it is the control of the freeze wall deformation during installation of the lining which is the most important criteria, and not strength.

The whole approach to freeze wall design evolves from initially designing the shaft lining to withstand the long term ground loading. An economic balance must then be struck between providing a thicker freeze wall on its own for deformation control over the unsupported sump depth or using a thinner freeze wall with additional temporary support where necessary to control deformation to an extent which is compatible with both the additional temporary support and the permanent lining. A 5m thick freeze wall could be acceptable for the freeze wall deformation example in this paper if fabricated steel tubbing were to be installed as temporary support at the 200m level prior to casting the concrete lining.

Unfortunately, the elastic and creep data used in the freeze wall deformation example were not results from the same samples. Such information was not available to the author at the time the paper was prepared and therefore combining the two could be suspect. A similar problem was experienced in choosing suitable available data for the sand creep properties in the example. The ULS value of 1.7 N/mm^2 (Klein 1985, Walsumer Meeressand) used is a long term (several years) value whereas a higher value could be expected in the short term (installation period). It also produced greater deformations in the example than the clay, which is contrary to expectation. However, using a larger value would increase the deformation further when substituted into Equations (6), (7) and (9) which is also questionable. These aspects need further review.

Two points need to be noted with reference to Figure 5. Only the lining inside face deformation has been shown, and not that for the outer face which is in contact with the freeze wall. However, as the lining is constructed from a relatively rigid material, the difference between the two would be insignificant. Secondly, the freeze wall deformation curves are based on constant support pressures. Until the concrete lining reaches its 28-day design strength, the support pressure varies with time, which would lead to higher deformation values than those shown using the constant support pressures. The magnitude of this increase, however, is not considered to be significant with regard to changing the basic principles adopted for checking the compatibility of the lining and the freeze wall.

From the exercise carried out in this paper, an indication of what the main principles are in arriving at a satisfactory freeze wall design can be gleaned. The originality of the paper is in its attempt to use the principles with reference to the direct needs of shaft sinking construction and its overall detailed engineering. A substantial amount of further work still needs to be carried out using improved data to confirm the principles and additionally to check the design concepts with measurements taken in situ during the construction of such projects.

REFERENCES*

Auld, F.A. 1985. Freeze wall strength and stability design problems in deep shaft sinking. Is current theory realistic? Proc. of The Fourth International Symposium on Ground Freezing, Sapporo, Japan. 5-7 August: 343-349.

Domke, O. 1915. Über die Beanspruchungen der Frostmauer beim Schachtabteufen nach dem Gefrierverfahren (Stresses affecting the ice wall in shaft-sinking by the ground-freezing method). Gluckauf. 51, No.47: 1129-1135.

Hegemann, J. and Jessberger, H.L. 1985. Deep frozen shaft with gliding liner system. Proc. of The Fourth International Symposium on Ground Freezing, Sapporo, Japan. 5-7 August: 357-373.

Jessberger, H.L. 1980. State-of-the-art-report. Ground freezing: mechanical properties, processes and design. The Second International Symposium on Ground Freezing, Trondheim, Norway. 24-26 June, Preprints Volume: 1-33.#

Klein, J. 1980. Die Bemessung von Gerfrierschächten in Tonformationen ohne Reibung mit Berücksichtigung der Zeit (Structural design of freeze-shafts in frictionless clay formations taking account of the time factor). Glückauf-Forschungshefte. 41, No.2: 51-56.

Klein, J. 1980a. Die Festigkeitsberechnung von Frostwänden im Gefrierschachtbau. Glückauf-Forschungshefte. 41, No.5. October: 3-8.

Klein, J. 1981. Die Bemessung von Gefrierschächten in Sandformationen vom Typ B=2 mit Berucksichtigung der Zeit (Structural design of freeze-shafts in Type B=2 sand formations taking account of the time factor). Glückauf-Forschungshefte. 42, No.3: 112-120.

Klein, J. 1985. Influence of friction angle on stress distribution and deformational behaviour of freeze shafts in non-linear creeping strata. Proc. of

*See page X concerning ISGF papers

The Fourth International Symposium on Ground Freezing, Sapporo, Japan. 2, 5-7 August: 307-315.

Ladanyi, B. 1982. Ground pressure development on artificially frozen soil cylinder in shaft sinking. Special Volume in honour of Professor De Beer. Institut Géotechnique de l'Etat, Brussels, Belgium.

Link, H. Lütgendorf, H.O. and Stoss, K. 1976. Richtlinien zur Berechnung von Schachtauskleidungen in nicht Standfestem Gebirge (Instructions for design of shaft linings in unstable ground). Verlag Glückauf GmbH, Essen.

Morgenstern, N.R. and Wittebolle, R.J. 1985. Laboratory testing of specimens from Jining No.2 Project, China. Report prepared for Cementation Mining Limited.

Roesner, E.K., Poppen, S.A.G. and Konopka, J.C. 1982. Stability during shaft sinking (a design guideline for ground support of circular shafts). Proc. of the Conference on Stability in Underground Mining, Vancouver, Canada: 749-769.

Timoshenko, S. 1976. Strength of materials. Part 2. Advanced theory and problems. 3rd Ed. Van Nostrand Reinhold Company, New York: 205-210.

5th International Symposium on Ground Freezing, Jones & Holden (eds)
 ISBN 90 6191 824 3

Ground freezing with liquid nitrogen

P.Capitaine & D.Rebhan
Linde AG, Höllriegelskreuth, FR Germany

ABSTRACT: This paper is concerned with the propagation action of frost in freezing processes using either one or several freeze pipes. The propagation of frost around a single freeze pipe in sandy soils of different saturation has been examined in a number of tests. Calculations of the progagation of frost were based upon the function $R = k \cdot \sqrt{\tau}$. k-values ranging from 0,028 $\frac{m}{\sqrt{h}}$ to 0,072 $\frac{m}{\sqrt{h}}$ were obtained by plotting R versus $\sqrt{\tau}$. For the examination of the propagation of frost in soil surrounding an excavation several freeze pipes were employed. A number of thermocouples were installed and the propagation of frost was recorded prior to and following the closing point. The closing-time values recorded at different levels were compared with those obtained in calculations based upon the k-factors of experimental results.

1 INTRODUCTION

Owing to its physical properties liquid nitrogen is an ideal refrigerating medium. At atmospheric pressure it has a temperature of ebollution of -195.7 °C and a heat of evaporation of 250 $\frac{kJ}{m^3}$. At a temperature increase up to -110 °C another 110 $\frac{kJ}{m^3}$ are available. In ground freezing liquid nitrogen is conveyed through a down-pipe to the bottom of a freeze pipe installed in the ground. As a result of liquid nitrogen evaporating in the space between down-pipe and jacket and the subsequent temperature rise of the gaseous cold nitrogen to a set gas temperature, heat is withdrawn from the soil, thus permitting the freezing process to set in. Consequently, the heat exchange along the freeze pipe is not constant, due to fluctuations of the state of aggregation, the temperature and the gas flow rate along the freeze pipe.
Nevertheless, in order to be able and estimate the closing-time in sandy soils in a simple and time-saving manner - yet, with a precision sufficient for practical applications - the propagation action of frost around a single freeze pipe installed in sandy soil of different saturation was examined. Moreover, the individual stages of freezing - prior to and following the closing point - were examined within the scope of a project.

2 PROPAGATION OF FROST AROUND A SINGLE PIPE

The experimental set-up is shown in schematic Figure 1.
From tank 2, located on the inclination balance 1, liquid nitrogen streams into down-pipe 8a, evaporates along the inner wall of the jacket and finally escapes into the atmosphere through the freezer head 8c. Freeze pipe 8 is in a central position in the test vessel 14 installed in the ground. The resistance thermometer 9, mounted at the gas outlet of the freezer head, indicates the gas temperature. Control element 10 permits a comparison between the actual temperature registered and the set nominal temperature. If the temperature falls below the set value, the solenoid valve closes, if, however, the set temperature is exceeded, the solenoid valve opens. The soil temperature is registered by thermocouples 11 installed at 3 different levels and connected with the y-t-recorder, where the soil temperature is recorded constantly.

Vessel 14:
A 200 l oil barrel serves as receptacle. The oil barrel is provided with cross struts arranged at 3 different levels to ensure accurate positioning of the 30 thermocouples (fig. 2).

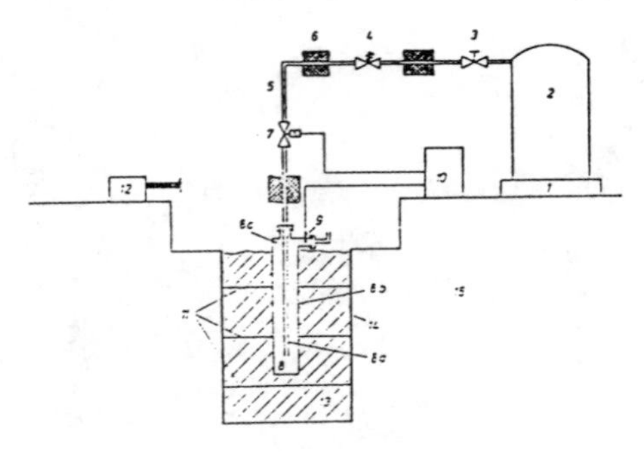

Legend

1 inclination balance
2 heat-insulated tank
3 isolating valve
4 safety valve
5 feeding pipe
6 insulation
7 solenoid valve
8 freezing tube consisting of:
   8a down-pipe
   8b heat-pipe
   8c freezer head
9 resistance thermometer
10 control element
11 thermocouple (30pcs arranged on 3 levels)
12 y-t recorder
13 medium to be frozen (silicic sand)
14 test vessel
15 surrounding soil

Fig.1 Test set-up for ground freezing tests

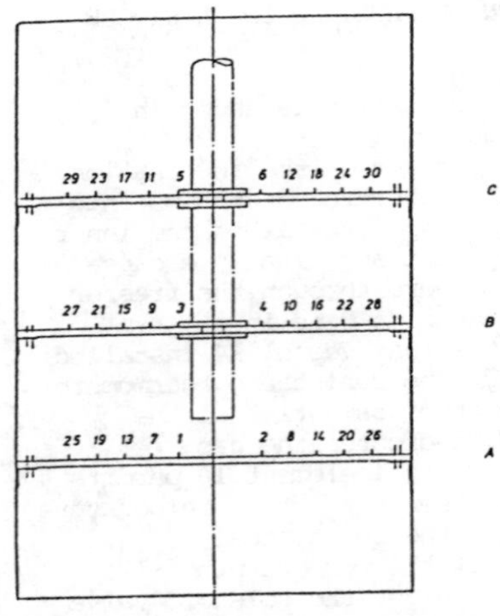

Fig.2 Position of the thermocouples in the test vessel.

2.1 Propagation of frost as a function of time

The freezing properties of silicic sand 0/4 with a differing moisture content were examined with the moisture content being varied from 4 % above 14 % up to a value of 21 %. The temperature between frozen and not frozen ground is 0 °C. Thus, the precise position of the 0°C interface at the cross strut levels is defined, as soon

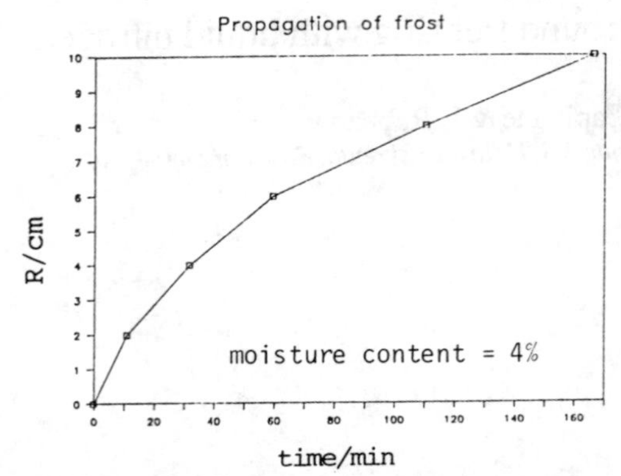

Fig.3 Wall thickness of ice mass as a function of time

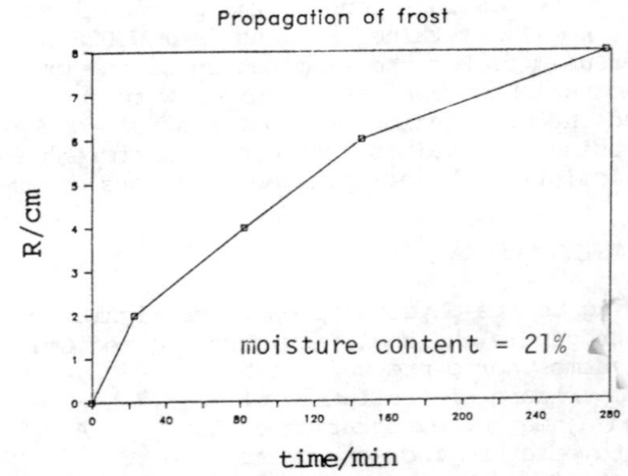

Fig.4 Wall thickness of ice mass as a function of time

as freezing has spread over the radial space, bringing the 0 °C line up to the level of the thermocouples. The expansion of the 0 °C interface as a function of time is a root function with the equation (Ständer, 1967).

$$R = k \cdot \sqrt{\tau}$$

If, therefore, the wall thickness of the ice mass is plotted versus the root of time, a straight line with a slope equal to the k-factor is obtained. Figures 3 and 4 specify - as an example - the wall thickness of the ice mass as function of time at a gas escape temperature of -100 °C in grounds with 4 %, respectively 21 % moisture content at the B level. Figures 5 and 6 depict plotting R versus $\sqrt{\tau}$ chosen for the determination of the k-factors. The result was a k-value of 0.072 $\frac{m}{\sqrt{h}}$ at a moisture content of 4 % and a k-value of $0.038 \frac{m}{\sqrt{h}}$ at a moisture content of 21 %.

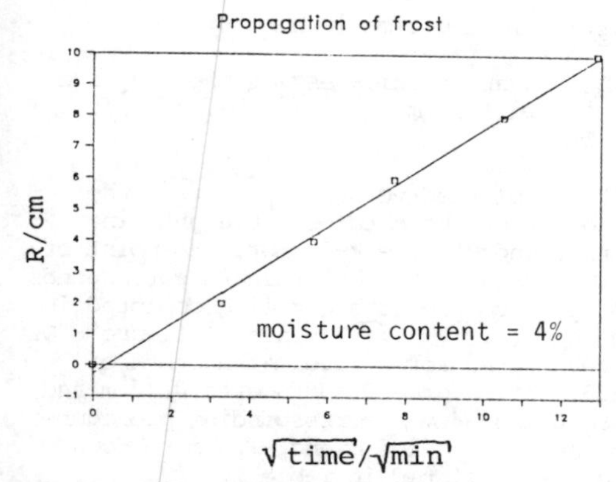

Fig.5 Wall thickness of ice mass as a function of root of time

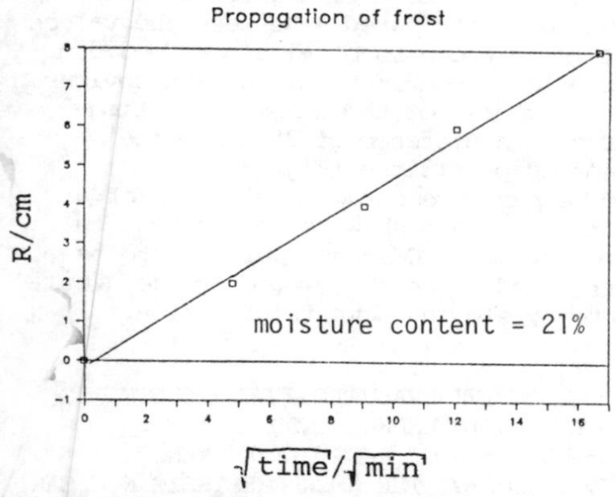

Fig.6 Wall thickness of ice mass as a function of root of time

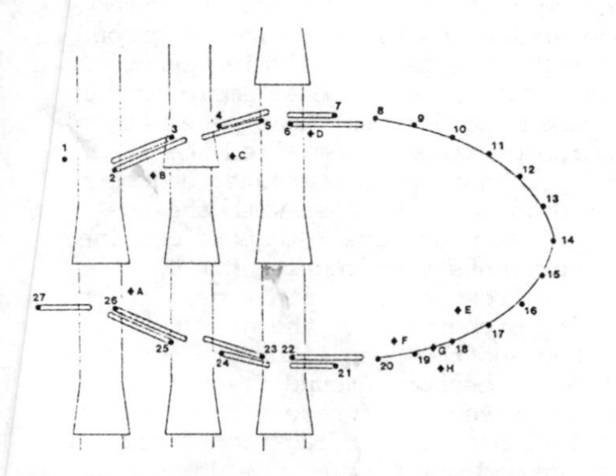

Fig.7 Arrangement of bores for freeze pipes and thermocouples in ground freezing project Neckarwestheim.

3 PROPAGATION OF FROST OF AN ELLIPTICAL TRENCH ENCLOSURE AT THE SITE OF THE NECKARWESTHEIM NUCLEAR PLANT

For repair works on a damaged pipe-line a trench 4 m deep and approx. 3 m wide had to be excavated. The decision to support the trench with an elliptical ice mass was mainly based upon the two subsequent reasons:

1. Low space requirements. The pipe rupture was in the immediate vicinity of an arterial road which could not be closed to traffic.
2. In addition to the damaged pipe-line other pipe-lines crossed the trench at different levels. These pipelines had to be prevented from sinking during repair works.

To limit the volume of the ground to be frozen to a minimum, soil to a depth of up to 1.2 m below ground level was excavated in a first step. In a second step the remaining 3 meter soil were excavated by placing a protective ice mass girdle around the trench. Freezing was carried out with 27 freeze pipes (pipe spacing 0.5 m, pipe lengths 1.9 m up to 3.2 m). Composition of ground: homogeneous gravelly sand of high compactness, having a moisture content of approx. 8 %. As the ground condition had been defined, the frost action could be recorded accurately by installing several temperature control points (Fig. 7).

The closing-time was determined through temperature shaft G, in a central position between pipes 18 and 19. The temperature feelers installed at a depth of -1.5 m, -2.5 m and -3.5 m indicated $0^{o}C$ after 16, 21 and 12 hours. The corresponding k-values are 0.063, 0.055 and $0.072 \frac{m}{\sqrt{h}}$. Temperature shafts E, F and H allowed recording of the frost action subsequent to the closing point. Because of the large radius of curvature in the zone of the temperature shafts, the freeze pipes are in a straight position in the first approximation. On this condition the frost action following the closing point can be reduced to the propagation of frost on a plane wall (Ständer 1967).

From

$$D = k \cdot \sqrt{\tau}$$

2 D: ice wall thickness in m at freezing time

τ: Freezing time in h

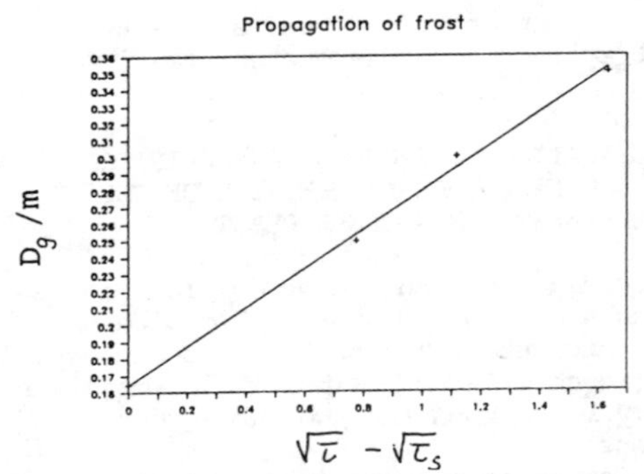

Fig.8 D_g as a function of $(\sqrt{\tau} - \sqrt{\tau_s})$

Table 1: D_s and k as function of depths below ground surface

depth below ground surf. in m	$\frac{D_s}{m}$	$\frac{k}{m/\sqrt{h}}$
- 1.5	0.2	0.1
- 2.5	0.17	0.06
- 3.5	0.16	0.12

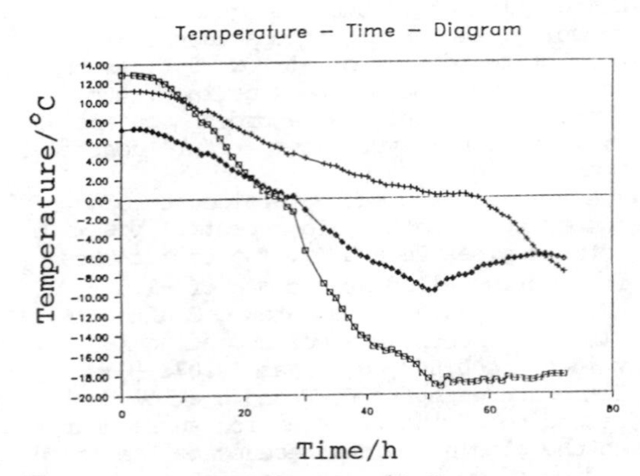

□3.5 m +2.5 m ◊1.5 m

Fig.9 Temperature-time-curve at a distance of 35 cm to the freeze pipe ellipse.

follows by approximation

$$D_g = D_s + k \cdot (\sqrt{\tau} - \sqrt{\tau_s})$$

$2\ D_s$: avarage thickness of the wavy ice-wall at closing time

τ_s : closing time in h

$2\ D_g$: avarage thickness of the wavy ice wall in m

With D_g plotted versus $(\sqrt{\tau} - \sqrt{\tau_s})$ the gradient of the obtained straight line corresponds to the k-factor, its point of intersection with the ordinate corresponds to D_s. From the straight line depicted in the D_g, $(\sqrt{\tau} - \sqrt{\tau_s})$ scale of figure 8 D_s is evaluated for a level at a depth of -3.5 m below ground surface to 0.16 m and k to 0.115 $\frac{m}{\sqrt{h}}$. A corresponding procedure for depths of -1.5 m and -2.5 m renders the values listed in Table 1.
At a depth of -2.5 m the rate of frost action was at a minimum - both, prior to and following the closing point. As a consequence, the k-values do not lie - as would be anticipated - between the values obtained at depths of -1.5 m and -3.5 m but below these instead. This is obvious in the course of the temperature-time-curve at a distance of 35 cm to the freeze pipe ellipse (Fig. 9).
The Temperature curve development at a depth of -2.5 m shows a long retention time in the 0 °C range, which is to be attributed to an increase in moisture content, slowing down frost action.

4 COMPARATIVE EVALUATION OF RESULTS OBTAINED IN EXPERIMENTS ON THE PROPAGATION OF FROST IN A TRENCH ENCLOSURE OF THE NECKARWESTHEIM NUCLEAR PLANT

It was the main object of the ground freezing tests to be in a position to predetermine the closing time in ground freezing projects in a simple manner. As a short freeze pipe had been employed for the chosen conditions, heat penetrating from the environment impaired the rate of frost propagation to a greater degree than would have been the case with the use of a long freeze pipe. The k-factors obtained in experiments thus represent the lower limit of frost action with regard to the freezing properties of the grounds examined and the specified set temperature of the gas escape. Ground freezing with the use of several freeze pipes permits calculation of the propagation action of frost around one single pipe up to the closing point of the ice rings surrounding the individual pipes, taking into account that in the precooling stage the heat is taken up by two pipes. A closing time - predetermined by the k-factors obtained in experiments - is, therefore, despite the fact that the k-values represent the lower limit of the propagation action of frost, to be regarded the maximum required closing time. The ground at the nuclear plant had a moisture content of 8 %. If the k-factors obtained in experiments are taken into consideration for calculating the closing time, a duration of approx. 15 hours is computed. Closing times at depths of -1.5 m. -2.5 m and -3.5 m were 16, 21 and 12 hours. If the closing time at a depth of -2.5 m is disregarded for reasons lined out above, the closing time obtained with the experimental k-factor is equal to the one actually recorded.

REFERENCES

Ständer, W. Mathematische Ansätze zur Berechnung der Frostausbreitung in ruhendem Grundwasser im Vergleich zu Modelluntersuchungen für verschiedene Gefrierrohranordnungen im Schacht- und Grundbau. Veröffentlichung des Instituts für Bodenmechanik und Felsmechanik der Technischen Hochschule Fridericiana in Karlsruhe, Heft 28, Karlsruhe 1967.

5th International Symposium on Ground Freezing, Jones & Holden (eds)
© 1988 Balkema, Rotterdam. ISBN 90 6191 824 3

Effect of frozen soil creep on stresses at the moving thaw front around a wellbore

P.Huneault & B.Ladanyi
Northern Engineering Centre, Ecole Polytechnique, Montreal, Quebec, Canada

L.N.Zhu
Lanzhou Institute of Glaciology and Geocryology, Academia Sinica, People's Republic of China

ABSTRACT: An analytical solution for the radial stress variation at the moving thaw front around a permafrost production wellbore is obtained based on the ageing theory of creep as applied to frozen soils. A complete formulation is shown to lead to a Stefan- type free boundary problem.

1. INTRODUCTION

The exploration and exploitation of petroleum resources in Arctic frontier regions are beset by problems mostly unique and substantial in character. An interesting overview can be found in Koch (1971) and a comprehensive report on drilling and completing arctic wells in Goodman (1978).

In particular, prolonged extraction of warm hydrocarbons through arctic production wellbores results in local melting and thaw subsidence of permafrost sediments. As stated by Mitchell et al. (1983), the pore-ice to pore-water phase change can induce two commonly accepted soil straining mechanisms. The first is termed stiffness reduction and pertains to the decrease of the load carrying capacity of the soil upon thaw. The second is related to the 9% pore-ice to pore-water volume reduction which in effect may induce lateral thaw strains through volumetric contraction of the soil skeleton or a decrease in pore pressures and the establishment of a vertical gradient in pore pressure change bringing about a gravity like loading, or both.

Several attempts have been made in the past to assess theoretically the extent of expected soil deformations and their potential detrimental effects on the structural integrity and stability of production casings and surface facilities (Merriam et al. (1975), Mitchell (1977), Mitchell and Goodman (1978), Ruedrich et al. (1978), Smith and Clegg (1971)). These studies concentrated on the behaviour of the thawing soil column while neglecting both the nature of the stresses at the thaw front and the viscoelasticity of the surrounding permafrost. In this paper, only one aspect of the whole problem is investigated, that of the effect of creep closure of frozen ground on the stresses at the thaw front when the thawing soil column experiences a full lateral thaw strain.

1.1 PROBLEM DEFINITION

A wellbore of external radius "a" is being used for extraction of warm oil through a thick permafrost sediment (fig. (1)). This results in permafrost thawing around the wellbore forming a long vertical unfrozen soil column of radius "b". As the soil contracts and experiences a reduction in stiffness upon thawing, the frozen ground tends to move inward both by its elastic response and by creep. It is required to find the resulting radial stress at the moving thaw front.

The elements of a solution, i.e. the governing equations for characterizing the various phenomena that act simultaneously, were outlined previously by Ladanyi (1985). Specifically one must consider:

1) The propagation of the thaw front,
2) The soil contraction due to thawing,
3) The time dependent response of the surrounding frozen ground.

1.2 ASSUMPTIONS

In the following, it is assumed that:

1) The heat flux around the well is two-dimensional, axisymetric and constant
2) The initial temperature, ice content, unfrozen water content and thermophysical properties of the permafrost layer are

constant

3) The soil is initially uniform in the radial direction

4) The far-field stress in the frozen ground is constant

5) The plane strain conditions are satisfied.

By totally disregarding any axial component in the above, the problem at hand has obviously been greatly simplified. In reality, the temperature in the circulating fluid, along the casing surface and in the permafrost varies with depth as does also the unfrozen water content, the permafrost lithology, etc. Thermal modelling in the axial and radial dimensions is best carried out numerically (Lin and Wheeler (1978)) and is difficult to account for analytically so it will not be attempted here.

2. PROPAGATION OF THE THAW FRONT

Under certain simplifying assumptions, an acceptable analytical expression between the radius "b" of the thaw front and the time t can be deduced. It has the form (Carslaw and Jaeger (1959))

$$\alpha^2 \ln \alpha - ((\alpha^2 - 1)/2) = C_1 t / H \qquad (1)$$

where the radius ratio of thaw front to wellbore is given as

$$\alpha = b/a \qquad (2)$$

and C_1, a constant, as

$$C_1 = 2 (T_w - T_s) K_{th} / a^2. \qquad (3)$$

Equation (3) refers to the differential in temperature existing between that of the wellbore, T_w and the thaw temperature of the soil T_s, to the thermal conductivity K_{th} of the thawed soil and to the total amount of heat H required for thawing the soil unit and heating it to the average temperature in the thawed annulus. The latter may be written as

$$H = C_2 + C_3 \frac{\alpha^2}{(\alpha^2 - 1)} \qquad (4)$$

with C_2 and C_3 being constants defined by

$$C_2 = C_f \gamma_f (T_0 - T_s) + Q + (C_{th} \gamma_{th} (T_w - T_0) / 3), \qquad (5)$$

$$C_3 = 8 C_f \gamma_f (T_0 - T_s) / 3. \qquad (6)$$

Here, Q refers to the latent heat whereas

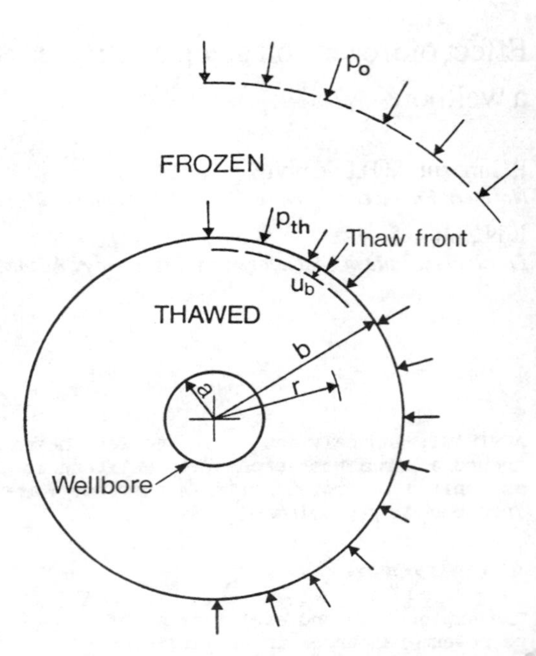

Figure 1: Notation in thawing around a wellbore problem.

C and γ denote the specific heat and the bulk unit mass, respectively, the subscript "f" applying to the frozen state and "th", to the thawed state.

Inserting eq. (4) into eq. (1) and solving for the time t yields

$$t = [\alpha^2 \ln \alpha - \frac{(\alpha^2 - 1)}{2}] \cdot [C_{21} + C_{31} (\frac{\alpha^2}{\alpha^2 - 1})] \qquad (7)$$

with

$$C_{21} \equiv C_2 / C_1, \qquad (8)$$

$$C_{31} \equiv C_3 / C_1. \qquad (9)$$

Finally, the rate of propagation of the thaw front is found from eq. (7) to be

$$\frac{d\alpha}{dt} = [t,_\alpha]^{-1} \qquad (10)$$

where

$$t,_\alpha \equiv \frac{dt}{d\alpha} = C_{31} [\frac{\alpha}{\alpha^2 - 1} + \frac{2 \alpha^3 (\alpha^2 - 2) \ln \alpha}{(\alpha^2 - 1)^2}] + 2 C_{21} \alpha \ln \alpha \qquad (11)$$

3. THAW STRAIN

It can be shown that the unit volumetric strain v, due to thawing of soil, is given by

$$v \equiv \frac{\Delta V}{V} = \frac{-0.09\ (W - W_u)}{(\rho_w / \rho_s) + (W / S_r)} \qquad (12)$$

where W and W_u denote the total and unfrozen water content of frozen soil, respectively, ρ_s is the density of solid particles, ρ_w is the density of water, S_r is the degree of saturation and V is volume. Note that $(v < 0)$ corresponds to contraction.

A kinematic relationship between volumetric strain v and radial displacement u of the boundaries "a" and "b" also exists for the plane strain case in the form

$$\frac{\Delta V}{V} = \frac{\alpha^2 \ [(\frac{u_b}{b})^2 + 2\ (\frac{u_b}{b})] - [(\frac{u_a}{a})^2 + 2\ (\frac{u_a}{a})]}{\alpha^2 - 1} \qquad (13)$$

Considering the wellbore to be sufficiently rigid (i.e. $(u_a \approx 0)$), eqs. (12) and (13) combine to give

$$\frac{u_b}{b} = [1 + (\frac{\alpha^2 - 1}{\alpha^2})\ v]^{1/2} - 1 \qquad (14)$$

4. PRESSURE VARIATION AT THE THAW FRONT

As the thaw front progresses radially outward and the central core of unfrozen soil undergoes continuous contraction, the surrounding outer zone of permafrost, cylindrical in shape with a cavity boundary congruent with that of the thaw front, responds both elastically and through creep. The elastic response is immediate and equivalent to the unloading of the cavity boundary leading to stress relaxation at the thaw front. In contrast, the creep response, by definition delayed in time, will tend to repressurize the thaw front interface.

Eq. (14) suggests that the stress variation at the thaw front may be determined by investigating the tangential strain-rate condition at that location, the total rate of strain being made up of an elastic and creep component:

$$\dot{\epsilon}_\Theta = \dot{\epsilon}^e_\Theta + \dot{\epsilon}^c_\Theta \qquad (15)$$

where Θ denotes the usual tangential cylindrical coordinate, the superscripts e and c refer to elasticity and creep, respectively, and the dot denotes the time derivative.

From eq. (14) the total strain-rate is found to be

$$\dot{\epsilon}_\Theta = \frac{\dot{u}_b}{b} = [1 + (\frac{\alpha^2 - 1}{\alpha^2})\ v]^{-1/2}\ \frac{v}{\alpha^3}\ (\frac{d\alpha}{dt}) \qquad (16)$$

with eqs. (10) and (11) being explicitly referred to here.

According to Lamé's theory for a very thick cylinder undergoing plane strain and loaded at its external boundary by a constant pressure (here the natural frozen ground pressure (fig. (1)), the elastic strain-rate component at the thaw front is simply

$$\dot{\epsilon}^e_\Theta = \frac{\dot{p}_{th}}{2G} \qquad (17)$$

where p_{th} is the pressure acting at the frozen-unfrozen soil interface and G is the shear modulus of the frozen soil.

The creep strain-rate can be derived by invoking the constitutive model

$$\dot{\epsilon}^c_{ij} = \frac{3}{2}\frac{\dot{\epsilon}^c_e}{\sigma_e} S_{ij} \qquad (18)$$

with

$$S_{ij} = \sigma_{ij} - \frac{1}{3}\delta_{ij}\sigma_{ii} \qquad (19)$$

and

$$\sigma_e = [\frac{3}{2} S_{ij} S_{ij}]^{1/2} \qquad (20)$$

Briefly, we recognize that the flow-rates $\dot{\epsilon}^c_{ij}$ are subject to hardening, as related by the equivalent creep strain-rate formulation $\dot{\epsilon}^c_e$, and are mobilized by deviatoric stress S_{ij}, the basis of the equivalent Von Mises stress σ_e. Presently and for the plane strain cylindrical case, eqs. (18) to (20) reduce to

$$S_{\Theta\Theta} = (\sigma_\Theta - \sigma_r) / 2, \qquad (21)$$

$$\sigma_e = \frac{\sqrt{3}}{2} |\sigma_r - \sigma_\Theta|, \qquad (22)$$

and

$$\dot{\epsilon}^c_\Theta = -\frac{\sqrt{3}}{2}\dot{\epsilon}^c_e \qquad (23)$$

considering that $(p_{th} < p_o)$ and having invoked incompressibility. The hardening law will now be taken in the power law form

$$\dot{\epsilon}_e^c = K\,\sigma_e^n\,b\,t^{b-1} \quad (24)$$

where $(K > 0)$, $(n \geq 1)$, $(0 < b \leq 1)$ are creep parameters such that eqs. (22) to (24) along with Lamé theory lead to

$$\dot{\epsilon}_\Theta^c = -\frac{(\sqrt{3})^{n+1}}{2} K\,b\,|p_{th} - p_o|^n\,t^{b-1} \quad (25)$$

The final forms of the relevant equations for $\dot{\epsilon}_\Theta$, $\dot{\epsilon}_\Theta^e$ and $\dot{\epsilon}_\Theta^c$ are obtained by substituting eqs. (10) and (11) into eq. (16) and eq. (7) into eq. (25), all the while maintaining eq. (17). Reconsidering eq. (15), we then obtain the differential equation

$$\dot{P} + C_4\,P^n + C_5 = 0 \quad (26)$$

where

$$P \equiv p_{th} - p_o\,, \quad (27)$$

$$C_4 = -\,(\sqrt{3})^{n+1}\,K\,b\,G\,[A_1\,A_2]^{b-1} \quad (28.a)$$

$$A_1 = \alpha^2 \ln\alpha - \left(\frac{\alpha^2 - 1}{2}\right) \quad (28.b)$$

$$A_2 = C_{21} + C_{31}\left(\frac{\alpha^2}{\alpha^2 - 1}\right) \quad (28.c)$$

$$C_5 = \frac{-\,2\,G\,v}{\alpha^3\,[\,A_3 + A_4 + A_5]\,\left[1 + \frac{(\alpha^2 - 1)}{\alpha^2}\,v\right]^{1/2}} \quad (29.a)$$

$$A_3 = 2\,C_{21}\,\alpha\ln\alpha \quad (29.b)$$

$$A_4 = \frac{\alpha}{\alpha^2 - 1}\,C_{31} \quad (29.c)$$

$$A_5 = \left[2\,\alpha^3\,\frac{(\alpha^2 - 2)}{(\alpha^2 - 1)^2}\ln\alpha\right] C_{31} \quad (29.d)$$

recalling that C_{21} and C_{31} follow from eqs. (3), (5), (6), (8) and (9), v is defined in eq. (12) and that, to be consistent, the incompressible form for G is now referred to in eqs. (28) and (29).

Expression (26) may lead to stress relaxation or repressurization at the thaw front depending on the C_4 and C_5 values. The latter are function in particular of the thaw front location α such that expression (26) effectively constitutes a Stefan-type free boundary problem whose solution may only be attempted numerically. Nevertheless, an analytical solution, albeit less rigourous, can still be arrived at based on the ageing theory for creep.

5. AGEING THEORY

In order to take into account the creep response of the frozen ground, the ageing theory simply considers that the shear modulus G of frozen soil is now a function of time and independent of the stress condition at the thaw front. Here, we adopt the form

$$G = \frac{G_o}{(1 + t)^\beta} \quad (30)$$

where G_o denotes the initial shear modulus and $(\beta > 0)$ is a creep parameter. Equation (30) is consistent with previous analytical and experimental proposals, for example those of Schapery (1965) and Tinawi and Gagnon (1984). In that the long term shear modulus tends to zero, eq. (30) is best applied to ice-rich soils. A slightly modified version of eq. (30) may be more appropriate for modelling the asymptotic behaviour in time of ice-poor soils.

The ageing solution is obtained by simply neglecting the creep contribution figuring in eq. (15). As a result, eqs. (16) and (17) can then be equalized in order to arrive at

$$\dot{p}_{th} = 2\,G\,v\,\left[1 + \left(\frac{\alpha^2 - 1}{\alpha^2}\right) v\right]^{-1/2} \frac{1}{\alpha^3}\,\frac{d\alpha}{dt}\,, \quad (31)$$

the solution of which is readily found to be

$$p_{th} - p_o = 2\,G\,\left[\,\frac{\{(1 + v)\,\alpha^2 - v\}^{-1/2}}{\alpha} - 1\right] \quad (32)$$

under the condition $(p_{th} = p_o)$ for $(\alpha = 1)$. Then by first substitutiing eq. (7) for the time t figuring in eq. (30) and thereafter inserting the result into eq. (32), the ageing solution is

$$p_{th} - p_o = \frac{2\,G_o}{[\,1 + A_1 A_2]^\beta}\,.\,\left\{\frac{[(1 + v)\,\alpha^2 - v]^{1/2}}{\alpha} - 1\right\} \quad (33)$$

where, once again, the constants C_{21} and C_{31} are defined through the set of eqs. (3), (5), (6), (8) and (9), A1 and A2 are given in eq. (28) and v is found from eq. (12).

Figure (2) illustrates the main aspects of the pressure variation function defined in eq. (33). Implicit to figure (2) are the following parametric values:

Table 1. Values used in establishing the pressure variation at the thaw front around a wellbore (figure (2)).

a	= 0,122	m		T_s	= - 3,5°	C
C_f	= 1.41	kJ / Kg C		T_w	= 10°	C
C_{th}	= 1.50	kJ / Kg C		W	= 0,286	
G_o	= 400	MPa		W_u	= 0,150	
K_{th}	= 0,0017	kJ / m C s		ρ_w	= 1000*	
Q	= 33800	kJ / m³		ρ_s	= 2700*	
S_r	= 0,98			γ_f	= 1940*	
T_o	= - 2,1°	C		γ_{th}	= 2040*	

* Kg / m³

Figure (2) indicates dramatic pressure drops at the thaw front when the latter is in the immediate vicinity of the wellbore. However, in time, the creep response tends to repressurize the thaw front interface. The delay and magnitude of the creep response is greatly influenced by the value of the β parameter. The lower the β value, the less susceptible the frozen soil is to creep. In the limit, creep is absent for (β = 0). Higher β values in effect tend to minimize the initial decay in pressure and accelerate the repressurization process at the thaw front.

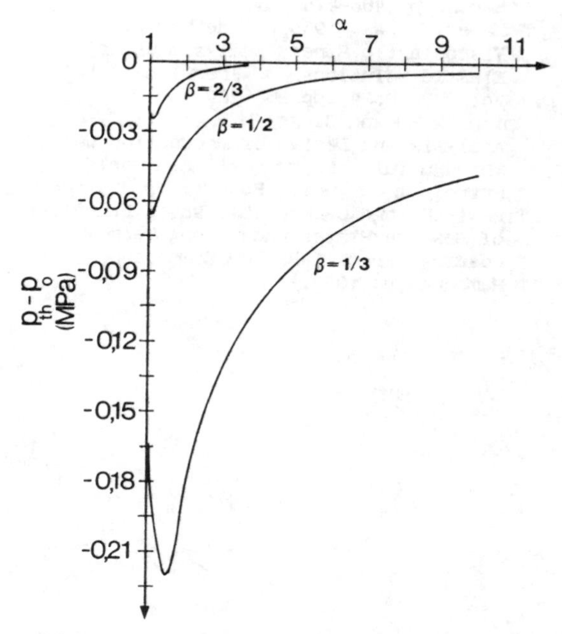

Figure 2: Pressure drop at the thaw front based on the ageing theory for creep.

6. CONCLUSION

The radial stress variation at the moving thaw front around a wellbore is shown to constitute a Stefan-type free boundary problem whose solution may only be attempted numerically.

By reverting to the ageing theory, an analytical solution was nevertheless arrived at which illustrates well the main aspects of the problem, specifically, an elastic response of the frozen soil which leads to stress relaxation, in contrast with the creep closure response which brings about repressurization.

7. REFERENCES

Carslaw, H.D. and Jaeger, J.C. (1959), Conduction of Heat in Solids, 2nd ed., Oxford University Press, Oxford, England

Goodman, M.A. (1978), Handbook of Arctic Well Completions, reprinted from "World Oil", Gulf Publ. Co., 52 pp.

Koch, R.D. (1971), The Design of Alaskan North Slope Production Wells, Proc. Instn. Mech. Engrs., Vol. 185, 73/71, pp. 989-1002

Ladanyi, B. (1985), Creep Closure of Frozen Ground around a Thawing Soil Column, Proc. of the Colloquium on "Free Boundary Problems: Applications and theory", Bossavit, A., Damlamian, A. and Frémond M. (eds.), Pitman Advanced Publishing Program, Vol. IV, pp. 347-354

Lin, C.J. and Wheeler, J.D. (1978), Simulation of Permafrost Thaw Behaviour at Prudhoe Bay, J. of Petroleum Tech., March, pp. 461-467

Merriam, R., Wechsler, A., Boorman, R. and Davies, B. (1975), Insulated Hot Oil-Producing Wells in Permafrost, J. of Petroleum Tech., March, pp. 357-365

Mitchell, D.E., Curtis, D.D., Laut, S.W., Pui, N.K. and Burgess, A.S. (1983), Well Casing Strains due to Perma+rost Thaw Subsidence in the Canadian Beaufort Sea, Proc. 4th Int. Conf. on Permafrost, Fairbanks, Nat. Acad. Press, Washington, Vol. 1, pp. 855-860

Mitchell, R.F. (1977), A Mechanical Model for Permafrost Thaw Subsidence, J. of Pressure Vessel Tech., February, pp. 183-186

Mitchell, R.F. and Goodman, M.A. (1978), Permafrost Thaw Subsidence Casing Design, J. of Petroleum Tech., March, pp. 455-460

Ruedrich, R.A., Perkins, T.K., Rochon, J.A. and Christman S.A., (1978), Casing Strain resulting from Thawing of Prudhoe Bay Permafrost, J. of Petroleum Tech.,

March, pp. 468-474

Schapery, R.A. (1965), A Method of Viscoelastic Stress Analysis using Elastic Solutions, J. Franklin Inst., Vol. 279, No. 4, pp. 268-289

Smith, R.E. and Clegg, M.W. (1971), Analysis and Design of Production Wells through Thick Permafrost, 8th World Petroleum Congress, Moscow, pp. 379-388

Tinawi, R. and Gagnon, L. (1984), Behaviour of Sea Ice Plates under Long Term Loading, Proc. IAHR Ice Symposium, Hamburg, pp. 103-112

5th International Symposium on Ground Freezing, Jones & Holden (eds)
© 1988 Balkema, Rotterdam. ISBN 90 6191 824 3

Tunnel displacements under freezing and thawing conditions

H.Meißner
Institute for Soil Mechanics and Foundation Engineering, University of Kaiserslautern, FR Germany

ABSTRACT: In tunnel construction the artificial freezing process is a well known practice. To predict underground movements numerical computations are performed. In this approach dimensionless equations are obtained to determine both the convergence of the tunnel and the settlement curves. Comparisons with measured displacements, obtained by model tests are presented. Furthermore, the effects of thawing on displacements and stress redistributions are shown.

1. INTRODUCTION

Under certain conditions the artificial freezing process in tunnel construction presents an economical method. During its application, attention must be given to ground movements differing from those using other driving methods. A reciprocal interaction, from the initial heading state onwards, exists between the frozen shell and the subsoil. After excavation ground movements arise due to creep of the frozen soil. The tunnel is driven in a saturated sand with an assumed fixed density. In this paper results of parametric studies are dealt with to predict both tunnel convergences and settlements of the ground surface. Some results are obtained by model tests in the laboratory.

This paper extends earlier investigations Meißner, 1985) by varying the tunnel dimensions as well as applying a surface load. For the parameter studies the finite element method is used. The material model for unfrozen soil is of the plastic type; for frozen soil, an elastic viscoplastic law is introduced. Both models are described elsewhere (Meißner, 1985). Therefore, in this paper the expressions of the two laws are dealt with in the necessary range to recognize the physical meaning of the models. Finally, in numerical analysis thawing of the frost shell is simulated. Underground movements are presented after both the placement of the lining and the thawing of the soil.

2. NUMERICAL MODEL

An underground section of the considered tunnel as well as the used discretization through a finite element mesh is presented in Fig.1. Only plane deformation problems will be treated.
The ranges of the different model parameters are chosen as follows:

Tunnel radius: $2\ m \leq r_a \leq 8\ m$
Overburden heigth: $12\ m \leq l_m \leq 28\ m$
wall thickness: $0{,}75\ m \leq d \leq 4\ m$
Surface load: $0 \leq p_o \leq 655$ KPa
Creep time: $0 \leq t \leq 1000$ h
Average temperature in the freezing pipe circle: $-35°C \leq T_m \leq -10°C$.

In all calculations the two distances l_u and b remain constant that is

$$l_u = 40\ m$$
$$\text{and } b = 40\ m$$

The excavation is simulated by

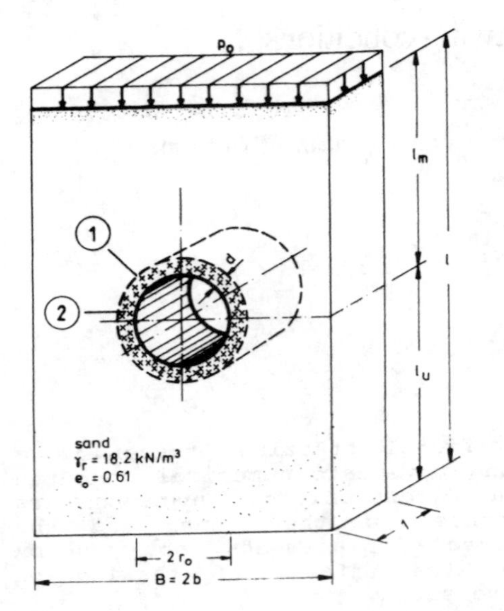

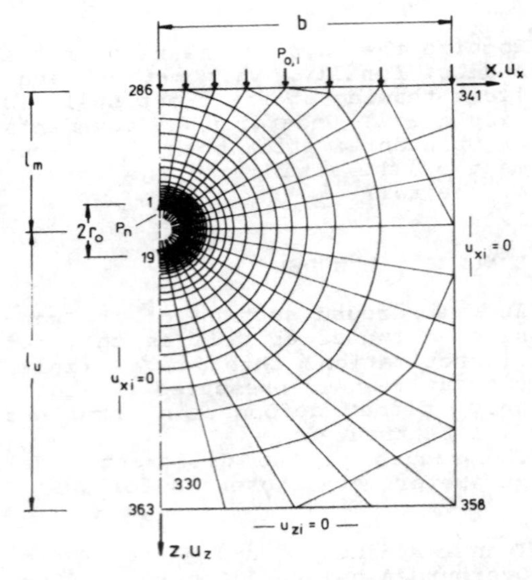

Fig. 1 Physical and numerical model of the tunnel.

stepwise decrease of the balance loads P_n at the tunnel surface, Fig. 1. For computation purposes compressive stresses and contractive strains are negative. Notations used in this paper are defined as follows:
Deviator components of both the stress tensor and the strain tensor:

$$s_{ij} = \sigma_{ij} - I_\sigma \delta_{ij}/3$$

$$de_{ij} = d\varepsilon_{ij} - dI_\varepsilon \delta_{ij}/3$$

where σ_{ij} is the Kronecker delta, while I_σ and I_ε, are first invariants of the corresponding tensors. Invariants of the tensors:

$$II_{ep} = e^p_{ij}\, e^p_{ji} \quad \text{(plastic components)}$$

$$I_\sigma = \sigma_{ii}\ ;\ I_a = I_\sigma/p_a,$$

p_a: atmospheric pressure

$$II_s = s_{ij}\, s_{ji}$$

$$III_s = s_{il}\, s_{lj}\, s_{ji}.$$

The sand has the initial void ratio e_o.

To describe the displacements of the frozen shell we introduce a dimensionless parameter δ through the equation

$$\delta = (u_{z19} - u_{z1})/r_o$$

where u_{z1} and u_{z19} are the vertical displacements of the marked points in Fig. 1. The parameter δ is denoted as the relative convergence.

The plastic material model used for unfrozen sand truly describes the stress strain behaviour of granular soils as investigated in biaxial and triaxial tests (Meißner, 1983). Softening at large strains, as well as changes of soil stiffness for curved stress paths from the at rest earth pressure, are considered. To describe both the elastic and the consolidation volumetric strains a variable bulk modulus K is introduced.
The corresponding shear modulus G is a function of the two values I_a and e_o. The yield function of the plastic model reads

$$f = II_s^{1/2} - A\, I_\sigma \left(1 - \frac{B}{6^{1/2}} \cos 3\alpha_\sigma\right)^{-m} \quad (1)$$

where A,B and m are parameters determined from sand samples subjected to triaxial test conditions. To determine the direction of the deviatoric plastic strain increments, the plastic potential

$$g = II_s^{1/2} - \chi\,(1 + 0{,}47 \cos 3\alpha_\sigma)^{-0,2} \tag{2}$$

is introduced, where χ is found by equating g to zero. Besides, dilatancy strains are presented by a volumetric flow rule.

The total incremental strain for frozen soil yields

$$d\varepsilon = d\varepsilon_e + d\varepsilon_{vp}$$

where $d\varepsilon_e$ is the elastic part and $d\varepsilon_{vp}$ the viscoplastic part. In good agreement with test results the volumetric strains are assumed to be zero. A characteristic strain rate-time curve of loaded samples is presented in Fig. 2. The time is denoted as t.

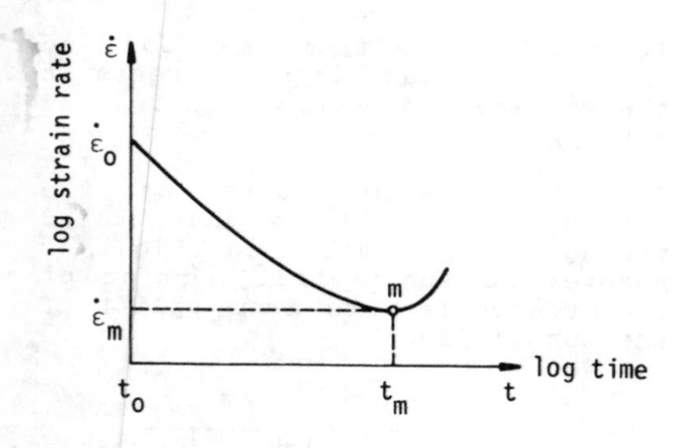

Fig. 2 Strain rate-time curve

The point m in Fig. 2 coincides with the point of inflexion of the corresponding creep curve.

The basic equation for creep strains is given by:

$$II_{\dot{e}}^{1/2} = II_{\dot{e}m}^{1/2} \cdot \exp(\zeta - 1)/\zeta \tag{3}$$

where the parameter ζ is defined as

$$\zeta = t/t_m$$

and t_m is the time at point m in Fig. 2.

From frozen soil samples subjected to triaxial tests we find after an approximation of test results:

$$t_m = \theta^{3,55}(II_s^{1/2}/I_a)^{-8,3}(650 I_a)^{-5,64} \ [h] \tag{4}$$

and

$$II_{\dot{e}m}^{1/2} = \theta^{-4,4}(II_s^{1/2}/I_a)^{11} \cdot (45{,}5\ I_a)^{7} \ [\%/h] \tag{5}$$

where $\theta = T/T_{abs}$ and $T_{abs} = -273\,°C$.

Equation (3) gives the second invariant of incremental viscoplastic strains

$$\Delta II_{e,vp}^{1/2} = II_{\dot{e}}^{1/2} \cdot \Delta t$$

Using g from Eq. (2) as potential forincremental deviatoric viscoplastic strains we get

$$dv_{ij} = \frac{\partial g}{\partial s_{ij}} - \frac{\partial g}{\partial s_{kk}} \cdot \frac{\delta_{ij}}{3}$$

and

$$\Delta e_{ij}^{vp} = \frac{II_{e,vp}^{1/2}}{|dv_{ij}|} \ dv_{ij}$$

The quasi elastic behaviour of the frozen soil may be described by the two parameters G and ν. From test results we find that the shear modulus G is expressed as follows (for $G \geqslant 5$ MPa)

$$G = -12{,}2\ p_a\,(20 + 1300\,\theta + \frac{II_s^{1/2}}{p_a}) \cdot (1{,}63 - \frac{II_s^{1/2}}{I_a}) \tag{6}$$

In the presentation of numerical results we will use G for $T = T_m$ and $II_s^{1/2} = 0$. Then Eq. (6) gives

$$G = 2000\ (20 + 11{,}35\,|T_m|) \tag{7}$$

The corresponding Poissons ratio may be expressed by

$$\nu = 0{,}45\ (0{,}1 + \frac{II_s^{1/2}}{I_a})^{0,2} \tag{8}$$

In good agreement with test results the creep rate is assumed to be only a function of the actual stress state.

3. NUMERICAL RESULTS

Using the finite element method the effect of distinct parameters on ground movements was studied. In the computations the distribution of the temperature T_r in the frozen shell is assumed to be

$$T_r = T_m \frac{\ln r/r_k}{\ln r_m/r_k}, \quad r_k = \begin{cases} r_o & \text{for } r<r_m \\ r_a & \text{for } r>r_m \end{cases}$$

where r_m is the radius of the freezing pipe circle and r_a the radius of the outer freezing front. The initial stress state in computations is the at rest earth pressure. Within 15 steps the balance loads at the tunnel surface are decreased to zero. The creep time is chosen to be about 800 hours. It should be noted that in all computations, the reference strain state is the state after the ground surface is subjected to p_o. In this study we use a certain medium sand. The initial void ratio remains constant. Therefore, the general expression for ground movements may be simplified and presented in dimensionless form as

$$u/l_m = f(r_o/l_m, d\gamma_s/G, \theta, p_o/G, \tau) \quad (9)$$

where $\tau = t/t_1$ and $t_1 = 1$ h, γ_s: unit weight of grains, G: shear modulus of the frozen soil in Eq. (7). f is a function which has to be determined by approximation of numerical results.

3.1 Tunnel Convergences

The relative convergences are defined by Eq. 1. The total value of δ is obtained through the following two steps:

1. determination of the convergence δ_1. The ground surface is subjected to p_o, the time parameter yields $\tau = 1$.

2. The time is increasing and the convergences become

$$\delta = \delta_1 \, \tau^m \quad (10)$$

In Figs.7 and 8 relative convergences obtained from parameter studies are plotted versus the relative overburden load.

We obtain a good qualitative approximation of numerical results by the following expression

$$\delta = 1,2\cdot 10^4 \frac{l_m \gamma_s}{G} \left(\frac{r_o\theta}{d}\right)^{0,5} \cdot \Big[1 + 0,37 \frac{p_o}{\gamma_s l_m} \theta^{0,3} \left(\frac{G}{\gamma_s l_m}\right)^{0,1} \left(\frac{d}{r_o\theta}\right)^{0,5} \exp\left(0,2\frac{r_o}{d}\right)\Big] \cdot \tau^m \quad [\text{o/oo}] \quad (11)$$

where the exponent m depends on the different parameters according to the following.

$$m = 0,2\left(\frac{l_m\gamma_s}{G}\right)^{0,5} \left(\frac{r_o}{d}\right)^{0,2} \exp\left(290 \frac{p_o r_o}{d\,G}\right) \quad (12)$$

It can be seen from Eqs (11) and (12) that a significant increase in the values of δ occurs with increasing p_o.

The displacement curve of the frozen shell wall is illustrated through an example in Fig. 3. However, a substantial portion of convergence results from uplift of the tunnel floor.

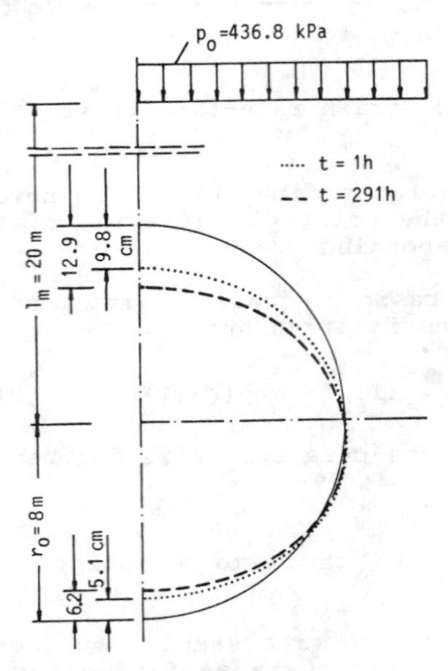

Fig. 3 Deformed frozen shell, T_m = - 20°C

For the same cases as those taken in Fig. 3 the distribution of tangential stresses in three cross sections of the frozen wall are illustrated in Fig. 4. Consequently, at the crown and the floor of the tunnel changes in stress distribution are not significant during the creep phase. On the other hand a remarkable redistribution of stresses from the frozen shell to the unfrozen sand occurs at the sidewall.

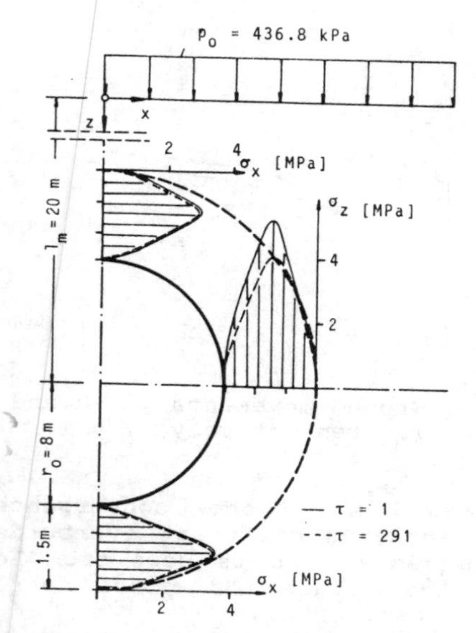

Fig. 4 Stress distribution in the frozen shell

3.2 Settlements of the ground surface

The concept to determine the function of the settlement curves is based on five distinct steps:

1 determination of the settlements s_{1o} of point 1 in Fig. 5. The overburden pressure p_o is zero, the time yields t = 1h.

2 The ground surface is subjected to p_o. The settlements $s_{1p} = s_1 - s_{1o}$ are determined, where s_1 are total settlements. The time is t = 1h.

3 The settlements in point 1 at the times t > 1 h are approached by a function. Then, s_1 is denoted by s_{1t}.

4 The settlements s_{2t} at a distance of 3 r_o from point 1 are expressed as function of s_{1t}.

5 Similar to a formula proposed by Kany (1974) the settlement curves are functions of s_{1t} and s_{2t}.

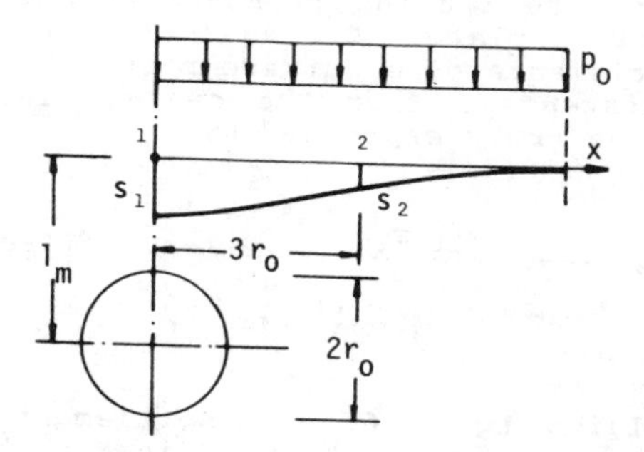

Fig. 5 Notations for the settlement curves

A detailed presentation to derive the functions is given elsewhere (Meißner, 1988). Approximation of numerical results yielded the following expression for point 1 in Fig. 5.

$$\frac{s_{1t}}{l_m} = 5,7\cdot 10^{-6}\left(\frac{r_o}{d}\right)^{0,25}\left(\frac{r_o}{l_m}\right)^2 \theta^{-2/3} \cdot \left(\frac{l_m}{\kappa}\right)^{1,6}\cdot\left[1+4,39\cdot 10^6\frac{p_o\,\theta}{G}\cdot\left(\frac{l_m}{\kappa}\right)\right]^{-1,6}\cdot \tau^m \quad (13)$$

in which

$$\kappa = \frac{G}{2,73\cdot 10^5\,\gamma_s\,\theta}$$

Again, m is found as a function of the distinct parameters. The approach of the computation results gives:

$$m = 10^{-5}\frac{r_o l_m}{\theta\, d^2}\left(1+1,5\cdot 10^6\frac{p_o\,\theta}{G}\right) \quad (14)$$

The settlements at point 2 in Fig. 5 are given by the expression

$$s_{2t} = s_{1t} \exp(0,3 \frac{p_o}{\gamma_s l_m} - 16,3 \frac{r_o}{l_m} (\frac{d}{l_m})^{0,5}) \qquad (15)$$

Finally, the settlement curves may be presented by both settlements s_{1t} and s_{2t}. Therefore, a similar expression is intro duced as Kany (1974) used it in his model to describe the bearing pressure beneath a base plate. Consequently, the settlements of a surface point i at a distance x from the center line may be truly expressed by

$$s_{it} = \frac{s_{2t}\, s_{1t}}{s_{2t} + (s_{1t} - s_{2t}) (\frac{x}{3r_o})^2} \qquad (16)$$

Applying Eq. (16) the settlements at the two points 1 and 2 in Fig. 5 are s_{1t} ans s_{2t} respectively.

The dimensionless expression (11) until (15) are those of the ground movements for a wide class of tunnels with frozen shells. It is to be noted, that the expressions are valid for saturated medium sand with a fixed and known density.

4. EFFECTS OF PARAMETERS ON GROUND MOVEMENTS

In a given tunnel project the two parameters, the thickness of the frozen shell and the freezing temperature may be altered. Therefore, to gain some insight into the effectiveness of these two parameters, displacements are determined for a given tunnel project. In Fig. 6 both settlements and convergences are plotted versus d and T_m respectively. With increasing values of d as well as θ the rate of displacements decreases. It is interesting to note that the convergences yield about twice to four times the maximum value of the settlement curves. The presented settlement curve is obtained for T_m = - 20°C and d = 1,5 m.

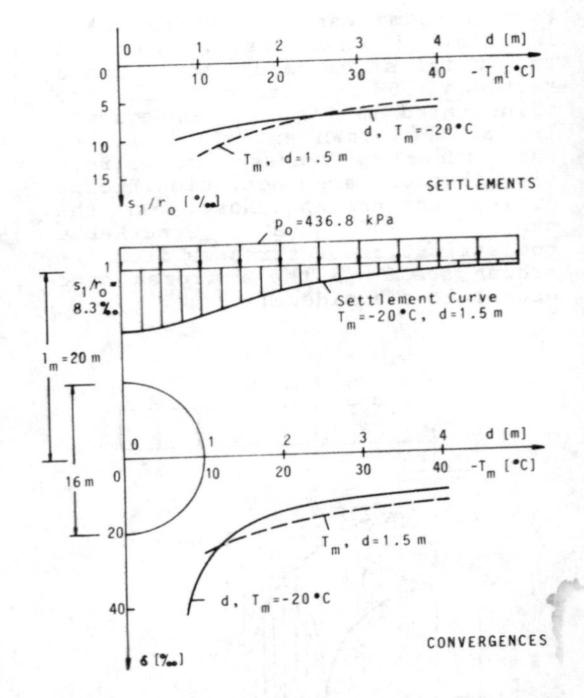

Fig. 6 Ground movements vs. d and T_m, respectively, τ = 1

Figures 7 and 8 show convergences when applying different overburden loads. As it can be seen from Eq. (11) the values for δ are nearly independent of l_m.

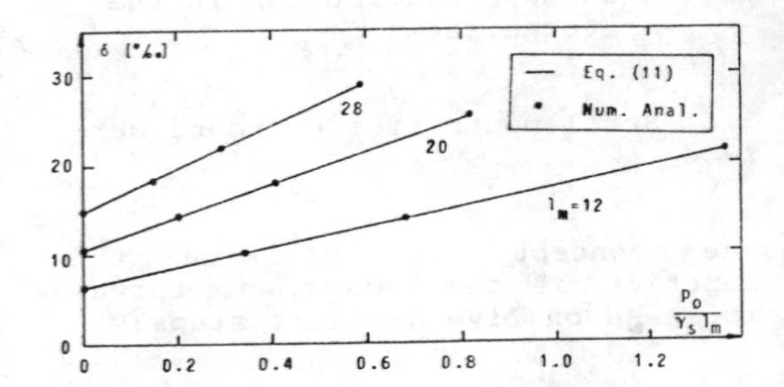

Fig. 7 Convergences vs. relative overburden load, τ = 1, T_m= -10°C, r_o/d = 5,33.

The convergences are linear functions of the overburden load. Fig. 7 shows the effects of varying the values of l_m. while Fig. 8 shows those of r_o/d on δ.

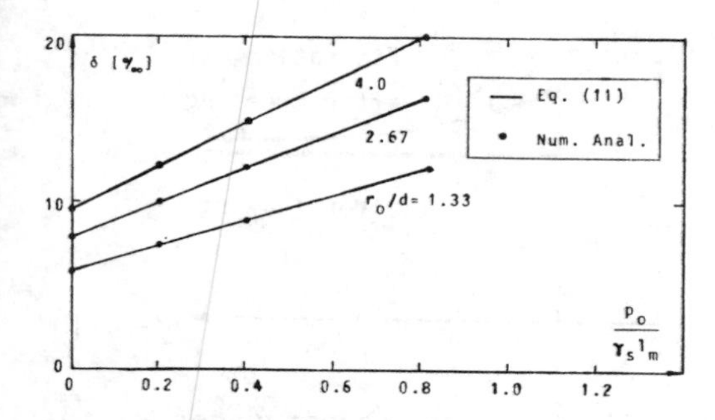

Fig. 8 Convergences vs. relative overburden load, T_m = - 10°C, l_m = 20 m, τ = 1.

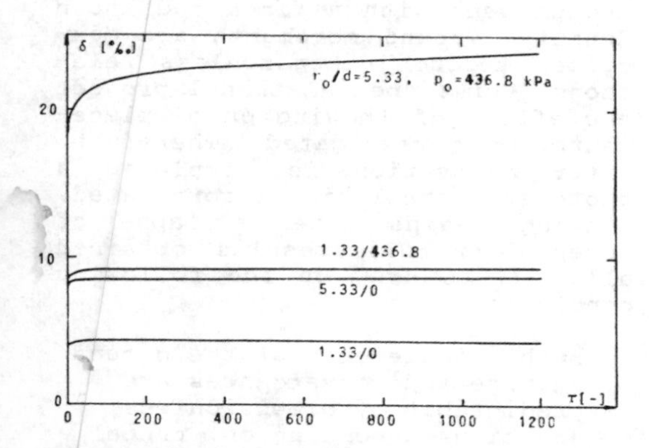

Fig. 9 Time dependence of convergences, l_m = 20 m, T_m = -20°C

Finally, the magnitudes of some creep convergences are presented in Fig. 9. The increase in the duration of the excavation process leads to a remarkable increase in convergences for both large diameters and high overburden loads.

5. MODEL TESTS

In order to check the reliability of the numerical model, tests on instrumented tunnel models were performed. The test process is reported elsewhere (Orth and Meißner, 1985) but the values of the parameters used were as follows:

$0,45 \text{ m} \leqslant l_m \leqslant 0,60 \text{ m}$, r_o = 15 cm, d = 10 cm, $T_m \approx -$ 2°C to - 3°C
$0 \leqslant p_o \leqslant 250$ KPa.

In numerical analysis the same dimensions are applied as those in

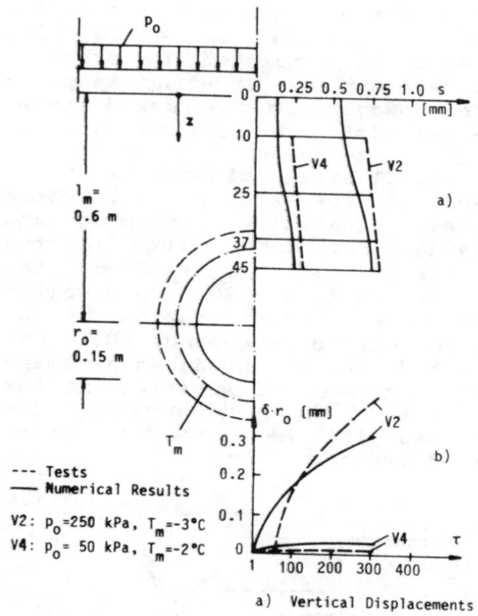

Fig. 10 Displacements from both tests and numerical analysis

the tests. However, it should be noted that in model tests the surface is subjected to a rigid load plate. Figure 10 illustrates the soil displacements above the tunnel crown as well as the convergences beginning from the excavation state. During the tests convergences were observed from this state onward. Comparison revealed a good agreement between empirical and numerical results. Thus, in well defined conditions, the numerical model is a reliable tool for describing the bearing behaviour of frozen shells in tunnel driving processes.

6. THAWING OF THE FROZEN SHELL

The effects of thawing on the ground movements is studied through an example. After excavation the creeping phase lasts for about seven days. Thereafter, a thin shotcrete shell is provided to the system. The material parameters for shotcrete are chosen as follows:

E = 5000 MPa, ν = 0,3

The ultimate compressive stress for shotcrete yields σ_V = 20 MPa. Thawing is simulated by stepwise increase of the freezing temperature. However, in the thawed state, the material law for sand must govern.
In Fig. 11 stress distributions are shown at the end of all three stages: excavation, creeping and thawing. For both values of the overburden load, p_o = 0 and the temperature t_m = - 20°C no significant stress redistribution is obtained during creeping. But, as expected, thawing produces a remarkable increase of stresses in the shotcrete shell, whereas the stresses in the formerly frozen soil decrease.

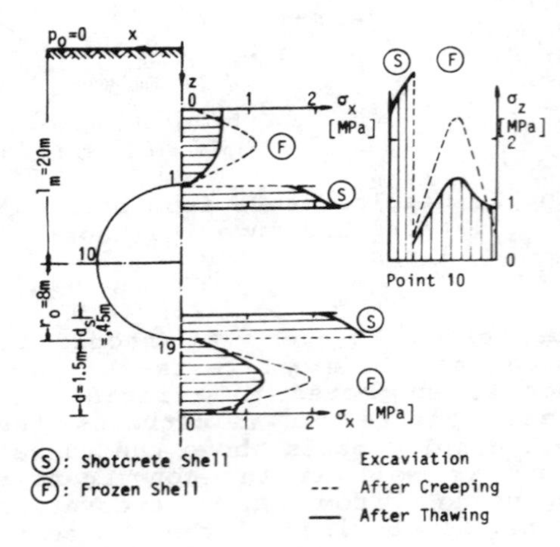

Fig. 11 Stress distribution at the end of the stages: excavation, creeping, thawing

The effects of different shotcrete thicknesses on convergences are illustrated in Fig. 12. For shotcrete thickness less than d=10 cm severe increase in convergences is indicated.

7. CONCLUSION

Results of a study on ground movements due to tunnel driving by means of frozen shell protection are presented herein. The Underground consists of a saturated

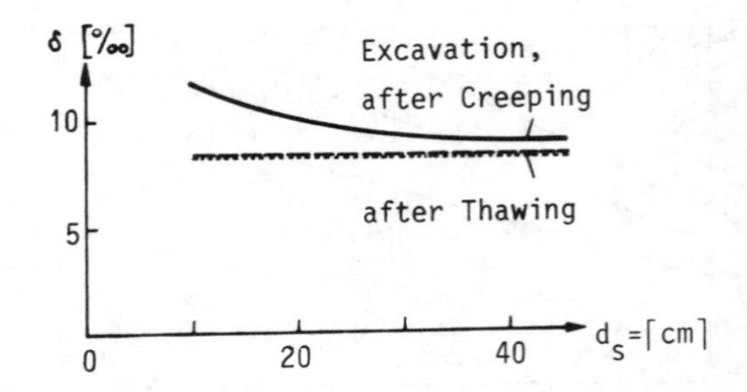

Fig. 12 Relative convergences vs. wallthickness d_s of the shotcrete shell.

medium sand with a fixed and known density. Ground movements are described through dimensionless relations. For a special tunnel project the effect of thawing on displacements is investigated. Therefore, after excavation is completed a shotcrete shell is incorporated. Thawing begins after a lapse of seven days. The results obtained allow us to draw up the following conclusions:

1 Both, settlements of the ground surface and convergences are predictable by dimensionless equations depending on: tunnel dimensions, temperature, time and overburden load.

2 For special tunnel projects the thickness of the frozen shell and the freezing temperature may be optimized without affecting the restrictions on the settlements.

3 The minimum thickness of the frozen shell depends on the distances of the freezing pipes. For thin frozen shells, large tunnel diameters and high loads acting on the ground surface rupture of the tunnel arises according to numerical studies.

4 Parameter studies will be carried out for different sand densities. Furthermore, criteria for dimensioning the shotcrete shell are developed.

8. ACKNOWLEDGEMENT

The investigations presented in this paper were supported by the Deutsche Forschungsgemeinschaft (DFG), Bonn, which is gratefully appreciated. Computations are performed both on the Univac 1100 of the University of Karlsruhe and the Siemens 2000 of the University of Kaiserslautern, West Germany.

REFERENCES*

Kany, M., 1974
Berechnung von Flächengründungen, Berlin: Ernst & Sohn

Meissner, H., 1983
Tragverhalten axial oder horizontal belasteter Bohrpfähle in körnigen Böden. Veröffentl. des Inst. fur Bodenmechanik und Felsmechanik Karlsruhe, Heft. 93.

Meißner, H., 1985
Bearing behaviour of frost shells in the construction of tunnels. Sapporo: Fourth Int.Symp.o.Ground Freezing, Vol. 2, P. 37-45

Meißner, H., 1988
Ground movements by tunnel driving in the protection of frozen shells. 6th Int.Conf.on Num.Meth. in Geomech., Innsbruck

Orth, W., Meißner, H., 1985
Experimental and numerical investigations for frozen tunnel shells. Sapporo: Fourth Int.Symp.o.Ground Freezing, Vol. 2.P. 259-262.

*See page X concerning ISGF papers

5th International Symposium on Ground Freezing, Jones & Holden (eds)
© 1988 Balkema, Rotterdam. ISBN 90 6191 824 3

Some results of in-situ measurement of freezing pressure and earth pressure in frozen shafts

Su Lifan
Beijing Research Institute of Mine Construction, CCMRI, Beijing, People's Republic of China

ABSTRACT: Artificial ground freezing has largely been applied to water-bearing overburden in China. Some results of in-situ measurements are discussed in this paper. The cause of freezing pressure and effect of soil character, the type of lining and the depth on freezing pressure have been analysed. Two curves of the increment and three forms of distribution along a section of freezing pressure are given. An empirical formula for estimating freezing pressure in clay soil is presented. In the paper, the different mechanism of earth pressure in sand and in clay is discussed according to the measured data, and that practical values of earth pressure were less than that given by P=0.013H is pointed out. Using statistical methods and probability theory, a standard value of earth pressure and hydrostatic pressure for design is proposed.
Keywords: deformation pressure, frost heave, freezing pressure, soil pressure, earth pressure

1 INTRODUCTION

In China, freezing method is largely applied in water-bearing sediments of the Quaternary Period. More than 280 freezing shafts with the total depth of 44km have been constructed so far. The maximum frozen depth was 415m with 358.5m in the Quaternary Period. When the thickness of clay strata exceeds 6 to 8 m and its depth is more than 120-150m, the outer lining may be damaged and fracture of freezing pipes may occur. Most linings in freezing shafts were composed of outer and inner lining. The inner was only made of reinforced concrete with the maximum thickness 1.1m. The outer lining was made of precast block, or reinforced concrete with compressible material seam, or compound construction, or only reinforced concrete. A flexible lining with asphalt was set up in 1985.

In order to investigate the problems encounted in design and construction of the deep freezing shafts in the Quaternary Period, in-situ measurement has been performed since 1964. Vibrating wire gauge, containing pressure cell, reinforcement strees gauge, concrete strain gauge and other probes were applied. The investigation involved freezing pressure, earth pressure, displacement of side wall, stress and strain of the lining and temperature. This paper dicusses some results of the freezing pressure and earth pressure on the lining.

2 THE RESULTS OF FREEZING PRESSURE MEASUREMENTS

Based on the measurements, the load on the lining could be divided into four stages as shown in Fig.1.

The term ''freezing pressure', which isn't the sole result of the frost heave of soil, refers to the contact pressure between the lining and the icewall during the construction of freezing shaft. The freezing pressures during stage I (construction of outer lining/), stage II (construction of inner lining), stage III (after turning off the refrigeration plant) are shown in Fig.1; stage IV is the earth pressure.

2.1 The causes of freezing pressure

The measurement has shown that there mainly were two causes of freezing pressure, the deformation of strata-icewall system after sinking and frost heave of soil.

As a rule, digging the frozen shaft immediately give rise to stress redistribution and displacement of strata-icewall

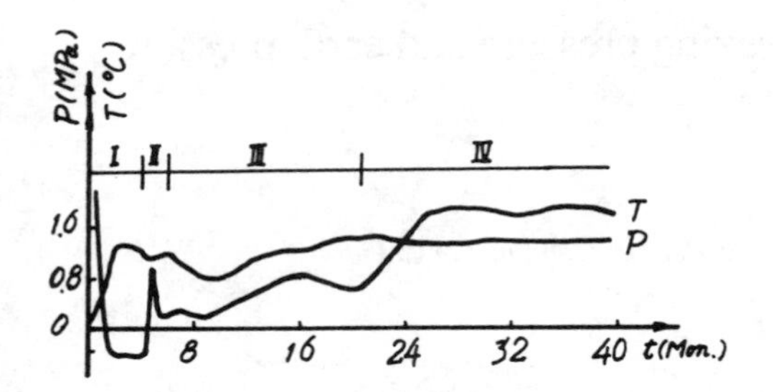

Fig.1 The curve of the load on the lining in frozen shaft (from main shaft of Xing Long Zhuang, H=142m, sand)

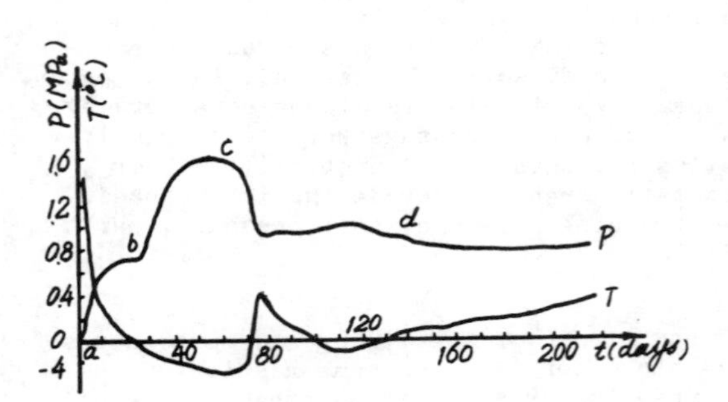

Fig.2 The curve of freezing pressure on the outer lining of reinforced concrete (from fan shaft of Ban Dian, H=142m, clay)

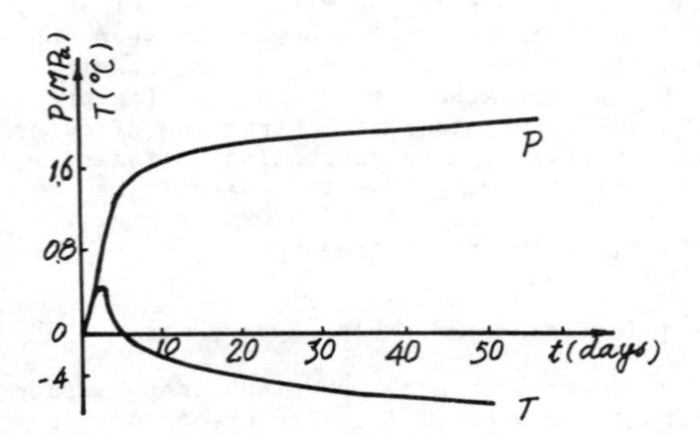

Fig.3 The curve of freezing pressure on the outer lining of precast block (from main shaft of Hai Zi, H=166m, clay)

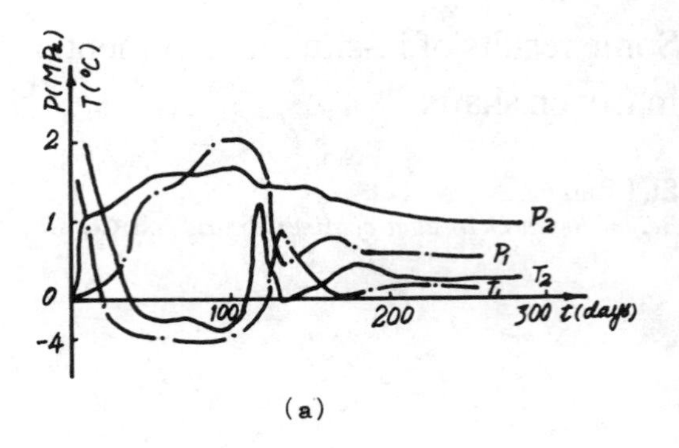

(a)

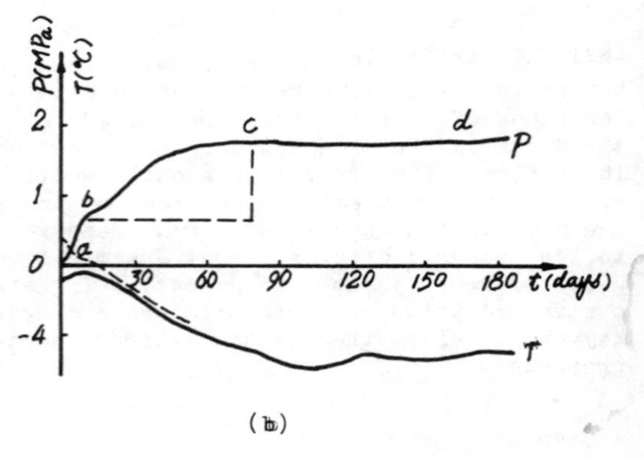

(b)

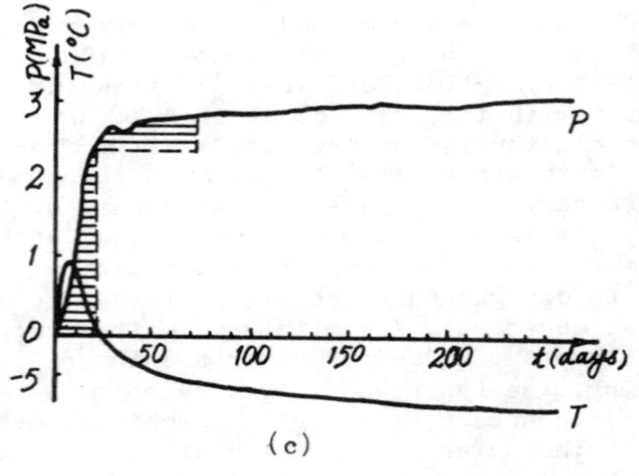

(c)

Fig.4 Two patterns of the curve of the frezing pressure
(a) The comparison between two patterns of the curve (The curve P_1, T_1 from main shaft of Xing Long Zhuang, H=157m, clayey sand; The curve P_2, T_2 from main shaft of Xing Tai, H=165m, clay)
(b) The curve of slowly increasing freezing pressure (from fan shaft of Pan San, H=152m, sand)
(c) The curve of rapidly increasing freezing pressre (from fan shaft of Pan San, H=323m, clay)

system. The process is related to depth, original condition of strata and icewall, creep of icewall, the location of face and the sinking speed. After the outer lining is set up, the boundary of the system will be changed, resulting in the interruption of displacement of the sidewall and then producing a pressure on the lining. In fact, no matter how hard the icewall is, the pressure on the lining will always occur and increase continously. This is a sort of 'deformation pressure'. When reinforced concrete was used, a thawing soil seam, 200 to 300mm in thickness, would be foumed behind the outer lining. After the release of hydration heat of concrete was over, the thawing soil would be frozen again, resulting in frost heave of soil and then producing a frost heave pressure acting on the lining.

Fig.2 gives the curve of the freezing pressure on the lining made of reinforced concrete. In Fig.2, part ab is the deformation pressure. The tempreture curve in Fig.2 illustrate the frozen soil behind the lining was being thawed in the period. Part bc is the frost heave pressure after the thawing soil was frozen again. Part cd is the pressure after the inner lining was set up. In the period, the soil behind the lining was thawed and frozen again.

Fig.3 gives the pressure curve of precast block lining. In this case, there was no frost heave pressure corresponding to the part bc in Fig.2, because the frozen soil behind the lining hadn't been thawed.

2.2 Two patterns of freezing pressure

Using the measured data, the freezing pressure can be classified into two patterns, the slowly increasing and rapidly increasing pattern as curve a and b in Fig.4.

If the icewall had high strength and stability, and the outer lining had certain flexibility, the freezing pressure would increase slowly, producing a curve of slowly increasing pattern. In this case, the amount of pressure during the initial 5-7 days after the lining was set up, was approximately 30-40 percent of the maximum pressure. As soon as the thawing soil behind the lining was frozen again and the flexibility of the lining disappeared, the pressure increased faster. The effect of frost heave of soil on the curve of freezing pressure was appearent.

In the second case, the pressure increases rapidly, exceeding 0.3--0.4 Mpa per day. The amount of pressure exceeded 60--70 percent of the maximum in 5--7 days after the outer lining set up. The deformation pressure was the main part. In this case, the outer lining of reinforced concrete was often damaged due to the low strength of concrete.

Tne curve of slowly incresing pressure occured in sand soil and the rapidly increasing pressure occured in clay soil, especially in calcareous heave clay. The damage to the outer lining in frozen shaft took place in clay in more than 90 percent of the cases.

2.3 The forms of distribution of the freezing pressure along a lining section

The distribution of the freezing pressure along a section is closely related to the properties of the soil.

The distribution of the freezing pressure and displacement of the sidewall along a section located in the same soil (clay soil) is given in Fig.5.

We can find out that the freezing pressure on the lining is inversely proportional to the displacement of sidewall before the outer lining sets up. Therefore, the form of distribution is conical. The maximum pressure is located at the foot of the section, which was confirmed by the fact that the damage to the lining in a section always began at this part.

When there were different soils in a section, the form of distribution of the pressure along the section became bilinear as Fig.6.In Fig.6, the upper soil was clay and the lower soil was sand. Though their depth was almost same in the section and the sand soil was located at the foot, the pressure in the sand was less than that in the upper clay. This fact illustrated that the amount of freezing pressure at the same depth was mainly dependant on the properties of soil. According to the above, three forms of the distribution of freezing pressure along a section have been obtained as shown in Fig.7. It is apparent that the lining will be subject to damage easily when the form of pressure is similar to Fig.7(c).

2.4 The estimation of the freezing pressure in clay

The experience and measurements have shown that the damage of the lining occured mainly in clay. It is essential to determine the freezing pressure in clay layer for reasonable design. But, as shown above many factors, such as depth, property of soil, the condition of icewall, can influence the freezing pressure. It is difficult to calculate it by means of a simple theoretical formula. Now, an empirical formula

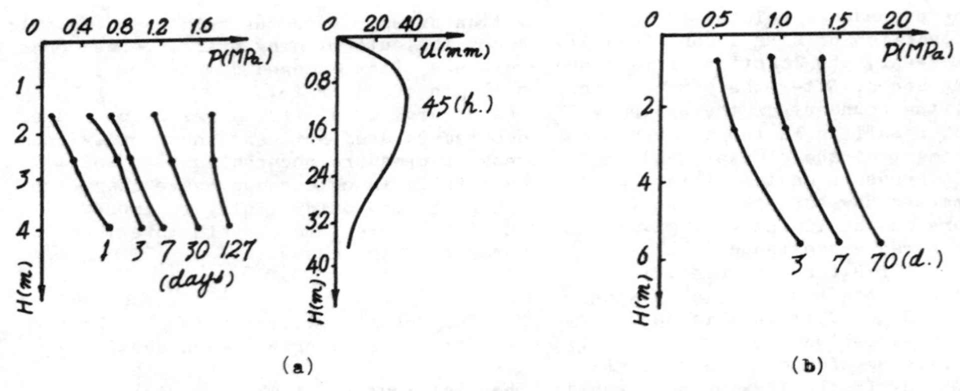

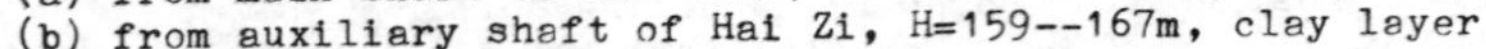

Fig.5 The forms of freezing pressure along a section in clay layer
(a) from main shaft of Lin Huan, H=126.6--133.8m, clay layer
(b) from auxiliary shaft of Hai Zi, H=159--167m, clay layer

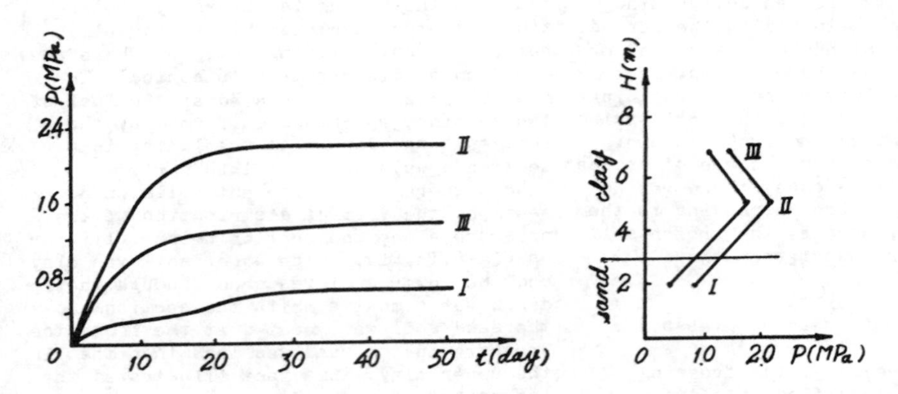

Fig.6 The forms of freezing pressure along a section in diefferent soil layers (from fan shaft of Hai Zi, H=203--212m)

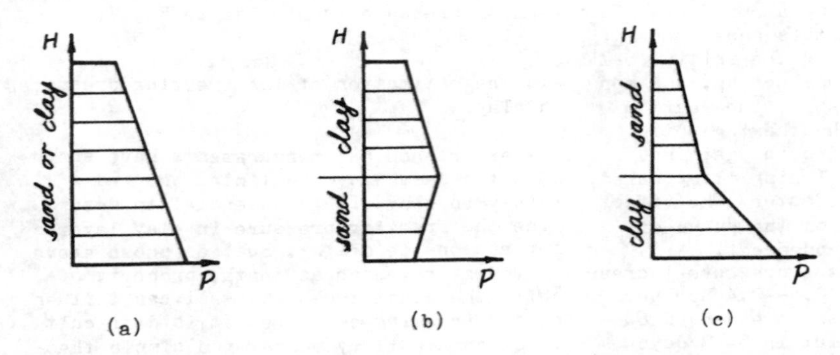

Fig.7 Three forms of the distribution of freezing pressure along a section

for determining the maximum freezing pressure P has been developed based on the measured data as follows:

$$P=K(1.36\log H-1.22) \qquad H\leqslant 100m$$
$$P=K(0.007H+0.8) \qquad H\geqslant 100m \qquad (1)$$

H--calculated depth (m): K--empirical coefficient associated with property of soil, sinking diameter etc. K=1--1.2.

When the outer lining is made of reinforced concrete, the strength of concrete at 5--7 day after the outer lining set up should be examined. At that time, 60 to 70 percent of the pressure determined by formula (1) can be applied as the load on the lining so that the increasing strength of concrete can be checked against the increment of the frezing pressure.

3 THE RESULTS OF EARTH PRESSURE MEASUREMENTS

In general, earth pressure P of shaft in water-bearing strata contains two parts, the soil pressure P_s and hydrostatic pressure P_w

$$P=P_s+P_w \qquad (2)$$

where $P_w=\gamma H$, γ--unit weight of water: H--height of water table. In the aquiclude, e.g. thick clay stratum, P_w should be zero.

Before the 1970's earth pressure for designing linings was calculated by П.М.Цимбаревич formula. On the assumption that average unit weight of soil was 2t/m^3, the average earth pressure would be about 0.0077H(Mpa) in aquiclude and about 0.015--0.017H in aquiferous sand. Since 1973, the following formula has been applied to calculate earth pressure

$$P=0.01H+0.003H=0.013H(MPa) \qquad (3)$$

where P_s=0.003H, soil pressure and P_w=0.01H, hydrostatic pressure, which indicates that earth pressure increases with depth linearly.The other formulas such as K. Terzaghi,В.Г.Березанцев, Н.С.Бульцев, which considered space effects require that the soil pressure on shaft should tend to a limiting value. So,it should be smaller in a deep shaft. That those formulas are founded on the limit equilibrium theory should be pointed out.

3.1 The measured data of the earth pressure

Table 1 and Table 2 gave the data of earth pressure measured in sand and clay layers. The Tables show that hydrostatic P_w' and the total pressure P could be measured out in sand layers, which indicated that the earth pressure in sand layer was comprised from hydrostatic and soil pressure. In the aquiclude, there only was the total pressure which was equal to the soil pressure and no hydrostatic pressure.

In Table 1, the soil pressure: $P_s=P-P_w'$; K=P/H; $\Phi_w=P_w'/P_w$, that is a coefficient characterizing the ratio between measured and theoretical hydrostatic pressure. In the Table 2, Kn= P/H is a coefficient characterizing the ratio of measured pressure to the depth.

3.2 Discussion of the measured data

The measured values of the earth pressure and the line given by P=0.013H are marked out in Fig.8. From here we see that all the measured values were less than that calculated by (3), which indicated that the lining designed by (3) should be safe.

The data also show that the measured values of hydrostaic pressure were less than the theoretical calculation. That was because water seepage of the lining and drainage resulted in the formation of a cone of depression around the shaft. The curves of hydrostatic pressure in the strata near the sidewall and at the interface between the outer and the inner lining are given in Fig.9. Two curves almost coincide, illustrating that the outer lining was waterlogged.

The measured results have indicated that the mechanism of earth pressure in water-bearing sand layer is different from that in clay aquiclude. In sand layers, the earth pressure resulte largely from hydrostatic pressure (soil pressure only was equal to about 13 percent of the total pressure), which illustratesthat the soil pressure in sand layers was consistent with the limit equilibrium theory qualitatively. In aquiclude of clay, the soil pressure calculated by the limit equilibrium theory should be smaller, e.g. equal to 0.0077H by means of П.М.Цимбаревич formula. But the measured earth pressure was a considerably higher value (see Table 2). Therefore, the mechanism of earth pressure in aquiclude is at variance with the limit equilibrium theory. In fact, the cause of the soil pressure in clay is complicated, and may result from deformation of strata, heave of soil and be related to character of the lining. It is difficult to determine the earth pressure in clay by a simple theoretical formula.

3.3 The standard value of earth pressure based on in-situ measured data

The earth pressure on a shaft may be considered to be a random varible. Hence, its standard value for design may be determined by statistical methods. Practically, a simple formula calculating earth pressure may be applied as follow

$$P=\psi H \qquad (4)$$

ψ--coefficient determined by the measured data.

The relevant statistics we derived from the follow formula

mean value $\bar{x}=\frac{1}{n}\sum_{i=1}^{n}x_i$

standard deviation $\sigma_{n-1}=\left[\sum_{i=1}^{n}(x_i-\bar{x})^2/(n-1)\right]^{1/2}$

devistion coefficient $\nu=\sigma_{n-1}/\bar{x}$

where x_i--the measured data; n--number of data points.

The statistics concerning K, Kn and ϕ_w have been derived from the data in Table 1 and Table 2, and shown in Table 3.

According to probability theory, the standard value for design under a allowable guarantee probability p is obtained on the follow formula

$$x=\bar{x}+t\sigma$$

where t--coefficient of guarantee probability, which is related to p one by one, e.g. t=1.645, p=95%; t=2, p=97.5%; t=3, p=99.9%, when the random varible is subject to normal distribution.

Now, the standard value of K, Kn, Kw for designing the lining in associated with different guarantee probability p have been obtained in Table 4. In general, the guarantee probability may assume p=95%, and then the standard value of the earth pressure for design would be

P=0.01107H in sand soil

P=0.01161H in clay soil (MPa)

and the hydrostastic pressure would be

P=0.00953H (MPa)

CONCLUSION

The paper outline some of results of in-situ measurement of freezing and earth pressure. Two patterns of the increment of the freezing pressure and three forms of its distribution along a section have been given. An empirical formula estimating maximum freezing pressure in clay layer

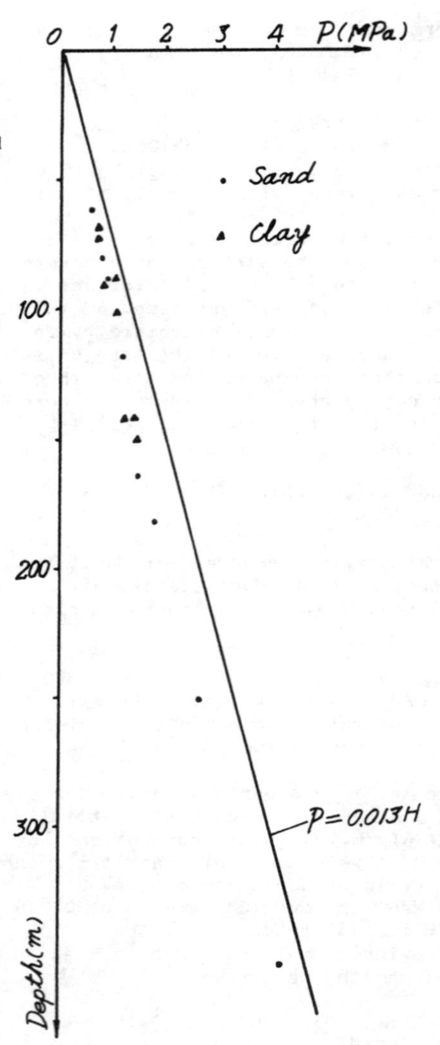

Fig.8 The measured values of earth pressure

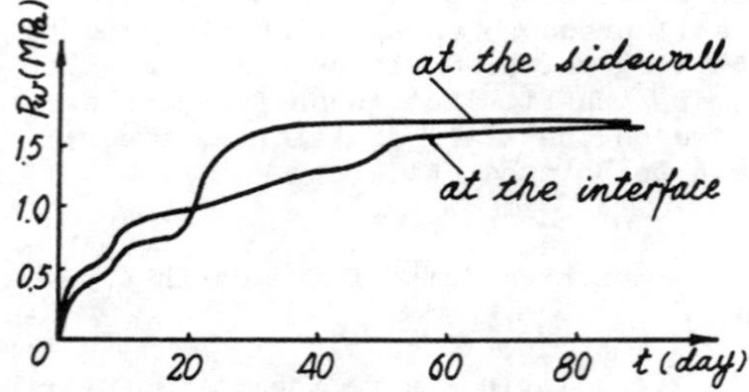

Fig.9 The curve of hytrostatic pressure (from fan shaft of Bao Dian, H=164.5m, sand)

Table 1 The measured data earth and hydrostatic pressure in sand layer and the relevant coefficient

Name of Mine	sinking diameter(m)	depth (m)	P (MPa)	Kx10 (MPa/m)	P_w' (MPa)	Φ_w (%)	P_s (MPa)	P_s/P (%)
A	4.4	61	0.65	0.92	0.46	81	0.10	18.0
A	4.4	80	0.77	0.96	0.65	85	0.12	15.6
A	4.4	88	0.83	0.94	0.75	89	0.08	9.6
B	4.35	118	1.16	0.98	0.96	84	0.20	17.2
C	3.4	164	1.45	0.89	1.40	86	0.06	4.0
D	4.65	192	1.73	0.901	--	--	--	--
D	4.65	250	2.55	1.02	2.32	92.8	0.23	9.9
D	5.3	329	--	--	2.72	82.9	--	--
D	5.05	356	4.06	1.14	3.26	91.6	0.80	19.7

A--Hong Yang; B--Xing Long Zhuang; C--Bao Dian; D--Pan San

Table 2 The mensured data of earth pressure and coefficient kn in clay layer

Name of Mine	A	A	B	A	C	B	C	B
diameter(m)	4.4	4.4	4.35	4.4	3.4	4.35	3.4	4.35
depth(m)	67.0	74.0	88.0	90.0	103.0	142.0	142.5	151.0
P(MPa)	0.72	0.71	1.06	0.80	1.01	1.39	1.25	1.45
Knx10(MPa/m)	1.07	0.96	1.20	0.88	0.98	0.98	0.88	0.96

Table 3 The statistical data of K, Kn, Φ_w

	K	Kn	Φ_w
$\bar{x}$	0.00969	0.00989	0.8650
σ	0.00812	0.01050	0.04207
ν	0.0838	0.1062	0.0486

Table 4 The standard values of K, Kn, Kw

P(%)	95.0	97.7	99.0
K	0.01107	0.0113	0.0116
Kn	0.0161	0.01198	0.01236
Kw	0.00935	0.00950	0.00964

(MPa/m)

is presented. The main cause of freezing pressure is analysed, Regarding earth pressure, the paper discuss the different mechanism of sand soil and clay aquiclude upon the measured data. An important result, that all the measured earth pressure is less than that calculated by P=0.013H(MPa), is gained. Finaly, the standard value of earth pressure and hydrostatic pressure for design are presented according to statistical method and probability theory.

REFERENCES

Цимбаревиц, П.М. 1951. Рудничное Крепление, Углемехиздат.

Terzaghi, K. 1953. Theoretical Soil Mechanics. Wiley.

Березанцев, В.Г. 1952. Осесимметричная Задача Теории Предельного Равновесия Сыпучей Среды, ГИТТЛ.

Бульчев, Н.С. 1982. Механика Подземных Сооружений Москва, "Нкдра"

Merritt, F.S. 1970. Applied Mathematics In Engineering Practice, McGraw-Hill.

5th International Symposium on Ground Freezing, Jones & Holden (eds)
© 1988 Balkema, Rotterdam. ISBN 90 6191 824 3

The cause and prevention of freezetube breakage

Wang Zhengting
Central Coal Mining Research Institute, Hepingli, Beijing, People's Republic of China

ABSTRACT: This paper describes the situation of freezetube breakage in shaft sinking using the freezing method, the parts easily broken,& the relative position between the broken freeztube and sinking section of the shaft. The main reason, which was analysed for the increasing freezetube beakage in Huainan Coal Mining Area, is the low strength of ice wall in clay and the large plastic deformation under deep ground pressure. In order to prevent freezetube breakage at the main shaft in the Xieqiao Colliery the freezing was strengthened and the height of the sinking section was reduced. This increased the strengthof ice wall and reduced the forces applied on the freezetube resulting in the disappearance of the freezetube beakage.
KEYWORDS: freezetubes, insitu testing, deformation

1 THE SITUATION OF FREEZETUBE BREAKAGE

In loose and soft water-bearing ground the freezing method is adopted to sink many shafts. Freezetube breakage is the biggest problem for the construction. Light breakage influences the advance of shaft sinking and increases the engineering cost. Heavy breakage endangers the safety in construction and may even flood the shaft resulting in heavy economic losses.

Since China adopted the freezing method in 1955, there have been 58 shafts, in which accidents of freezetube breakage happened in various degrees. 294 freezetubes broke in total and 5 freezetubes per shaft in average. From 1977 to 1984 in Huainan Coal Mining Area there were 10 shafts in which 140 freezetubes broke, which is more than a half of the total in China. The number of freezetube breakages was increasing gradually and the problem was getting more and more serious, until the auxiliary shaft of the Xieqiao Colliery was flooded due to freezetube breakage.

2 THE BROKEN PARTS IN A FREEZETUBE

A freezetube consists of several seamless steel tubes, which are connected by threads or welding. Breakages normally occur at the connections. The broken shape has two types as following.

Splits occur at the edge of welding seam. The main shaft in the Gequan Colliery of Xingtai Coal Mine used steel tubes with 127mm diameter and 7.5mm wall thickness. Welding connection was still used for installing the freezetubes. When the shaft was sunk to 104m, freezetube breakage was found. After digging out and checking them, a ringlike split was found at the top end of the tube fitting on welding seam which was welded at installing time. See Fig.1.

Strips of the taper thread occur. The service shaft in the Xieqiao Colliery of Huainan Coal Mining Area used 5A-J55 type steel tube with 139.7mm diameter and 7.72mm wall thickness. Taper thread connection was used. No.21 freezetube deviated into the excavated diameter of the shaft at depth of 220m. After digging out and chedking the tube, it was found that the freezetube had bending deformation, but no breakage and the thread connection was stripped. See Fig.2 and Fig.3.

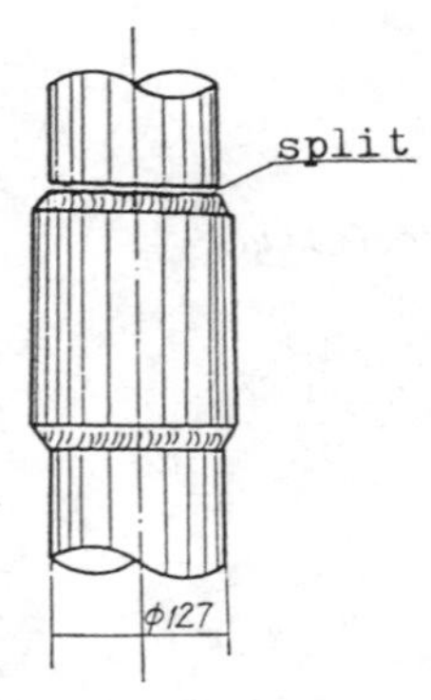

Fig.1 The split at the edge of welding seam on freezetube

Fig.2 Bending deformation and strip of the thread connection on freezetube

3 RELATIVE POSITION BETWEEN BROKEN PARTS AND SINKING SECTION OF THE SHAFT

After freezetube breakages happened at the East air shaft of the Pansan Colliery, the service shaft and the auxilliary shaft of Xieqiao Colliery the relation of the depth along the shaft between the broken parts of the freezetube and sinking section of the shaft was investigated and analysed. The results of the investigation for these three shafts is as following.

Fig.3 Strip of the thread connection on freezetube

The East air shaft of the Pansan Colliery was 6.5m in finished diameter and 415m in freezing depth. The freezetubes were 5A-J55 type steel tube with taper thread connection. Quarternary clay is from 252m to 328m along the shaft depth. When the shaft was sunk through this clay, freezetube breakage happened 4 times in a total of 22 tubes. During sinking to 325m 16 freezetubes were broken at same time. According to observation the locations of the broken parts were not in the sinking section but were in the supported section. The maximum height from broken part to the bottom of the shaft wall was 269.9m.

The service shaft of the Xieqiao Colliery was 6.6m in finished diameter and 330m in freezing depth. In April of 1984 the shaft was sunk to depth of 224.3m in clay. The brine leaked out to the working face. 10 broken freezetubes were checked and breakage continued of 33 broken freezetubes, the position of 5 brokon tubes which is 15% of the total was in the sinking section, the position of 4 broken tubes which is 12% was under the working face and the position of 24 broken tubes which is 73% was in the supported section.

The auxilliary shaft of the Xieqiao Colliery was 8m in finished diameter and 360m in freezing depth. The thickness of the Quarternary alluvium, which was loose and soft water-bearing unstable strata including quick sand, clay and so on, was 299.3 m. Below 226m there were two thick strata of clay (Fig.4). The freezing method using freezetubes on two concentric circles was adopted. The distance between the two circles was 1.5m. The diameter of inner circle was 17m. In order to form the designed ice wall thickness as quick as possible, two circles were frozen simultaneously to the whole depth. Freezing began on February 24, 1984. Shaft sinking began on August 1. When the shaft was sunk to depth of 230.6m in clay on December 13, at the evaporator of the refrigeration plant it was found that 40 cubic meters of brine had leaked away. After checking, 5 broken freezetubes were found. Brine was not found on the working face and shaft sinking continued. In order to prevent freezetube breakage again the height of the sinking section was reduced from 4.3m to 3.3m. When the shaft was sunk to a depth of 239.6m near a quick sand stratum on Dec. 21, the working face was bulging and brine was leaking. Crevices appeared and continued to enlarge in the northeast and northwest part of ice wall. Water and sand were flowing into working face at an increasing rate. According to measurement water and sand rose 3.6m in 8 minutes. The shaft was flooded. By analysing it was found that the broken parts of the freezetubes located under the working face. After the 8 days, under the clay stratum the ice wall in sand melted resulting in the accident. After flood and checking 27 more freezetubes were broken.

Of 59 freezetubes in the above three shafts which broke, the position of 5 broken tubes which is 9% of the total was in sinking section, the position of 45 broken tubes which is 76% was in the supported section, and the position of 9 broken tubes which is 15% was under working face (Table 1).

4 ANALYSING THE CAUSE OF FREEZETUBE BREAKAGE

In the East air shaft of the Pansan Colliery, service shaft and auxilliery shaft of the Xieqiao Colliery the freezetube breakage happened while the shaft was sunk through deep and thick clay stratum. Before breakage some abnormal phenomena appeared in varying degrees, such as bulging of working face, breakage of hydrologic tube, large displacement of the ice wall, breakage of shaft lining and so on (Fig.5). In the auxilliary shaft of the Xieqiao Colliery when the working face stopped sinking for 24 hours the maximum height of bulging was 850mm. The maximum displacement of the ice wall was 25mm in 6 hours. The lower part of every shaft lining section split and some broken parts dropped down. From the above phenomena the plastic deformation of ice wall can be divided into three parts: deformation under working face, deformation in sinking section and deformation in supported section. The plastic deformation causes the freezetubes to move towards the center of the shaft (Fig. 5). When displacement exceeds the allowable deformation of the freezetube breakage happens. This is the main reason for freezetube breakage. The height of the sinking section, material properties of freezetube, strength of tube connection and the strength of shaft lining influence the freezetube breakage. During shaft freezing and sinking the freezetubes carry tension, pressure, shearing and bending loads. Analysing the data in Table 1 when the broken position is located in the sinking section, the freezetubes broke by bending. When the broken position is located in the supported section and under working face the freezetube broke by tension.

5 EFFECTIVE WAY OF PREVENTING FREEZETUBE BREAKAGE

Strengthening freezing is the first effective way of preventing freezetube breakage. The thickness of ice wall is normally calculated according to freezing strength of sand. In shaft sinking it is asked that frozen soil goes into the excavated shaft diameter the less the better, to keep a soft center for easy digging. The strength of the ice wall in clay is about one quarter of the strength in sand. Strengthening freezing increases the strength of ice wall in clay. In construction active freeing time is prolonged. The ice wall in clay develops into the excavated diameter, so that the whole shaft is completely frozen. The main shaft of the Xieqiao Col-

Stratum	Column	Thickness (m)	Accumulaten deptn (m)
			201.0
Sandy clay and sand		15.0	216.0
Sand		10.3	226.3
Clay		13.9	240.2
Clayey sand		1.7	241.9
Sand		2.1	244.0
Clayey sand		9.4	253.4
Clay		12.7	266.1
Sand		18.5	284.6
Clay and sand		14.7	299.3
Lower strata is rock			

a. Auxiliery shaft

Stratum	Column	Thickness (m)	Accumulaten deptn (m)
			210.6
Sandy clay		13.9	224.5
Clay		17.2	241.7
sandy clay		11.1	252.8
Clay		12.7	265.5
Sandy clay and sand		19.0	284.5
sandy clay		14.4	298.9
Lower strata is rock			

b. Main shaft

Fig.4 Geologic column in alluvium of the main and auxiliery shaft of the Xieqiao Colliery

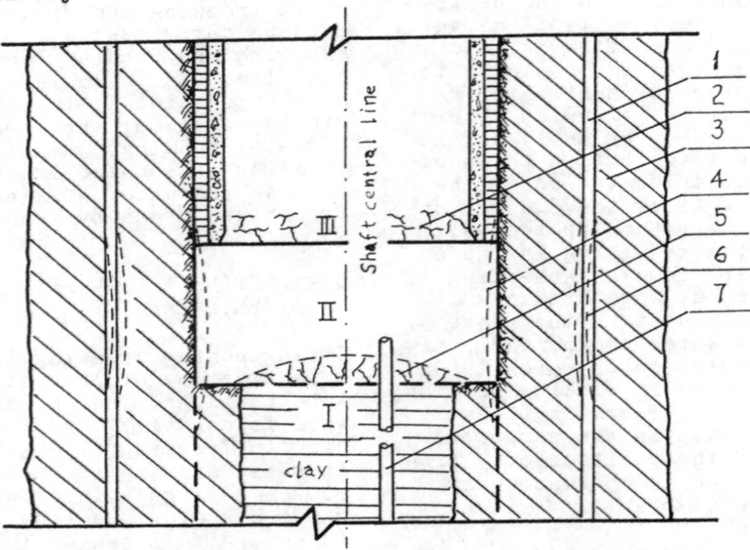

Fig.5 Deformation of ice wall and freezetube in clay stratum

I. Deformation under working face
II. Deformation in sinking section
III. Deformation in supported section
1. Freezetube
2. Crevices in shaft lining
3. Ice wall
4. Deformation line of ice wall
5. Shape of bulging
6. Deformation line of freezetube
7. Hydrologic tube

Table 1. Broken position of freezetubes

Name of shaft	Number of broken freezetubes	Broken position		
		In supported section	In sinking section	Under working face
1. East air shaft of Pansan Colliery	21	21		
2. Service shaft of Xieqiao Colliery	33	24	5	4
3. Auxiliery shaft of Xieqiao Colliery	5			5
4. Total	59	45	5	9
5. Percentage		76	9	15

liery was the last deep feezing shaft in Huainan Coal Mining Area. Its finished diameter was 7.2m. Thickness of alluvium was 298.9m (Fig.4). Freezing depth was 363m. The diameter of the circle of freezetubes was 17m.

Shaft freezing began on August 1, 1984 and shaft sinking began on December 5. When the shaft was sunk to depth of 200.2m on July 23, 1985, sinking stopped and two thick clay strata under the working face were frozen for 102 days. Shaft sinking was continued on November 1. When the shaft was sunk in clay under depth of 224.5m the whole shaft was completely frozen. The effective thickness of the ice wall was 6.3m. The average temperature was -15°C. The temperature of the shaft wall was -13°C. The soil temperature at shaft center was -1.5°C (Fig.6). The height of sinking section was 4.3m. The average moving speed of shaft wall was 0.83mm per hour, which was one fifth of the speed in the auxiliary shaft (Table 2). During the concreting of the shaft the maximum height of the ground swelling at the working face was 200mm. The average speed of swelling was about one fifth of the speed in the auxiliary shaft (Table 3). During construction there was no freezetube breakage, and the shaft was successfully sunk through the clay strata.

The main reason that the strengthening of freezing got such a good result was as follows. Because the freezing time was prolonged, the frozen soil developed to the shaft center, and the actual thickness of the ice wall in the next section and its load-bearing capacity were increased.

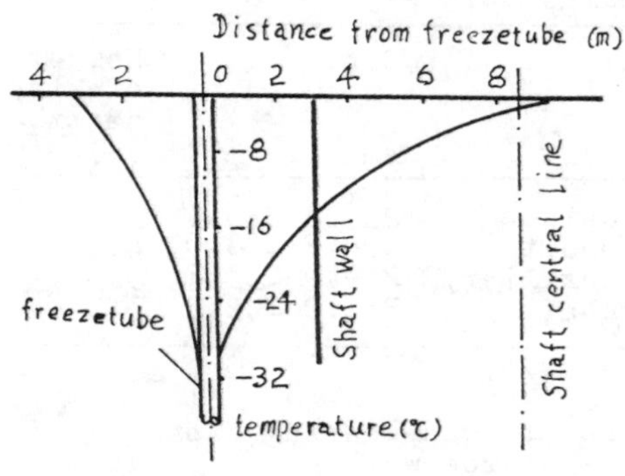

Fig.6 Extending of ice wall in clay at main shaft of Xieqiao Colliery

Considering the ice wall as a thick walled tube of infinite length according to the formula, its designed thickness at the main shaft of the Xieqiao Colliery was 4.76m. The average temperature was -9.5°C. At the shaft depth from 224.5m to 265.5m there were two thick clay strata. According to heavy liquid ground pressure formula:

$$P = 1.3\gamma H \qquad (1)$$

where γ: Density of water
H: Shaft depth
P: Ground pressure

the maximum ground pressure was 34.5 kg per sq cm.

The load-bearing capacity at the designed and actual ice wall thickness of the main shaft was calculated according to the following formula:

$$2.3\frac{P_0^2}{K^2} + 0.29\frac{P_0}{K} - \frac{E_0}{R_0} = 0 \qquad (2)$$

Table 2. Displacement of shaft wall at main and auxiliary shaft of Xieqiao Colliery

Shaft	Depth (m)	Height of sinking section (m)	Soil	Measuring time (hour)	Displacement (mm)	Moving speed (mm/h)
1. Main shaft	225.8-230.1	4.3	Clay	19	13	0.7
	262.8-267.1	4.3	Clay	36	30	0.8
2. Auxiliary shaft	231-	3.3	Clay	6	22	3.7
	237-	3.3	Clay	6	25	4.2

Table 3. Ground swelling of working face at main and auxiliary shaft of Xieqiao Colliery

Shaft	Depth (m)	Height of sinking section (m)	Soil	Measuring time (hour)	Height of ground swelling (mm)	Ground swelling speed (mm/h)
1. Main shaft	225.8-230.1	4.3	Clay	36	200	5.6
	254.2-258.5	4.3	Clay	36	200	5.6
2. Auxiliary shaft	229.9-233.2	3.3	Clay	30	750	25.0
	233.2-236.5	3.3	Clay	30	850	35.4

where P_o: Load-bearing capacity of ice wall
K: Unconfined compressive strength of frozen soil
E_o: Actual thickness of ice wall under the face
R_o: Radius of not frozen soil in shaft

The relation between the frozen soil thickness developing into excavated diameter and the load-bearing capacity is illustrated on Fig.7. From Fig.7 the bigger the frozen soil thickness developing into excavated diameter the bigger the load-bearing capacity of the ice wall. When the ice wall thickness was designed its load-bearing capacity was 8.8kg per sq cm, only one quarter of the ground pressure. When the frozen soil thickness developing into the excavated diameter was 4.6m, the load-bearing capacity of ice wall was equal to ground pressure. When the whole shaft was basically frozen the load-bearing capacity was bigger than ground pressure. Ground swelling at working face was small, so that the deformation under working face was eliminated.

Secondly, the effective thickness of the ice wall in the sinking section and its strength were increased. The load-bearing capacity was

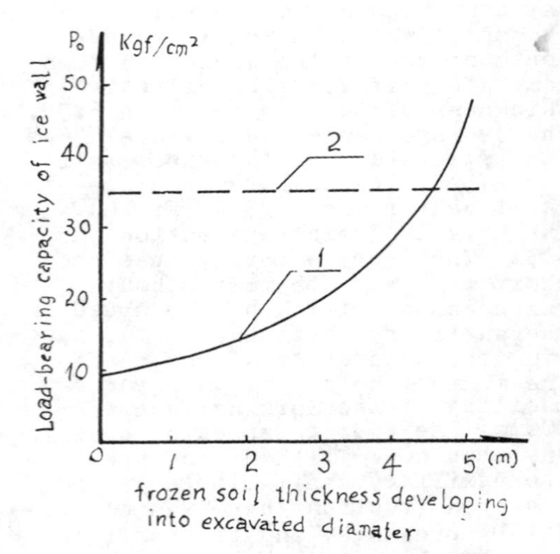

Fig.7 Relation between frozen soil thickness de eloping into excavated diameter and load-bearing capacity of ice wall
1. Relation curve
2. Maximum ground pressure in clay stratum at depth of 265.5m

calculated using the theory for a thick walled tube of finite length, the designed ice wall thickness and effective ice wall thickness.

When the designed ice wall thickness was calculated it was supposed that in the sinking section the upper end was fixed and the lower end was not fixed. The load-bearing capacity is:

$$P_1 = \frac{E_1 \sigma_1}{\sqrt{3}\, h} \qquad (3)$$

When the effective ice wall thickness after strengthening freezing was calculated, it was supposed that both ends were fixed. The load-bearing capacity is:

$$P_2 = \frac{2 E_2 \sigma_2}{\sqrt{3}\, h} \qquad (4)$$

where E_1: Designed ice wall thickness
E_2: Effective ice wall thickness
h: Height of sinking section
σ_1, σ_2: Long term compressive strength of frozen clay

According to formulas (3) and (4) the calculated results were arranged in Table 4. From Table 4, when the heights of the sinking section were the same, the load-bearing capacity of effective ice wall thickness was 4.1 times of that of the designed ice wall thickness and higher than the ground pressure. The deformation of the ice wall in the sinking section was controlled, after concreting the deformation was eliminated, so during shaft sinking through clay no freezetube breakage happened.

Cutting down the height of the sinking section is the second effective way of preventing freezetube breakage. From formulas (3) and (4) the height of the sinking section is inversely proportional to the load-bearing capacity of ice wall. It is reasonable to reduce the height of the sinking section. The height of sinking section in the main shaft and several shafts in Which freezetube breakage happened, were listed in Table 5. From Table 5, when the height of sinking section was equal to or higher than that of the main shaft in the Xieqiao Colliery, more freezetubes broke during the shaft sinking. This explained the height of the sinking section related to freezetube breakage.

6 CONCLUDING REMARKS

In clay strata at main shaft of the Xieqiao Colliery strengthening freezing and cutting down the height of sinking section were adopted. The strength of the ice wall was increased and the deformation of ice wall

Table 4. Load-bearing capacity of ice wall

Item	Designed thickness	Effective thickness
1. Thickness of ice wall (m)	4.76	6.30
2. Temperature of shaft wall (°C)	0	-13
3. Average temperature (°C)	-9.5	-15
4. Strength of frozen soil (kg/sq cm)	16.0	24.8
5. Height of sinking section (m)	4.3	4.3
6. Load-bearing capacity (kg/sq cm)	10.2	42.0

Table 5. Height of sinking section

Name of shaft	Height of sinking section (m)	Number of broken freezetube
1. Main shaft of Xieqiao Colliery	4.3	0
2. Auxiliary shaft of Xieqiao Colliery	3.3-4.3	5
3. Service shaft of Xieqiao Colliery	4.4-7.3	34
4. East air shaft of Pansan Colliery	4.3-8.4	22

was effectively controlled. The situation of the forces applied to the freezetube was improved. This provided a way for solving the problem of freezetube breakage Which had existed in Huainan Coal Mining Area for a long time.

In clay strata the thickness of ice wall below the face and the height of sinking section can be calculated according to formulas (2) and (4). But the height of sinking section must be controlled whithin 2 to 3 meters.

Strenthening freezing does not require an increase in the number of freezetubes. A single freezetube circle and a prolonging of the freezing time can properly be used. When the upper part of frozen soil develops to the position of shaft wall, sinking begins. When sinking to the clay stratum at large depth, the whole shaft is frozen and under this condition the shaft can be sunk safely and economically.

REFERENCES

Chen Wenbao 1986. The freezing sinking method. Mine construction engineering handbook, Vol.4. Beijing: The China Coal Industry Publishing House.
Chen Minghua 1986. The experience and moral for three shafts using freezing sinking method in Xieqiao Colliery. Coal Science and Technology 11.

5th International Symposium on Ground Freezing, Jones & Holden (eds)
© 1988 Balkema, Rotterdam. ISBN 90 6191 824 3

Structure and stress analysis of seepage resistant linings in shafts sunk with the freezing method

Yu Xiang
Central Coal Mining Research Institute, Beijing, People's Republic of China
Wang Changsheng
Beijing Research Institute of Mine Construction, CCMRI, Beijing, People's Republic of China

ABSTRACT: Ground freezing is a major method for sinking shafts through alluvium in China. To suit different ground conditions and to prevent water seepage various types of lining structure have been adopted. Based on a great number of data obtained in site observations and stress analysis, the paper concentrates on mechanism of both plastic sandwich and foamed plastic layer as a constituent part of shaft lining. It describes briefly the flexible lining and a new type of lining in China.

KEYWORDS: earth pressure, freezing pressure, gliding layer, lining,seepage, shafts

1 INTRODUCTION

Freezing method for sinking shaft has been introduced to China since 1955. Up to now, 280 shafts have been sunk or being sunk. The total length is in excess of 44 Km. Freezing method has already become a major special shaft sinking method in China, especially in North and East China (Table 1).

Table 1

Year	No. of shafts	Total freezing length m	Max. freezing depth m
1955-1959	17	1,473.9	162
1960-1969	45	5,077.2	330
1970-1979	129	22,048.3	415
1980-1986	89	15,519.6	363
Total	280	44,119	-

With the increase of thickness of Quarternary alluvium, through which the shafts pass, the structure of shaft lining has experienced a process developing from simple to complicated one (Fig1).

From 1955-1963 single wall structure (Fig.1,a) was used because of thin alluvium. Except brick masonry used in a few shafts, in-situ concrete or reinforced concrete were used. Heavy water seepage occurred in the shaft lining after thawing of ice wall.

Double-wall lining (Fig.1,d) began to be used in 1963.Starting from 1966, double wall lining with an intermediate plastic sheet(Fig.1,e) was employed. It was proved to be an effective method for controlling water seepage from the shaft lining.

Since 1978, another type of double wall lining was applied in Huainan and Huaibei mine areas in East China, as shown in Fig. 1,g. The outer wall is made of pre-cast concrete blocks and a plastic sheet is placed in between. This type of structure has some effect on preventing the ice wall and shaft lining from being damaged in thick clay layers, and on prevention of water seepage.

In 1983, compound bitum-steel-concrete flexible lining (Fig.1,h) was used in west air shaft of Kongji Colliery,Huainan, with the purpose to recover shaft pillars in the future.

In addition, to prevent water seepage, plastic sheets were used at the junctions of two sections of single wall (Fig.1,b). when the sinking and walling stage was high, shotcreting was applied as temporary support to control excessive deformation

Table 2. Examples of water seepage in shafts built in the 50s with freezing method

Name of shaft	Net dia. of shaft m	Thickness of overburden m	Quantity of water seepage m^3/hr
Linxi air shaft	5.0	60.5	100.0
Tangshan air shaft	5.0	148.0	75.0
Fangezhuang main shaft	5.5	82.7	67.0
Lüjiatuo main shaft	5.5	61.7	117.0
Xingtai main shaft	5.0	248.0	94.3

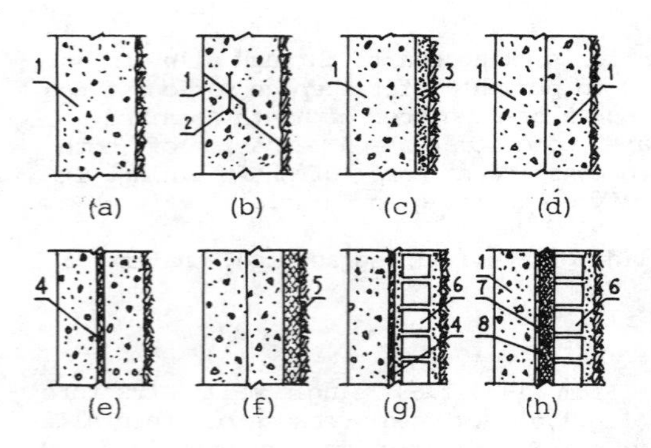

1 - concrete lining,
2 - water sealing plastic sheet,
3 - wire-mesh and shotcreting
4 - plastic plate,
5 - foamed plastic sheet,
6 - pre-cast blocks,
7 - steel plate,
8 - bitumen

Fig.1 Longitudinal section of various types of shaft lining structures

of ice wall (Fig.1,c).In the recent years, a layer of foamed plastic sheet was filled between the ice wall and the lining in some shafts (Fig.1,f). This can not only reduce the freezing pressure on lining, but also reduce the heat loss of the outer wall, so that the strength of concrete can go up rapidly.

In the 60s a number of measurements of external and internal forces exerted on the lining were made. Based on investigation, the mechanism of two structures of shaft lining,namely,the double wall lining and sandwich lining was understood. The research project on using foamed plastic sheet between ice wall and shaft lining gained prominent engineering results.

2 SHAFT LINING WITH PLASTIC SHEET AND ITS MECHANISM

In the initial period of sinking shafts with freezing method, many cracks were found in the lining, and junctions of sections of lining were not properly treated. Water seepage happened in most shafts when the ice wall was thawed.Water-make in some of the shafts reached scores of cubic meter per hour,even in excess of 100 cu m per hour (Table 2). Although double wall structure was applied, water seepage problem was not fundamentally solved.

In general, backfilling with grout was necessary, or even to add another wall inside the shaft, as a result, the effective cross section of the shaft was reduced.

In 1966, plastic sandwich lining, i. e. a sheet of polyvinyl-chloride was laid between the outer and inner reinforced concrete walls, was first adopted in an auxiliary shaft, Suoli Colliery, Huaibei, with satisfactory water-sealing result. The Quarternary alluvium was 61 m thick, and it contained 4 layers of running sand with a total thickness of 10 m. The calculated max. water- make was 108 m^3/hr, and the freezing depth was 67 m. The outer wall, 300 mm thick, was placed in shaft sinking; and then a sheet of 2 mm thick PVC was applied on the outer wall from shaft crib as a water-proofing layer; and then an inner wall was placed (Fig.2).

Cement grout was injected between plastic layer and concrete walls, so that the inner and outer walls can stand in common the forces as one solid body, and can cut off the water feeder in the void between two layers and seal off the water. Trials on nail ejector and binding method were made to bind the plastic sheet on the outer wall. Plastic sheets were connected by binding, welding and overlapping methods. By the end of the 70s and in the

Table 3. Examples of shaft lining with plastic layers

Name of shaft	Year	Net dia. of shaft m	Freezing depth m	No. of plastic layers
Suoli auxiliary shaft	1966	6.0	67	1
West air shaft of Panji No.2 mine	1978	6.5	327	2
Auxiliary shaft of Panji No.2 mine	1978	8.0	327	2
South air shaft of Panji No.2 mine	1979	6.5	320	2
Main shaft of Haizi mine	1979	6.5	285	1
Central air shaft of Haizi mine	1979	4.5	275	1
West air shaft of Haizi mine	1979	6.0	286	1
Main air shaft of Panji No.3 mine	1979	7.0	280	2
East air shaft of Panji No.3 mine	1981	6.5	415	1

early 80s, this kind of sandwich lining was extensively used in a dozen of shafts in Huaibei and Huainan mine areas. As a result, no water seepage was found(Table 3).

Most of the shafts mentioned in Table 3 passed through 300 m, or even thicker alluvium. These strata mainly consisted of clay, calcareous clay, water-bearing sand, consolidated sand and gravel layers.Of which, clay and calcareous clay layers constituted over 60% of the total thickness. Heavy deformation of ice wall or floor heave occurred in these strata when freezing method was applied. For example, at -250 m level a 77 m thick clay layer was found in east air shaft in Panji No.3 mine. Evident displacement up to 250 mm and serious floor heaving up to 1.3 m at the centre of shaft face happened in process of shaft sinking. Under such bad conditions to overcome the shortcoming of in-situ concrete, namely, low strength of lining in the initial period, which may be damaged by heavy freezing pressure, the thickness of the outer wall was increased to 700 to 900 mm. And pre-cast blocks, which have high strength and do not produce hydration heat, were used to build outer wall. Some of shafts even employed foamed plastic sheet to buffer the freezing pressure acting on the outer wall. With these improvements the sandwich lining has become perfect day by day and can accommodate different geological and operational conditions.

Plastic sandwich lining has solved the practical problem of water seepage in shafts. At the same time, different mechanism of sandwich lining and general double wall lining was understood based on a great number of field measurements and investigations. This is an important theoretical contribution.

We can take main shaft of Xinglongzhuang Colliery in Yanzhou mine area and eastern air shaft of Panji No.3 Colliery as examples to compare the mechanism of double wall lining and plastic sandwich lining. The observation results of stresses on the shaft lining at certain level of two shafts are shown respectively in Fig.3 and Fig.4.

The net diameter of Xinglongzhuang Colliery main shaft is 6.5m; the thickness of Quarternary alluvium, 189 m; freezing depth, 220 m. The stress curve at level of -88 m shown in Fig.3 illustrates that the longitudinal reinforcements in the inner wall were in tension for long periods, and the maximum stress reached 75 N/mm^2. The tensile stress lasted 8 to 9 months. Results of observations made at other levels and those made in other shafts using double wall lining were identical. This fully indicates that in-situ double wall reinforced concrete lining has to stand the tensile stress in a considerably long time, which is a most unfavourable stress to the concrete. Because the strength of concrete at initial period was low, generally only 2.4 N/ mm^2. When the tensile stress was greater than its strength, the lining was damaged. Fissures and even cracks were found, which led to water seepage after thawing.

In east air shaft of Panji No.3 Colliery a 1.5 mm thick PVC plastic sheet was laid between the inner and the outer reinforced concrete lining. Fig.4 shows a stress curve of reinforcement in the inner wall at level of -328 m. Because the plastic sheet allowed a relative gliding movement between the inner and outer walls, the surrounding confinement of the outer wall to the inner wall could be eliminated. The inner wall was in tension for only 20 days, and the tensile stress of longitudinal reinforcements was 7.8-12.6 N/mm^2. This greatly reduced the possibility of producing cracks,

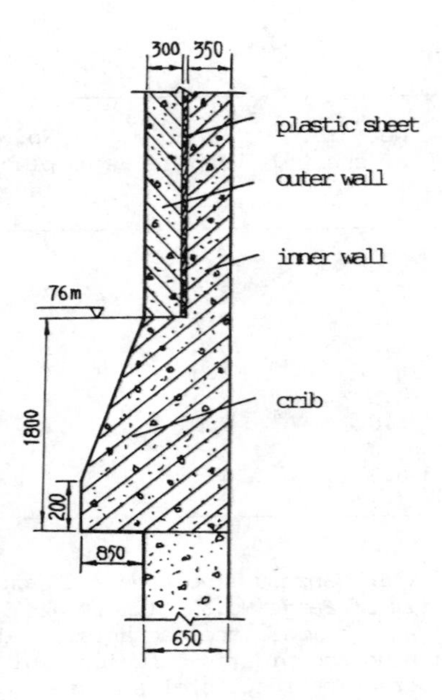

Fig.2 Plastic sandwich shaft lining (unit mm)

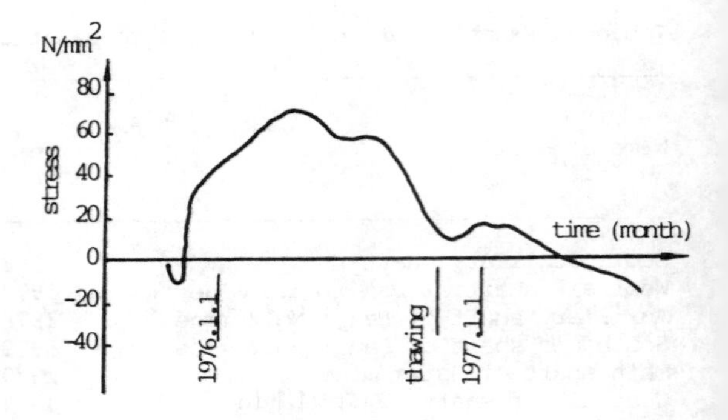

Fig.3 Curve of stress on shaft lining at level of -88 m in the main shaft in Xinglongzhuang Colliery

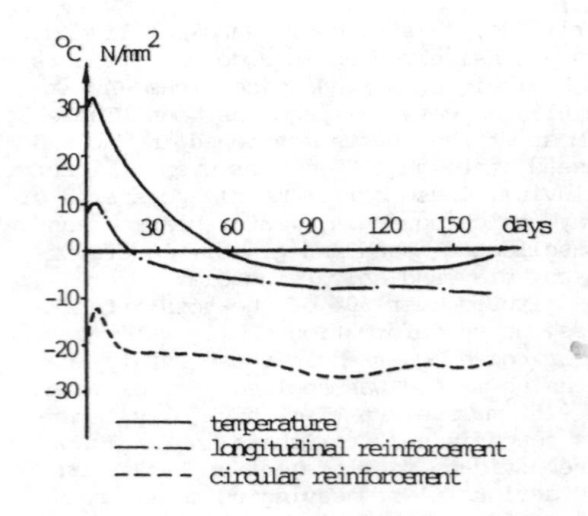

Fig.4 Curves of stresses in reinforcements in inner shaft lining at level of -328 m in east air shaft, Panji No.3 Colliery

thus ensuring the water-proof performance of the shaft lining. At the same time, we gained enlightenment from this fact. As a result, we introduced a plain concrete structure as the inner wall (see section 5). The above-mentioned conclusions were also proved by the measured data obtained in other shafts, e.g. Xingtai main shaft in Hanxing mine area, Haizi main shaft and Luling west air shaft in Huaibei mine area, Nantun main shaft in Yanzhou mine area.

3 FOAMED PLASTIC CUSHION LAYER AND ITS MECHANISM

Since the 70s to avoid damage caused by freezing pressure acting on shaft linings in thick alluvium a layer of concrete blocks has been added on double wall lining or sandwich lining. Thus, the total thickness of the lining reached 1.1-1.2 m or even greater. However, the results were not ideal. From 1985, a new type of structure was introduced in several freezing shafts in Huainan and Huaibei mine areas, which consisted of a plastic sandwich lining and a foamed plastic cushion layer. The polystyrene foamed plastic sheet was 25 mm thick. Depending on the rock type, one to three layers of plastics were laid betweem the outer wall and the shaft wall.

Curves in Fig.5 show variation of freezing pressure exerted on lining based on the measured data.

The pressure reducing effect was quite obvious. The loading tests made in the laboratory showed that the pressure began to go up when one layer of foamed plastics was compressed by 21 mm; and the pressure started to rise when two or three layers of plastics were compressed by 40 or 60 mm respectively.

Since the foamed plastics has heat isolation effect, it can greatly increase the curing temperature of outer wall concrete. The temperature measurement records show that a layer of foamed plastics can protect heat about 10°C, and three layers of foamed plastics can cure the outer wall at posi-

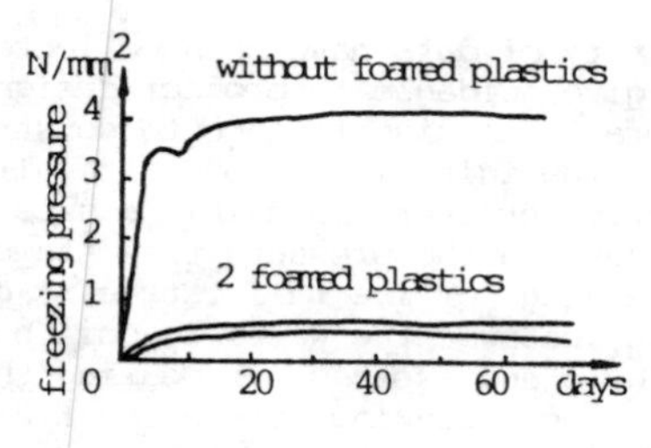

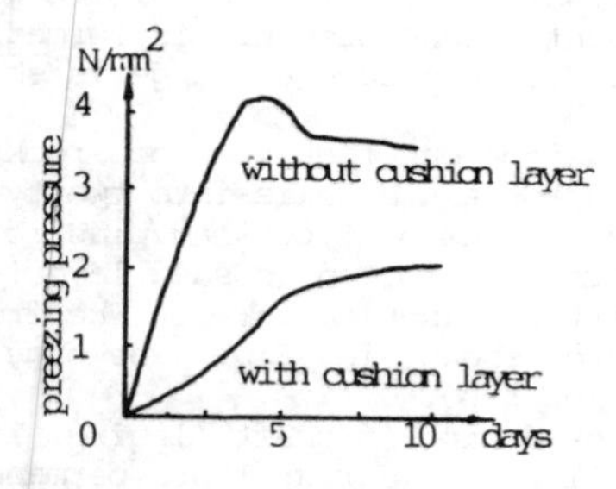

Fig.5 Comparison of freezing pressure on lining with or without foamed plastics

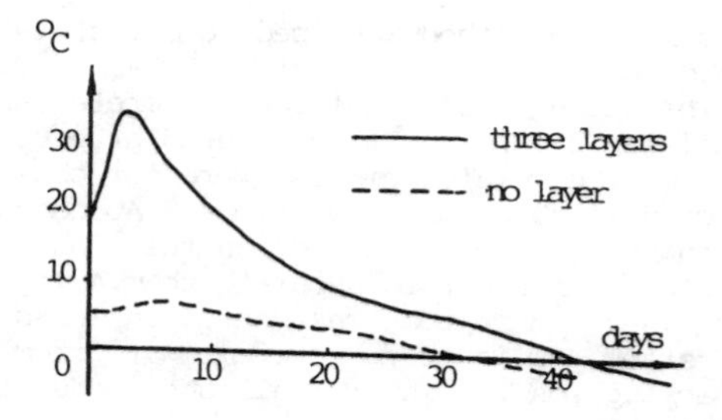

Fig.6 Heat isolation effect of foamed plastics

tive temperatures over 44 days (Fig.6).

In addition, since the foamed plastics will reduce migration of moisture content in the concrete, which can ensure the necessary water for hydration of concrete and avoid the rise of freezing pressure resulted from increased moisture content due to refreezing of the thawed soil. Since the surface of foamed plastics is smooth, the temperature stress of the outer wall reduces to 15-25%, thus the number of cracks due to temperature stress is reduced remarkably.

4 FLEXIBLE SHAFT LINING AND PRELIMINARY STRESS OBSERVATIONS

The flexible shaft lining, also called the gliding lining, is extensively used in Western Europe and North America. This type of lining is flexible, because it has a layer of bitum, which can withstand axial deformation with a curvature of 3,000 to 5,000 m. So, when safety pillars of the shaft are extracted, the shaft lining will not fail due to deformation.

From 1983 to 1985, China built the first shaft with flexible lining--the west air shaft of Kongji Colliery, Huainan mine area. With this shaft lining 6.6 Mt of coal could be recovered from shaft pillars.

The shaft with a net diameter of 4.5 m was 270 m deep in total. Freezing method was applied. Thickness of alluvium it passed was 65.5 m, freezing depth, 112 m. The

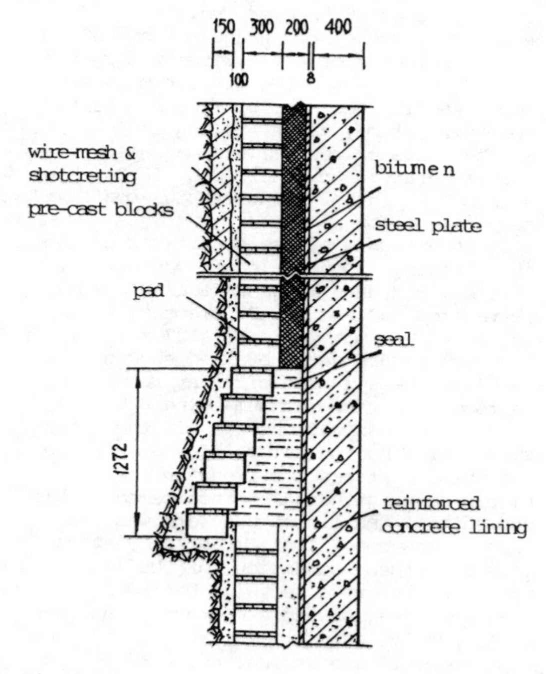

Fig.7 Structure of flexible lining used in west air shaft of Kongji Colliery

structure of flexible shaft lining is shown in Fig. 7. The height of shaft collar and sections of flexible lining was 114 m. Below it were sections of rigid lining, 21 m, which consisted of pre-cast blocks, steel cylinder, concrete lining, crib and supporting rings.

The height of each section of steel plates was 3 m, and the total height was 122.95 m. They were jointed together by hand welding vertically and horizontally. The vertical groove angle was 30°, horizontal groove angle, 45°. Ultrasonic nondestructive method was used to detect the weldings.

The strength of reinforced concrete inner wall was 25 N/mm^2. It was poured at sections of 3m. The reinforcements were not connected at junction of sections. Abandoned machine oil was applied on the contact surfaces in order to separate them.A layer of 0.5-0.7 mm pitch was coated on inside of steel cylinder to prevent it from sticking on the inner wall and made it rust-resistant.

The bitum layer served as a sealing material and it was used to equalize the pressure and allowed a relative gliding movement between the inner and outer walls. It was a mixture of No. 100 bitum and slacked lime powder with finess of 0.99 mm, volume weight 915 Kg/m^3, void rate 67.3%. Pouring temperature ranged from 120 to 130°C, height 3 or 6 m. Bitum was poured into junctions and the set bitum in the lower section became fluidized and melted together. It may cause thawing of ice wall in the range of 200 mm, and may also raise the temperature of concrete by 12°C, which ensured curing of concrete in positive temperature for 28 days.

To investigate stress conditions of the shaft lining when the shaft was put into service, and when it was affected by coal extraction, altogether 37 elements were buried at -36 m, -71m, and -127 m levels. The preliminary data obtained showed that tensile stress in the circular reinforcements was low,the maximum being 8-9 N/mm^2. Tensile stress was observed in the outer circle of longitudinal reinforcements,and compression stress was found in the inner circle of longitudinal reinforcements.This showed that the lining concrete was in a good stress condition. Further investigation on stress condition of the lining, when coal is extracted will be made.

Since wire-mesh and shotcreting was used as temporary support, the height of precast concrete blocks reached 30-40 m. And since the cribs were improved and their size was reduced, the pace of operation was quickened.

5 FREEZING PRESSURE AND NEW TYPE OF SINGLE WALL LINING

The so-called freezing pressure refers to the lateral pressure acting on the lining due to ground freezing in shaft sinking. It is a temporary operational loading in the period starting from walling, thawing and refreezing of the ice wall. Beginning from 1964 observations have been made in more than 20 freeze shafts, and large quantity of data have been collected,which can give guidance to project design and is of practical significance to construction.

In the initial period of walling, freezing pressure is mainly a deformation pressure in the pressure rise stage. In refreezing of ice wall stage it goes up again, because the water in soil becomes ice and expands. In pouring the inner wall, the freezing pressure fluctuates slightly due to rethawing and refreezing of ice wall. After thawing of ice wall,the permanent ground pressure is formed. The general rule of such a change is shown in Fig.8.

Depending on the type of rock, the freezing pressure falls into two types, namely, expansion type and plastic type. The former is formed in sand layer, and the latter, in clay layer. The freezing pressure shown in Fig.8 belongs to expansion type.

In the design of shaft lining it is generally considered that permanent pressure $P = 1.3\ \gamma H$. Of which water pressure, $P_1 = 1.0\ \gamma H$, which is supported by inner wall; the earth pressure $P_2 = 0.3\ \gamma H$, is supported by the outer wall. In fact, the outer wall is not designed on the basis of earth pressure, because it shall also stand the freezing pressure in operation. The measured value of freezing pressure is about 3 to 4 times higher than the earth pressure. Obviously, if the design is based on the earth pressure, the outer wall will certainly be damaged by the excessive freezing pressure. However, if the design is entirely based on freezing pressure, the lining will be too thick.

With the aim of reducing thickness of lining for freezing trials on new type of lining structure are made in several shallow shafts, that is, wire-mesh and shotcreting are applied as temporary support in the process of shaft sinking until it reaches the stage height, say, 30 to 40 m, depending on the rock type. Then a layer of gliding material is laid on the sprayed concrete layer to prevent the shaft lining from sticking on it. Then, the concrete wall is poured. Fig.9 is one of the examples of this type of lining.

The reason why thickness of this type of lining can be reduced is based on the following major factors: plasticity of the temporary support,consisting of wire-mesh and shotcreting, allows certain displacement deformation of the ice wall, thus reducing the freezing pressure; less thick sprayed concrete produces less heat from hydration, and less pressure occurs in refreezing of thawed soil.

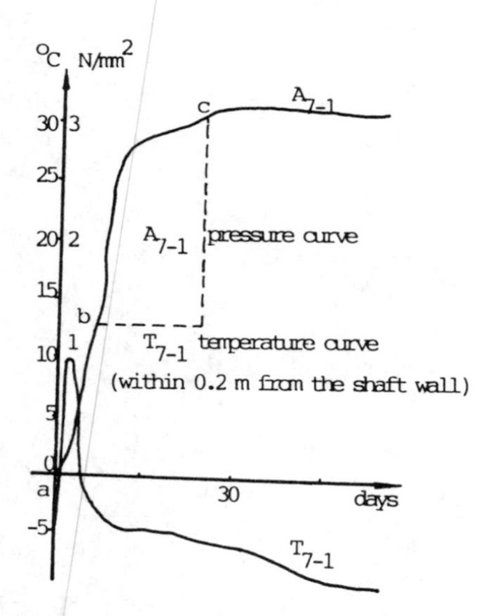

Fig.8 Curves of freezing pressure at level of -335.6 m in east air shaft of Panji No.3 Colliery

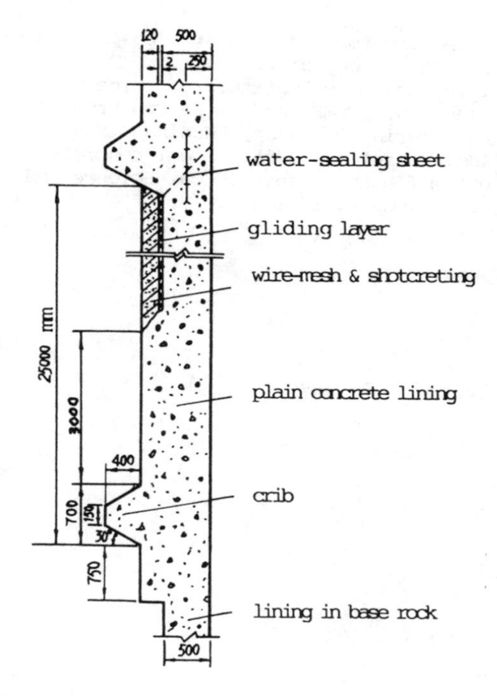

Fig.9 New type of lining structure of auxiliary shaft Weicun Colliery, Jiaozuo mine area

The role of a gliding layer is similar to the plastics layer between the two walls.

In accordance with large quantity of measured data in the past a plain concrete lining is adopted to replace the reinforced concrete in view of the shallow depth and thin alluvium.

Further observations shall be made to obtain some convincing results with its gradual expansion of application.

6 CONCLUSIONS

The conclusions are as the following:

1) with the increase of the shaft depth and of the alluvium thickness, the trend of development of shaft lining for freezing is the structure becoming from simple to complicated, from thin to thick. However,with the appearance of new structure and new materials, its thickness can be controlled and properly reduced, and in certain cases, even plain concrete inner wall can be employed.

2) laying foamed plastics on the shaft wall has an effect of easing and equalizing the pressure, heat isolation and sealing off water. Care should be taken in selecting conditions to avoid excessive deformation of shaft wall and collapse of outer wall. It is necessary to search for a cushion material that can stand greater freezing pressure.

3) plastic sandwich lining is currently an effective and rational type of lining which can stop water seepage. Flexible lining should be adopted, when it is necessary to recover shaft safety pillar; or when the shaft is located at a place where subsidence takes place due to other factors.

4) the characteristics and application scope of a temporary support consisting of wire-mesh and shotcreting remain to be further studied in practice.

REFERENCES

Yu Xiang, Wang Zhengting and Zhong Faying. 1985. Thirty years of shaft sinking by freezing method in China. Coal Science and Technology 7: 2-6.

Wang Changsheng. 1987. Freeze sinking. Coal Science and Technology 4: 28-34.

Yu Gongchun. 1983. Water seepage from cracks in permanent shaft lining in sinking a vertical shaft through thick overburden with freezing method and its prevention. Mine Construction Technology 4: 9-15.

Sun Qikai. 1987. New type of shaft lining

structure for freeze shaft. Coal Science and Technology 5: 26-28.

Beijing Research Institute of Mine Construction, CCMRI and Mine Construction Engineering Division, Huainan Mining Administration. 1986. Steel-concrete-bitum flexible lining. Coal Science and Technology 4: 17-20.

4. Case histories

5th International Symposium on Ground Freezing, Jones & Holden (eds)
© 1988 Balkema, Rotterdam. ISBN 90 6191 824 3

Vienna Subway construction – Use of brine freezing in combination with NATM under compressed air

Franz Deix
Vienna Subway Construction, Vienna, Austria

Bernd Braun
Frontier-Kemper Constructors, Inc., Evansville, Ind., USA
(Formerly: Deilmann-Haniel GmbH, Dortmund, FR Germany)

ABSTRACT: The most difficult portion of Section U6/3 of the Vienna Subway System required construction of two single-track tunnels under a telecommunications facility, where strict ground settlement limitations had been set to protect sensitive installations. Local geology, hydrology, and site constraints led to the decision to use the New Austrian Tunneling Method (NATM) supplemented by compressed air and partial dewatering (depressurization). Additional measures had to be taken under the telecommunications building (TCB) where the tunnel crown was either in or close to the alluvial deposits and only 1.5 m under the foundation of the TCB. Ground freezing was considered to be the best construction approach to deal with all problems and constraints. Extensive laboratory testing of frozen soil and verification of design data by freezing of two test sites took place before the final design for crossing under the sensitive building could begin. Thus, a freezing and excavation scheme could be adopted that ensured minimal heaving magnitude during active freezing and minimal subsidence during subsequent tunnelling and thawing of soil.

KEYWORDS: Tunnels, Underpinning, Compressive Strength, Creep, Frost Heave, In Situ Testing.

1 INTRODUCTION

The second construction phase of the Vienna Subway was started in 1983 and will add another 25 km by 1993 to the 32 km already in operation. Substantial portions of this phase will necessitate underground construction -- without surface disturbance -- in waterbearing, unstable soils. Findings of extensive economic and technical investigations led to the decision to use the NATM rather than modern tunnelling shields employed in Phase I construction. Total groundwater control is the most important pre-requisite for problem-free use of the NATM in loose soils in inner-city underground construction. The compressed air method -- successfully used in Phase I for groundwater control -- was adopted in conjunction with the NATM for difficult tunnelling in densely urban areas in Phase II. To ensure success with the NATM in the inner city, given existing loose soil conditions, additional construction methods are required which allow for adaptation to local geology, hydrology, and site constraints. The methods used for this job included stabilizing poorly graded gravels and uniform sands, dewatering water-bearing formations, and depressurizing perched aquifers. Water control measures involved a combination of dewatering wells, depressurization wells, horizontal drains, sump-pumping, compressed air, grouting, and -- in special cases -- ground freezing. The costs of these additional methods comprised between 5% and 20% of tunnel construction costs, the actual amount depending on existing geological and hydrological conditions.

2 GEOLOGY AND HYDROLOGY

Difficult ground conditions, a densely built-up area, and complicated intersection requirements characterize Section U6/3 of the Vienna Subway System (Deix and Braun 1987). This section has a total length of about 893 m consisting of twin-track and single-track tunnels, a transition zone, a station, and various auxiliary shafts for access and grouting purposes. The general geology is fill, underlain by alluvial deposits consisting of sands, gravels, and boulders, followed by tertiary clays with embedded sand layers. The hydrology is characterized by two aquifers, one in the alluvial deposits with a seasonally varying water level and one in the sand layers within the tertiary formation. The second one was perched.

3 SPECIFICATIONS

Tunnelling under the TCB with a street on either side was one of the most challenging parts of Section U6/3. The facility was not only old, but its function as housing for sensitive equipment used to handle 110,000 telephone connections made tunnelling below a very delicate undertaking.

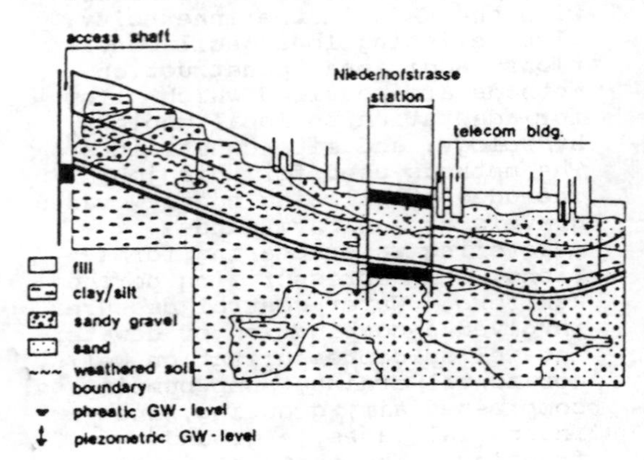

Fig. 1. Longitudinal section U6/3 - geology, hydrology

In this location, the crown of the tunnel was either in or close to the alluvial deposits and only 1.5 m under the foundation of the building.

A disturbance of the telephone communications system had to be avoided under all circumstances. Additional construction techniques had to be used to fulfill the building Owner's requirements, of which the main ones were:

1) adjacent bearing walls differential settlement should not exceed five mm.

2) maximum angular distortion not to exceed 1:500.

Measures taken to meet these requirements included:

1) use of the NATM with compressed air.

2) simultaneous and continuous driving of both tunnels.

3) reduced excavation length from 1.0 m to 0.8 m.

4) use of pre-stressed, closed-circle lattice girders.

5) securing of heading with shotcrete.

6) construction of a one meter thick frozen slab over the tunnel.

Using the grouting method for stabilizing the alluvial deposits under the TCB was precluded in the Specifications due to sound transmission concerns. The ground freezing method was chosen by the Owner as the best alternative to provide groundwater control and structural support as well as to prevent air losses and blow-out conditions in the alluvial deposits. Two test sites were frozen prior to final design of the ground freezing system under the TCB to verify design criteria and behavior of the frozen soil. A value engineering proposal for conventional brine freezing -- developed by a joint venture of Gruen & Bilfinger, Vienna, Austria and Deilmann-Haniel, Dortmund, FRG -- was accepted as the best and most cost-effective solution.

4 PRE-CONSTRUCTION TESTS AND ANALYSES

Undisturbed soil samples were taken from the tertiary soil layers, and disturbed soil samples

Frostkörper (freeze wall)
Gefrierrohre (freeze pipes)
Keller (basement)
Fundament (foundation)
Gleis 1, Gleis 2 (tracks)
Kies (gravel)
Schluff (silt, clay)
Sand (sand)

Fig. 2. Cross-section under TCB with freeze wall

were taken from the alluvial deposits during excavation of the subway station for laboratory testing.

4.1 Uniaxial compression tests of frozen alluvial deposits

The alluvial deposits prevalent in this area can be described as a well-graded gravel-sand mixture (GW) with no fines, 37% sand, and 63% gravel. Other index parameters were as follows:

Bulk density (Mg/m^3): 2.67
Dry density (Mg/m^3): 2.03
Moisture Content (%): 10.5
Degree of Saturation (-): 0.89
Porosity (-): 0.24

Samples were frozen at -10°C and tested to obtain the short term uniaxial compressive strength (q_g) and the modulus of elasticity (E_{vg}). The results from tests performed at -10°C were used to calculate values at -3°C and -5°C (Jessberger 1986).

Table 1. Uniaxial compressive strength and modulus of elasticity (tangent modulus at 50% of strength) of frozen alluvial deposits.

Temperature [°C]	Compressive Strength q_g [MPa]	Tangent Modulus E_{vg} [MPa]
-10	5.08	770
- 5	2.54	385
- 3	1.52	230

4.2 Uniaxial creep tests of frozen alluvial deposits at -10°C

The uniaxial creep behavior of frozen soils can be approximated -- not including initial elastic deformation -- using the following creep equation:

$$\varepsilon = A * \sigma^{B} * t^{C} ,$$

where ε is the strain, σ the axial

Table 2. Creep parameters and modulus of elasticity of frozen alluvial deposits

Temperature T [°C]	A [$(m^2/MN)^{B}*hr^{-C}$]	B [-]	C [-]	E_{vg} [MN/m^2]
-10	$7.21*10^{-4}$	2.78	0.30	240
- 5	$2.56*10^{-3}$	2.78	0.30	150
- 3	$1.07*10^{-2}$	2.78	0.30	90

Table 3. Index parameters of soils for heave tests

Sample-# [-]	Soil [-]	Clay [%]	Grain Size Silt [%]	Sand [%]	Gravel [%]	ρ_d [Mg/m^3]	w [%]	LL [%]	I_p [%]
1977	CH	55	45	--	--	1.71	20.5	52	27
1978	CI	42	42	13	3	1.76	20.2	48	25
2041	CL	20	67	13	--	1.84	18.2	30	11
2039	CH	44	56	--	--	1.70	23.7	55	26
2046	GW	--	--	37	63	2.02	8.4	--	--

stress, t the time, and A,B,C the creep parameters obtained from the frozen soil tests.

Uniaxial creep tests were performed at -10°C. The creep behavior of the frozen gravels for other than the tested temperature were approximated by comparing the uniaxial compressive strength and assuming that only the creep parameter A is time-dependent (Jessberger 1986).

4.3 Frost heave behavior

The frost heave behavior of the various formations was evaluated using the Segregation Potential (SP) (Konrad, Morgenstern 1983). The SP quantifies the ability of a soil to sustain the formation of ice lenses during the freezing process. The SP is defined as:

$$SP = \dot{h} / \text{grad } T \quad [\ mm^2/s\ °C]\ ,$$

where $\dot{h}$ is the frost heave rate [mm/s] and grad T the temperature gradient within the sample [°C/mm].

Approximate heave amounts can be calculated with the SP when the freeze time and the temperature gradient at the freeze front is known. The SP was evaluated in the tests as a function of the surcharge load SP = f($_1$).

The tertiary formation on this

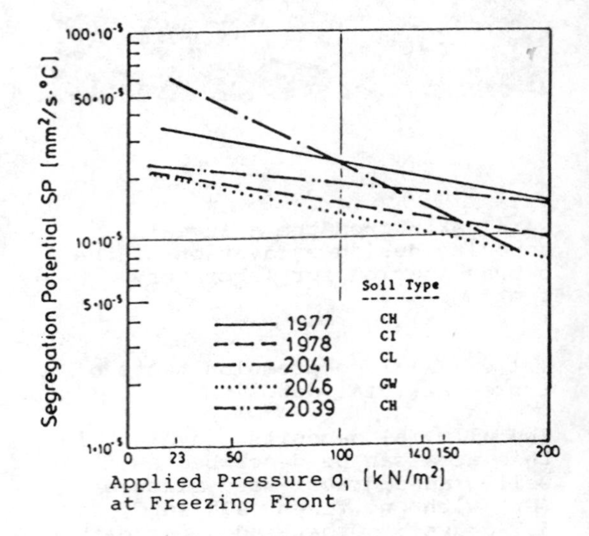

Fig. 3. Segregation Potential as a function of applied pressure

site can be divided into three groups consisting of clays with low (CL), medium (CI), and high plasticity (CH).

The heave test results are presented in Fig. 3. A detailed description of the testing procedures as well as the discussion of the test results are being presented at this symposium by Jessberger, Jagow, Jordan (1988).

4.4 Thermal analyses

The heave tests revealed that the tertiary formation was frost susceptible. Accordingly, it was necessary to find a suitable freezing scheme that would minimize heave. Past experience has shown that this can best be done by intermittent freezing during maintenance of the freeze wall. Therefore, a thermal analysis using FEM was performed for the frozen test site V_1. The objective of this analysis was to determine the pre-freezing time to form a one meter thick freeze wall with various freeze pipe spacings and to determine the temperature distribution within the freeze wall during intermittent freezing. A cycle of two hours freezing and twelve hours shutoff was used for the analysis. During intermittent freezing, the freeze wall core temperature rose from -9°C to -5°C, concurrent with a slow growing of the freeze wall. This procedure would result in a lower temperature gradient at the freeze front, reducing the heave amount. A paper on the model used and the results is presented at this symposium by Jessberger, Jagow, Jordan (1988).

4.5 Groundwater flow measurements

Lateral groundwater flow in the alluvial deposits was a concern during design of the ground freezing system. Groundwater flow measurements using the single borehole method and a radioactive tracer showed that the groundwater flow was less than half a meter per day, a magnitude not detrimental to freezing.

Furthermore, a chemical analysis revealed that the groundwater did not contain any substances that would suppress the freezing point of the water.

5. FROZEN TEST SITES V_1 AND V_2

A structural analysis -- using data obtained from the frozen soil tests -- indicated that a continuous one meter thick frozen slab with an average core temperature of -4°C would satisfy the structural requirements. To prevent air losses in the alluvial deposits, it was necessary that the freeze wall tie into the tertiary clay formations. It was desirable to place the freeze pipes approximately 0.5 m above the tertiary formation so that the freeze wall would barely tie into this layer. Two test sites to be frozen in the vicinity of the TCB were designed to obtain field data during freezing, tunnelling, and thawing Test site V_1 was to provide thermal data, frost propagation, optimal intermittent freezing cycles, heave, and thaw consolidation not influenced by tunnel excavation. Test site V_2 -- incorporating the experience from the first test site -- was to provide information on drilling accuracy, heave with optimal intermittent freezing, settlements during tunnelling, and thaw consolidation.

The test sites were equipped with horizontal and vertical temperature monitoring pipes to observe frost propagation in the soil. Surface settlement markers were installed to measure surface movements. Furthermore, PVC-pipes containing special measuring rings were placed in the ground at one meter intervals to observe deformations at various depths using the sliding micrometer deformation techniques. All data were collected by computer at a central data acquisition center near the refrigeration plant station using a PC with a 20 MB hard disk. Special software allowed immediate graphic display of desired data.

The double-head overburden drilling technique was used to install the freeze pipes on 0.9 m centers with an alignment tolerance of 1% according to the Specifications and a further restraint of adjacent freeze pipe spacing not to exceed 1.2 m. The drill pipe was left in place and preliminarily sealed by a drop-off drill bit. The drill pipe served as the freeze pipe and was finally sealed with a mechanical packer.

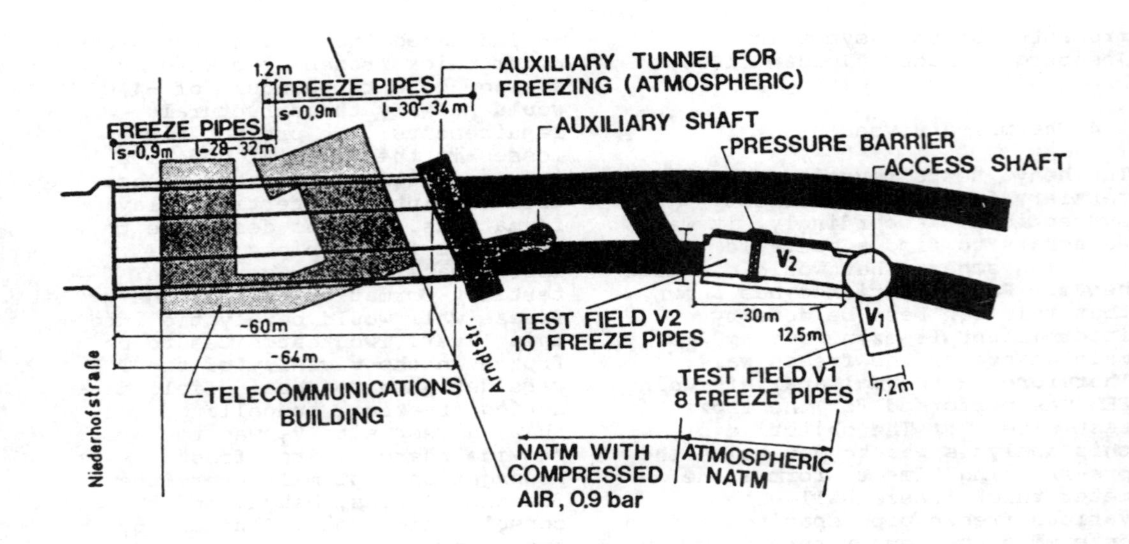

Fig. 4. General layout of freezing system for the TCB and the test sites V_1 and V_2

The alignment of all holes were checked using a multi-shot camera. No additional pipes had to be installed due to excessive deviation. The temperature monitoring pipes were placed at locations with maximum freeze pipe spacing.

Two refrigeration plants equipped with reciprocal compressors, each with a capacity of 735,000 kJ/hr at -25°C, were installed for the freezing, one serving as a standby unit.

The results from the sliding micrometer measuring soil deformations above, within, and below the freeze wall were interesting. Approximately 65% of the entire soil movements during the freezing period could be attributed to heave above the freeze wall, while approximately 35% resulted from compression of the soil below the freeze wall.

The main test site results showed the following:

1) The optimal intermittent freezing operating time was two to five hours per day, depending on the freeze pipe spacing.

2) Freeze pipes could reliably be installed within the specified tolerances.

3) Heavings could be minimized through optimal intermittent freezing.

4) TCB-Owner's requirements on settlement and differential settlements could be fulfilled.

5) Frozen soil design criteria were verified.

The ground freezing approach was finally accepted by the consulting engineer hired by the postal office, and final design of the ground freezing system under the TCB could proceed.

6. GROUND FREEZING UNDER THE TCB

On the south side (SS), freeze pipes could easily be installed through the diaphragm wall of the partially constructed station. On the north side (NS), it was initially planned to drive an auxiliary tunnel under compressed air from the main tunnel for freeze pipe placement. After further study, it was decided to move the access tunnel four meters further back from the TCB and construct this tunnel from an additional access shaft under atmospheric conditions using local dewatering. By separating the actual tunnel driving from the freeze pipe placement, driving of the main tunnels could proceed without any interruptions,

Table 4. Main data for test sites V_1 and V_2

Description	Test Site V_1	Test Site V_2
# of freeze pipes / length (m)	8 / 12.5	10 / 30.0
# of horizontal temperature pipes / # of temperature sensors	3 / 4	4 / 10
# of vertical temperature pipes / # of temperature sensors	1 / 8	2 / 8
# of surface settlement markers	19	24
# of sliding micrometer pipes	3	5
Pre-freezing time (days)	5	6
Maintenance freezing time (days)	23	85
Surface heave predicted (mm)	5 to 14	5 to 15 *
max. actually measured (mm)	6	5
Max. measured soil movements (mm)	10.5	9.5
Thaw consolidation (mm)	2	-- **

* Prediction was only for a maintenance freezing time of 30 days.
** Reliable data could not be obtained due to other construction activities.

and the total freeze time could be reduced, resulting in lower risk of potential heave. A total of 25 freeze pipes were installed in the north field (N-field), while 28 were placed in the south field (S-field). The freezing process in the ground was constantly monitored through five horizontal and five vertical temperature monitoring pipes. A total of 146 temperature sensors were installed. Twenty-four heave markers were placed at strategically important locations within the TCB. A special warning system was installed to indicate any brine losses during freezing, but fortunately this system was never activated.

Freezing of the N- and S-fields was started simultaneously to avoid differential heave and early enough so that the freeze wall was formed at the time the tunnel reached the TCB.

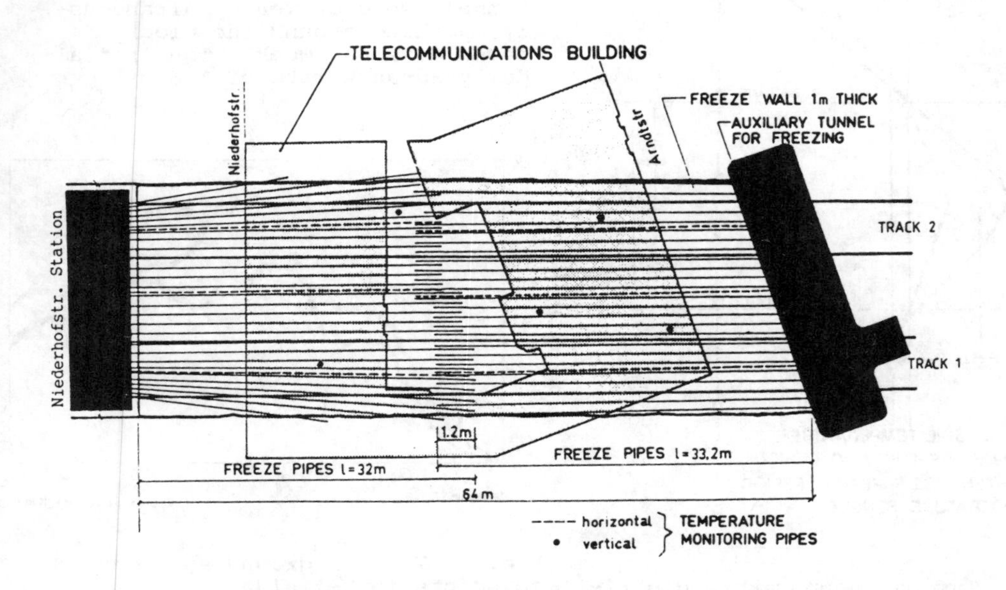

Fig. 5. Layout of ground freezing system under the TCB

The S-field showed a more uniform freezing behavior which can be attributed to the fact that mostly alluvial deposits had to be frozen. After nine days of freezing the required freeze wall was formed, and intermittent freezing started with three hours per day operating time. After a few days, the time was further reduced to 2.5 hours. A typical temperature distribution within the freeze wall after pre-freezing and after 20 days of maintenance freezing is given in Fig. 6.

The N-field could be divided into two parts. The E-side was frozen after 10 days and intermittent freezing was done with a daily operating time of three hours. On the W-side, concern was caused by an exhaust channel under the base slab connected to the TCB's heaters. Temperatures close to +100°C were measured there, resulting in a soil temperature of +40°C approximately 0.5 m above the proposed freeze wall location. It was obvious that the freeze scheme had to be modified in this area to achieve a closed freeze wall. A water cooling system was installed within the exhaust channel dropping the temperatures to less than +40°C. For a period of five days, continuous freezing was applied, until the freeze wall closed. Then intermittent freezing was used, with a daily operating time of five to 7.5 hours until tunnelling had passed this area. At that time the daily freeze operation for the entire N-field was reduced to two to three hours. Heave of bearing walls were typically less than 10 mm in the S-field and between 10 mm and 13 mm in the N-field. Maximum heave of 24 mm occurred in the basement where the heaters were installed and the overburden pressure was lowest. Heave markers SP 277 (W-side) and SP 281 (E-side) are typical examples of the heave development in the N-field and are presented in Fig. 7.

The heave/settlement developments can be divided into distinct periods and are presented in Tables 5 and 6.

Fifteen days after the start of freezing, actual tunnelling under the TCB commenced. The two tunnels were driven simultaneously, working around the clock, seven days per week. The initial daily advance rate of 1.6 m

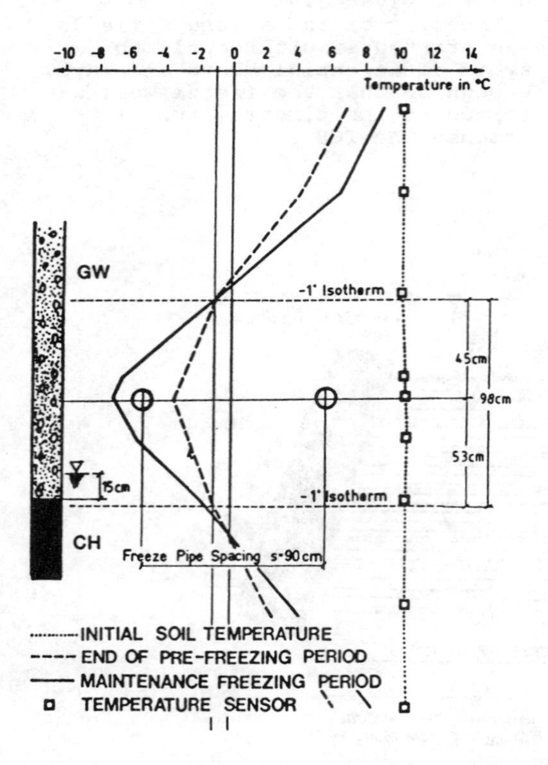

Fig. 6. Typical temperature distribution in S-field freeze wall

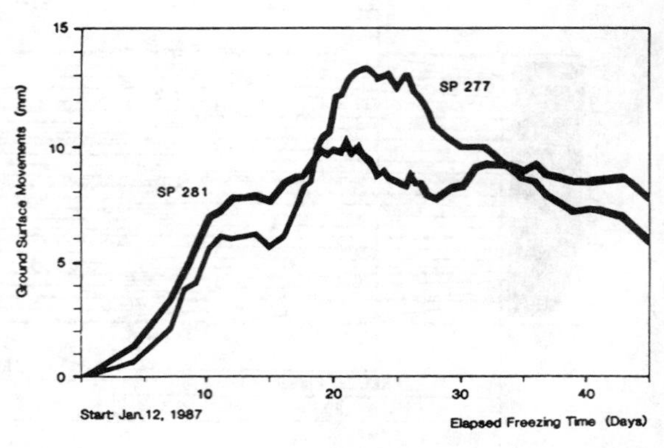

Fig. 7. Typical ground surface movements in N-field

Table 5. Soil movements at heave marker SP 277 (N-field, W-side)

Time (days)	Activity	Result
1 - 10	pre-freezing (24 hrs/day)	continuous large increase
10 - 14	intermittent freezing (4 hrs/day)	no increase
15 - 20	continuous freezing (24 hrs/day)	continuous large increase
21 - 26	intermittent freezing (5 hrs/day)	no increase
27 - 30	intermittent freezing and tunnelling (7.5 hrs/day)	initial settlement due to tunnelling
31 - 42	intermittent freezing and tunnelling (2 to 3 hrs/day)	initial settlement due to tunnelling
> 42	settlements due to tunnelling and thaw consolidation	thaw consolidation and secondary settlement due to tunnelling

Table 6. Soil movements at heave marker SP 281 (N-field, E-side)

Time (days)	Activity	Result
1 - 10	pre-freezing (24 hrs/day)	continuous large increase
10 - 14	intermittent freezing (4 hrs/day)	no increase
15 - 18	intermittent freezing (8 hrs/day)	smaller increase
19 - 27	intermittent freezing and tunnelling (3 hrs/day)	initial settlement due to tunnelling
28 - 30	intermittent freezing and tunnelling (6 hrs/day)	small increase
31 - 42	intermittent freezing and tunnelling (2 to 3 hrs/day)	initial settlement due to tunnelling
> 42	settlements due to tunnelling and thaw consolidation	thaw consolidation and secondary settlement due to tunneling

Table 7. Total soil movement under TCB

Cause	Soil Movement (mm)
Heave during freezing	10 to 13
Primary settlement due to tunnelling	4 to 6
Thaw consolidation, secondary settlement due to tunnelling, depressurization of tunnel	16 to 18
Final total subsidence	10 (max. 14)

increased steadily to 2.4 m, with an average rate of 2.0 m. The freeze wall in the N-field was verified by probe holes from within the tunnel excavation, while it encroached the excavation in the S-field. The warm compressed air in the tunnel had a negligible influence on the freeze wall in the N-field (the tertiary clays acted as an insulator) and only a minor influence on the freeze wall in the S-field, with the effect that the daily intermittent freezing time had to be increased to three hours for the last few days. Tunnelling proceeded without any difficulties and reached the diaphragm wall at the station after approximately five weeks. Freezing was terminated three days prior to that date. A total of 5,100 m^3 of soil were excavated, 1,000 m^3 of shotcrete placed, 30 tons of reinforcement and 120 tons of lattice girders installed.

The total soil movements -- after thawing had taken place, the tunnel was depressurized, and the final lining installed -- are summarized in Table 7.

The maximum differential settlement measured was 3.7 mm and well below the specified 5.0 mm. The maximum angular distortion amounted to 1:1,285, also well below the specified 1:500.

7 CLOSING REMARKS

With this project it was demonstrated that tunnels can be driven under settlement-sensitive buildings with a minimum cover using the NATM in combination with compressed air and ground freezing. It was shown, that under the given geologic and hydrologic conditions, a thin freeze wall can be formed and maintained over an extended time period without much growth. Heave under the foundations could be limited to 13 mm in the frost susceptible soils. Pre-requisites for the successful completion of this project were:

1) extensive soil investigations including laboratory testing and site trials of freezing.

2) thermal analysis.

3) continuous and extensive monitoring of the freezing process with daily adjustments to changing conditions.

4) close cooperation among owner, designer, and contractors.

REFERENCES

Deix, F. & B. Braun 1987. The use of NATM in combination with compressed air and ground freezing during Vienna Subway construction. Rapid Excavation and Tunneling Conference (RETC), Vol. 1, New Orleans:488-506.

Jessberger, H.L. 1986. Untersuchung der Festigkeitseigenschaften und des Frosthebungsverhaltens an Proben aus dem Bereich der geplanten Frostkörper, U6/3 in Wien, Vivenotgasse (Frozen strength and frost heave tests with samples from the planned freeze wall at U6/3, Vienna). Report prepared by Jessberger + Partner, Bochum.

Jessberger, H.L., R. Jagow & P. Jordan 1988. Thermal design of a frozen soil structure for stabilization of the soil on top of two parallel metro tunnels. Proceedings 5th International Symposium on Ground Freezing (ISGF 88), Nottingham.

Konrad, J.M. & N.R. Morgenstern 1983. Frost susceptibility of soils in terms of their Segregation Potential. Proceedings 4th International Permafrost Conference, Fairbanks:660-665.

ACKNOWLEDGEMENTS

Owner: City of Vienna, Magistratsabteilung 38, Subway Construction.

Construction Supervision: Magistratsabteilung 38, Subway Construction.

Contractors: Joint Venture "ARGE U6/3 Vivenotgasse", comprised of Neue Reformbau, Gruen + Bilfinger, Hinteregger & Söhne, Teerag-Asdag AG, Alpine, Hazet, Hochtief AG, Tiefbau.

Subcontractor Freezing: Gruen + Bilfinger, Deilmann-Haniel.

5th International Symposium on Ground Freezing, Jones & Holden (eds)
© 1988 Balkema, Rotterdam. ISBN 90 6191 824 3

The comparison between the results of heat transfer analysis and the observed values on a refrigerated LPG inground tank

Sadao Goto & Masafumi Shibuya
Tokyo Gas Co., Ltd, Japan
Takashi Nakajima & Masanobu Kuroda
Shimizu Corporation, Japan

ABSTRACT: In the Tokyo Gas Sodegaura LNG Terminal facing the Tokyo Bay, a 60,000KL refrigerated LPG inground tank was completed in October, 1986. It is one of the largest refrigerated LPG tanks in the world and it has been functioning successfully since its inauguration. In this paper, an analysis of heat transfer in the design stage and a comparison between its results and the values observed during tank operation are described. The observed values in the main parts of the tank agreed well with the values computed in the design stage of the tank, and the accuracy of the initial analysis was confirmed.

Also in this paper, the results of follow-up research of the frozen soil earth pressure on other LNG tanks, which had been mentioned in our 4th ISGF proceedings paper, are described. The measured values of the frost heaving pressure agreed well with the computed values, and the accuracy of the design method was affirmed.

1 INTRODUCTION

The Tokyo Gas Sodegaura Terminal is the largest LNG (Liquefied Natural Gas) receiving terminal in the world, and supplies natural gas, clean energy as it called, as town gas to the Tokyo metropolitan area. Refrigerated LPG (Liquefied Petroleum Gas) is used for calorie control of natural gas, and the inground tank was constructed for the storage of refrigerated LPG. As LPG is stored at a low temperature and as a result the sides and the bottom of the tank are gradually frozen, heating equipment was provided in order to control this freezing process. In this paper, an observation of the freezing and an analysis of heat transfer are described from chapters 2 to 6. The measurement of the frost heaving pressure on other LNG tanks is described in chapter 7.

2 OUTLINE OF THE TANK

This inground storage tank was constructed in a reclaimed area facing the Tokyo Bay. As shown in Figure 1, the bottom slab is founded on diluvial sandy soil. The tank itself is 64 m in diameter and 18.7 m in liquid depth. The tank construction is; from the inner side, a stainless steel membrane to obtain a gas tightness, rigid polyurethane foam panels to prevent heat transfer into the tank and a cylindrical outer shell made of reinforced concrete to resist liquid, soil and water pressures as well as an earthquake pressure. A steel roof is fixed on the top of the side walls, and the side wall membrane is connected to the roof so as to obtain a gas tightness. Insulation mats are fitted inside the dome-shaped roof, so that the heat transfer through the roof is negligible.

Since the tank is near the sea, the ground water level is high. In this tank, a cylindrical slurry wall was placed in the impermeable layer under the ground so as to reduce the penetration of water beyond the slurry wall. The small amount of water that collects in the gravel layer under the bottom slab is pumped up and out of the ground.

3 OUTLINE OF THE CONTROL OF FREEZING

The refrigerated LPG is stored at -45°C in this tank. Because of this low temperature, the soil surrounding the tank gradually freezes. If the freezing zone were to become too large, the frost heaving pressure on the tank as well as on the foundations of other structures might

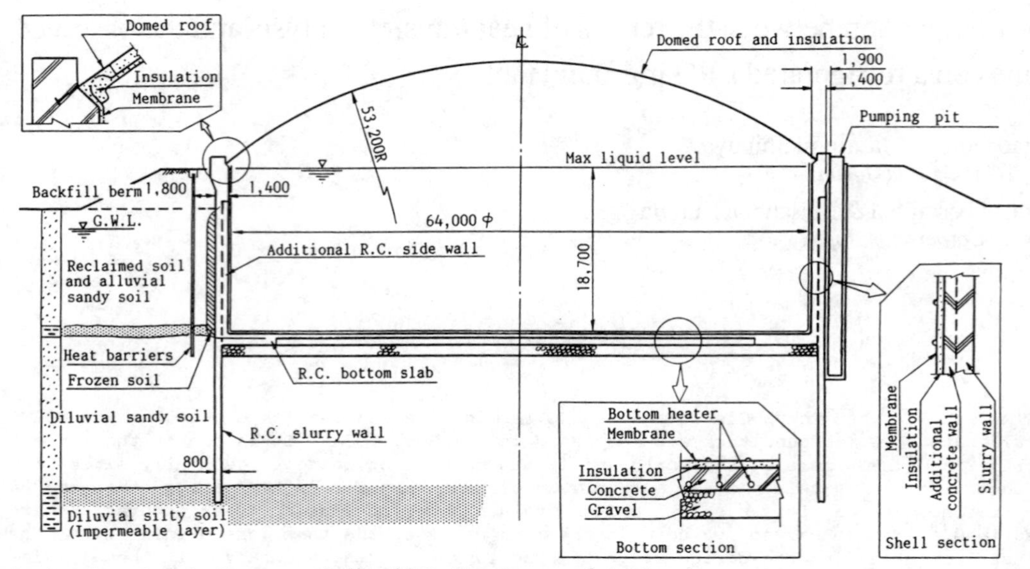

Fig.1 Structural diagram of 60,000KL LPG inground tank

reach harmful proportions.

Therefore heating equipment is needed to prevent the tank from being affected by frost heaving pressure. For this purpose, heat barriers were fitted vertically around the tank 1.8 m away from the outer surface of the side wall. The horizontal freezing of the soil is controlled by these barriers. These heat barriers are illustrated in Figure 2. Thus the thickness of the frozen soil is limited to a distance of 0.5 m from the outer surface of the side wall. This 0.5 m range was adopted in the design to satisfy two incompatible conditions. They are, one - to increase the watertightness of the tank by means of a thicker frozen soil barrier, and two - to control the frozen soil to a lesser degree from the viewpoint of its influence on the other structures.

The heat barriers are composed of 60 heat pipes, each of which is a double tube. Heat is supplied via circulating water in these pipes.

In regard to the situation under the bottom slab of the tank, if the soil there froze, the bottom slab would be affected by heaving pressures. Heat pipes therefore were also installed in the bottom slab so that the freezing front would be controlled within the bottom slab. Figure 3 shows the configuration of the bottom heater.

Two series of vortex heating systems are provided to promote reliability : one system acts as a main heater ; the other acts as a subheater. Heat is supplied via circulating brine (soluble 76% ethylene glycol water) in these pipes.

4 ANALYSIS OF HEAT TRANSFER OF THE TANK

4.1 Analysis method

The time-depended temperature distribution in the tank and the ground was obtained by an axial symmetry cylindrical model of the finite element method.

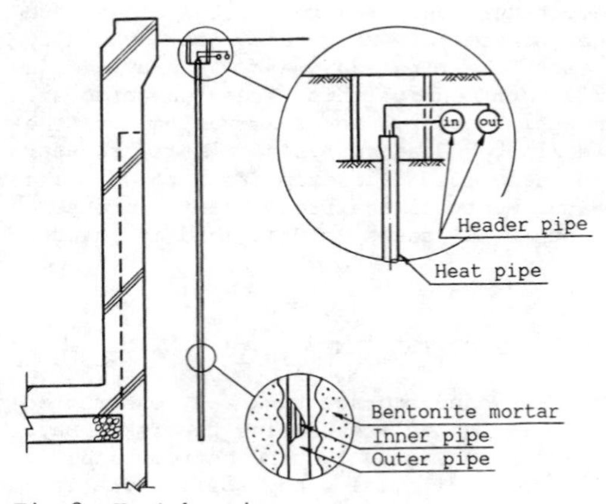

Fig.2 Heat barriers

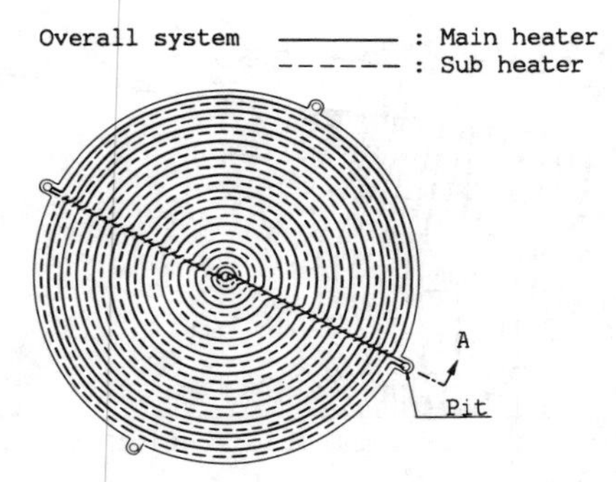

Fig.3 Bottom heater

The design conditions are as follows and are shown in Figure 4.

4.1.1 Boundary conditions

The boundary conditions of the ground surface and the inner surface of the tank were defined as heat transfer coefficients. The boundary conditions at the side and bottom edges of the ground model were defined as adiabatic boundaries. The distance between the tank and the adiabatic boundaries were adopted to be sufficiently far so that the result of analysis was not significantly affected by these boundaries. The heights of refrigerated LPG in the design stage were set at 30%, 50% and 100% of the full capacity height. In this paper, the height of refrigerated LPG was set at 50%.

4.1.2 Original temperature

The original temperatures of the atmosphere, the ground and the refrigerated LPG were determined as +15°C,

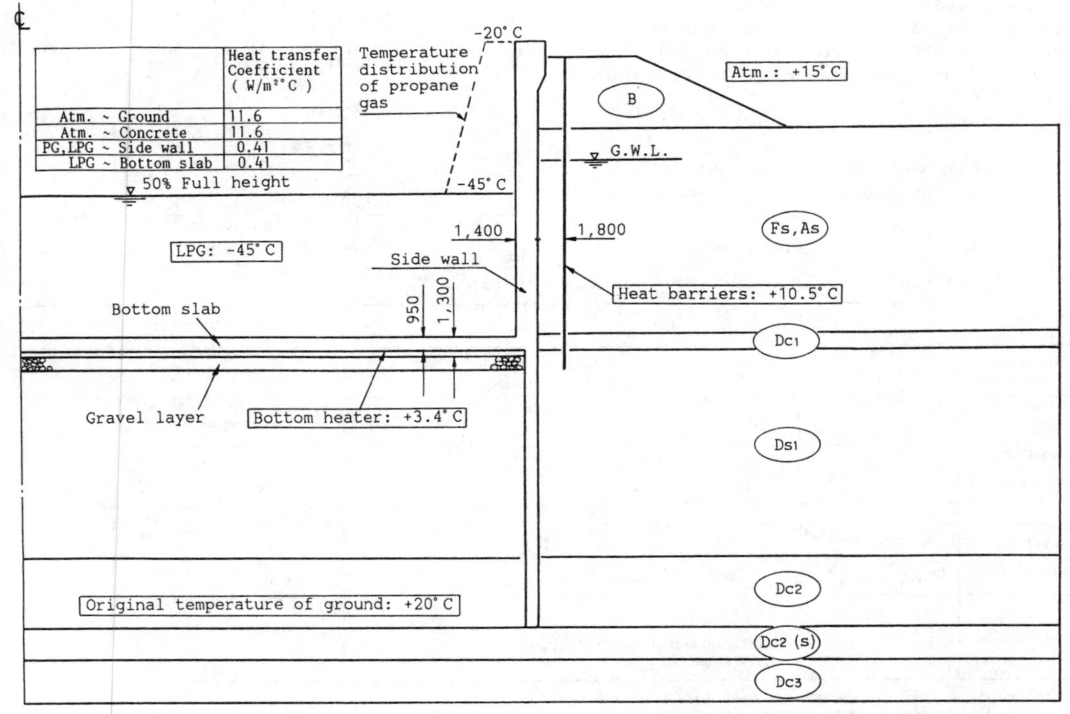

Fig.4 Conditions of analysis

+20°C, -45°C which had been obtained in previous observations of the other tanks. The temperature of the propane gas was determined as -45°C at the surface of the refrigerated LPG and -20°C at the top of the side wall. The distribution of the temperature was taken as being linear.

4.1.3 Heat temperature

The temperature of the heat barriers was accepted as being +10.5°C so that the thickness of the frozen soil would be controlled up to about a 0.5 m range from the outer surface of the side wall.

In the heat transfer analysis using the axial symmetry cylindrical model, the heat pipes installed at intervals of an angle of 6° radiating from the center of the tank were represented as a thin continuous cylindrical heat film. The validity of this supposition was confirmed by a previous two dimensional horizontal model of the finite element method which justified such a representation of each heat pipe. A smooth freezing front line was confirmed when each heat pipe radiating at an interval of an angle of 6° from the center of the tank. (See Figure 5)

As to the bottom heater, the temperature of the heater, set at a point 0.95 m from the inner surface of the bottom slab, was set at +3.4°C so that the freezing front would be confined within the bottom slab.

4.1.4 Thermal characteristics

The thermo-physical properties of concrete, insulation and soil were

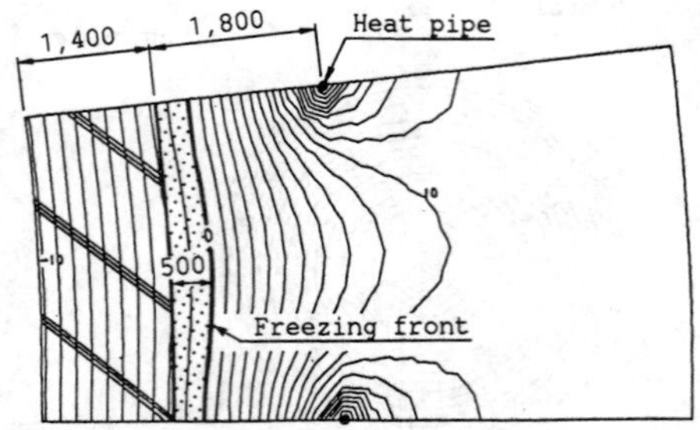

Fig.5 Temperature distribution in horizontal model

determined by using "Recommended Practice for LNG Inground Storage" (Japan Gas Association, 1979). The thermal constants of the soil were calculated using the calculation formulas shown in Table 1 and the results are shown in Table 2.

Table 2 Thermal characteristics

			Density γ (kg/m³)	Specific heat C (J/kg°C)	Thermal Conductivity λ (W/m°C)	Latent heat of freezing L (J/kg)
Concrete			2,400	840	2.33	——
Insulation t= 65 mm	Side		——	——	0.027	——
	Bottom		——	——	0.027	——
Gravel	in air	BF	1,800	750	1.16	——
	in water	BF	2,160	1,420	2.12	——
Soil	B	BF	1,940	1,670	1.77	——
		AF	1,890	1,170	2.88	78,700
	Fs	BF	1,940	1,670	1.77	——
	As	AF	1,890	1,170	2.87	78,700
	Dc1	BF	1,710	2,010	1.49	——
		AF	1,640	1,300	2.72	116,000
	Ds1	BF	1,940	1,670	1.77	——
	Dc2	BF	1,640	2,130	1.40	——
	Dc2 (s)	BF	1,820	1,840	1.59	——
	Dc3	BF	1,640	2,130	1.40	——

[Remarks] BF : Before freezing
AF : After freezing

Table 1 Methods of calculating thermal characteristics

Item	Formula	Physical properties [1]
Water content in volume φ_ω	$\varphi_\omega = n \cdot \frac{S_r}{100}$ n : Porosity; S_r : Degree of saturation (%)	
Density γ (kg/m³)	$\gamma_1 = (1-n)\gamma_s + \varphi_\omega \gamma_i$ $\gamma_2 = (1-n)\gamma_s + \varphi_\omega \gamma_\omega$	γ_i = 881.6 (Density of ice containing air bubbles)
Specific heat C (J/kg°C)	$C_1 = \frac{C_s\gamma_s(1-n) + C_i\gamma_i\varphi_\omega}{\gamma_1}$ $C_2 = \frac{C_s\gamma_s(1-n) + C_\omega\gamma_\omega\varphi_\omega}{\gamma_2}$	C_s = 920 C_i = 2,090 C_ω = 4,190
Thermal conductivity [2] λ (W/m°C)	$\lambda_2 = \frac{1}{3}\left[\lambda_\omega\left\{\frac{1-(1\frac{3\lambda_s}{2\lambda_\omega+\lambda_s})(1-\varphi_\omega)}{1+(\frac{3\lambda_\omega}{2\lambda_\omega+\lambda_s}-1)(1-\varphi_\omega)}\right\}+\lambda_s\left\{\frac{1-(1-\frac{3\lambda_\omega}{2\lambda_s+\lambda_\omega})\varphi_\omega}{1+(\frac{3\lambda_s}{2\lambda_s+\lambda_\omega}-1)\varphi_\omega}\right\}+\frac{3\lambda_s\lambda_\omega}{3\lambda_\omega+\lambda_s\varphi_\omega}\right]$ λ_1 is calculated by substituting λ_ω for λ_i in the respective formulas of λ_2.	λ_s = 3.49 (Average of feldspar, granite, quartzite, and basalt) λ_i = 2.26 λ_ω = 0.58
Latent heat of freezing L (J/kg)	$L = \frac{L_\omega\varphi_\omega\gamma_\omega}{\gamma_1}$ (The latent heat of freezing of soil grains is neglected.)	L_ω = 333,000

[Remarks] [1] T. Takashi: Refrigeration, Vol, 36, On Soil Freezing Working Method [I] (1961).
[2] Applied to saturated soil of $\varphi w = n$

Suffixed letters: 1: After freezing 2: Before freezing S: Mother rock of soil grains i: ice w: water

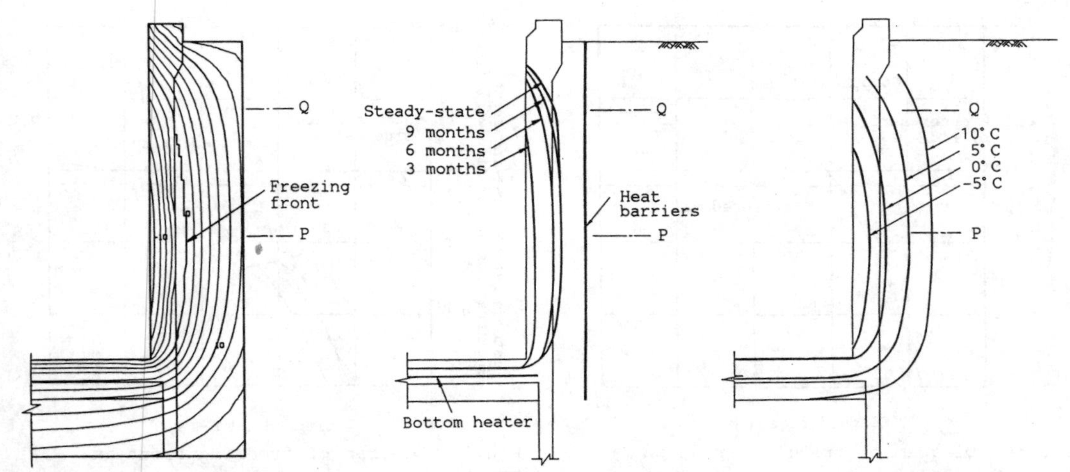

Fig.6 Temperature distribution in analysis (9 months after the start of tank operation)

Fig.7 Progress of freezing front

Fig.9 Temperature distribution in observation (9 months after the start of tank operation)

4.1.5 Heating operation conditions

The temperature distribution in the tank and in the ground was computed in an unsteady-state condition.

In this analysis, the starting period of the heating system was set as follows.

The bottom heater was started at the same time the tank started operation. If the bottom heater were not in operation just after the tank began operation, the freezing front would immediately extend under the surface of the bottom slab and the gravel layer would freeze.

The heat barriers were started 9 months after the tank had started operation. This was because, the progress of the freezing front at the side of the tank was slow and, at that time, it extended to only about 0.5 m from the outer surface of the side wall without the heat barriers in operation.

4.2 Result of the analysis

The temperature distribution in the tank and the ground 9 months after the start of tank operation is shown in Figure 6. At this time, the heat barriers were not yet in operation. The progress of freezing front from the start to the steady-state of tank operation is shown in Figure 7.

5 MEASUREMENT OF THE TEMPERATURE IN THE TANK AND THE GROUND

The arrangement of thermometers (C-C thermo-couples) in the tank and the ground is shown in Figure 8.

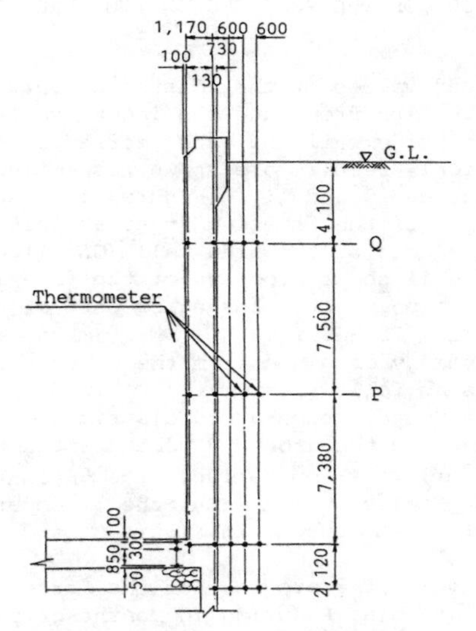

Fig.8 Arrangement of thermometers

The temperature distribution obtained by these thermometers in August, 1987, is shown in Figure 9. This figure shows the temperature distribution 9 months after the start of tank operation. The thickness of the frozen soil at this time was about 0.35 m from the outer surface of the tank.

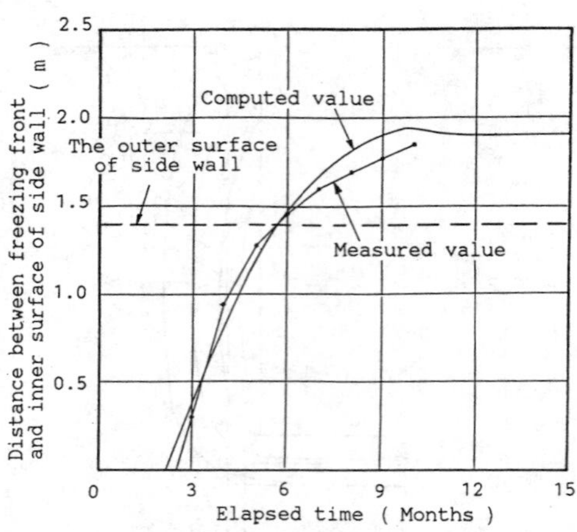

Fig.10 Diagram of freezing front in point P

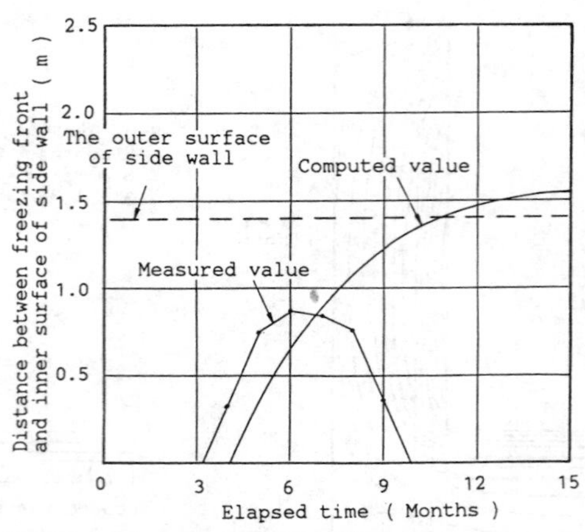

Fig.11 Diagram of freezing front in point Q

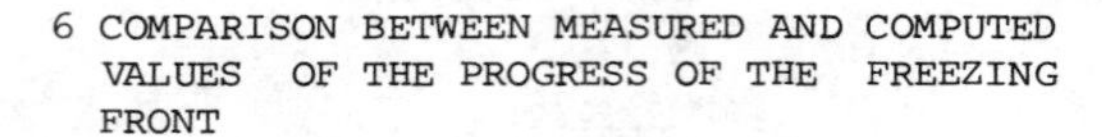

6 COMPARISON BETWEEN MEASURED AND COMPUTED VALUES OF THE PROGRESS OF THE FREEZING FRONT

Measured values of the distance between the freezing front and the inner surface of the side wall and the corresponding analytical results are shown respectively in Figures 10 and 11. Figure 10 shows progress of the freezing front at point P in the middle of the side wall (GL -11.6m).

Figure 11 shows progress of the freezing front at point Q in the upper part of the side wall (GL -4.1m).

A summary obtained from these comparisons is as follows.

1. As to the temperature distribution in the tank and the ground 9 months after the start of tank operation, the measured values agree well with the computed values. (As shown in Figure 6 and in Figure 9)

2. As to the progress of the freezing front at point P during 10 months of tank operation, the measured values agree well with the computed values. (Shown in Figure 10) Only in the 7th to 10th month period from the start of tank operation, the measured values are slightly smaller than the computed values. A reason for this phenomenon is that the temperature at point P was actually warmer than the temperature in the design condition due to the hotter summer. As a result, the thickness of the frozen soil did not reach 0.5 m by this time and actually the heat barriers were not in operation 10 months after the start of tank operation, although in the analysis the heat barriers started operation 9 months after the start of tank operation.

3. In regard to the progress of the freezing front at point Q, the measured values tend to agree with the computed values during the first 7 months but after that time the measured values disagree with the computed values. (Shown in Figure 11) The reason for this discrepancy at point Q is considered the same as that for the discrepancy at point P. Point Q was however more influenced by weather conditions than point P, since it is nearer the ground surface and above the ground water level.

As a result of these actual field measurements of the freezing phenomenon, we can conclude that the analysis of heat transfer in the main parts of the tank in our design stage are valid.

7 RESULTS OF THE FROZEN SOIL EARTH PRESSURE ON OTHER LNG TANKS

In this section, the additional results of frozen soil earth pressure on other LNG tanks, which had been mentioned in our 4th ISGF proceedings paper (Goto et al 1985), are described below:

The frost heaving pressure was measured by a new type of earth pressure gauge which had a hard stiffness diaphragm (diameter= 500mm) supported by three load cells. These gauges were installed on the side walls of Tank A and Tank B : The tank A gauge measured the frost heaving pressure of cohesive soil; The tank B gauge measured that of sandy soil.

Figure 12 and Figure 13 show the results

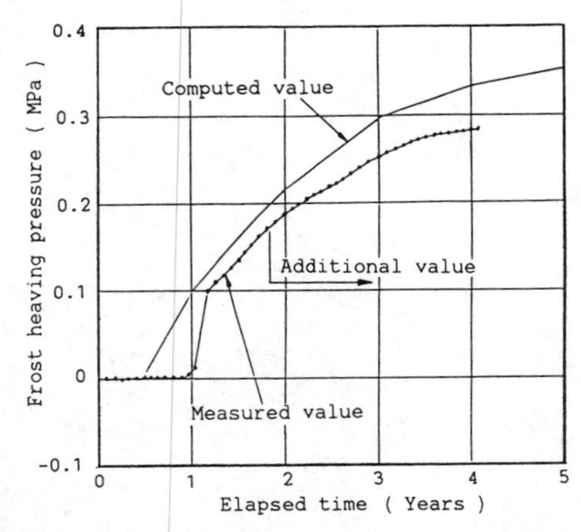

Fig.12 Diagram of frost heaving pressure in Tank A

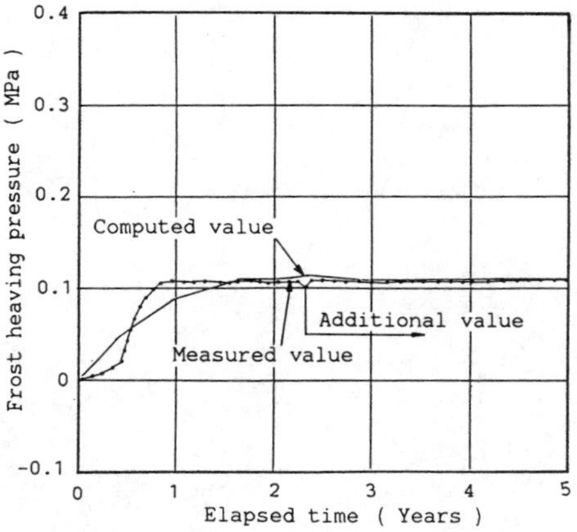

Fig.13 Diagram of frost heaving pressure in Tank B

of these measurements.

A summary of the results of these frost heaving pressures is as follows.

1. A delay in the increase of frost heaving pressure was observed in the measured values when compared to the results of the analysis for Tank A. A reason for this phenomenon is that the effectiveness of the insulation of Tank A was slightly better than predicted in the design value.

The frost heaving pressure, measured as P= 0.28 MPa for the Tank A about 4 years after the start of tank operation, agrees well with the predicted value taking account of the delayed start of freezing.

2. The frost heaving pressure, measured as P= 0.11 MPa for Tank B about 5 years after the start of tank operation, agrees well with the predicted value.

According to these additional field measurements of the frost heaving pressure, we can conclude that the value of frost heaving pressure used in the design of the concrete walls of LNG tanks were valid.

8 CONCLUSIONS

In this paper the analysis of heat transfer, observations of the temperature distribution and the progress of the freezing front in the refrigerated LPG tank were described.

The measured values of the progress of the freezing front agreed well with the computed values in the main part of the tank, so that we can conclude that the analysis of the heat transfer in the design stage was reliable and valid.

According to the results of additional actual field measurements of frost heaving pressure in LNG tanks, the analysis of frost heaving pressure used in the design stage was confirmed as also being reliable and valid.

REFERENCES*

Goto,S. Watanabe,O. Iguro,M. and Nakajima,T. 1985. The measurement of frost heaving pressure on LNG inground tank, Proceedings of the fourth International Symposium on Ground Freezing, p.337-341

Sugawara,Y. and Minegishi,K. 1984. Design of an inground storage tank for refrigerated propane, GASTECH'84 LNG/LPG conference, p.466-480

Japan Gas Association Committee on LNG Inground Storage. 1979. Recommended Practice for LNG Inground Storage, p.17- 37

*See page X concerning ISGF papers

5th International Symposium on Ground Freezing, Jones & Holden (eds)
 ISBN 90 6191 824 3

Construction of the Asfordby mine shafts through the Bunter Sandstone by use of ground freezing

S.J.Harvey
Ground Freezing Division, British Drilling and Freezing Company Ltd, UK

C.J.H.Martin
Cementation Mining Ltd, UK

Keywords. Borehole Surveying, Drilling, Economics, Excavation, Shafts, Thawing.

ABSTRACT: This paper reviews the techniques involved in the construction of two 7.32 metre finished diameter shafts through the heavily water bearing Bunter Sandstone formation encountered at the Asfordby Mine.

The operation entailed a sub-surface freeze from a level 275 metres below surface and extended to 405 metres depth. This approach required precise freeze hole drilling and surveying, construction of an underground working chamber to locate and furnish the freeze holes, and a comprehensive monitoring system.

1 INTRODUCTION

British Coal is investing in the new Asfordby Mine, near Melton Mowbray in the N.E. Leicestershire coalfield, to provide a low cost, high productivity, replacement capacity for the 1990's and beyond.

The mine will be the largest in the Midlands with a planned 3 million tonnes/year capacity of low cost coal for the supply to nearby power stations.

Twin vertical shafts, one to carry men and materials and one for skip winding of coal, are being sunk to gain access to 146 million tonnes of reserves in the Deep Main, Parkgate and Blackshale seams.

Work on site commenced in 1985 and will take approximately 10 years to achieve maximum production levels. Specialist contractors are employed for the shaft sinking, ground freezing and pit bottom development works.

2 GEOLOGY

To prove the geology and quality of the coal some 25 deep exploration boreholes were drilled around the area between 1974 and 1983. The geological section of the shaft site is reproduced in Fig. 1.

The presence of the Keuper waterstones and in particular the Bunter (Sherwood) Sandstone were proven to be heavily water bearing. Pump tests carried out have verified that the possible water make into an unlined shaft, without prior treatment, would be approximately 1000 gpm and with virtually full hydrostatic head present.

This problematic zone was identified some 330 metres below surface and extends to a depth of some 390 metres. This section is overlain by the Keuper Marls and Lower Lias, and is situated above the Westphalian Coal Measures, all predominantly dry formations.

3 CONSTRUCTION BACKGROUND

Several methods of ground treatment were considered by British Coal to enable the shafts to be sunk efficiently and safely through the problematic Bunter Sandstone formations. The final decision was made in favour of the Ground Freezing technique. This decision allowed confident programming and budgeting to be carried out from the beginning of the project. On the projected mine development cost of £412 million (1984 prices), if the use of ground freezing had eliminated an 8 week or more overrun to the overall programme then the total cost of freezing would have been covered by the interest charges alone.

It was decided to carry out a sub-surface freeze from an underground chamber constructed above the water bearing zone. The chambers were constructed at a depth of approx. 275 metres below surface which provided a minimum 45 metres of cover to any known waterbearing zone. The freeze was extended into the Coal Measures and terminated at a depth of 405 metres to form an impervious cut off. This would prevent any possibility of ground water permeating under the projected ice wall and into the excavation.

Construction work on the Shaft Sinking Contract commenced in July 1985 with the construction of a reinforced concrete drill pad. A series of three skid rails were set into the concrete to form a level base to support the drill rig subframes and to assist with rig movement between adjacent hole positions. The foundation was designed to with stand a maximum 40 tonne point loading on the inner rails. Thirty nine freeze holes were planned for each shaft on a 16.1 metre P.C.D. together with three monitoring holes. Each freeze hole surface location was boxed out through the concrete foundation with 375 mm diameter polystyrene plugs. The monitoring hole positions were decided upon after a substantial number of the holes were completed and located at positions of largest spacings between adjacent freeze holes.

4 FREEZE HOLE DRILLING

4.1 Drilling

The mobilisation of drilling equipment to site commenced on 5 August 1985. A drilling mud farm was erected in a position central to both shafts. This consisted of a series of mixing and circulation tanks, mud cleansing cyclones, vibrating screens and high pressure/high volume circulation pumps.

Three drilling rigs were brought to site on 19 August 1985, two Failing 2000's and one Failing 2500.

Drilling commenced using a 305 mm dia. rock bit through the surface deposits to a depth of approximately 50 metres. A steel conductor casing 220 mm dia. was then lowered in and cemented into position. This casing prevented erosion of the soft surface deposits and interconnection between adjacent holes being drilled.

After the cement had set, normally a period of 8 hours, drilling then recommenced using a 200 mm mill tooth bit in an open hole.

The drilling mud type and hydraulic design were selected with consideration to the strata type and the expected rate of penetration. The fluid hydraulics had to be sufficient to keep the drill bit cool and clean, whilst quickly removing all the cuttings being produced to prevent the bit "balling-up". Failure to keep the bit clear of cuttings results in substantially reduced penetration rates and the cuttings being ground into much finer particles, which are then more difficult and costly to separate. A circulation rate of 45 to 75 litres/min per 10 mm diameter of drill bit was required to achieve the above.

The drilling mud used was a dispersed water based gypsum and bentonite mixture with additives to control viscosity, pH and inhibitors for the Triassic Mudstones. This mud type was chosen to prevent contamination by the many gypsum and anhydrite bands present, for its long operational life qualities, and for its economics. Very few problems were experienced with the mud system. A 150 mm dia. centrifugal vacuum type pump was used to return the drilling fluid from each drill rig to the mud farm for cleaning. These units initially gave very serious problems with the float control gear and vacuum pumps failing after a short period of pumping. The reason for the problem was the drilling fluid becoming caked-up around the floats which then allowed drilling mud to find its way into the vacuum pumps, thus causing failure. After very limited success by the pump manufacturer, a modification was made by site which greatly improved the reliability.

A drilling target area of 0.5 metres radius around each hole's planned position was specified by the Client. Reservations were made prior to tender on the necessity and practicality of such a restrictive target zone.

The initial drilling operations on site reinforced these views with substantial difficulty being experienced in maintaining the holes within the target. An enlargement of the target zone was agreed with the Client after some 8.5 weeks drilling. The revised target zone, (Fig. 2), had a similar maximum hole spacing but allowed for more deviation to occur, especially below the Freeze Chamber level. This additional target area was used to carry out necessary corrections to the direction of the hole when a survey indicated that it was deviating towards the target limit.

It should be emphasized that a constant angle in the drill hole of 0 deg. 9 min. 0 sec. (0.15 degrees) would still deviate the hole out of the revised target area before it reached the Freeze Chamber level 275 m B.G.L.

4.2 Surveying

Surveying of the drill holes was carried out by lowering a self contained high precision gyroscopic instrument on a wire line within the bore of the drill string. Two main types of instrument were used, single shot and multi shot.

The single shot units were used to record the angle and direction of the drill bit onto a photographic disc every 5-10 metres to evaluate the holes progress. When a single shot indicated that the hole was beginning to deviate, a multi shot survey was carried out to determine the exact course and position of the hole. This instrument is lowered on a monoconductor wire line and the angle and direction changes are indicated instantaneously at surface in analogue and digital format. The accepted accuracy of these instruments is 0.05 degrees per 100 metres, i.e. at 275 metres depth the expected instrument accuracy should be +/- 240 mm.

The average survey error actually recorded at the 275 metre level was 233 mm. A plot showing the surveyed positions of the freeze holes in relation to the located positions is shown in Fig. 3. All the holes were found within their respective target zones at this level. This information was then used to recalculate the surveyed positions of the holes from 275 metres to 405 metres thus eliminating the accumulated error from surface to 275 metres.

4.3 Correctional Drilling

Where a survey result confirmed that a hole was on a course to deviate outside the target zone, corrective measures were employed to redirect the course of the drill bit back on line. This was achieved by the use of a bent sub placed on top of a down hole motor. The bent sub causes deflection to occur by pushing the bit to one side. Angle and/or direction change was achieved by orientating this assembly in the desired direction and holding that orientation whilst drilling until the desired result was obtained.

The down hole motors used were multi stage positive displacement Monieau screw type units consisting of two main parts, the motor body and the rotating bit sub. Rotation of the bit sub is caused when the drilling fluid is pumped under pressure into the cavities formed between the rotor and stator. The power produced by these motors is equivalent to approximately 45 kw with rotational speeds up to 500 revs/min.

When orientating these assemblies, account had to be taken of the torque produced by these motors twisting the drill string, which was locked at surface. Even using the heavy steel drill pipes and collars 15 to 20 degrees of torque induced twisting was experienced.

Due to the relatively high rotational speeds of the motors and the hard bands of strata experienced, it was necessary to resort to polycrystalline diamond compact (PDC) drill bits for the corrective runs. By modifying the setting of the cutters around the periphery, bit life was increased from 300 metres to in excess of 1000 metres as compared to conventional rock bits which were worn out in less than 10 metres.

Of the 85 holes drilled, 18 were drilled without any corrective measures being required whilst the remaining 67 holes required some 300 corrective runs.

Due to numerous surface limitations, e.g. pile foundation positions etc., 15 holes were directionally drilled from an adjacent holes surface position and top hole section (Fig. 4). The majority of these diversions proved to be time efficient in comparison to drilling a hole from surface. This also proved that generally, directional drilling with some angle in the hole to measure, is faster and can be controlled easier than trying to drill vertically where the probability of survey error is greatest.

4.4 Freeze Tube Installation

Upon completion of drilling each hole to 405 metres depth and verification of the survey results, the hole was conditioned with clean mud and the drill string removed. A cementing string was then lowered to the hole bottom and a predetermined quality of Class B cement slurry was injected. The cement was mixed to very accurate proportions with a slurry weight of 1.68 kg/litre. Fluid loss additives and retarders were included in the slurry to give a 24 hour working period before the initial set. The cementing string was then withdrawn and the 125 mm O.D. freeze casing was assembled, each joint being torqued up to specific limits and externally pressure tested to 2700 p.s.i. (18.62 MPa).

The freeze casing, approximately 130 metres long, had a sealed top and bottom plug arrangement. A steel inner pipe 65 mm dia. was also installed at this time and the casing string filled with water to provide negative bouyancy.

When this arrangement was completed, a detatching mechanism was connected to the top plug and the whole casing string was gently lowered into the cement. When the casing was at the correct horizon, important for later operations, the casing was detatched from the lowering string. The lowering string was fitted with a circulation port at the bottom and, after confirmation that the casing was detatched, mud was slowly circulated and the surface returns monitored to ensure that cement was noted. This then confirmed that the freeze casing was

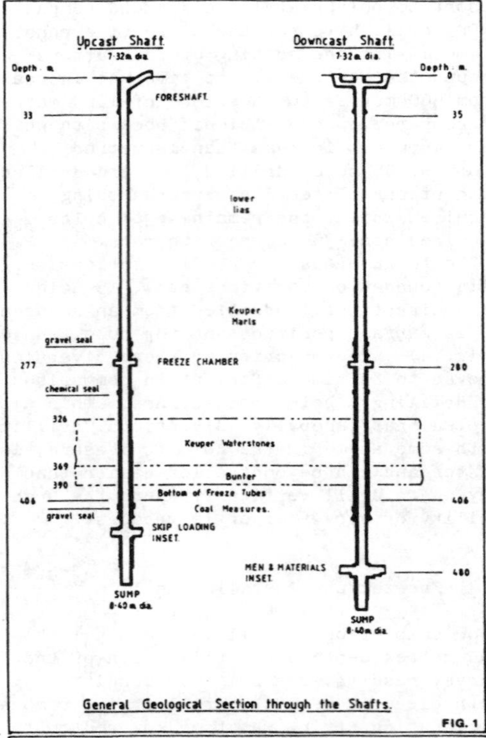

General Geological Section through the Shafts.

FIG. 1

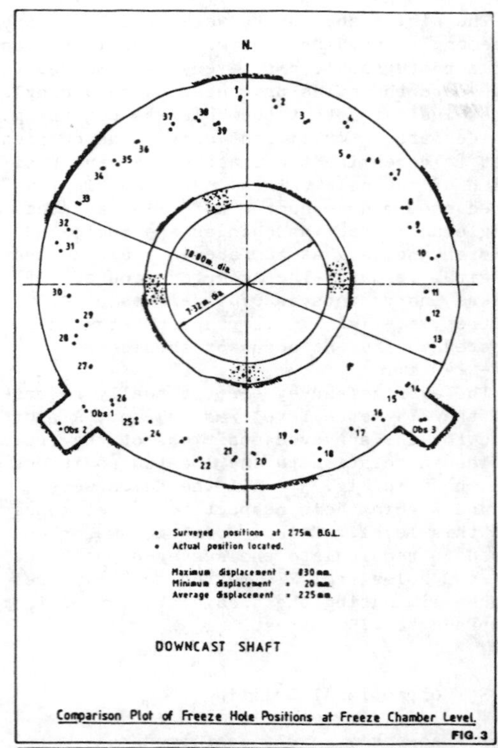

Comparison Plot of Freeze Hole Positions at Freeze Chamber Level.

FIG. 3

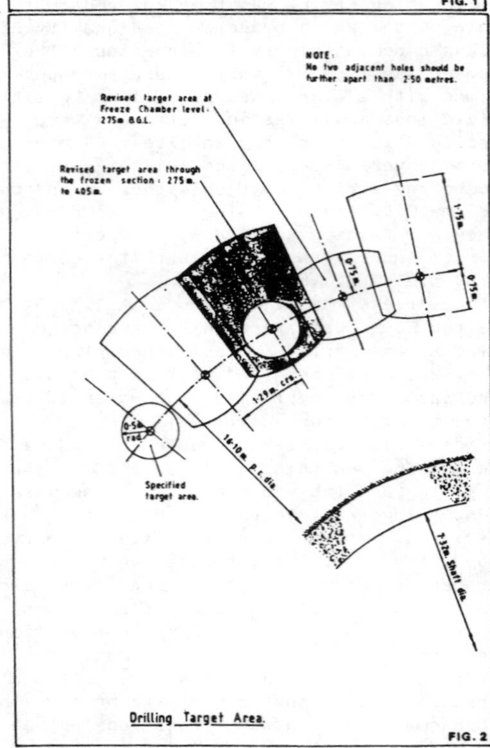

Drilling Target Area.

FIG. 2

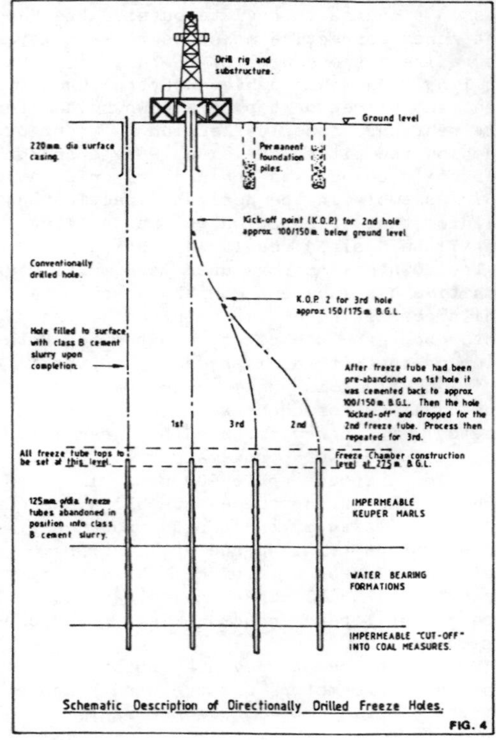

Schematic Description of Directionally Drilled Freeze Holes.

FIG. 4

totally cemented and no possibility existed of water being able to find its way up the drill hole annulus to flood the shaft during Freeze Chamber construction.

Immediately the above was confirmed, the hole above was then cemented to the desired level using a Class B cement slurry with accelerator as required.

Drilling and abandoning works were completed on schedule to the Upcast Shaft on 26 January 1986 and 5.5 weeks ahead of schedule to the Downcast Shaft on 16 April 1986. The use of a fourth rig was necessary on the Upcast Shaft for a period of 6 weeks to achieve the programme due to the directional drilling difficulties experienced in the first 8.5 weeks of drilling. The total drilling carried out for the two shafts amounted to 34425 metres with the installation of 10220 metres of freeze tube casings.

5 CONSTRUCTION OF THE SHAFT AND FREEZE CHAMBER

5.1 Construction

When the freeze hole drilling was completed to each shaft, piling rigs installed the large diameter concrete piles to form the foundation for the permanent headgear.

The shafts were then sunk by conventional means to the Freeze Chamber level.

Construction of the Freeze Chamber then began by driving a top heading in and opening out. Roof support was by radial beams supported at the shaft from hanging rods cast into the concrete lining and steel upright legs at the back of the Chamber (Fig. 5). Roof bolts and mesh provided supplementary support together with temporary square work. The freeze hole positions were located in the roof and as the Chamber was further opened up, care was exercised during blasting to prevent damage to the freeze casings. Spoil was mucked away by Eimco 630 into the main hoist kibbles.

Exposed strata was also protected from long term weathering by the application of gunite.

5.2 Furnishing of Freeze Tubes

Once all the freeze casings had been located at the correct horizon and identified, the top plugs were backed off using an hydraulic power tong to expose the male thread on the casing. Class 600 weld neck flanges which had been specially threaded were then screwed on and torqued up. An hydraulic pressure test of 1125 p.s.i. (7.76 MPa) was then carried out for a 10 minute duration to confirm the competency of the joint and casing. The 65 mm dia. inner pipes were then connected onto the flow head (candlestick) arrangement and the flange faces bolted down to 350 lbsf/ft (48.4 Kgf/m) torque.

The shaft sumps were then extended down to approx. 290 metre depth to accommodate the stage during construction of the permanent headgear, and the backwall grouting operations were completed to surface.

Installation of the brine delivery and return ring mains was then carried out by suspending them from the radial roof support girders to minimise any effect from creep which may occur in the floor, both prior to and during freezing operations. Flexible connections between the ring mains and freeze tubes were made using high pressure, low temperature hydraulic hoses, the flow being controlled by valves. (Fig. 6).

To protect the system from total loss of brine in the event of a burst hose etc., remotely operated main valves were fitted to the shaft range connections. When all this system had been assembled, it was then subjected to a hydraulic pressure test of 1125 p.s.i. (7.76 MPa) for 2 hours. Insulation was effected by applying a 50 mm thickness of British Coal approved phenolic foam and vapour sealed.

5.3 Shaft Mains

Design of the shaft mains had to take account of a number of situations:-

1. The hydrostatic head and working pressure of the circulation medium - $CaCl_2$ brine @ S.G. 1.29;
2. The operating temperature range of the brine +20°C to - 35°C;
3. The contraction of the range on cooling and subsequent expansion when warmed up creating a movement of 165 mm;
4. The shaft operational activities, e.g. blasting.

The grade of steel used for all the underground pipework was chosen to take account of the above situations and the required wall thicknesses calculated. It was decided to support the shaft mains from a purpose made stool constructed 10 metres above the Freeze Chamber. This support was designed to take the total load of the pipe and brine to surface. Special sliding brackets were installed at regular intervals to provide horizontal support, but allow vertical movement during contraction/expansion. All movement thus being accommodated at surface by a flexible hose connection to the surface delivery/return mains.

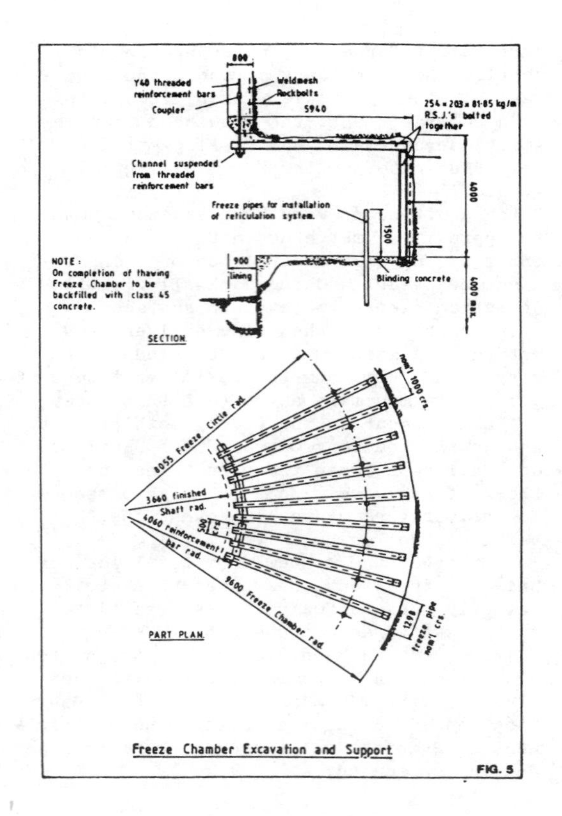

Freeze Chamber Excavation and Support.

FIG. 5

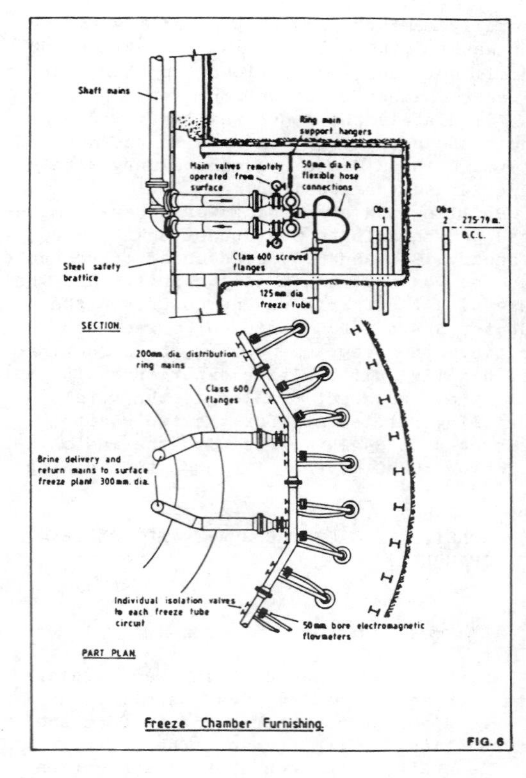

Freeze Chamber Furnishing.

FIG. 6

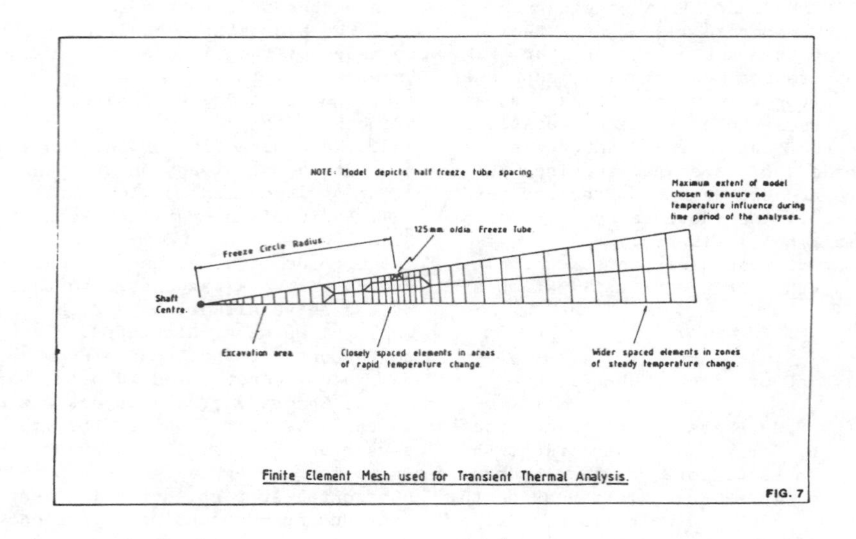

Finite Element Mesh used for Transient Thermal Analysis.

FIG. 7

5.4 Centre Pressure Relief Hole

During the Freeze Chamber installation works, a small hydraulic rig was positioned in the shaft sump and a conductor hole drilled to 36 metres depth. A standpipe 65 mm dia. was grouted in and pressure tested to 400 p.s.i. (2.76 MPa) to confirm its competence. Drilling then continued in open hole at 50 mm diameter through the water bearing formations and into the top of the Coal Measures.

After the drill rig had been removed from the shaft, the standing pressure on the standpipe was 326.34 p.s.i. (2.25 MPa) indicating a 229.8 metre head of water i.e. 59 metres from surface. A pump test was then carried out which confirmed that the Bunter Sandstone was a major aquifer capable of producing flows up to 1000 gals/min. (76 litres/sec) into an unlined, untreated shaft.

6 GROUND FREEZING

6.1 Ice Wall Design and Formation

The ice wall, when formed, must provide a structurally sound support for the prevailing strata conditions and methods of working. In order to design such a frozen structure, specific qualities and behaviour of the strata are required for the input of such a calculation. Very little relevant data for this site was available, therefore assumed data for the Bunter Sandstone, based upon past experiences, was applied with the following:

1. Full hydrostatic pressure was present P_w = 3.9 MPa;
2. Vertical pressure P_v = 5.9 MPa;
3. Horizontal pressure P_h = 1.1 MPa;
4. Frozen compressive strength of the Bunter Sandstone formation = 22 MPa
5. Unfrozen compressive strength = 14 MPa.

A finite element model (Fig. 7) was then analysed and indicated that the worst case would require an ice wall 1.5 metres thick. Allowing for the practical aspects and a factor of safety, the minimum desired ice wall thickness was set at 3.0 metres.

A finite element thermal analysis, using a model shown in (Fig. 8) was also carried out using pessimistic data. The results indicated that the desired 3 metre thickness could be achieved within 15 weeks of freezing. In actual fact, the time to achieve the 3 metre thickness was 10 weeks for the Upcast Shaft and 12 weeks for the Downcast Shaft. This, however, was brought about by programme necessity. The time lag between the two shafts was reduced to 10 weeks, therefore, the Upcast Shaft was required to be frozen ready for when the Downcast freeze was switched on. This then allowed the Downcast freeze to be expedited.

6.2 Freeze Plant

The installed freeze plant capacity was determined on the basis that both shafts may have to operate concurrently. Its total rated output of some 600 tonnes/refrigeration/day consisted of three screw type compressors and two reciprocating type with a combined input capability of 2500 h.p. Each compressor is mounted on its own self contained unit together with evaporator, condenser, control circuits etc. The primary refrigerant used is Ammonia (NH_3) with shell and tube or evaporative type condensers rejecting the heat to atmosphere.

Circulation of the brine to each shaft was carried out by a bank of four pumps arranged in parallel. The flow rate for the prefreeze period was approximately 2500 gals/min. (190 l/sec.)

6.3 Freezing Operation and Monitoring

During the freezing operations, the performance of the freezing systems and progress of the ice wall build up was constantly monitored by a sophisticated logging system. Temperatures, flows and levels, both at surface and underground, were collected and analysed by computer. All important points which were monitored were also protected by an audio visual alarm system and, in the case of the main brine tank levels, should a major brine loss be detected, the underground valves were programmed to close automatically within 15 seconds, thus limiting any major loss.

The individual brine flows to each freeze tube were balanced at the commencement of the freeze and then regularly monitored together with the return temperatures to ensure optimum operation.

During the freezing operations, the permanent winding tower construction works were carried out. Access to the Freeze Chamber was by a temporary winder arrangement installed in the fan drift/air inlet to the shafts. Access underground was, therefore, very limited and could take up to 30 minutes to achieve in an emergency. Such a comprehensive monitoring system was deemed necessary to counteract the unavailability of immediate access. The system was required to be intrinsically safe and British Coal approved.

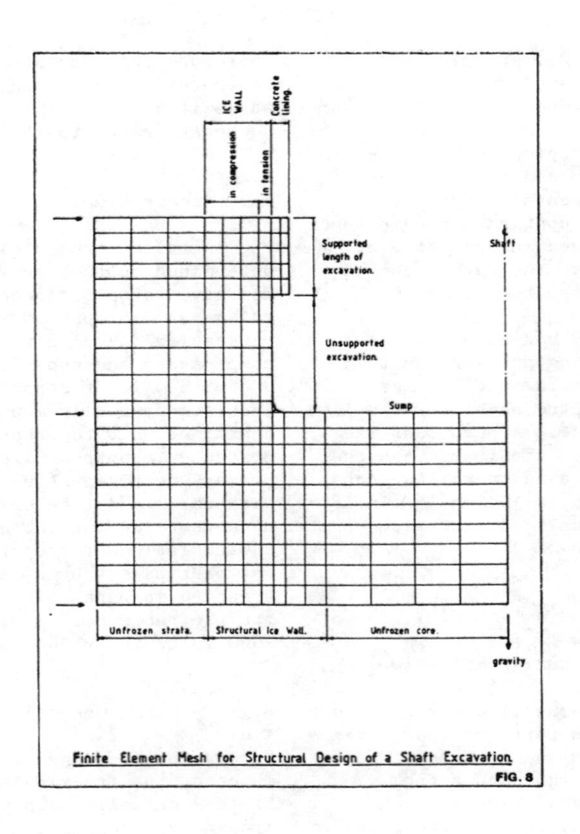

Finite Element Mesh for Structural Design of a Shaft Excavation.

FIG. 8

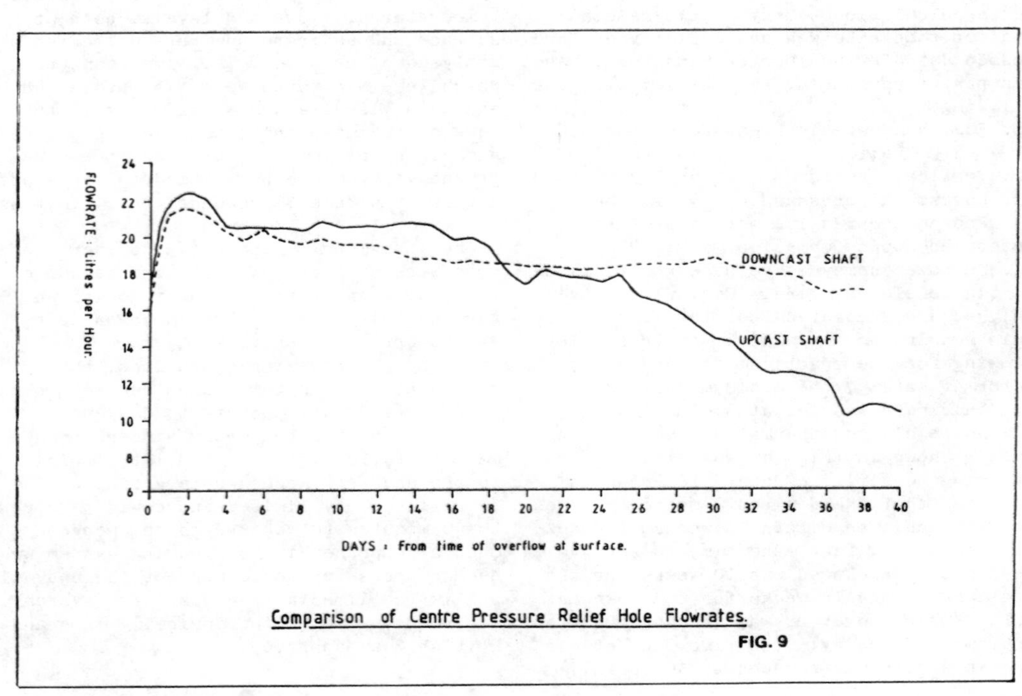

Comparison of Centre Pressure Relief Hole Flowrates.

FIG. 9

Freezing to the Upcast Shaft commenced on 3 April 1987. The centre pressure relief hole was monitored closely and indications that ice wall closure was occurring were first noted on 1 May when the pressure indicated a 0.5 bar increase. Steady small increases continued until 17/18 May when a rapid rise in pressure expelled water at surface. This continued to flow reaching a peak of 21 litres/hour and steadily reducing with time.

Fig. 9 shows the pressure relief hole flows recorded for each shaft.

Freezing to the Downcast Shaft commenced on 16 June 1987 and continued in much the same manner as the Upcast Shaft.

Brine circulation temperatures to both shafts were maintained in the order of -27 to -30°C during the prefreeze periods. During maintenance of the ice walls, the temperatures were kept in the order of -20 to -22°C.

Creep of the Keuper Marls in the Freeze Chambers was monitored regularly by means of levelling fixed reference points in the floor and collating same with shaft survey stations. The floor of the Upcast Freeze Chamber heaved some 10 mm before the commencement of freeze and some 17 mm during the freeze. It can therefore be assumed that any movement that did occur was not freeze induced but rather the natural relaxation of stress in the surrounding formations.

7 CONSTRUCTION OF SHAFT THROUGH FROZEN GROUND

7.1 Excavation

Excavation of the shafts was by conventional drill and blast methods.

A pattern of shotholes to a predetermined depth (see 7.2 below) was drilled in the shaft floor, charged and blasted. After clearance of fumes an Eimco 630 compressed air operated rocker shovel was lowered down onto the blasted muck pile. This machine excavated the blasted rock and loaded it into 3.5 cu.m. capacity kibbles which were hoisted up out of the shaft and tipped. When all the blasted rock had been removed and the floor cleaned up, the sequence was repeated.

Rates of progress when sinking and lining through the frozen ground averaged between 12 and 15 metres per week depending on ground conditions. This was slower than in the upper sections of the shaft because of blasting restrictions and the increased thickness of the concrete lining.

7.2 Blasting

It was imperative that blasting operations in the frozen ground did not cause damage to:

1. The shaft brine mains going into the Freeze Chamber.
2. The Freeze Chamber installation.
3. The freeze tubes in the ground.
4. The ice wall itself.

Care was taken before sinking recommenced below the Freeze Chamber to install steel and mesh protection around the shaft brine mains and to build a barrier enclosing the Freeze Chamber. This eliminated the danger of damage to the installation from flying rock but still left the worry of damage caused by vibration from shock waves.

To minimise this danger the following precautions were taken:

1. Depth of shotholes was limited to 1.5 metres (this was increased to 2.4 metres in certain strata with experience).
2. The amount of explosive being initiated at any time was minimised by maximising the number of delay detonators used in the blast. This was done by using a mixture of both half second and millisecond delays so that in all 16 different delays were used per blast.
3. The outer shotholes, (i.e. those nearest to the freeze tubes and the ice wall), were charged with 25 mm diameter explosive sticks as opposed to the 32 mm diameter explosive used in the rest of the holes.

The above precautions proved adequate and no vibration damage was caused. Although some vibration monitoring was carried out in the Upcast Shaft Freeze Chamber it was not possible to predict at what levels damage to either the installation or the ice wall would occur without testing the system to destruction. This was considered unadvisable!

7.3 Lining

The shaft concrete lining was formed in 6 metre lengths behind the advancing shaft sump. The distance between the lining and the shaft bottom varied between 3 metres to 15 metres depending on strata conditions.

8 THAWING AND ABANDONMENT

The thawing and abandonment operations are an equally important stage of the freezing technique. The main objective of this operation is to ensure that no artificial water paths are allowed to exist between the aquifers, the coal measures and the Freeze Chamber.

The ice wall maintenance was terminated to each shaft upon satisfactory completion of the hydraulic lining through the frozen section and into the Coal Measures.

8.1 Thawing

The frozen mass was allowed to warm up naturally with the main input of heat being by means of the forced ventilation from within the shaft, thus the change of state of the frozen mass from a solid to a liquid occurred primarily from the lining outwards.

Within a period of 2 to 3 weeks after switching off the freeze, the ground temperature around the freeze tubes was generally between -1 and -2 deg. C. Brine circulation was maintained throughout this period and had formed an unfrozen annulus around each freeze tube which allowed them to be permanently abandoned before the ice wall had receded and allowed ingress of the ground water.

8.2 Abandonment

It was necessary to comply with British Coal's Production Instruction 1977/9 for the abandonment of boreholes. The pre-abandonment procedures previously carried out formed the first stage which was then completed after thawing as follows: Once thawed the brine was replaced by water and Class B cement slurry was tremmied into each freeze hole via the 65 mm inner pipe until the hole was full.

The cement slurry was mixed at surface to a weight of 2007 kg/cu.m. and supplied to the Freeze Chamber via a 38 mm dia. shaft range.

The water displaced by the grout from the freeze tubes was piped into a tank and later pumped out of the shaft to the settling lagoons.

8.3 Backwall Grouting

Backwall grouting of the shaft lining through the frozen section took place in two phases.

Phase one was a low pressure filling operation to ensure that the joints between the lengths of concrete lining were completely filled.

The second phase of backwall grouting through the frozen zone was carried out some 3 weeks after the freeze had been switched off when temperature monitoring probes indicated that the rock concrete interface had thawed. This exercise was carried out from a gravel filled gravel seal below the frozen section of shaft and progressed upwards to the Freeze Chamber.

Injection pressure was limited to 1.25 times the theoretical maximum hydrostatic head.

9 CONCLUSIONS

The construction of the shafts in this unusual fashion has allowed concurrent construction of the shafts and permanent head towers to occur for a substantial period of time. This has provided the Client with several advantages in respect of overall programme time and budget. The freezing technique has also accommodated weaker than anticipated water bearing strata which may have resulted in serious delays had a method other than freezing been adopted.

It is noteworthy that after some 160 weeks of the contract the works are still achieving the original programme in respect of the shaft construction and budget. This achievement is partially due to the use of the ground freezing technique. Its dependability in respect of programme and cost has again been demonstrated.

10 ACKNOWLEDGEMENTS

The authors would like to thank

Mr. J.C. Boyle, Chief Mining Engineer, Asfordby, British Coal Corporation,

Mr. P.J. Wheelhouse, Managing Director, British Drilling & Freezing Company Limited,

Mr. J.C. Black, Managing Director, Cementation Mining Limited,

for their permission to present this paper.

Any opinions expressed in the paper are those of the authors and not necessarily those of their respective company's.

5th International Symposium on Ground Freezing, Jones & Holden (eds)
© 1988 Balkema, Rotterdam. ISBN 90 6191 824 3

Thermal design of a frozen soil structure for stabilization of the soil on top of two parallel metro tunnels

H.L.Jessberger & Regine Jagow
Ruhr-University Bochum, Bochum, FR Germany

Peter Jordan
Jessberger & Partner Ltd, Bochum, FR Germany

ABSTRACT: Within the section U6/3 of the new Vienna metro system the telecommunication office Meidling has to be underpassed. The allowable inclination of the foundation is restricted to 1:500 with respect to sensitive relay stations within the building and to numerous transmission lines passing by.
The tunnels lay in tertiary formation. The tunnel cap is located only 1.6 m underneath the foundations of the telecommunication office in quaternary formation. For tunnel construction the New Austrian Tunneling Method was chosen in connection with air pressure stabilization. In order to avoid air pressure loss through quaternary layer and to minimize vertical deformations of the building it was decided to build a 1.0 m thick frozen soil plate between tunnels and building foundations. The intermittent soil freezing method was chosen with brine as coolant. The precalculation of the ground freezing project is based on the analysis of time dependent temperature fields and on the results of frost heave tests.
In this paper the results of the tests and the calculation are reported and discussed. Also the observations and field measurements during the execution of two field tests and during the performance of the tunnel construction are stressed.

1 INTRODUCTION

The line U6 of the new Vienna metro system is composed of a remodelled city railway and a new part with four sections. The 950 m long U6/3 section starts from an access shaft at the Vivenotgasse and leads up to the junction with the line U4 at the Langenfeldgasse (see Figure 1). The telecommunication office Meidling is located between Niederhofgasse and Arndtstrasse, which has to be underpassed by two single track tunnels.

The tunnels are mainly embedded in tertiary silt and clay formations and they are driven using the New Austrian Tunneling Method in connection with air pressure stabilization in order to prevent water migration into the tunnels during construction work. The air pressure is kept at 0.2 to 0.3 bar higher than the water pressure at the tunnel base. In some locations grouting is necessary to reduce the permeability of the overburden. In addition partial dewatering is carried out. The underpassing of the telecommunication office

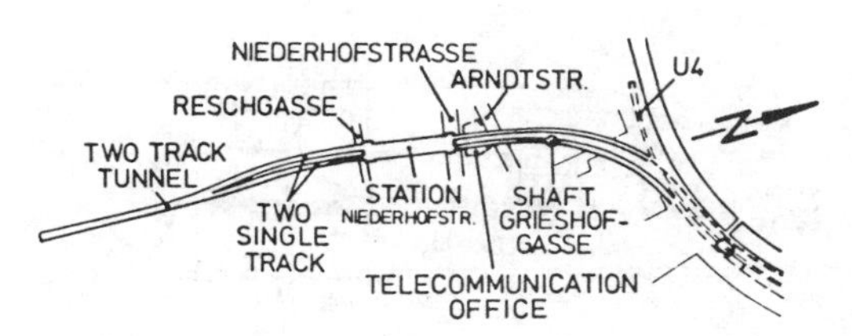

Figure 1. Plan of section U6/3 (Deix,1987).

Meidling leads to special requirements. The vertical differential deformation is restricted to 5 mm because of the sensitive relay station within the building and the allowable inclination of the foundations is restricted to 1:500. It should be mentioned that the distance between the tunnel cap and the building foundation is only about 1.6 m. In order to avoid air pressure loss and to minimize construction related vertical deformations the decision is made to build-up a 1.0 m thick frozen soil plate between tunnels and foundations. Grouting is not permitted with respect to the risk of resonance effects.

2 DESIGN OF THE FROZEN SOIL PLATE

2.1 General

The situation in the vicinity of the telecommunication office Meidling is described in Figures 2 and 3. The frozen soil plate underneath this building is designed with an area of 65 x 22 m. The plate has to be impervious against air pressure. To prevent detrimental frost heave of the frost susceptible tertiary layer the thickness of the frozen soil plate is restricted to 1 m and intermittent freezing is chosen. Furthermore the frozen soil plate should serve for uniform load and deformation distribution during tunneling work.

In finite element calculations the development of time dependent temperature fields during build-up of the frozen soil plate is analysed. The cycles for the intermittent freeze process during maintenance of the frozen soil mass are preinvestigated also with FE calculations. This intermittent freezing will be achieved by switching off and on the refrigeration plant in certain intervals.

In advance of the freezing work underneath the telecommunication office two in-situ field tests are performed in order to adapt the calculated freeze cycles to the real ground conditions. The location of the two test fields is given in Figure 3.

This paper concentrates on the calculations and measurements for the heavily instrumented test field 1, but also some data of the underpassing work are given.

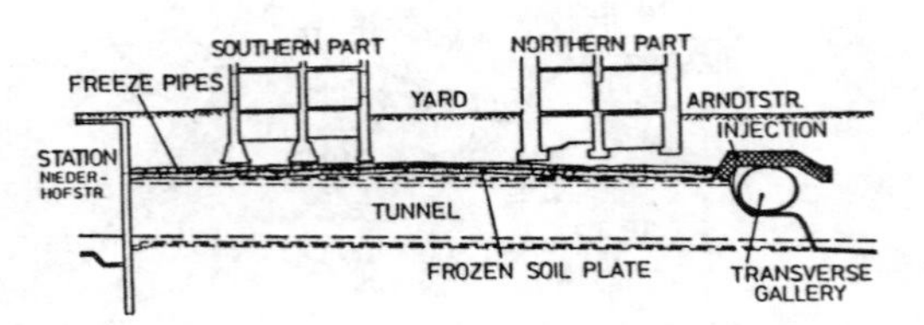

Figure 2. Cross section of telecommunication office "Meidling".

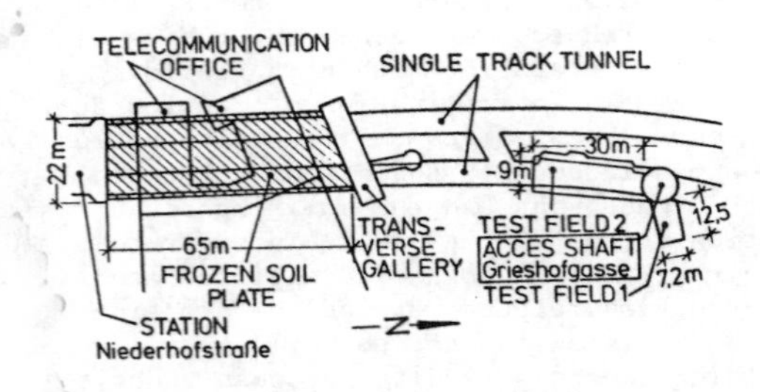

Figure 3. Location of the telecommunication office and the test fields (Fischer, 1987).

2.2 Development of temperature field

Thermal calculations are performed using the finite element method. In the first stage the time for the build-up of the frozen soil plate with 1.0 m thickness is calculated. In the second stage the development of the temperature field during intermittent freezing for chosen freeze cycles is evaluated with the aim of maintaining the frozen soil plate thickness and to minimize frost heave. The parameters of the soil materials (Table 1) which are used for the thermal calculations are given by the City Authority or taken from other projects respectively.

The geometry of the area with the frozen soil plate is reproduced by FE mesh with 337 nodal points and 610 elements as shown in Figure 4. The mesh includes 5 freeze pipes in distinct nodal points indicated by black circles. The planned distance of the freeze pipes is between 0.9 and 1.2 m. In the initial stage the soil temperature is assumed as 12 ℃ and the air

Table 1. Soil parameters.

DIN 4022	gravel, sandy G,s	gravel, sandy/silty G,s,u	clay silty T,ū
ρ_d [kN/m³]	21.9	20.5	16.8
w [%]	6.9	11.2	21.1
S [%]	0.62	1.0	1.0
c [kJ/m³·℃]			
unfrozen	141	158	166
frozen	123	125	123
k [W/mK]			
unfrozen	1.89	1.89	1.20
frozen	2.49	2.49	1.98

ρ_d: dry density
w: water content
S: degree of saturation
c: heat capacity
k: heat conductivity

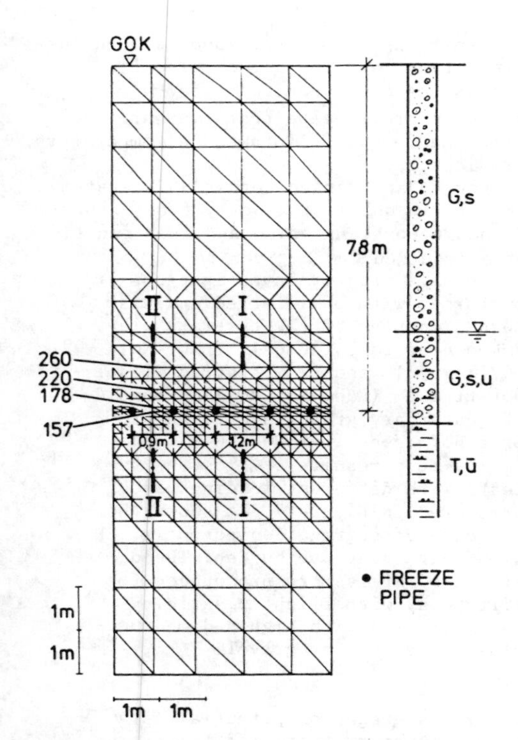

Figure 4. FE-mesh of test field 1.

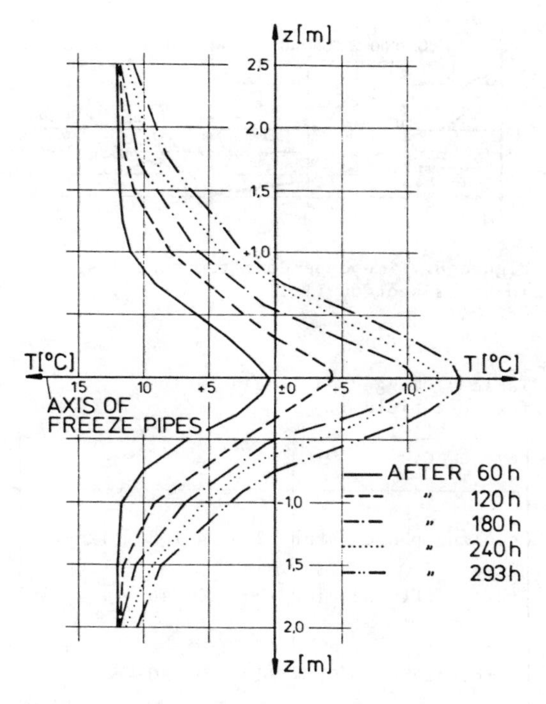

Figure 5. Time dependent temperature distribution for test field 1 section II-II.

Table 2. Freeze time as function of freeze pipe distance.

freeze pipe distance [m]	0.9	1.2
time period for frozen soil plate build-up of 1.0 m [days]	7.5 to 8.5	10 to 12

temperature as 15 °C. The brine temperature starts with 12 °C and is reduced to -30 °C within one day and then kept constant during the build-up of the frozen soil plate.

The result of the thermal calculations for the build-up stage is shown in figure 5, where the time dependent temperature distribution in cross section II-II is plotted for various freeze periods. According to the design the calculation is concentrated on the frozen soil plate thickness of 1.0 m, but including the worst condition of a freeze pipe distance of 1.2 m. For this distance the build-up time is 10 to 12 days, whereas for the designed freeze pipe distance of 0.9 m during a freeze period of 10 to 12 days the thickness of the frozen soil plate reaches 1.5 m (see Figure 5). The calculated time period for build-up of the frozen plate with a thickness of 1.0 m related to -1 °C isotherms is shown in Table 2.

In the calculation the intermittent freeze process starts after 10 days, when the thickness of the frozen soil plate reaches d = 1.0 m between two freeze pipes with the distance of a = 1.2 m. The cycle of the intermittent running of the cooling plant results from precalculated energy consumption and from experiences of other projects. For the calculation 10 intermittent freeze cycles are chosen with 2 hours cooling phase and 12 hours turned-off phase of the cooling plant.

In Figure 6 the calculated time dependent temperatures at three points of cross section II-II (see Figure 4) are plotted. The figure indicates that during intermittent freezing an increase

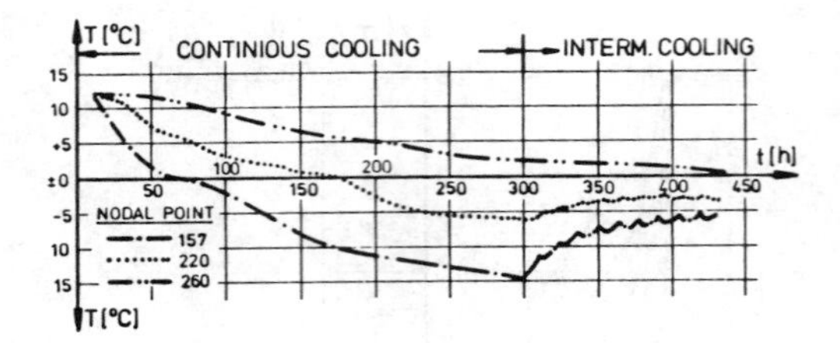

Figure 6. Time dependent temperatures in cross section II-II.

Table 3. Suggested freeze cycles for test field 1.

cycle No.	1	2	3	4
cooling phase [h]	2	4	8	12
turned-off phase [h]	10	20	16	12
brine temperature [°C]	-30	-30	-30	-30

of the temperatures in the frozen soil takes place, but with respect to a certain frozen soil plate thickness a quasi steady state can be reached.

For investigating the influence of different freeze cycles on the frost heave behaviour in test field 1 the cycles of Table 3 are suggested.

2.3 Investigations concerning frost heave

Based on thermal calculations and laboratory tests the frost heave behaviour is estimated using the approach of Konrad and Morgenstern (1981). According to this theory the heaving characteristics of a soil can be represented by the segregation potential SP which is defined as

$$SP = h/\text{gradT} \quad [\text{mm}^2/\text{s}\cdot{}^\circ\text{C}] \quad (1)$$

with h: frost heave rate [mm/s]
gradT: temperature gradient across the frozen fringe [°C/mm]

Frozen fringe means the zone between the frost front and the warmest ice lens. The values of SP can be determined in frost heave tests measuring frost heave, water intake and temperature gradient.

Konrad and Morgenstern (1982) stated that for a quasi stationary frost front SP has a constant value and is a function of overburden for a given soil.

In this project investigations are performed with undisturbed and disturbed samples of the tertiary clay taken near test field 1 (Lab. No. 1977, 1978) and from station Niederhofstrasse (Lab. No. 2041). Samples of quaternary gravel are taken from Grieshofgasse (Lab. No. 2046).

The test results of frost heave tests with different loadings are presented in Figure 7. It shows the decrease of segregation potential SP with increasing overburden. The decrease of SP is more pronounced for silty clay with slight plasticity than for clay with high and medium plasticity and for gravel.

With segregation potential data from laboratory tests and temperature gradients from thermal calculations for quasi steady temperature conditions the time dependent frost heave can be described by equation (1). The heave caused by volume increase during freezing of soil water has to be added to the frost heave during build-up of the frozen soil plate and to the frost heave during maintenance of the frozen soil mass:

$$h(t) = d(t) \cdot n \cdot (0 \text{ to } 0.09) + SP_1 \int_0^{t_1} \text{gradT}_1(t_1) \cdot dt_1 + SP_2 \cdot \text{gradT}_2 (t - t_1) \quad (2)$$

with h(t): time dependent frost heave
d(t): time dependent thickness of the frozen soil
n: porosity
SP: segregation potential (overburden)
gradT: temperature gradient
Index 1: indicates the time period of build-up of frozen soil plate

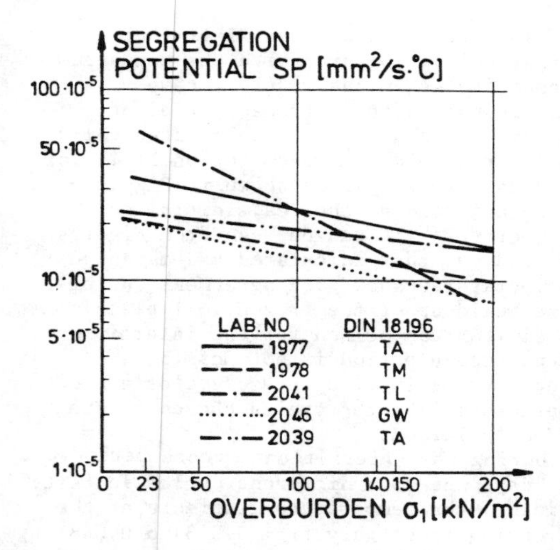

Figure 7. Segregation potential for different soils as a function of overburden.

Index 2: indicates the time period of maintenance of frozen soil mass

The laboratory test results are evaluated in a way that frost heave caused by volume change of freezing water and by ice lensing during build-up phase are summarized to only one factor which is controlled by frost penetration into the soil.

From the tests it is found that this heave rate during frost penetration is of the order of 0.03 mm/cm for quaternary layers and of 0.2 mm/cm for tertiary layers.

From Figure 7 two lines are selected for characterising the segregation potentials of quaternary and tertiary layers, respectively. With an overburden load of $\sigma = 140$ kN/m² it is found

- for quaternary sandy, silty gravel (Lab. No. 2046) $SP = 10 \cdot 10^{-5}$ [mm²/s · °C]

- for tertiary clay with slight plasticity (Lab. No. 1978) $SP = 15 \cdot 10^{-5}$ [mm²/s · °C]

As the surface of the tertiary layer is not known precisely, it is assumed that the frozen tertiary layer has a thickness of 5 cm to 30 cm. For test field 1 the frost heave is calculated for following conditions:

- freeze period 29 days (5 days for build-up of frozen soil plate and 24 days intermittent freezing)

- frozen soil thickness a = 1.5 m

- mean temperature gradients (taken from the temperature measurements in test field 1)
  gradT = 0.065 [°C/cm] in the quaternary layer
  gradT = 0.069 [°C/cm] in the tertiary layer

The frost heave depending on the thickness of the frozen gravel and clay respectively will be of the order of $h_{min} = 8.9$ mm to $h_{max} = 11.4$ mm.

Considering the influence of the soil layer between frozen soil plate and soil surface, it is assumed that the measured surface heave will take 50 to 75 per cent of the calculated frost heave.

3 FIELD OBSERVATIONS

For test field 1 eight horizontal freeze pipes are drilled from the 8 m deep shaft Grieshofgasse (see Figure 3). The freeze pipes with the length of 10 m and a distance of 0.9 m are placed in a depth of 6 m. For temperature measurements three horizontal tubes and one 8 m deep vertical tube are drilled. To control the temperature field 38 temperature probes are placed in the tubes. The surface deformations are measured at 14 points. The vertical inground displacements are measured by 3 sliding micrometers down to a depth of 8 m (Fischer 1987.

In May 1986 the freezing of test field 1 was started. During the build-up of the frozen soil plate a temperature influence of the shaft wall became obvious. The nearer the measuring point to the shaft wall the higher the initial soil temperature, but even in a distance of 2 m to the wall the temperature stays at 11 °C and does not reach 12 °C which is the initial temperature for the FE-calculations. The difference in the initial temperatures leads to an adjustment of the calculated time dependent temperature curves of nodal points 178, 220 and 260 (see Figure 4) which corresponds to the in-situ temperature probes of TH 13/1, TH 11/2 and TH 12/4.

In Figure 8 three measured and calculated temperature curves are presented. The result of the thermal calcu-

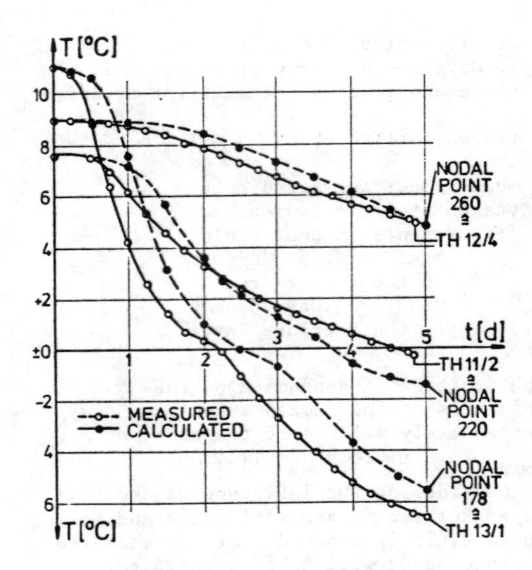

Figure 8. Comparison of measured and calculated time dependent temperatures.

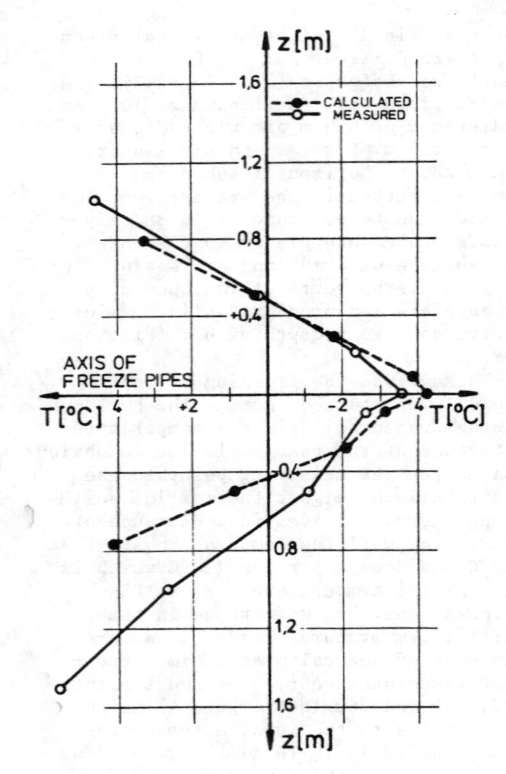

Figure 9. Measured and calculated temperature field in cross section II-II after 5 days.

lation for cross section II-II (see Figure 4) is compared with the measured temperatures in the vertical temperature hole after a freeze period of 5 days. In addition the calculated temperature field in cross section II-II for a 5-day-freeze period shown in Figure 9 is also adjusted to the real initial soil temperature. It can be seen from Figures 8 and 9 that calculated and measured temperatures show good agreement during the build-up of the frozen soil plate. A similar comparison for the intermittent freeze period is not possible, because the in-situ freeze cycles are not performed like the cycles chosen in the FE-calculation.

During the intermittent freeze period of 24 days and four different cycles in test field 1 the temperature gradients at the freezing front vary from 0.043 to 0.088 ℃/cm in the lower part and from 0.041 to 0.076 ℃/cm in the upper part of the frozen soil. Therefore the precalculation of frost movements in chapter 2.3 is carried out with an average gradient of gradT = 0.065 ℃/cm for the quaternary and gradT = 0.069 ℃/cm for the tertiary soil.

In Figure 10 the measured vertical deformations in test field 1 are shown with respect to sliding micrometers GM 51/2. These results are evaluated from sliding micrometers which indicate a maximum settlement underneath the frozen soil plate of 4 mm and a heave on top of the frozen soil plate of 5.6 mm, but for the frost heave effect the amount of both values together has to be taken into consideration. Table 4 gives the data for comparison with the measured and calculated frost heaves of sliding micrometers GM 51/2 and 3.

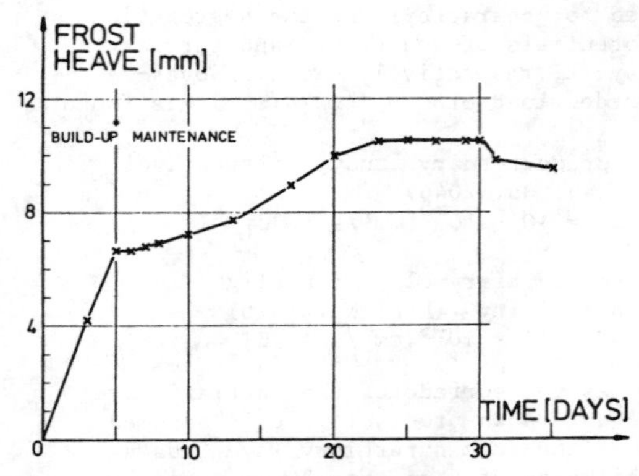

Figure 10. Accumulated frost heave in the vicinity of the frozen soil plate during test period.

In test field 1 the surface heave is 5.5 mm which is about 62% of the total vertical movement and which is in good agreement with the estimated data given in chapter 2.3. The measured inclination of the surface, caused by ground freezing is between 1:1,150 and 1:2,000 and well below the required limit of 1:500.

4 TEST FIELD 2 AND FROZEN SOIL PLATE FOR UNDERPASSING THE TELECOMMUNICATION OFFICE

After the termination of test field 1 the Griesgasse shaft was sunk for another length of 8 m and test field 2 is performed. Test field 2 lays on top of the tunnel which is excavated from the shaft in the direction to the telecommunication office. The observed vertical movements during this test field 2 are influenced by the irregular upper bound of the tertiary layer.

Based on the experiences of the two test fields the freeze concept for underpassing the telecommunication office is established. The frozen soil plate underneath the building is frozen in two sections. In the northern part 25 freeze holes are drilled from a transverse gallery (see Figure 3) with a length of 28 m and a cross section of 30 m². For the southern part 28 freeze pipes are installed from the station Niederhofgasse. In both sections the length of the pipes is 33 m. To control the temperatures 162 temperature measuring probes are placed. The surface displacements are measured at 30 points (Fischer 1987).

Table 4. Measured and calculated frost heave in test field 1 for a test period of 29 days.

	measured [mm]	calculated [mm]
Build-up of frozen soil plate	5.2 to 6.5	3.9 to 8.1
maintenance of frozen soil plate	2.5 to 5.6	3.3 to 5.0
total frost heave	9.0 to 10.8	8.9 to 11.4

11 days after the beginning of freezing (January 11, 1987) the thickness of the frozen soil plate reaches the designed value of 1 m. In the following intermittent freezing period the cooling plant is running between 2.5 and 7.5 hours per 24 hours. During tunneling with a speed of 2 m/d the thickness of the frozen soil plate is kept to 1.1 m and the measured heave reaches 13 mm. When the measuring points are passed by the excavation, the heave is reduced to values between 3 and 8.5 mm.

Based on precalculations the loadings on the frozen soil plate could be assumed to:

σ = 140 [kN/m²]
under the exterior foundation
σ = 190 [kN/m²]
under the interior foundation
σ = 60 [kN/m²]
under the basement

With these data and results of the frost heave tests the surface heave at the end of freezing period can be calculated to:

$$h_{min} = 6.5 \text{ mm to } h_{max} = 12.5 \text{ mm}$$

(see chapter 2.3). The temperature gradient is assumed to gradT = 0.06 °C/cm.

At the end of February 1987 the excavation passed beneath the telecommunication office. Any defects or damages at the telecommunication office and its sensible instrumentation did not appear.

5 SUMMARY

In this project ground freezing is used to strengthen the soil, but mainly to build-up a pervious layer against air pressure loss. For underpassing the telecommunication office Meidling the inclination caused by differences in vertical displacements are restricted to 1:500. FE-calculations and laboratory tests are performed to predict the time dependent development of the frozen soil plate and the frost heaves. In connection with the experiences made in two test fields the data are adapted to real ground conditions. With the obtained experiences the project "Meidling" is performed without any influence on the operation of the telecommunication office.

ACKNOWLEDGEMENT

The project was planned by Magistratsverwaltung, Vienna. The joint venture of Neue Reformbau, Grün und Bilfinger, Hinteregger und Söhne, Teerag-Asdag AG, Alpine Baugesellschaft, Hazet, Hochtief und Tiefbau, is the main contractor. The ground freezing work is done by the subcontractor Deilmann-Haniel, Dortmund. The permission to publish this paper is gratefully acknowledged.

REFERENCES

Deix, F. und Gebeshuber, J. 1987. Erfahrungen mit der NÖT beim U-Bahn-Bau in Wien. Felsbau 5 Nr. 3.
Fischer, P. 1987. Eisbären in Wien Gefrierprojekt U-Bahnlos 6/3 - Vivenotgasse, Deilmann-Haniel.
Konrad, J.M. & N.R. Morgenstern 1981. The segregation potential of a freezing soil. Canadian Geotechnical Journal 18, p. 482-491.
Konrad, J.M. & N.R. Morgenstern 1982. Effects of applied pressure of freezing soils. Canadian Geotechnical Journal 19, p. 494-505.

5th International Symposium on Ground Freezing, Jones & Holden (eds)

ISBN 90 6191 824 3

Freezing a temporary roadway for transport of a 3000 ton dragline

Derek Maishman
freezeWALL Inc., Rockaway, N.J., USA

J.P.Powers
Aquon Ground Water Engineering, Hackettstown, N.J., USA

V.J.Lunardini
Cold Regions Research and Engineering Laboratory, Hanover, N.H., USA

ABSTRACT: This unusual ground freezing operation - probably the biggest ever accomplished in the United States - enabled a giant dragline 24m wide to walk 700m across the alluvial flood plain of the Green River in Kentucky in one day. The paper describes the environmental constraints that made the procedure necessary and the special pipelaying and ground insulation methods employed. The thermal progress of the project is reviewed and appropriate design methods are elaborated.

1 INTRODUCTION

In an unusual application of ground freezing, a roadway of frozen silty clay, 700m 2300 ft) long and 25m (83 ft) wide, was formed to support the movement of a strip mining dragline weighing 3000 tons to a newly opened mine in Kentucky, U.S.A. A series/parallel array of horizontal freeze pipes, totalling 12,600m (42,000 ft) length, was connected to a central, multi-unit freeze plant with capacity of 1,800,000 kilocalories/hr (600 tons refrigeration). The unusual procedure was cost effective because environmental constraints made conventional methods unacceptable. The frozen roadway successfully supported the dragline during its move. The ground freezing system was designed by empirical methods. After the fact, a methodology was developed for more analytic design, which may be suitable for future projects involving horizontal freezing.

2 WHY GROUND FREEZING WAS SELECTED

To dismantle the dragline into sections that could be moved by conventional transport and then reassemble it, would have been prohibitive in both cost and the time required to accomplish it. It was possible to barge the machine to within seven hundred meters of the new mine, but the final distance was across farm land fields of relatively soft silty clay. Fig 1 is a diagramatic sketch of the walking dragline. Movement is accomplished by alternately advancing a 16.8m (55 ft) diameter base and two rectangular flat shoes each 3m by 17m long (10 ft x 56 ft). The total width across the shoes is 23.7m (77.7 ft). In motion the shoes exert a bearing pressure of 5200 p.s.f. (250 KN/m2) and the base or tub 2450 p.s.f. (118 KN/m2). A road enbankment could have been built by conventional means, but the environmental authorities required that the fields be restored to their original condition after the move was accomplished. When the costs of road construction, removal and restoration of the fields was compared with ground freezing, the latter alternative appeared the preferable solution. The move was completed on schedule and within budget.

3 CONTRACTUAL ARRANGEMENT

The owner's engineers specified the horizontal plan and the frozen thickness of the roadway (Fig. 1). The central 6m (20 ft) wide strip, together with portions of the edge strips, were required to support an estimated 60% of the tub bearing surface area, and to accommodate dynamic loads (up to 38% of the total dragline weight). The contractor was made responsible for the design of the freezing system, and was

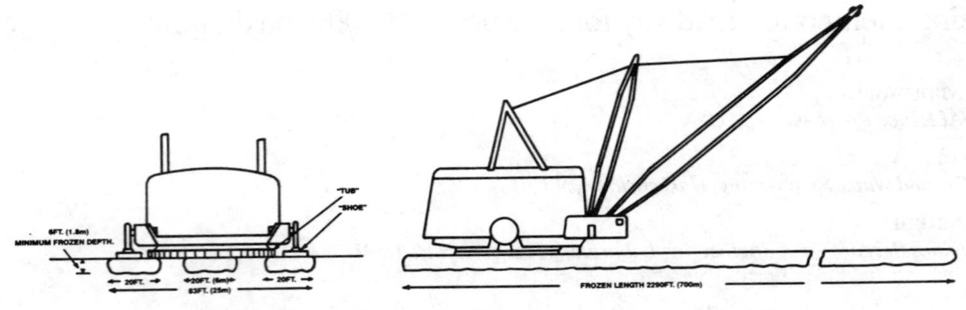

Figure 1: Walking dragline and frozen road requirement

instructed to provide thermal monitoring so that the engineer could confirm that the design thickness had been achieved. The schedule was extremely tight, allowing only ten weeks from notice to proceed until the dragline moved onto the roadway. Within this period the necessary equipment had to be mobilized and installed, and the earth had to be frozen. The owner planned to mobilize a diesel electric generating plant to furnish power for the dragline's move. This plant was mobilized early, to be available to power the freeze plants.

4 DESIGN CONSIDERATIONS

In vertical ground freezings, the frozen surface in the excavation is partly shaded from direct sunlight, and partly protected from the wind. The ambient heat load is moderate, and can be virtually eliminated by an insulating blanket. In tunnel freezings the ambient heat load is internal, from the ventilation air, and is usually moderate. But in this application, a horizontal surface of 15,300 square meters would be exposed to direct radiation from the sun, and would encounter heat load from wind and precipitation runoff. It was anticipated that this ambient load would be a much higher percent of the total freeze system capacity than is the case with more conventional freeze projects.

Two surface insulation systems were considered: sprayed polyurethane, and straw covered with polyethylene sheet. The latter was selected on the basis of cost, and because of environmental concerns over disposal of the polyurethane. The polyethylene sheet chosen to protect the straw was opaque and light colored, to minimize the absorption of heat of radiation.

The spacing and depth of the freeze pipes were chosen by empirical methods, with judgmental adjustment of the experience of conventional freezings to the peculiar geometry and ambient heat loads of this one. Figure 2 is a section through the roadway showing the depth and spacing of the freeze pipes. The design was intentionally conservative, because the tight schedule allowed a limited time for formation.

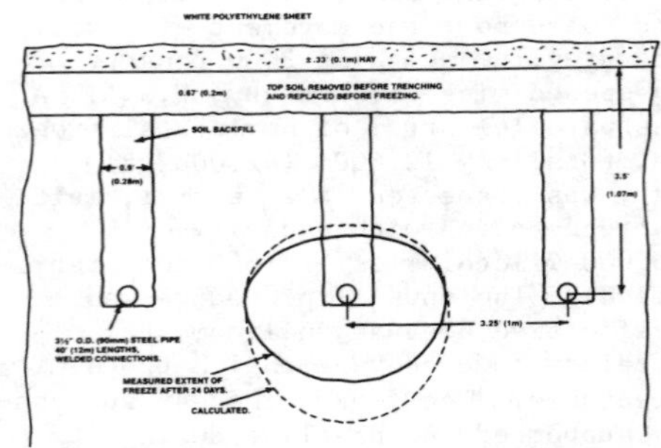

Figure 2: Section through part of roadway including results after 24 days of freezing

The central freeze plant included four plants of 120 nominal tons capacity, and two plants rated at 40 tons, all connected to a two-compartment brine tank and central pumping station. To provide compatibility between the brine circulation systems and the freeze pipe array, a series/parallel piping arrangement was used as illustrated schematically in Fig. 3. The total volume flowing through the freeze pipes was 6 cu. meters (1600 U.S. gallons) per minute, distributed through 36 equal underground loops each 1150 feet (350m) in length.

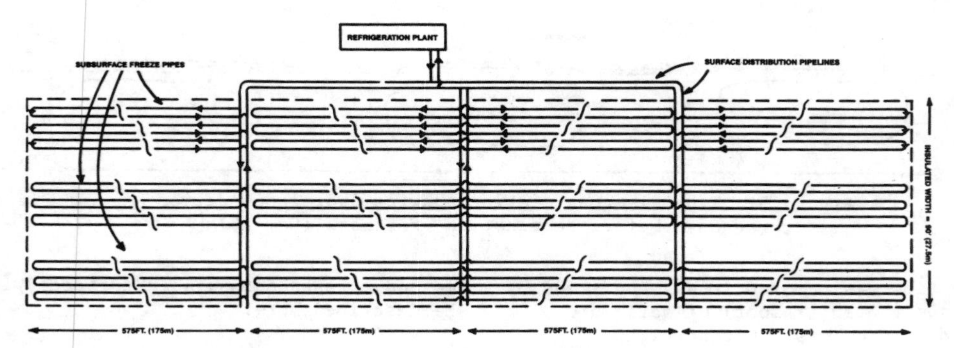

Figure 3: Brine piping circuitry

The operating plan was as follows: upon satisfactory formation of the frozen roadway the freeze plants would be shut down and all brine removed from the system, just before the dragline moved out. The insulation would be removed and disposed of, and surface piping removed. When the move was completed, vertical risers would be removed, the refrigeration plants and brine system demobilized and the surface regraded. The buried horizontal freeze pipes were to be abandoned in place, since they did not present a problem to the resumption of agricultural use of the land.

5 EXECUTION OF THE FREEZE

Promptly upon notice to proceed, mobilization of the freeze plants, pipe and fittings and associated equipment began. For the horizontal freeze pipes, ditching and welding techniques as used for cross country pipelaying were employed. At peak production 3,490m (11,500 ft) of 90 mm (3 inch) pipe were put in place per 10 hour shift. Concurrently the installation of refrigeration plants, brine tank and circulating pumps, surface piping and vertical risers took place. Three weeks after site entry, the refrigeration plants went on line, as the ground surface insulation was being laid down.

Approximately 4" of dry straw was distributed over the ground surface using an agricultural blowing applicator. The polyethylene sheeting was secured by stakes and weighted by metal bars. Some difficulty was encountered during high winds, and the system had to be reinforced. The insulation performed its function satisfactorily, although rain water seepage occurred in places.

Some difficulty was experienced with the diesel generating plant, and there were several shutdowns of the freeze system. The temperature of the brine fell to -15°C during the first eight days of freezing and thereafter decreased somewhat irregularly to -20°C. Ground temperatures were monitored inside 80 vertical pipes driven 2m (6 ft) into the ground. Temperature trends are shown in Fig. 4. Energy abstraction was in the range of 100 to 115 kilocalories per lineal meter of pipe per hour (120 - 140 BTU/ft/hr). Weather conditions were generally warmer than expected, the daytime levels varying between 8°C and 30°C.

Nevertheless, seven weeks after the freeze started, monitoring indicated that the roadway was equal or beyond its design thickness. A typical profile at six and one half weeks is shown in Fig. 5. Freezing was stopped on November 15th, 1985. When the

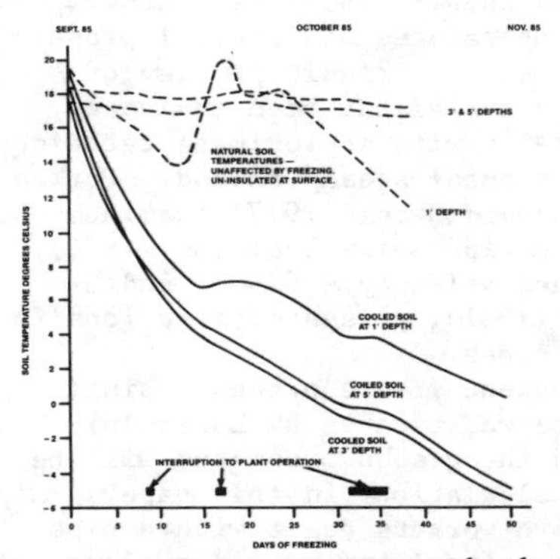

Figure 4: Temperature trends during the freezing period

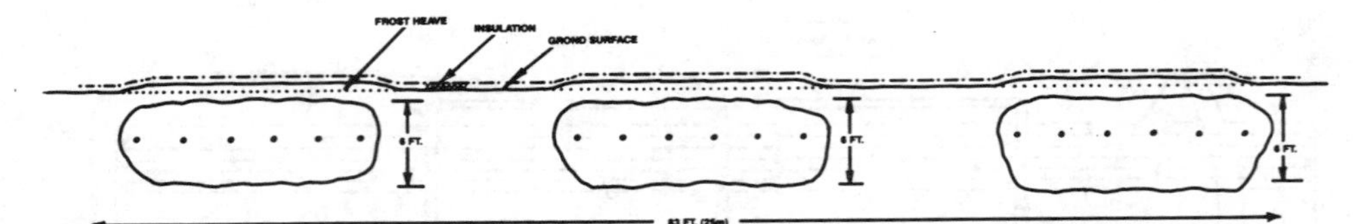

Figure 5: Typical cross-section showing average condition three days before termination of freeze (Day 46)

insulation was removed it was confirmed that about 0.3m (1 foot) of soils at ground surface was, in general, unfrozen. The dragline successfully walked on November 19th. When the dragline moved on to the frozen roadway, the unfrozen and very sticky clay adhered to the base (tub) and threatened to build up to the point where the walking mechanism would be unable to function. Two bulldozers had been mobilized as a contingency. They dozed the unfrozen soil to the sides of the roadway and the move resumed. The dragline reached its destination without further difficulty.

After ten weeks of preparation, the move of 700 meters across the fields took only seven hours.

6 SUGGESTED DESIGN METHOD

After completion of the project, one of the authors (3) undertook to develop a design methodolgy appropriate for this kind of work. No exact solution exists for phase change in the cylindrical geometry, however, the problem of a pipe surrounded by an infinite medium, i.e. infinite burial depth, has been evaluated numerically by Lunardini (1980), Sparrow et al. (1978), Tien and Churchill (1965), Carslaw and Jaeger (1959), and others, for various temperatures and thermal properties. The more difficult problem of finite pipe burial has been evaluated systematically with approximate techniques such as the quasi-steady method. Porkhaev (1963,1970) and Hwang (1977) examined the uninsulated pipe while Thornton (1976), Seshadri and Krishnayya (1980) and Lunardini (1981), presented solutions for insulated pipes.

The transient freeze around a single buried pipe was treated by Lunardini (1983) and the graphs presented will be used for calculations in this paper. All of the above workers dealt with a pipe beneath an uninsulated ground surface. A method to handle insulated surfaces will be described here.

7 STEADY STATE SOLUTION

Solutions are available for the steady state - after phase change ceases - using source-sink images or conformal transformations, Carslaw and Jaeger (1959) and Lunardini (1981a). The details will be given for freezing, as noted in Figure 6, but thawing is essentially the same. The thermal properties of the medium are constant, but different, for temperatures above and below the phase change value.

The temperatures and heat flow from a pipe are given by

$$q = 2\pi\, k_2 \frac{b(1+\beta)\,(T_o - T_f)}{\beta}\ , \tag{1}$$

$$\frac{T_1 - T_f}{T_p - T_f} = (1+\beta)\, f - \beta\ , \tag{2}$$

$$\frac{T_2 - T_o}{T_f - T_o} = \frac{1+\beta}{\beta}\, f\ . \tag{3}$$

The geometric function f is given by

$$f(\zeta,\xi) = \frac{b}{2} \ln \left\{ \frac{\zeta^2 + [\xi + a]^2}{\zeta^2 + [\xi - a]^2} \right\}. \tag{4}$$

The value of the function f, at any time, on the phase change interface is defined as f_o,

$$f_o = \frac{b}{2} \ln \left\{ \frac{\zeta_o^2 + (\xi_o + a)^2}{\zeta_o^2 + (\xi_o - a)^2} \right\}. \tag{5}$$

For the steady state the phase change interface is simply the location of the T_f isotherm, given by

$$f_{o\infty} = \frac{\beta}{1+\beta}\ . \tag{6}$$

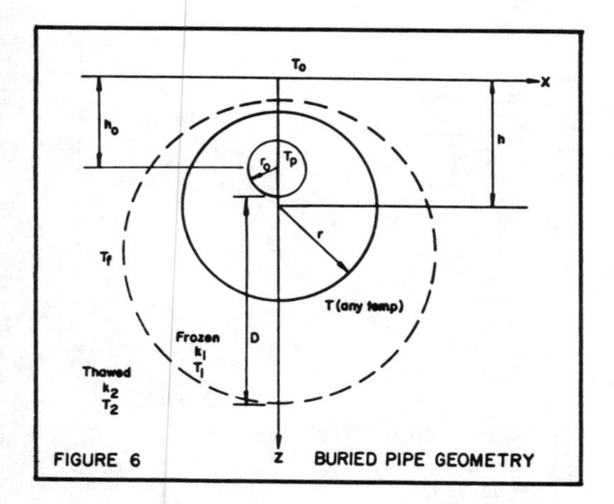

FIGURE 6 BURIED PIPE GEOMETRY

The limiting freeze depth can be evaluated from

$$\xi_{o\infty}/a = \frac{(\mu + a)^K + 1}{(\mu + a)^K - 1} = \frac{p + 1}{p - 1} , \quad (7)$$

where $K = \beta/1+\beta$, $p = \exp(f_{o\infty}/b)$.

In this equation $\xi_{o\infty}$ denotes the depth to the bottom of the freeze interface on the plane of symmetry where $\zeta_o = 0$. (It is also the steady state value of ξ_o, with phase change, after infinite time). The depth to the center of the T_f isotherm and its radius are

$$H_{o\infty} = a\left(\frac{e^{2K/b} + 1}{e^{2K/b} - 1}\right) = a\,\frac{p^2 + 1}{p^2 - 1} , \quad (8)$$

$$R_{o\infty} = \xi_{o\infty} - H_{o\infty} . \quad (9)$$

The heat flux at any point along the ground surface is

$$q/A = -k_2 \frac{(T_f - T_o)}{K r_o} \left(\frac{\partial f}{\partial \xi}\right)_{\xi=0} . \quad (10)$$

Thus

$$q/A = -\frac{2k_2\, ab\, (T_f - T_o)}{K r_o} \frac{1}{(\zeta^2 + a^2)} . \quad (11)$$

The normal heat flux at the origin ($\zeta=0$) is

$$q/A = \frac{2\, k_2\, b\, (T_o - T_f)}{K r_o a} . \quad (12)$$

7.1 Effect of Surface Insulation

The effect of surface insulation will be to decrease the effective ground surface temperature T_o to a value of T_o'. The heat flow through the ground, at the surface, is equated to the heat flow through the insulation. The total heat flow at the surface through a region of half-width ζ_s is found by integrating Eq (11) over the distance ζ_s and is

$$q = 2\, k_2 \frac{(1 + \beta)}{\beta}\, b \tan^{-1}\left(\frac{\zeta_s}{a}\right) (T_o - T_f) . \quad (13)$$

If the separation distance between the pipes is S then one-half of this value can be used to evalute ζ_s. The temperature T_o' is given by

$$T_o' = T_f + \frac{T_o - T_f - A\, k_{12}\, (T_f - T_p)}{1 + A} , \quad (14)$$

where $A = \dfrac{2a\, Q\, k_{21} \tan^{-1}(\zeta_s/a)}{\zeta_s}$,

$$Q = \frac{k_1}{k_i} \frac{d}{r_o} \frac{b}{a} .$$

It follows that

$$\beta' = \frac{\beta - A}{1 + A} . \quad (15)$$

The value of β' cannot be less than zero since this represents the situation when the soil is initially at T_f.

8 TRANSIENT SOLUTION

The preceding equations are for the steady-state, however, they are valid for intermediate times if the appropriate value of f_o is used. For any intermediate time the value of f_o is found from Eq (5). For convenience the equation can be evaluated at $\zeta_o = 0$, then

$$f_o = b \ln \left(\frac{\xi_o + a}{\xi_o - a}\right) . \quad (16)$$

From the geometry of the problem it is clear that

$$\xi_o = D/r_o + \mu + 1 . \quad (17)$$

The value of the total freeze depth, ξ_o, is found from the graphs of Lunardini (1983). These graphs, for a single pipe, can be used with β replaced by β'. Four of the graphs are reproduced here as Figures 7-10.

The theory presented up to this point neglects the sensible heat around the pipe. This can be accounted for by using an effective latent heat given by Lunardini (1981a) as

$$L_e = L\left(1 + C_{21}\beta\, S_T + \frac{S_T}{2}\right) . \qquad (18)$$

However, since each pipe of a series of pipes need freeze only the soil between any two pipes, the system of multiple pipes will freeze faster than a single pipe. To partially compensate for this the sensible heat will be ignored for multiple pipe systems.

The above equations and the graphs of Lunardini (1983) can now be used to desigr a soil freezing pipe system.

9 FREEZEWALL DESIGN STUDY

The task was to freeze a 6.1 m (20 ft) strip of silty clay to a depth of at least 1.83 m (6 ft) within a freeze time of 49 days. The pipe size chosen was 8.89 cm (3.5 in.) O.D. steel pipe with a wall thickness of 4 mm. The ground surface wac insulated with 10.2 cm (4 in.) of straw having an effective thermal conductivity, k_i, of 0.05-0.25 W/m-°C (see Appendix 1).

The soil parameters are: γ_t = 1.97 g/cc (120 - 125 lbm/ft^3), W = 20 - 27% (saturated soil), $\gamma_d = \gamma_t/(1+W/100)$ = 1.57.

The soil thermal properties are taken from Lunardini (1981a) for fine grained soils.

	Frozen	Thawed
k (W/m-°C)	2.04	1.40
C (cal/cm^3-°C)	.45	.65

It was assumed that 80% of the soil water freezes, since the soil is silty-clay, then $L = .8\, \gamma_d \ell\, W/100 = 26$ cal/cm^3.

This value of L is probably high since quite large values of soil density and water content were reported. Very likely the actual freeze depth will exceed the predictions based on a latent heat of 26 cal/cm^3. The soil ambient temperature wac 17.5 - 19.5°C and the soil surface temperature was assumed to be 18.5°C.

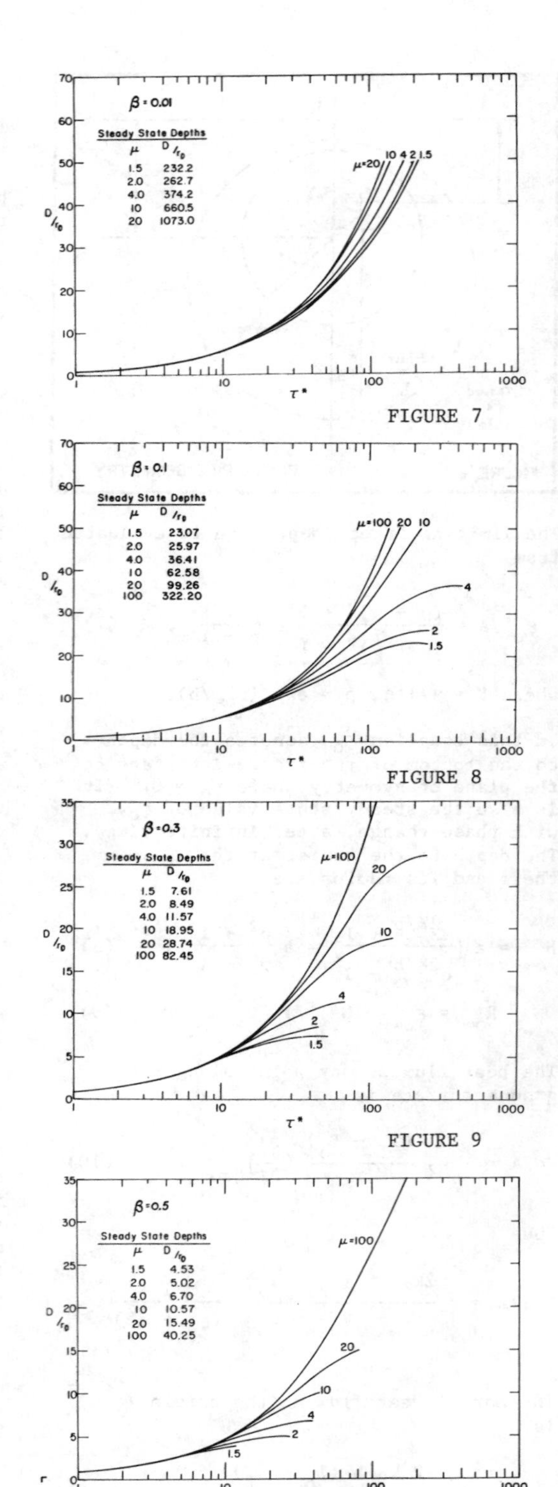

FIGURE 7

FIGURE 8

FIGURE 9

FIGURE 10

Weather data showed a mean air temperature of 15.7°C. Since this does not include solar effects, the value of 18.5 is probably reasonable. The pipe surface temperature will be on the order of -20°C. The following cases will be examined for r_o = 4.45 cm, L = 26 cal/cm^3.

Case	μ	T_p (°C)	T_o	k_i	τ	β
1	24	-20	18.5	.132	801.2	.6343
2	24	-15	18.5	.132	600.9	.8457
3	20.57	-20	18.5	.118	801.2	.6343
4	20.57	-15	18.5	.118	600.9	.8457

Using Figures (7-9) the calculated freeze quantities are given in Table 1.

Table 1. Calculated Freeze for Insulated Ground Surface After 49 Days.

Case	β'	ξ_o	y_o^*	r_f^*	Unfrozen layer*
1	.0978	48.67	216.4	82.0	52.3
2	.2399	43.77	194.6	68.1	58.4
3	0	46.67	207.5	83.6	40.4
4	.1091	43.23	192.3	74.4	43.4

* (cm)

Clearly any of these cases will satisfy the requirements. The actual burial depth used was 106.7 cm with a design pipe temperature of -20°C and the pipes separated by 99.1 cm (39 in.).

9.1 Predictions and Experimental Data

The ground freezing was monitored for freeze depths and temperatures. It was found that the average pipe temperature over the 49 day freeze was -14.8°C. The same values as in Table 1 (except as noted) were used to estimate the freeze depths. Table 2 gives the results obtained with various values of the straw insulation.

Table 2. Effect of Insulation on Freeze After 49 Days, μ = 24, T_p = -14.8°C

k_i	Max. Freeze Depth (m) Calc.	Max. Freeze Depth (m) Meas.	Unfrozen Layer (cm) Calc.	Unfrozen Layer (cm) Meas.
.270	1.71	2.36-2.41	66.5	43.4-52.8
.150	1.89	"	60.2	"
.075	2.11	"	53.8	"

The case for k_i = 0.075, h_o = 106.7 cm gives good predictions compared to the data. In any case it exceeds the required freeze depth of 1.83 m by a safe margin. The pipe burial depth mainly affects the total freeze depth and the unfrozen layer above the pipe as shown in Table 3. Based on the 106.7 cm burial depth a center-to-center pipe distance of 78.7 cm will have a nearly complete freeze overlap.

Table 3. Effect of Burial Depth on Freeze After 49 Days (k_i = .075).

h_o (cm)	Max. Freeze Depth (m)	Max. Freeze Radius (cm)	Unfrozen Layer (cm)
106.7	2.11	78.7	53.8
91.4	1.96	76.7	42.7
76.2	1.81	74.4	32.0

The effect of air temperature was evaluated, as shown in Table 4, for the same parameters as in the previous tables.

Table 4. Effect of Air Temperature on Freeze After 49 Days.

T_o	Max. Freeze Depth (m)	Max. Freeze Radius (cm)	Unfrozen Layer (cm)
18.5	2.11	78.7	53.8
20.5	2.04	74.4	55.6

The choice of an average air temperature of 18.5°C is quite reasonable since the solar input effect was excluded from the surface temperature used. Using the actual average air temperature does not significantly change the predictions. The data are from Evansville, Ind. (even these values differ from the nearby Evansville Airport data by .8-1.7°C), thus the Kentucky site can be expected to be somewhat warmer.

The freeze bowl was also estimated after the first 24 days with the values given in Table 5.

The agreement is fair especially since it is known that the quasi-steady method has the largest error during the early growth stages. The result can also be seen with Figure 2.

The quasi-steady predictions for a single buried pipe can be used for engineering design with some confidence. This avoids the need for numerical calculations and gives rapid estimates.

Table 5. Freeze Depth After 24 Days, T_p = -11°C, S_T = .1903, β = 1.1532, L = 26.0, k_i = .075.

Freeze Depth (m)		Mean Freeze Radius (cm)		Unfrozen Layer (cm)	
Calc.	Meas.	Calc.	Meas.	Calc.	Meas.
1.69	1.39	50.5	37.6	67.6	80.0

10 ACKNOWLEDGMENTS

The owner of the project was Peabody Coal Company. Mr. J. C. Turner was project manager. The freeze concept, and the design thickness of the roadway was developed by Law Engineering Testing Company, with Mr. N.G. Schmitt in charge. The design of the ground freezing system and its execution was by freezeWALL, Inc., a subsidiary of Moretrench American Corporation, Rockaway, NJ, USA. Mr. H. R. Seybold was project manager for freezeWALL.

APPENDIX 1. THERMAL CONDUCTIVITY OF WET STRAW

The surface insulation was 10.2 cm of straw. This presents some difficulties as to the thermal conductivity of the straw. The table below lists some materials and thermal conductivities.

Table 6. Thermal conductivities of common materials (Eckert and Drake 1959).

Material	Thermal Conductivity W/m-°C
Cardboard	.029
Cotton	.027
Cork	.020
Fiber	.209
Wood	.064 - .174
Air	.023
Water	.582

The straw solid material is assumed to have a thermal conductivity of 0.21 W/m-°C and the thermal conductivity of a mixture is estimated using the geometric mean

$$k_i = (k_a)^{x_a} (k_s)^{x_s} (k_w)^{x_w} \quad (19)$$

where x_i = volumetric fraction of each constituent, a,s,w - refer to air, solid (straw), water. The value of k_i for a range of volumetric fractions is given below.

Table 7. Effective thermal conductivity of 10.2 cm of straw.

x_w	x_a	x_s	k_i (W/m-°C)
.01	.5	.49	.071
.1	.41	.49	.094
.2	.31	.49	.130
.2	.1	.7	.209

NOMENCLATURE

A	$\frac{2a\, Q\, k_{21}}{S} \frac{\tan^{-1}(\zeta_s/a)}{\zeta}$
a	$\sqrt{\mu^2 - 1}$
b	$1/\ln(\mu + a)$
C	volumetric specific heat
d	thickness of surface insulation
D	phase change thickness beneath pipe
f, f_o	functions defined in text
h_f	depth to center of freeze bowl
h_o	depth to center of buried pipe
H_o	depth to center of phase change isotherm circle, h_f/r_o
k	thermal conductivity
k_{12}	k_1/k_2
k_i	thermal conductivity of insulation
K	$\frac{\beta}{1+\beta}$
ℓ	latent heat of fusion of water
L, L_e	volumetric and effective latent heat
p	$e^{f_o/b}$
q	pipe heat transfer rate

Q	$\frac{k_1}{k_i} \frac{d}{r_o} \frac{b}{a}$
r	radius of arbitrary isotherm
r_f	radius of freeze bowl
r_o	radius of pipe
R_o	phase change radius r_f/r_o
S	separation between pipes
S_T	$C_1 (T_f - T_p)/L$
t	time
T	temperature
T_f, T_o, T_p	freezing, ground initial, and pipe temperatures
T_o'	temperature between ground surface and insulation layer
W	water content of soil
x,y	Cartesian coordinate
x_i	volumetric fraction of ith component
y_o	phase change depth on plane of symmetry
β	$k_{21} \frac{(T_f - T_o)}{(T_p - T_f)}$
β'	$k_{21} \left(\frac{T_f - T_o'}{T_p - T_f}\right)$
γ_d, γ_t	dry and bulk densities of soil
ζ	x/r_o
ζ_s	$\frac{S}{2r_o}$
μ	h_o/r_o
ξ	y/r_o
ξ_o	y_o/r_o
τ	$\frac{k_1 (T_f - T_p)t}{r_o^2 \; L_e}$
τ^*	$2\sqrt{\tau}$

Subscripts 1,2: Frozen, thawed values.

REFERENCES

Carslaw, H.S. and J.C. Jaeger (1959) Conduction of Heat in Solids, 2nd ed. Oxford at the Clarendon Press.

Eckert, E.R.G. and R. Drake (1959) Heat and Mass Transfer, McGraw Hill Book Co., New York.

Gold, L.W., G.H. Johnston, W.A. Slusarchuk and L.E. Goodrich (1972) Thermal Effects in Permafrost, Proc. Can. Northern Pipeline Res. Conf: 25-45.

Hwang, C.T. (1977) On Quasi-static Solutions for Buried Pipes in Permafrost, Can. Geotechnical J. 14: 180-192.

Hwang, C.T., D.W. Murray and E.W. Brooker (1972) A Thermal Analysis for Structures on Permafrost. Can. Geotech. J. 9(2): 33-46.

Lachenbruch, A.H. (1970) Some Estimates of the Thermal Effects of a Heated Pipeline in Permafrost. U.S. Geol. Surv. Circ. 632.

Lunardini, V.J. (1983) Approximate Phase Change Solutions for Insulated Buried Cylinders, J. Heat Transfer 105(1): 25-33.

Lunardini, V.J. (1981) Phase Change Around Insulated Buried Pipes: Quasi-Steady Method, J. Energy Resources Tech. 103(3): 201-207.

Lunardini, V.J. (1981a) Heat Transfer in Cold Climates. Van Nostrand Reinhold, New York.

Lunardini, V.J. (1980) Phase Change Around a Circular Pipe. U.S. CRREL Report 80-27, Hanover, N.H.

Porkhaev, G.V. (1963) Temperature Fields in Foundations. Proceedings Permafrost Int. Conference, National Academy of Science, Washington, D.C. Pub. No. 1287: 285-291.

Porkhaev, G.V. (1970) Thermal Interaction Between Buildings, Structures, and Perennially Frozen Ground, Nauka Publisher, Moscow.

Seshadri, R. and A.V.G. Krishnayya (1980) Quasi-steady Approach for Thermal Analysis of Insulated Structures. Int. J. Heat Mass Transfer 23: 111-121.

Sparrow, E.M., S. Ramadhyani and S.V. Patankar (1978) Effect of Subcooling on Cylindrical Melting. J. Heat Transfer 100(3): 395-402.

Thornton, D.E. (1976) Steady State and Quasi-Static Thermal Results for Bare and Insulated Pipes in Permafrost. Can. Geotechnical J. 13(2): 161-170.

Tien, L.C. and S.W. Churchill (1965) Freezing Front Motion and Heat Transfer Outside an Infinite Cylinder. A.I.Ch.E. Journal 11(5): 790-793.

5th International Symposium on Ground Freezing, Jones & Holden (eds)
© 1988 Balkema, Rotterdam. ISBN 90 6191 824 3

Ground freezing in non-saturated soil conditions using liquid nitrogen (LN)

L.V.Martak
Municipality of the City of Vienna, Gruppe Grundbau, Austria

ABSTRACT: The new U3 Line of the Vienna Underground passes under the heavily-loaded foundations of a well-known department store. In view of the geotechnical situation, heave or settlements had to be avoided as much as possible. Just below the store one of the tunnels emerges from the silty clay and runs on in sandy or partially saturated gravel. Therefore, we decided to use ground-freezing to ensure the watertightness and strength of the surrounding soil. Laboratory tests showed that sandy gravels with a moisture content as low as 2.8% will have a strength of up to 200 KN/sq. m, with the temperature of the excavation surface at -10°. On the basis of these tests, freezing was carried out. This paper will discuss the difficulties of the drilling work, the supply of liquid nitrogen and the eventual success.

KEYWORDS: Borehole surveying, control, liquid nitrogen, partially saturated soil, tunnels, water content

INTRODUCTION

Ground freezing, although rather expensive, is increasingly used for supporting tunnel excavations, especially under difficult geotechnical conditions. When building an underground railway system in populous urban core areas today, it is necessary to give full consideration to environmental factors. In Vienna, these are in particular groundwater quality, the compatibility of the deformations caused by the tunnelling operation and the dynamic sensitivity of the buildings to the vibrations of the underground traffic. These geotechnical demands by and large rule out chemical grouting. Geotechnical methods which alter the void ratio by permanently compacting the soil are capable of intensifying the resonance between trains in the tunnel and modern reinforced concrete constructions (such as the well-known Herzmansky Department Store in Vienna).

The New Austrian Tunnelling Method (NATM) on the other hand requires that the soil which will later support the excavation be very compact so as to ensure a flat distribution of the settlements near the foundations above.

Ground freezing seems capable of meeting these conflicting demands, especially when liquid nitrogen (LN) is used because it makes it easy for us to control the compaction and sealing of the soil.

DESCRIPTION OF THE GEOTECHNICAL CONDITIONS

The Underground Line U3, connecting two populous Viennese districts, cuts across the city centre. The line is designed to pass through partially saturated quaternary sands and gravels as well as through fully saturated tertiary silts, clays and fine sands. Despite the more or less deep fill, the geological profile in Vienna's urban area generally shows quaternary loess, sand and sandy gravel and, below, tertiary silt, fine sand and clay. Just below the well-known Herzmansky Department Store the Line divides into two tunnels with different levels: one tunnel remains in the silty clay, while the other, twenty-five metres away, rises up into the sandy gravels.

The U3 Line continues along a well-known shopping street, with one tunnel running above the other. The top heading of the shallow tunnel remains 3 to 4 metres beneath the foundations of the department store. The foundations, which are heavily loaded, consist of flat beams and single slabs, with an estimated soil pressure of up to 450 kN/sq. m. Given a tunnel diameter of more than 6.5 m, the thin cover of soil requires artificial support so as to ensure that the building can cope with the settlements. Theoretically, the inspection of the modern concrete construction would allow a differential heave or settlement of 8 mm, distributed over an area of 20 m by 20 m. When looking for a suitable method of soil improvement, we had to take into consideration the deformations caused by the driving operation of the tunnel lower down and the expected settlements of the tunnel above.

The actual distribution of the quaternary and tertiary sediments was determined by a number of boreholes driven all around the department store. Geological analysis has shown that the top heading of the upper tunnel emerges from the silty clay directly under the house and runs on in sandy gravel. The clay forms a relief that is partially filled with groundwater-bearing gravels. The groundwater must be prevented from intruding into the tunnel. Experience has shown dramatically that small amounts of water in the tunnel top may lead to erosion combined with a relaxation of the covering soil, a process that can easily get out of control. At any rate, we were aware that the gravels did not have a continuous water table and might be non-saturated in some places when the tunnel roof penetrates them.

For the tunnelling concept, the following soil improvement measures were discussed, involving the NATM under free atmospheric conditions:

-- grouting with cement suspension or chemical mixtures:
chemical grouting was not used because of its environmental hazards; cement bentonite grouting in heterogeneously-graded sands and gravels may produce uncontrolled heaves and transmits the vibrations of the underground traffic to the house above;

-- horizontal jet grouting was not yet approved by the time of tendering the project;

-- vertical wells from outside and horizontal wells from inside the tunnel:
the vertical wells can dewater the artesian fine sand zones in the clay, but not the unknown troughs of the tertiary/quaternary relief. Wells inside the tunnel hamper the excavation work, especially when they are located in the top section;

-- ground freezing:
allows a controlled, non-permanent increase of the strength of the covering soil and of watertightness. In non-saturated sandy gravels negligible heave may be expected. However, the freezing reaction of gravels with a low water content (less than 6%) was not yet known so that tests were necessary.

We decided to require, in our invitation to tender, vertical vacuum wells to control the fine sands in

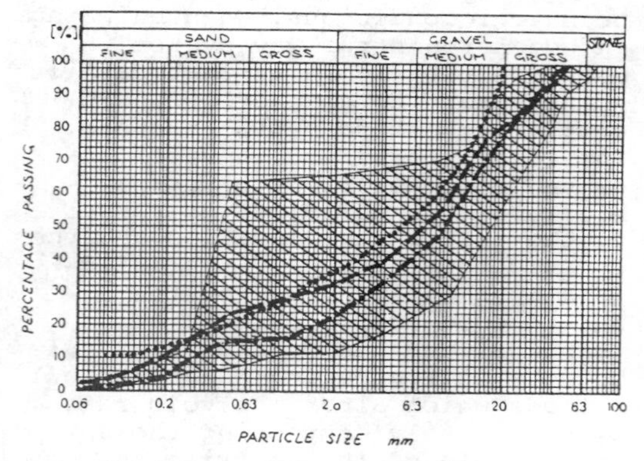

Figure 1: Grain size ranges of the tested soils for freezing suitability

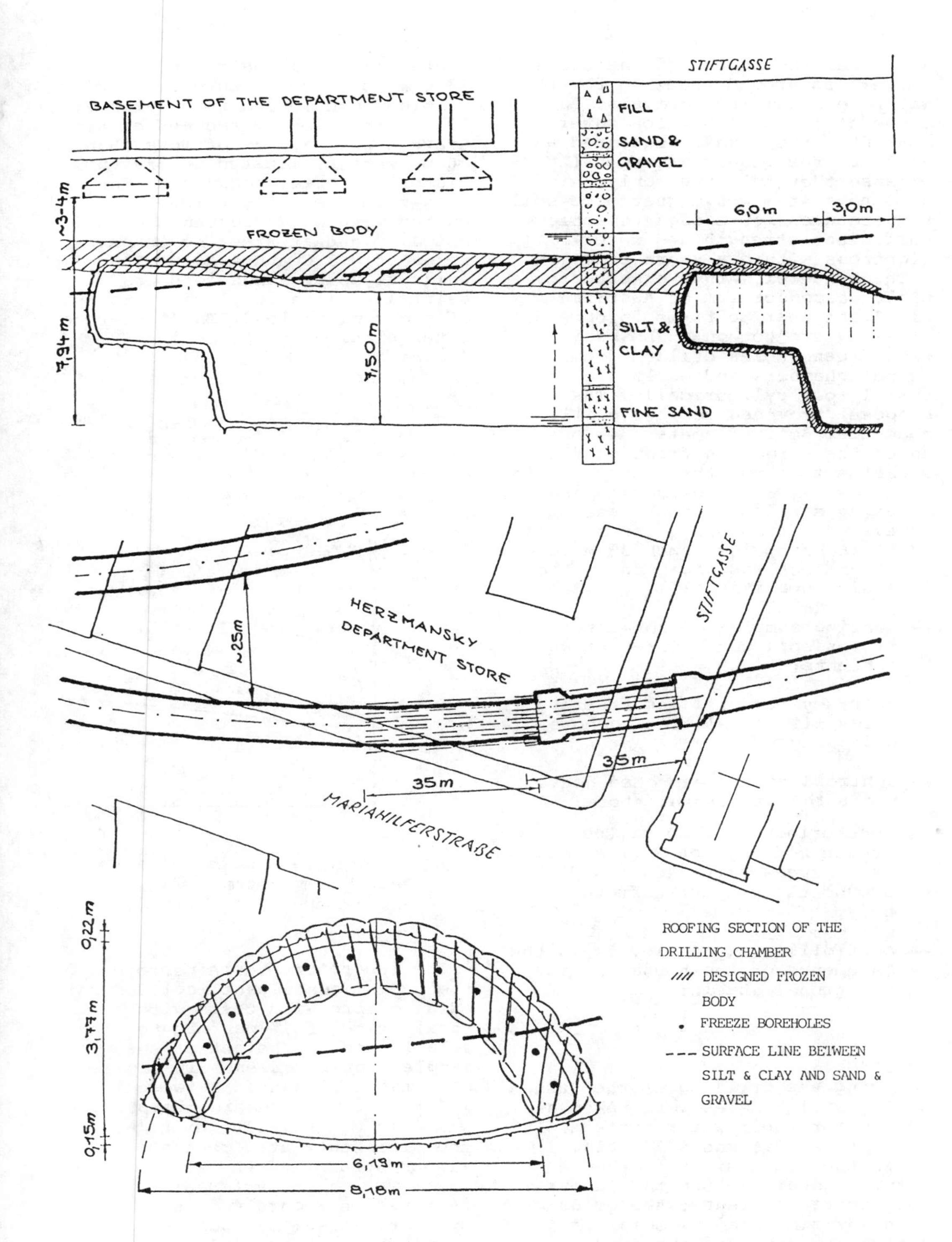

Figure 2: Longitudinal section, situation and cross section of the designed freezing body

and under the bottom of the tunnel as well as ground freezing for the water-bearing sandy gravels. Because of the possibly low water contents, the bidder requested a contract for liquid nitrogen (LN) because they felt that only this cooling system would guarantee soil temperatures low enough to achieve sufficient strength and watertightness within a short time.

The tenderer, Arge Volkstheater U3/10, a consortium of Austrian civil engineering firms, proposed a freezing project on an LN basis, with freeze holes drilled from tunnel chambers and enclosing the tunnel roof cylindrically. The proposal provided for two drilling chambers, 8m in diameter, -- one under the street in front of the department store, the other beneath the house -- and two overlapping freezing sections. The freeze-tube arrays were proposed to consist of 12 to 16 boreholes, each 35 m long.

The project was accepted on the following conditions:

-- testing samples of non-saturated frozen soil with different water contents;

-- accuracy for drilling freezing holes ±1% in relation to borehole length;

-- controlling the expected heave under the department store;

-- monitoring the size of the frozen body by means of temperature gauges situated in and around the projected frozen body;

-- controlling oxygen levels in the basements of the house and in the tunnel itself;

TESTING

During the drilling of the vacuum wells, soil samples were taken and tested for their water content. It was between 2.5 and 5.7%. Fig. 1 shows the grain sizes of the samples obtained. For the laboratory tests in Vienna, the graduation was simulated by means of a suitable mixture of standardized sand and gravel, as shown in Figure 1. The material was compacted to a unit weight of 18.0 kN/cu.m to 19 kN/cu.m. The samples were formed into cylinders, 200 mm in diameter and 400 mm high. Gravels with a size of more than 50 mm were eliminated to prevent distorting creep behaviour. The long-term creep tests were conducted with water contents of 2.5 and 4.5% and at freezing temperatures of - 5°C and - 10°C. The uniaxial compression load was maintained at a level of 50 to 60% of the rupture load. The pattern of interrelations is given in Figure 3.

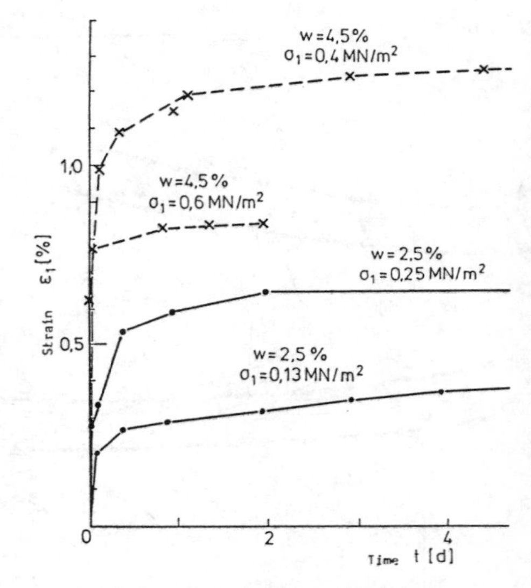

Fig.3: uniaxial creep tests with sandy gravel, temperature -10°C, water content 2,5%, 4,5%

The laboratory tests performed by the Bochum consultant included uniaxial compression tests with a strain rate of 1% per minute, in relation to the height of the sample. Water contents in similar soil material ranged between 1.8% and 10.7%. The freezing temperature was - 10°C. The relation between the different water contents, the linear strain and the strain rate, which was the same for all samples, is given in Figure 4. Also cylindrical samples were employed with a diameter of 100 mm and a height of 200 mm. Figure 4 clearly shows that there is a minimum water content of approximately 2.5% at which level you get a compressive stress of around 0.20 to 0.5 MN/m^2 with a strain of about 0.50 to 1.30%.

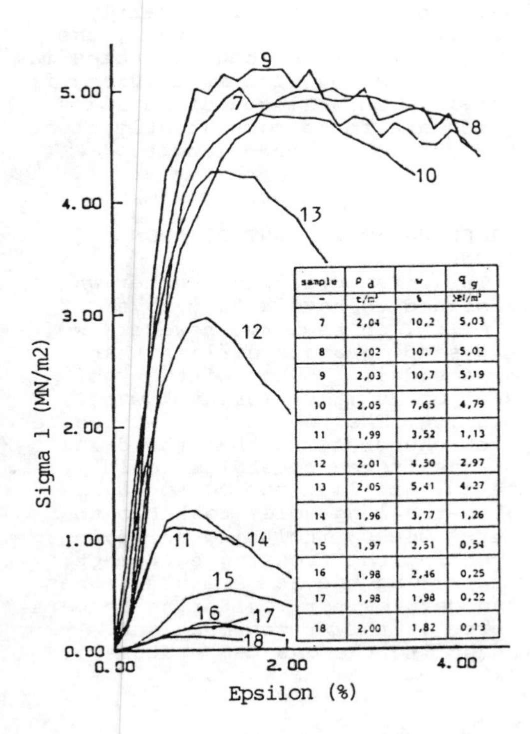

sample	ρ_d	w	q_g
-	t/m^3	%	MN/m^2
7	2,04	10,2	5,03
8	2,02	10,7	5,02
9	2,03	10,7	5,19
10	2,05	7,65	4,79
11	1,99	3,52	1,13
12	2,01	4,50	2,97
13	2,05	5,41	4,27
14	1,96	3,77	1,26
15	1,97	2,51	0,54
16	1,98	2,46	0,25
17	1,98	1,98	0,22
18	2,00	1,82	0,13

Figure 4: uniaxial compression test
temperature - 10° C
sandy gravel
strain rate 1%/min

Summing up the test results, we can draw the following conclusions:

-- a decrease in the water content of non-saturated soil is connected with a decrease in strength (uniaxial compression tests) and an increase in axial strain (creep tests);

-- an increase of the freezing temperature is related to an increase in uniaxial compressive strength and a decrease of creep deformation;

-- the realistic minimum water content necessary to obtain a considerable increase in compressive strength of the frozen soil is estimated at 2.5%;

-- for the actual freezing project, a temperature of at least - 10°C at the soil surface of the tunnel appeared necessary in order to support the excavation before it was sealed with shotcrete.

DRILLING WORK AND ACCURACY

The boreholes, each 34 to 35 m long, started from the two projected drilling chambers and continued in a line parallel to the tunnel direction enclosing the top heading.

Cased rotary drilling was used for the freeze holes, with the casing and pipe string rotating in opposite directions. The drilling fluid was water, the drilling diameter 108 mm.

In the casing a laser was mounted that fixed both the borehole face and the surveying markers on the shotcrete layer of the drilling chamber. According to West German experience, drilling tolerance required in the invitation to tender was no more than 1% of borehole length. Given this proviso, the theoretical distance between two freeze holes was 0.98 m. Including the tolerance of twice 0.35 (The total length of the hole was 35 m), a maximum distance of 1.70 from one hole to the next hole beside was estimated.

Monitoring drilling accuracy was effected by a so-called two-component INTERFELS deflectometer. It consists of two instrumented arms, each 1 m long, with a cardanic joint between them. One arm carries the inclinometer which measures in two directions, the other carries the electric pendulum. The data are transmitted via cable to the monitoring panel which is connected with a computer and a plotter. Metre by metre the inclinometer is advanced into the borehole with a calibrated rod. In the first freezing section, the boreholes ran through clay and silt for some metres, then continued in sandy gravels. The accuracy measured was satisfactory. The top of the holes was visible by laser ray. The second freezing section (total length 35 m) was situated entirely in layers consisting of sands and coarse gravels. Consequently, drilling was much more complicated. While the boreholes were equipped with freeze-tubes made of low-carbon steel, embedded in cement bentonite, the suspension penetrated deeply to the soil.

The borehole measurements documented that the drilling had produced a systematic torsion and displacement of the boreholes to the right and upwards. Repeated surveying produced different results, which made it clear that the tolerance of the two-component inclinometer itself is at least 1%.

The temperature measurement boreholes were drilled after surveying the freeze holes to find extreme control positions. Four to five temperature holes were required for every frozen body, crossing it from the inside to the outside or running parallel through it.

EXPERIENCES

Summing up the drilling and surveying experiences, we can say that the accuracy of the deflectometer system and of the surveying possibilities in boreholes more than 30 m long is approximately 1%. Rotary drilling in opposite direction, one of the most accurate methods nowadays, achieves the same accuracy in silt and clay without sand- or clay stone. In fairly heterogeneous sandy gravels, the accuracy of such long boreholes may also decrease. Greater accuracy is possible in well-graduated soils which are found consistently along the entire borehole length.

FREEZING WORK, CONTROL AND TUNNELLING

The first drilling chamber was designed to have a clay cover of 1 metre. In reality, however, we had sandy gravels earlier than expected in the top of the roof section but they were not water-bearing. The top heading was sealed with shotcrete so that the drilling of the freeze boreholes could begin. Drilling the 14 to 16 35-metre-long holes took two to three weeks, including surveying, mounting the freezing equipment, installing the temperature gauges and devices for monitoring temperatures in the frozen body and measuring the gas temperature of

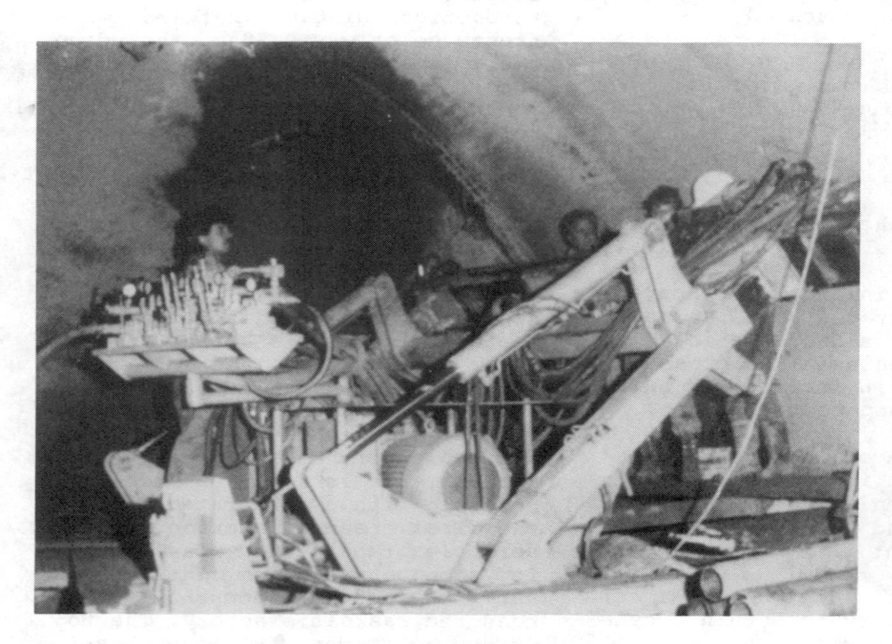

Figure 5: cased rotary drilling at the tunnel face of the drilling chamber

the nitrogen given off at the mouth of every freeze hole. Two 25,000-litre tanks for the LN were mounted on the pavement of Stiftsgasse, directly above the tunnel. Pipelines, which run from the surface to the top of the drilling chamber, ensure LN supply and conduct away the hot gases.

The freezing work started with an exhaust gas temperature of - 120°C. Hourly, the distribution of the falling temperatures was recorded automatically at the control centre. The temperature boreholes were equipped with six temperature gauges each, positioned at 13 m, 17 m, 21 m, 25 m, 29 m and 35 m from the borehole mouths. Within three days, the drilling chamber was completed with bench and bottom so that tunnelling could begin. The first five to seven metres of the frost-sheltered tunnel were in silty clay. Then the quaternary sediments appeared at the top of the roof section. A small cavity of perhaps 1 cu.m was dug to test the frozen gravels for strength and

Figure 7: frozen body in sandy gravels with 5% to 8% water content

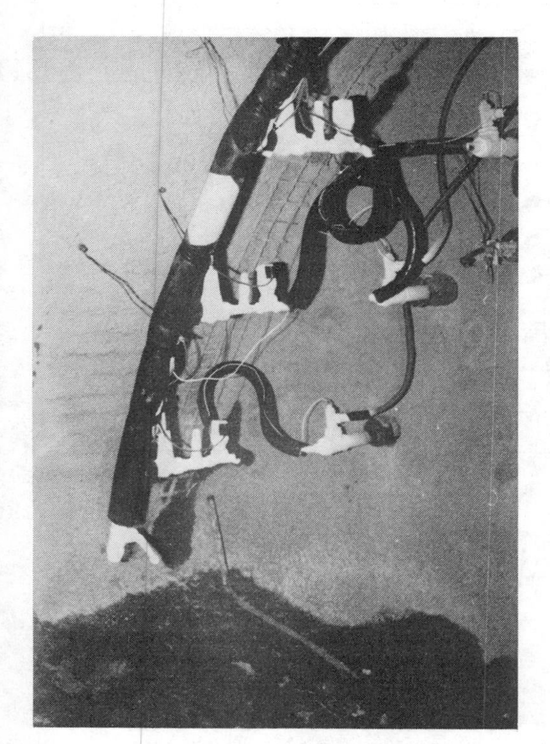

Figure 6: freezing equipment at the mouth of the freeze holes

determine the temperature distribution. The gravels were firmly held together by a thin film of ice. The temperature of the soil surface was - 25° C to - 50°C. On the basis of these positive results we decided to continue with an intermittent operation, i.e. freezing time from 3 p.m. to 9 p.m., followed by tunnelling work in two shifts from 9 p.m. to 3 o'clock the next afternoon. We avoided working in the tunnel during freezing time because of the danger of damaging a freeze tube under full nitrogen pressure.

An extensive control system was installed by Linde & Co. Austria, the main supplier of LN, to recognise a possible drop in oxygen levels in the tunnel or in the basements of the department store.

The Herzmansky Department Store, which continued to do business throughout, was inspected twice a day for which purpose measuring devices had been fixed in the concrete structure.

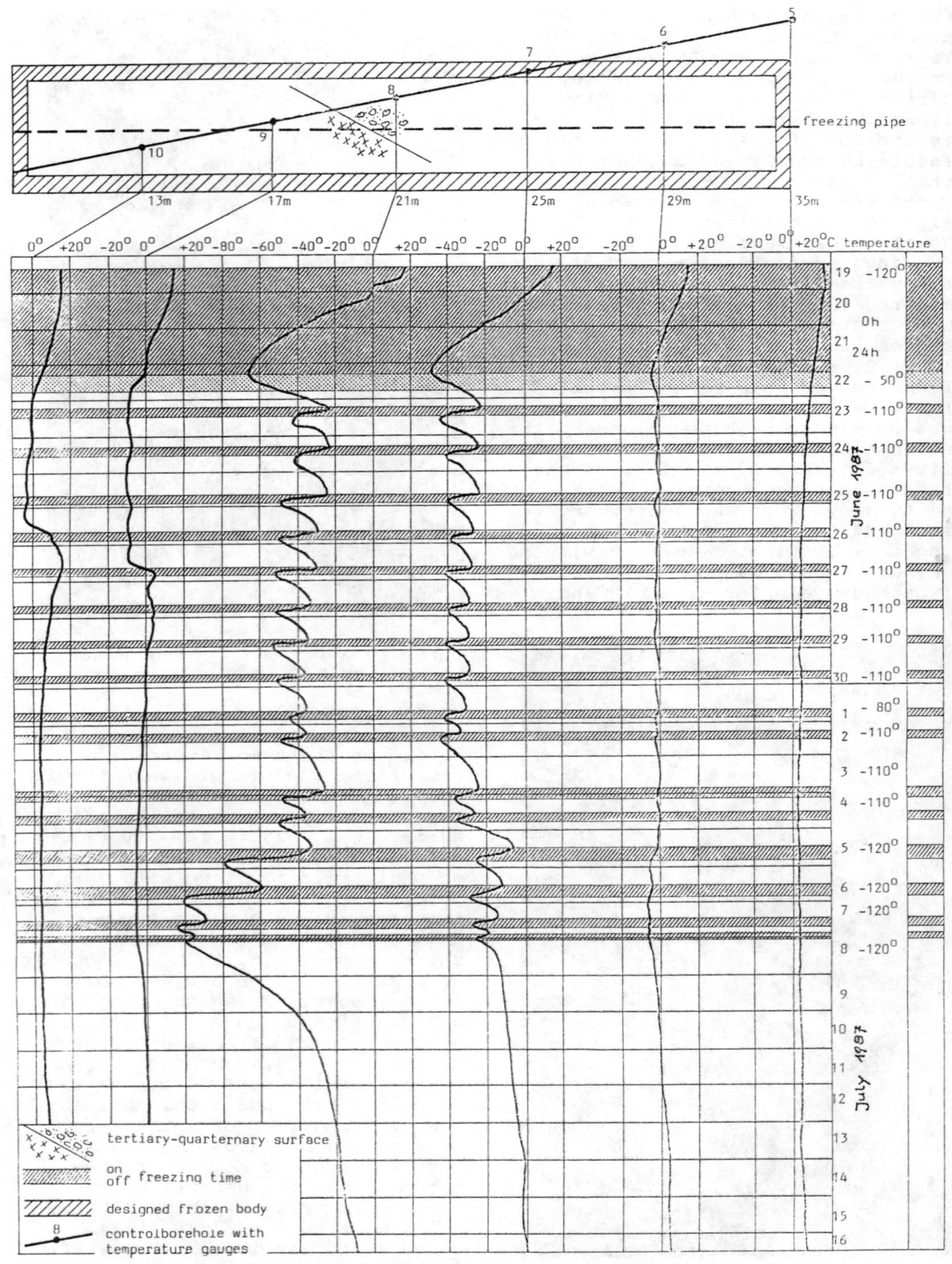

Figure 8: schematic distribution of freezing temperature in silt & clay and non saturated sandy gravel

The freeze-tubes in the silty clay were soon taken out of operation to stop unnecessary freezing in the cohesive material. As a result, heave at the department store was limited to 1 mm. Subsequent settlements caused by the excavation increased to 6 and 8 mm. After the thaw period the settlements evened out between 8 and 10 mm altogether.

Figure 8 shows the distribution of some typical temperature gauges. The schematic drawing indicates the positions of the temperature boreholes in relation to the freeze-tubes which are represented by the horizontal broken line. Surrounding them is the projected frozen body. The soil conditions are revealed by the outline of the frozen body.

Different temperature distributions are given:

The first two illustrate the freezing state in the silt and clay which form the side walls of the tunnel. The freezing process in the silt is characterized by a slow decrease of the soil temperature, especially in the range of ±0°C when the water freezes. The fully saturated clay and silt is easily distinguished from the non-saturated sand and gravel. Here the temperature fell below ± 0°C within an hour. Once the tunnel face is past the temperature gauge, the temperature increases again, due to the hydration heat of the shotcrete.

The following temperature distributions show the freezing process of the more or less sandy gravels in relation to their low water content. Samples taken from this area had an actual water content of 5% to 8%. These measurements are not representative, however, because the moisture content of the hot air in the tunnel causes the water to condense on the surface of the frozen soil particles.

The ups and downs of the freezing temperatures are related to the freezing cycles during which the exhaust temperature of the nitrogen gas was maintained at -90°C to -110°C. In the course of one prolonged period of tunnel driving in difficult soil conditions, during which the rhythmic freezing cycle had to be interrupted, some gauges registered an expected rise in the freezing temperature of up to -8°C about one day later. However, the intensive freezing periods at a gas temperature of - 120° C allowed tunnel driving to be continued and completed safely. The thaw period, represented by the temperature increase, extended over more than two weeks until the soil temperature reached 0°C. The distribution of the temperature measurements in the second freezing section was similar to that in the first one. Given the specific freezing conditions and borehole distribution, LN consumption averaged 900 l/cu.m of frozen soil.

The progress of the tunnel was dictated by the excavation time. As a rule, we took 4 to 6 hours to open the side parts of the roof section, which consisted mainly of highly-viscous frozen silt and clay, and 0.5 to 1 hour to open and shotcrete the tunnel face in the brittle-hard frozen sand and gravel.

CONCLUSION

Compared with many soil improvement measures, soil freezing as an auxiliary support for tunnelling involves low heave in sandy gravels. Especially under non-saturated conditions freezing with LN allows us to create low temperatures in the soil (-50°C to - 100°C), which results in extraordinary strength and water tightness. During the ensuing excavation and the thaw period only low settlements may be expected. LN is rapidly effective and easily controlled through exhaust temperature, even in soil with a water content of around 2.5%. During excavation the hot tunnel air flows into the soil and helps to raise the water content of the soil, which becomes obvious from ice condensation on the surface. Due to the low temperatures in the soil, the capacity of the frozen soil was so considerable that a safe opening of the tunnel roof was possible within 6 to 10 hours.

Figure 9: frozen body in the roofsection of the tunnel excavion in NATM

REFERENCES

Travnicek, R. (1986). Abschlußbericht über Dauerlastversuche an gefrorenem Grobboden, Bautechn. Prüf- und Versuchsges. m.b.H., Lanzendorf, unveröffentlicht

Jessberger, H.L., Jordan. (1986) Gutachtliche Stellungnahme zur Vereisungsmaßnahme des Kaufhauses Herzmansky im Zuge des U-Bahn-Baus, Bauabschnitt U3/10 Wien, Bochum, unveröffentlicht

ARGE Vereisung U3/10, Soletanche-Sonderbau (1987). Techn. Bericht Bohrlochvermessung der Vereisungsbohrungen und der Temperaturmessbohrungen der 1. und 2. Kaverne, Vienna, unveröffentlicht

5th International Symposium on Ground Freezing, Jones & Holden (eds)
© 1988 Balkema, Rotterdam. ISBN 90 6191 824 3

Ground freezing for the construction of a drain pump chamber in gravel between the twin tunnels in Kyoto

S.Murayama, T.Kunieda, T.Sato, K.Miyamoto, T.Hashimoto & K.Goto
Seiken Co. Ltd, Kawarayamachi Minami-ku, Osaka, Japan

ABSTRACT: The Ground freezing method was applied in the excavation work for constructing a pump chamber between two parallel shield tunnels in gravel ground with a large underground water flow. The following three points which constitute the main features of the method employed are described in this paper:
1) Frozen soil walls were constructed with the aid of (a) injecting a chemical liquid for decreasing the seepage flow rate in the gravel ground which had a high velocity of underground water and a high hydraulic pressure and (b) completely preventing water inflow by operating devices including attached freezing pipes.
2) A Three-dimensional frozen soil wall was constructed almost entirely in the circumference of the space , including the shield tunnels, the excavation for a pump chamber and shaft; then the pump chamber was built by excavation.
3) The freezing and excavating were controlled by the feedback of data measured at the actual site so that deformation of the shield tunnels caused by ground expansion resulting from freezing as well as the excavation work was minimized.

1 INTRODUCTION

The freezing method has been much used as a supplement for protection of during movement of a shield, strengthening of ground at a weak portion of construction, etc.

The present use of this method is a new application, in which ground of the upper, lower and lateral sides between the shaft and two shield tunnels were first frozen three-dimensionally, then the shaft and space between the tunnels was excavated, and the pump chamber was finally built by widening the excavated section.

This method was applied for the following reasons:
1) It ensures the prevention of water flow at the construction site near the Kamo River where the gravel ground has a large flow of spring water.
2) It achieves the stoppage of water and strengthening ground at the same time because both are required almost entirely in the circumference.
3) The area of the site for the pump chamber is too small to allow the use of large construction machinery.

2 AN OUTLINE OF CONSTRUCTION

The purpose of this construction is to build a pump chamber for discharging water from tunnels almost at the center of the shield area as a part of subway construction by Kyoto City. Two parallel Nos. 1 and 2 tunnels were constructed at a distance of approximately 5 m. The concrete secondary lining has not been established yet. The segmental outer diameter of the tunnel is 7,050 mm.

Plans for freezing are shown in Figs. 1 - 3. As illustrated in the figures, upper frozen soil was formed around the shaft, at the upper and the lateral sides of the two parallel shield tunnels using vertical freeze pipes drilled from the ground surface. Lower V-shaped frozen soil wall was built between two shield tunnels using inclined freeze pipes drilled from inside of each shield tunnel. The attached freeze pipes were installed into the inside of the segment to strengthen the adherence between the shield tunnels and the frozen soil wall and secure water stoppage. Two 150KW screw brine chillers were installed for cooling. During the freezing period the temperature of brine ($CaCl_2$) was -25 to -30 C, and was adjusted in accordance with excavating conditions.

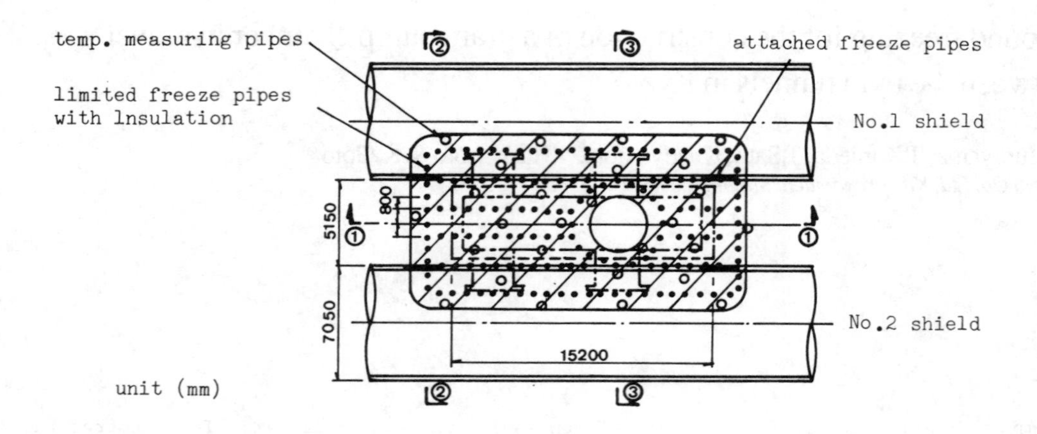

Fig.1 Plan View

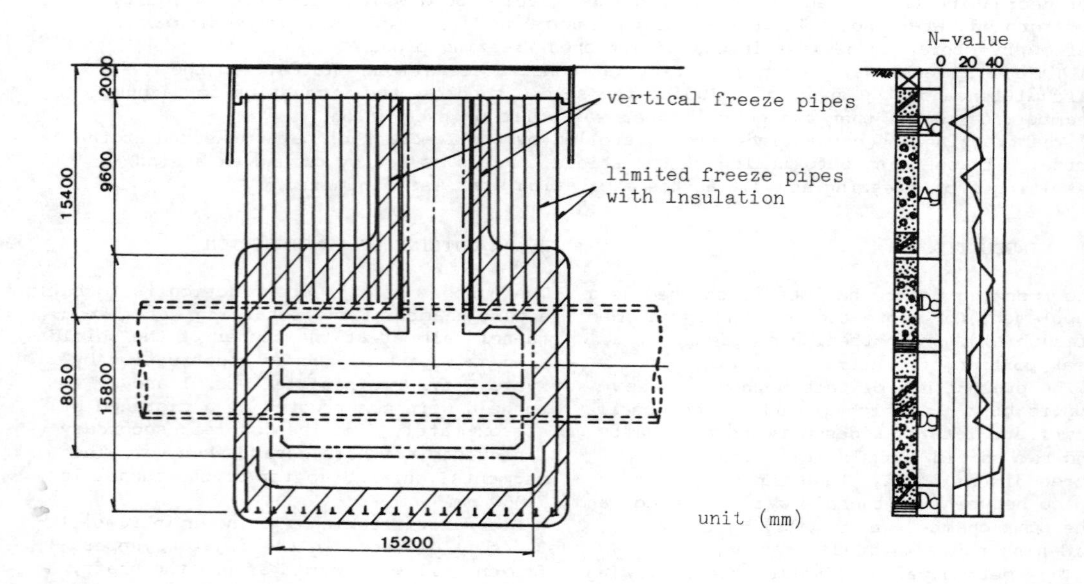

Fig.2 1 - 1 Section

Figure 4 shows a geological vertical section view. The ground subject to construction mainly consists of water-bearing and dense gravel with the N value of 30 - 50, but there is also a thick clay layer of approximately 2 m at a depth of 26 m, or 4 m below the bottom of the shield tunnels.

The period of construction was from July 1985 to August 1986. A total volume of 1,650 m^3 of soil mass was frozen and the total length of the freeze pipes was 2,516 m.

The frozen soil wall was designed so that it is securely protected from the excavation. Table 1 shows the strengths design utilized.

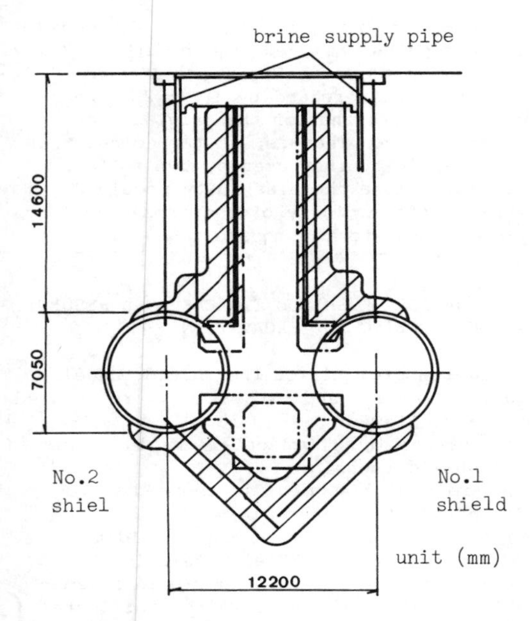

Fig.3 3 - 3 Section

Table 1 Design Strength of Frozen Soil at -10 C

	Cohesive soil (kgf/cm^2)	Sandy soil (kgf/cm^2)
Compressive	30	45
Bending	18	27
Shearing	15	22

3 COUNTERMEASURES AGAINST UNDERGROUND WATER FLOWS FOR DURING FREEZING

As the construction site near the Kamo River abounds in spring water, a high flow rate of underground water was expected. Therefore, measurements of the flow rate were made beforehand using a single-hole flow meter. Figure 5 shows the results of the measurements. As shown in the figure, the velocity of flow V was 5.6 m/day at a depth of 7 m, which exceeded the critical velocity of flow V_{crit} given by Takashi (1969). Accordingly, a chemical liquid was injected to reduce the flow rate. Generally, the temperature in the ground after the start of freezing decreased gradually at locations where the chemical liquid had been injected, thus showing the effect of lowering of the flow rate by the injection of the chemical liquid. As the temperature did not drop smoothly at the point of measurement at a depth of 12.1 m, however, the chemical liquid was injected again there. This made possible the formation of the frozen soil walls as designed.

4 CONFIRMATION OF STRENGTH OF FROZEN SOIL

The formation of frozen soil was judged generally by measuring the underground temperature. In this work we conducted uniaxial compression tests and P wave logging to examine whether the frozen soil wall had sufficient strength and rigidity as a retaining wall.

First, the uniaxial compression tests were conducted for samples of undisturbed frozen gravel collected in the frozen state at a depth of 21.8 m during the excavation. The result of the uniaxial compression tests was that ultimate compressive strength qu = 107

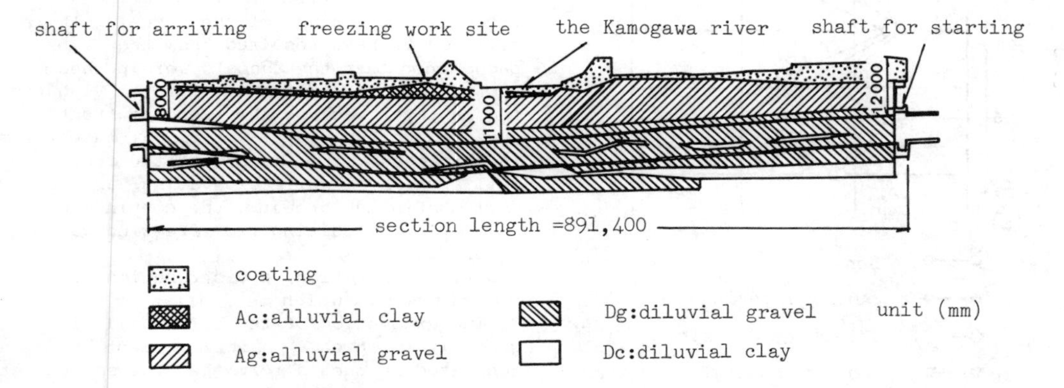

Fig.4 Geological Vertical Section View

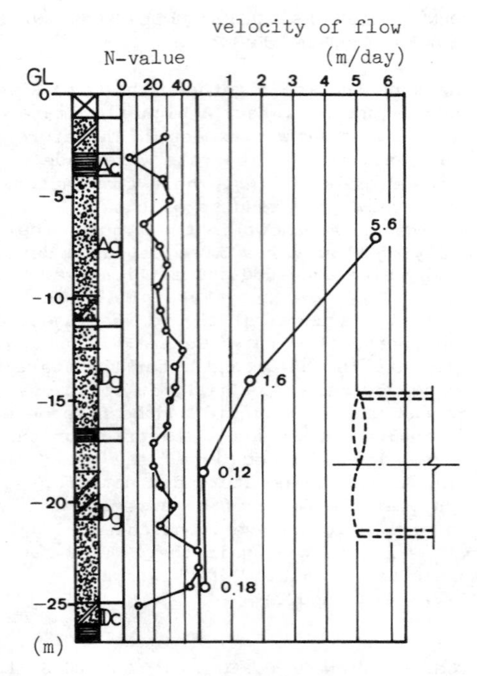

Fig.5 Velocity of Flow versus Depth

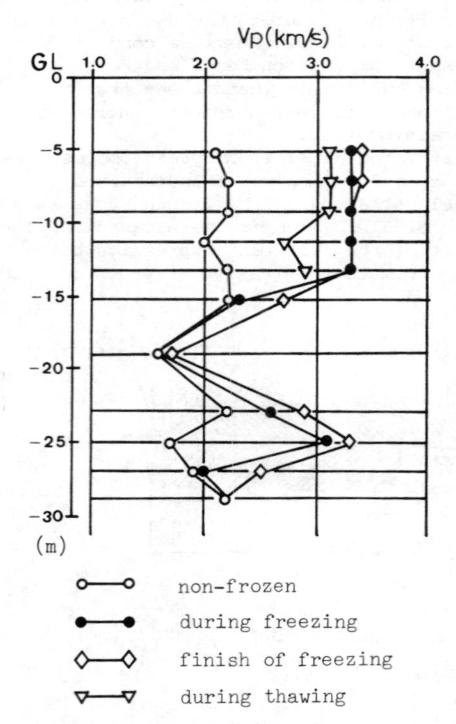

Fig.6 The Result of P Wave Logging

- 130 kgf/cm2 (at -15 C), which exceeded the strengths designed. The result of P wave logging is shown in Fig. 6, The P wave velocity V_p increased by 1.4 - 2.0 times compared with that of the original non-frozen ground. That is, it turned out that the elasticity increased by two to four times, because the elasticity modulus showing the rigidity of the frozen soil is proportional to the square of V_p.

5 BEHAVIOR OF SHIELD TUNNELS AS A RESULT OF FREEZING AND EXCAVATION

Field measurements of the shield tunnels during freezing and excavation were carried out as followed;

1) Vertical and horizontal displacements of the shield tunnels;
2) Strain of a segment in the sectional direction of the tunnels;
3) Inner deformation of the shield tunnels.

As shown in Fig. 7, the tunnels in the frozen area were pushed upward in a semi-elliptical shape, while in the horizontal direction the two tunnels were mutually pressed outward extending the distance between them. The maximum amount of displacements was approximately 25 mm upwards, and approximately 17 mm aside. As for the sectional deformation of the tunnels, the results of measurements of inner deformation were shown in Fig. 8 and the moment of the segment in the sectional direction is shown in Fig. 9. As for the mode of cross-sectional deformation, the section of the tunnels is irregularly pressed from both upward and downward directions due to compression from the V-shaped frozen soil wall at the bottom of the tunnels.

These deformations emerged even before starting the excavation. Almost similar deformation patterns occurred during excavating the sides of the tunnels. It is believed to have resulted from the phenomenon that the tunnels were pushed upward from the bottom through the V-shaped frozen layer because the freezing front advanced to the clay layer approximately 2 m in thickness at the bottom of the frozen soil area. Even while excavating, as frozen soil continued growing, the continuing deformation resulting from freezing was observed.

During excavation, a distribution of moments occurs which is similar to that during excavation of retaining walls in general, in which a positive moment is generated in such a form that the segment at the excavated portion expands toward the side of the excavation; and at the bottom of the excavation a negative moment is seen as a reactive force.

It was thought that the strain on the segment increased with the growth of frozen area, because the freezing front advanced to the clay layer at the bottom of the shield. As a countermeasure, the number of the freeze pipes in operation at the bottom was reduced by alternating operation of even-numbered piped and odd-numbered pipes as well as adjusting the temperature of brine. As a result an increase in amount of displacement generated was controlled successfully.

As for thawing, only the attached freeze pipes were subjected to forced thawing to remove the stress on the segments because the stress had to be rapidly removed; as a result, the segments approximately regained their original state. This was successful because the upward deformation of the segments due to the V-shaped frozen soil had stopped. The amount of subsidence of the entire frozen soil due to thawing was about 4 mm.

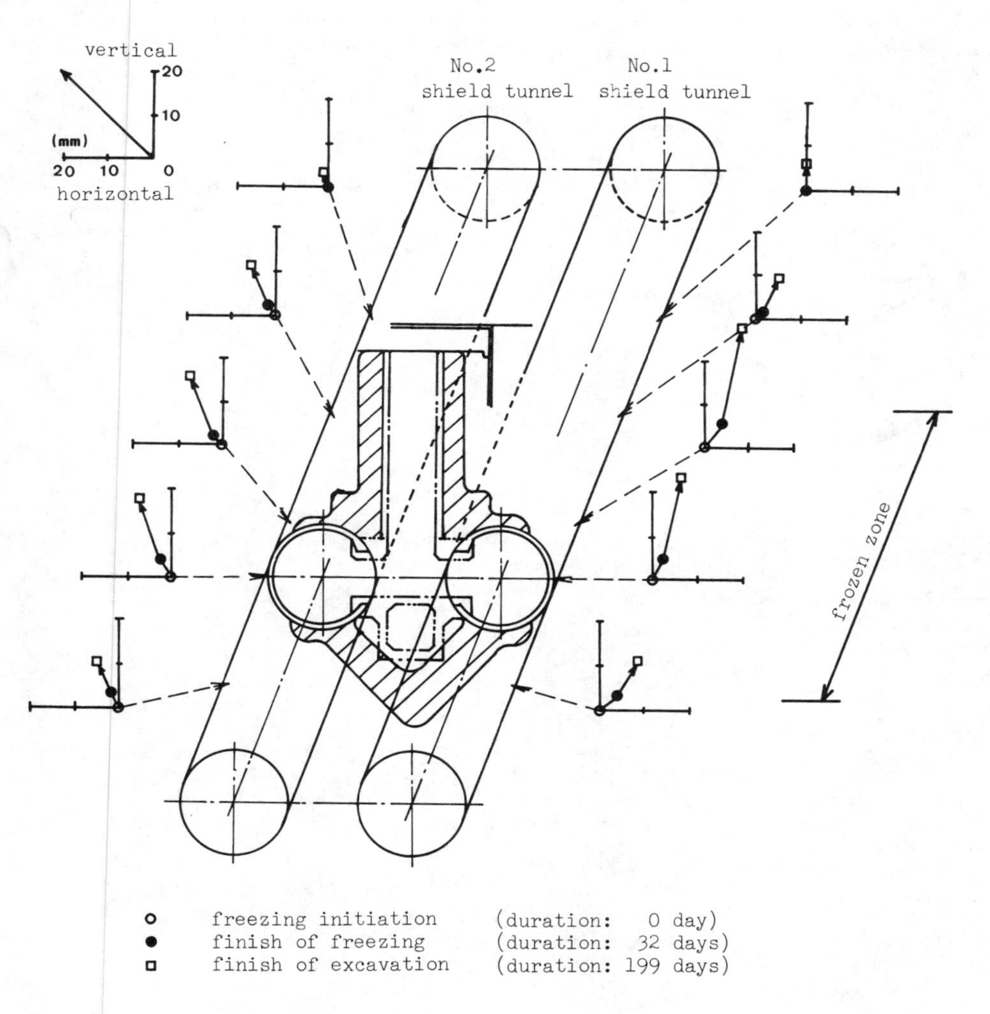

Fig.7 Tunnel Displacement Vector at the Depth of Tunnel Center

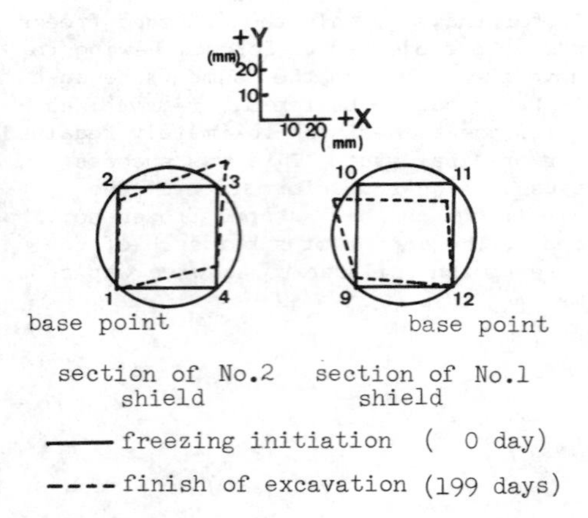

Fig.8 Inner Displacement

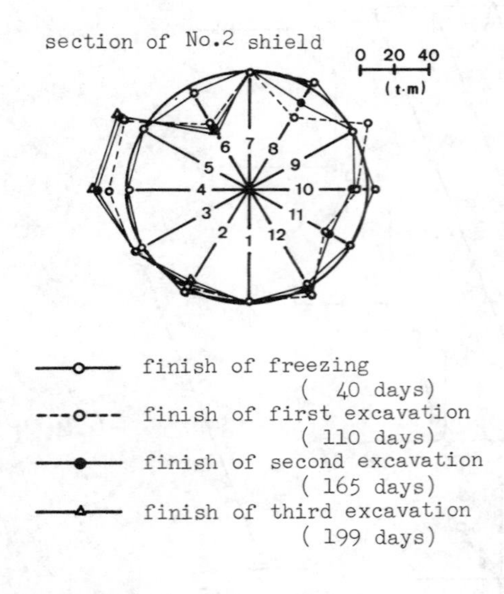

Fig.9 Moment of Segment

6 CONCLUSIONS

A pump chamber was successfully built at a water-bearing gravel layer with an underground water flow, using an unprecedented three-dimensional freezing method, which has made it possible to freeze the ground to form a perfect water-shield and a retaining wall which prevented spring water from leaking into the excavation increase of deformation was, however, restrained by reducing the number of freezing pipes in operation and adjusting the temperature of the brine, while maintaining the strength of frozen soil and the adherence between the frozen soil and the shield tunnels.

REFERENCE

Takashi, T. 1969. Influence of seepage stream on the joining of frozen soil zones in artificial soil freezing. Highway Research Board. Special Report. 103:273-286. Washington D. C.

5th International Symposium on Ground Freezing, Jones & Holden (eds)
 ISBN 90 6191 824 3

Application of the freezing method to the undersea connection of a large diameter shield tunnel

K.Numazawa
Tokyo Metropolitan Government Sewage Bureau, Japan

M.Tanaka & N.Hanawa
Kajima Institute of Construction Technology, Tokyo, Japan

ABSTRACT: The connection of two 9.7m diameter large scale shield tunnels at a depth of 22.5m under sea level (approximately 15m under the sea bed) with ground freezing was made in Japan.
In this method, 108 freezing pipes were installed radially from the inside of a tunnel, and a frozen soil ring formed around the tunneling connection.
However, the process of forming frozen soil with such a complicated arrangement of freezing pipes and the quality of the frozen soil were not fully understood.
Therefore, soil freezing model tests were conducted to obtain the design data.
This paper is a report on the outline of construction of tunneling connection and also describes the results of the model test.

INTRODUCTION

This project was to construct a 3.3km long, 9.7m diameter tunnel to connect the Morigasaki and the Nambu Sludge Treatment Plants. The tunnel will carry pipelines to transport raw sludge and plant treatment raw sludge and plant treatment water.

The greatest portion of the tunnel lies beneath the Keihin Canal at a depth of 22.5 m under sea level.

For construction purposes the tunnel was divided into two sections.

One tunnel section was driven by slurry shield working from a shaft at Nambu Sludge Treatment Plant to meet at the midpoint with the shield tunnel driven out from the Morigasaki end (photo -1).

When the two shield faces had approached to within about 80cm of each other, the freezing method was used to connect the two shields.

The construction work of the freezing section in the tunnel was completed successfully in 1985.

This paper describes both the full scale shield tunnel connection beneath the sea, and also the model freezing test which was performed before the actual full scale construction.

GENERAL OUTLINE OF WORK

The schematic of the connecting section of the two shields and also the soil profile are shown in Fig.-1.

For this case, in which the frozen soil is used for temporary support, we must examine its behavior from the mechanical and structural aspects.

However, no established design method existed, so we performed the basic design with the structural calculations as summarised Table-1.

The freeze pipes of the double tube type, which served for both drilling and freezing, were installed at three angles

Photo–1 Construction site and shield machine

of incidence (20°, 22° and 25°) through the 32mm thick steel skin plate into the surrounding ground (photo-2). The hydrostatic pressure in the ground was 0.2 to 0.3 MPa.

The spacing of the freeze tubes was about 80cm at their tips. There were 108 pipes in all, forming two cones at the connecting section. In addition, in order to maintain adfreezing between the frozen soil and the shield body, freeze pipes were attached to the inside wall of both shield bodies at the manufacturing stage.

Fig.-2 shows the arrangement of both the freeze tubes and the temperature measuring pipes.

Two refrigeration plants were installed about 50m from the freezing sites for both sections.

Each cooling unit had a cooling capacity of 96,000 Kcal/hr at an evaporation temperature of -27°C and a condensation temperature of +40°C.

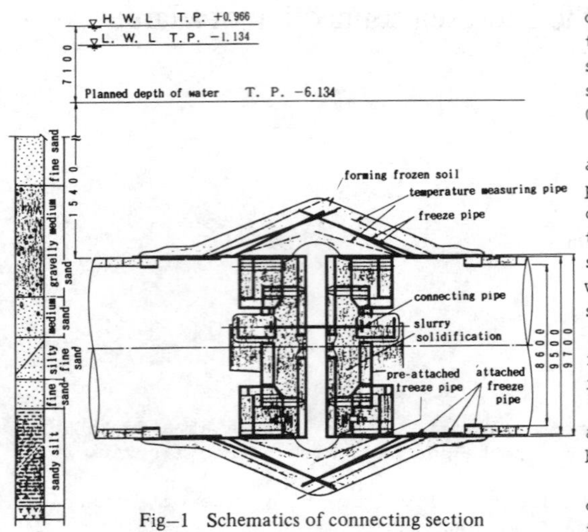

Fig–1 Schematics of connecting section of two shields and soil formation

Table–1 Structural design of frozen soil

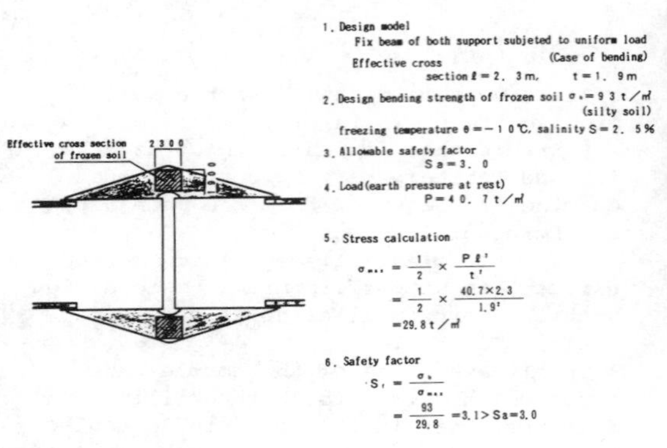

1. Design model
   Fix beam of both support subjeted to uniform load (Case of bending)
   Effective cross section $\ell = 2.3$ m, $t = 1.9$ m
2. Design bending strength of frozen soil $\sigma_b = 93$ t/m² (silty soil)
   freezing temperature $\theta = -10$℃, salinity $S = 2.5$%
3. Allowable safety factor
   $S_a = 3.0$
4. Load (earth pressure at rest)
   $P = 40.7$ t/m²
5. Stress calculation

$$\sigma_{max} = \frac{1}{2} \times \frac{P \ell^2}{t^2} = \frac{1}{2} \times \frac{40.7 \times 2.3}{1.9^2} = 29.8 \text{ t/m}^2$$

6. Safety factor

$$S_f = \frac{\sigma_b}{\sigma_{max}} = \frac{93}{29.8} = 3.1 > S_a = 3.0$$

Photo–2 Boring work of freeze pipe

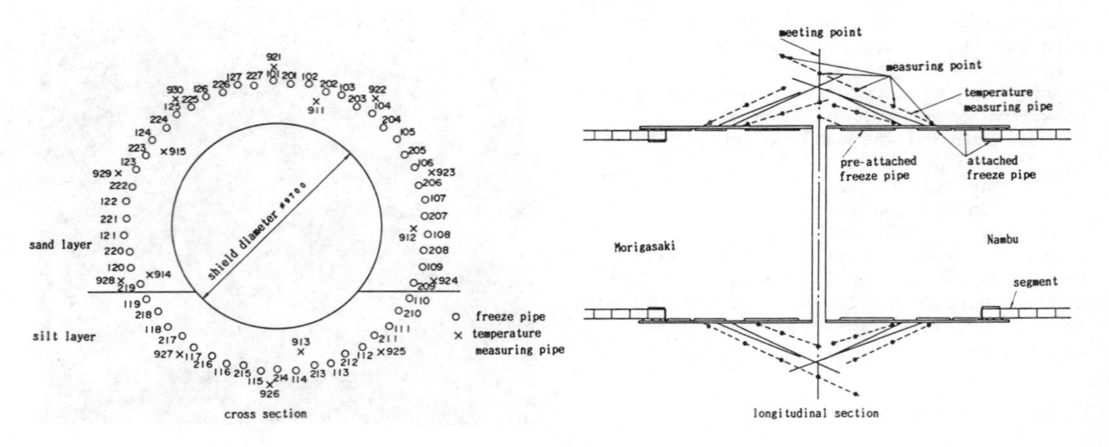

Fig–2 Arrangement of freeze pipes

Freeze tubes were connected with cooling unit by supply and return pipe-lines.

Calcium chloride brine at -25 to -30°C, was circulated through the freeze tubes to form the frozen soil ring.

The frozen zone was measured by 34 temperature-measuring pipes.

After the frozen soil ring with the design thickness had been formed, the steel plates of the shield front were cut off using acetylene, and then, the excavation of the frozen soil was carried out from both shields. The tunnel was lined temporary by welding together 6mm thick steel plates to prevent seepage at the frozen soil thawed (photo-3). Finally, the segments were erected and the two shield tunnels were connected.

The progress of the work is illustrated in Table-2.

Photo–3 First lining

Table–2 Progression chart of works

(hatched bar: execution; open bar: plan)

Works \ Period	1984: 10	11	12	1985: 1	2	3	4	5	6	7	8
Preparatory works	10/10 10/18										
Freeze pipe drilling works	10/18		12/10								
Refrigeration plant works		11/28	12/29								
Insulation works			12/10	1/10							
Freezing operation			12/15 12/28			3/6		5/7			
First lining						3/6	4/17	5/7			
Forced thawing							4/18	5/15	6/20	7/9	
Secondary lining										7/10	8/10 8/25
Withdrawal											[illegible]

PROPAGATION OF FROZEN SOIL

The cold brine was circulated first through the upper part and then through to the lower part of the freeze tube array.

This system was based on considerations of the propagation during preparation period and also to limit heaving.

Therefore, the freezing progress was dependent on the position of the freeze pipes.

Representative examples of the freezing progress in the soil at the centre of connecting section are shown in Fig.-3 and Fig.-4.

The freezing period for forming frozen soil, 1m in thickness, in anoutside direction from freeze tube at was 36 to 70 days, while the one for an inward direction from the freeze tube was at 14 to 32 days.

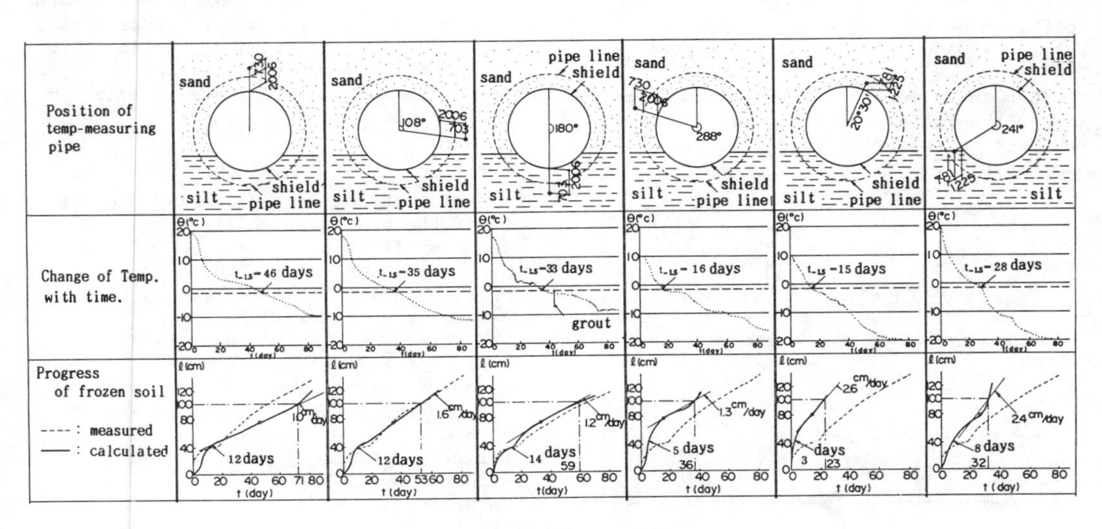

Fig–3 Progress of frozen soil (I)

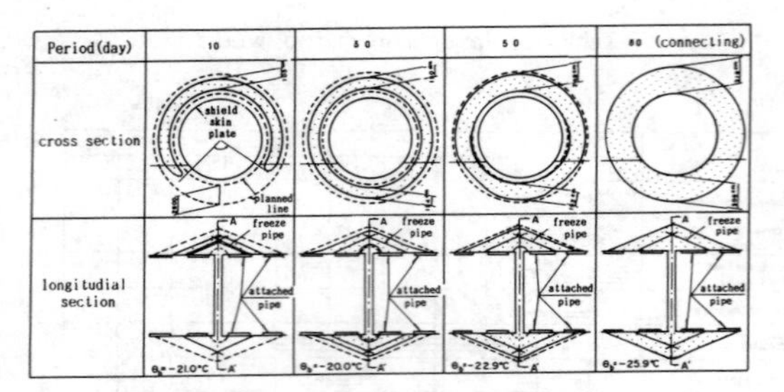

Fig–4 Progress of frozen soil (II)

The reason that the propagation in the inward direction faster than that of the outside, could be to the influence of cryogenic temperature of the pre-attached freeze pipes and the freeze tubes inclined to the shield side.

FROST HEAVING AND THAW SETTLEMENT

Since the frozen part beneath the tunnel was silty soil as shown in Fig.-1, frost heaving and thaw settlement were estimated by a heaving and thawing test considering both the freezing speed and the overburden pressure.

Predicted amounts of heaving and settlement were 70mm and 140mm, respectively. However, the maximum heave measured at the connecting section was 114mm by levelling as shown in Fig.-5.

Consequently, there were gaps between the segments, the influence of which extended about 15m from the centre of the freezing site, but this was no concern in relation to the connection work. The progress of heaving and subsidence is illustrated in Fig.-6.

After the completion of the connection, forced thawing and grouting were applied to the zone under the tunnel to prevent it subsidencing.

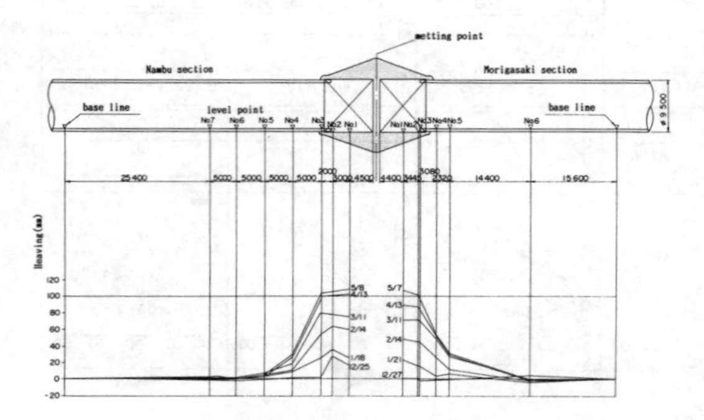

Fig–5 Heaving in the direction of longitude

The force thawing was executed by circulating the warm water at 60°C within freeze pipes in place of cold brine.

The injection method was step-back grouting from freezing front to tunnel body with liquid cement based grout.

After the grouting, the subsidence of connecting section was effectively zero.

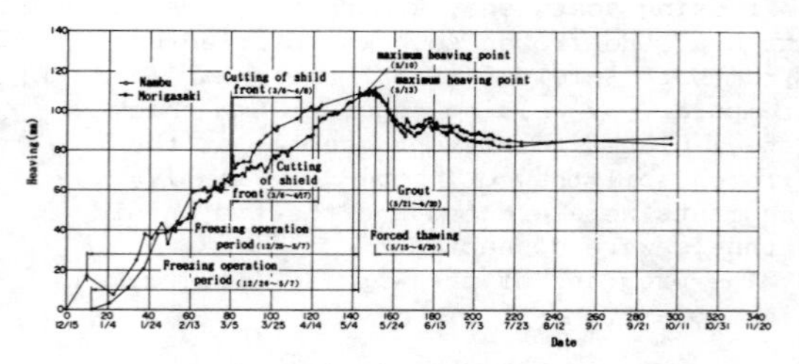

Fig–6 Heaving and subsidence change of tunnel

THE MODEL FREEZING TEST

The case in which the freezing method was applied to the full scale shield connection under the sea was outside our experience.

The quality of frozen mass formed simultaneously by the inclined freeze tubes driven from inner surface of shields together with the pre-attached freeze pipes on the inside surface of shield bodies was uncertain, since the complex process of forming the frozen soil was not completely understood.

Therefore, in order to complete the works safely and reliably, a model freezing test was performed prior to the actual construction. The arrangement of the freeze pipes at a portion of tunnel crown was modelled at half scale.

Test facilities and test procedure

The soil used for test resembled the sea sand of the construction site.

The test facilities consisted of a soil box, model freeze pipes, thermocouples supported on wire mesh, a brine bath and indicators, etc. (Fig.-7).

The soil is contained in a rigid box of 3*3*2m with a glass mesh window. The box was covered with insulation 5cm thick to insulate it from the ambient temperature (photo-4).

Firstly, 8 freeze pipes, 16 attached freeze pipes and 5 wire meshes of 216 thermocouples in all, were set in the soil box. Next, the box was filled hydraulically with the resemble sea sand to simulate the

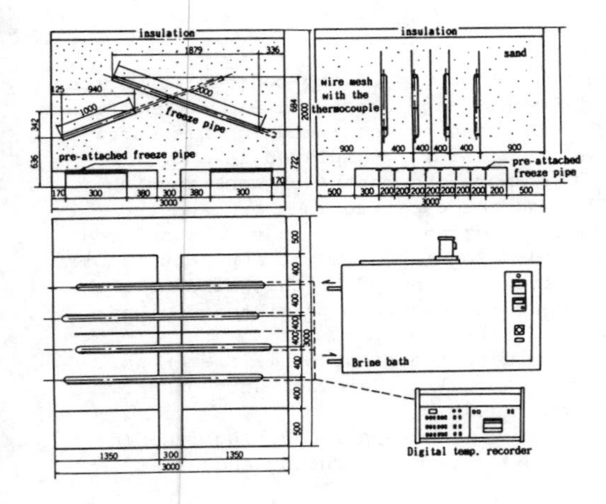

Fig–7 Schematics of model freezing test

Photo–4 View of freezing test

site conditions.
The predetermined temperature brine was then circulated within freeze pipes.

The change of temperature in the sand was measured until the predicted thickness of frozen soil was reached.

Test results

1.Propagation of the frozen soil

The propagation of frozen soil is shown in Fig.-8. From this figure, we can see that of the progress of inner frozen soil between freeze pipes and pre-attached pipes, is faster than that of the outer frozen soil because of the influence of the cryogenic temperature of pre-attached pipes.

Therefore, for an arrangement of this type, in order to create the intend thickness of frozen soil, the critical aspect is the progress of outer frozen soil at the the centre of connecting section where the spacing between freeze pipes is greatest.

The propagation of frozen soil may be predicted from existing theoretical equations for "single pipe freezing" and for "flat plate freezing", but theory has not developed for the "plural pipes freezing" proposed for actual condition.

To date, an approximate equation, based on the two theories mentioned above, has been used usually in Japan.

Fig.-9 shows the comparison of the propagation of frozen soil between the experimental values and the calculated values (for volumetric moisture content $0.4 m^3/m^3$, cooling temperature -25°C, ground temperature +16°C).

It is clear that the values for the outer frozen soil are approximately consistent except where values were measured near the tips of the freeze pipes.

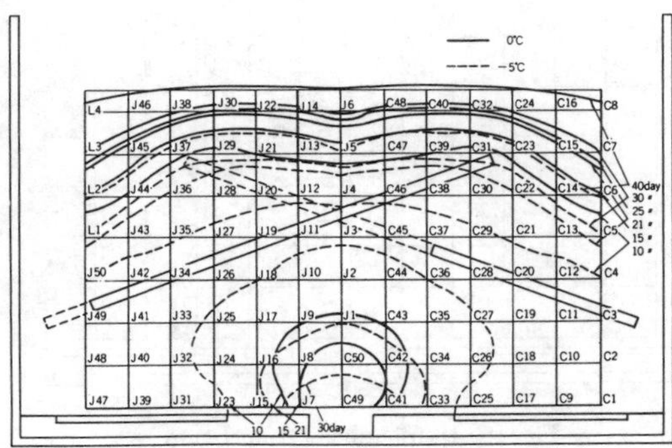

Fig–8 Progress of frozen soil

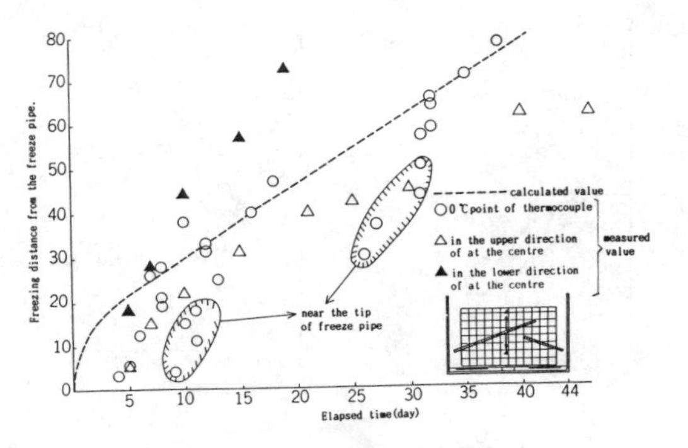

Fig–9 Progress of freezing line

2.Quality of the frozen soil

Generally it is known that the strength of frozen soil depends on its temperature. To estimate the strength of frozen mass, we must investigate the distribution of temperature.

Fig.-10 shows the distribution of temperature in the design range of frozen mass.

The state of freezing and the average temperature -10°C of frozen mass were determined by the evaluation of the distribution of temperature. The results as mentioned above were applied to actual construction work.

Fig.-11 shows the comparison of the change of measured values between the field and the laboratory.

From the agreement of the values, it may be implied that the propagation of frozen soil and its quality in field are similar to that in laboratory.

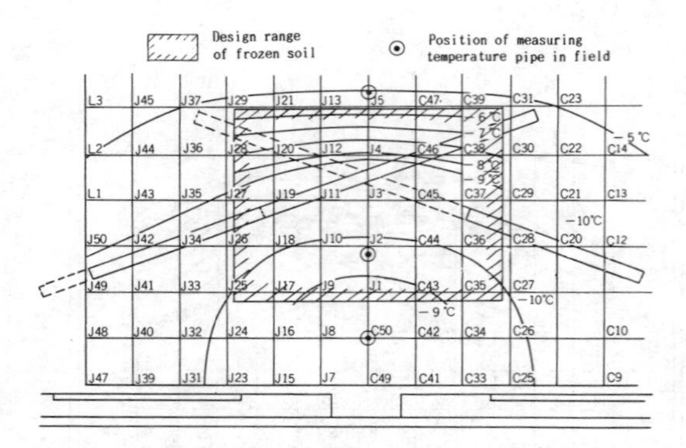

Fig–10 Distribution of temperature

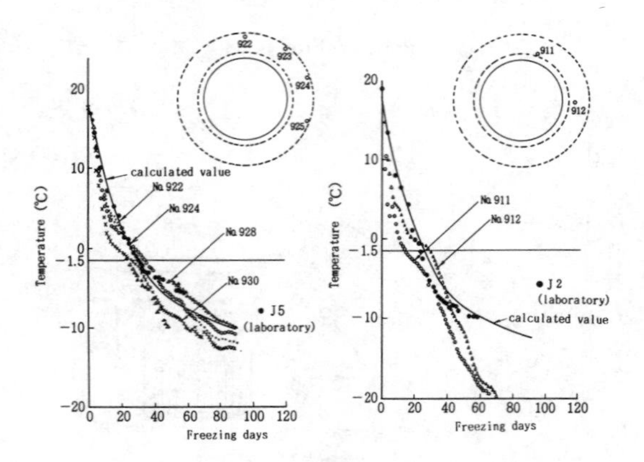

Fig–11 Comparison of change of measured values between the field and the laboratory

CONCLUSION

The construction of the connecting section in Nambu Sludge Treatment Plant Connecting Tunnel which is mentioned in this paper was successfully completed in Aug. 1985 after about ten months of construction. Through the construction work and the model freezing test, we have accumulated various experience on the application of the freezing method to this kind of work.

The work described permitted designer to:-

1. evaluate the progress of frozen soil and it's quality (average strength of frozen mass).
2. estimate the heaving during freezing soil and subsidence during and after thawing.
3. establish the rational design

ACKNOWLEDGEMENTS

Regarding the execution of tunnel connection, we are grateful to Tokyo Metropolitan Government Sewage Bureau, Japan Sewage Works Agency, Kajima J.V. and other persons concerned, for their support and advice.

REFERENCES

Japan Construction Machine Association, 1980, "Ground Freezing Method--Plan, Design and Execution--", (in Japanese).

M. Tanaka and N. Hanawa, 1986, "Stability of Frozen Caves (part 2)--Forming Frozen Soil--",Annual Report of Kajima Inst. of Construction Technology, Vol.34, pp.233-236, (in Japanese).

5th International Symposium on Ground Freezing, Jones & Holden (eds)
 ISBN 90 6191 824 3

Two practical applications of soil freezing by liquid nitrogen

Wolfgang Orth
Wibel & Leinenkugel, Karlsruhe, FR Germany

ABSTRACT: The first example given is of a small excavation in gravelly sand which was made to repair a leaking pipeline. Three other pipes extremely sensitive to displacement crossed the excavation at a higher level. In the second example, a concrete framework for an underpass was jacked through a railway embankment while the railway line was kept in operation. As the slope at the working face was not stable under the load from the trains, a one metre thick horizontal slab of frozen soil was prepared to distribute the loads and damp dynamic oscillations. The frozen soil was cut off in front of the framework as the jacking progressed. - The design for both cases was done with a formula which accounted for time effects and thus led to comparatively thin low-cost frozen bodies.

KEY WORDS: Embankment, excavation, gravels, liquid nitrogen, underground structures.

1 INTRODUCTION

The following two examples of soil freezing by liquid nitrogen (LN_2) were carried out to avoid conventional support of the soil. In the first case, the principal purpose of soil freezing was to reduce deformations in the subsoil, whilst in the second case it was mainly to improve stability. They both showed that soil freezing in general and especially by LN_2 can be an exceptionally versatile method and does not automatically lead to extreme cost.

2 A SMALL EXCAVATION

In an industrial area in southern Germany, a leaking pre-stressed concrete pipeline was to be repaired. The diameter of the pipe was 400 mm and its axis 3.6 m below ground surface, so an excavation of approximately 4 m depth was necessary. The subsoil consisted of 1 m of a silty gravel with some cohesion underlain by a dense gravelly sand with a water content around 7 %. There was no ground water in the excavation area.

Problems arose from three similar pipes which crossed 1.3 m above the leaking one. For these pipes, a maximum displacement of only 2 cm from their initial position was allowed. Further, an adjacent road was to be kept in service.

It was first thought that the solution would be a pit with slopes in the gravelly sand with the crossing pipes suspended from a steel girder (Fig. 1). This would have required much manual work during excavation

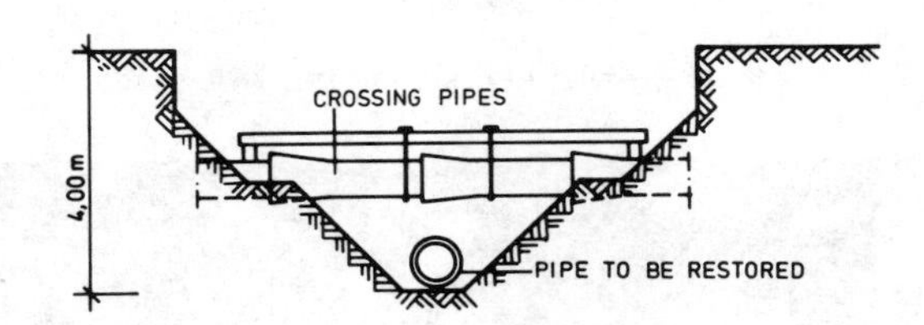

Fig. 1 Conventional pit

and later in compacting the soil, especially below the crossing pipes. Nevertheless, it was questionable whether the deformations could be kept low enough. Further, the adjacent road would have been cut off.

To avoid these disadvantages, a horizontal arch was frozen to stabilize the pit as well as the crossing pipes (Fig. 2).

To keep the deformations low, the frozen body was designed for earth pressure at rest. Within two days, the frost wall had been frozen 80 cm thick (Fig. 3) and after the same time the excavation was done with-

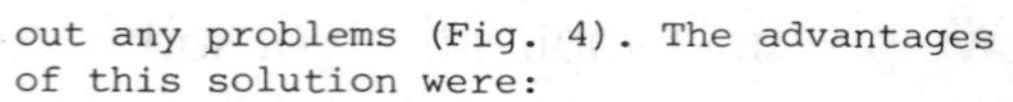

out any problems (Fig. 4). The advantages of this solution were:
- reduction of the excavated volume (most of which would have required manual labor) of about 140 m^3

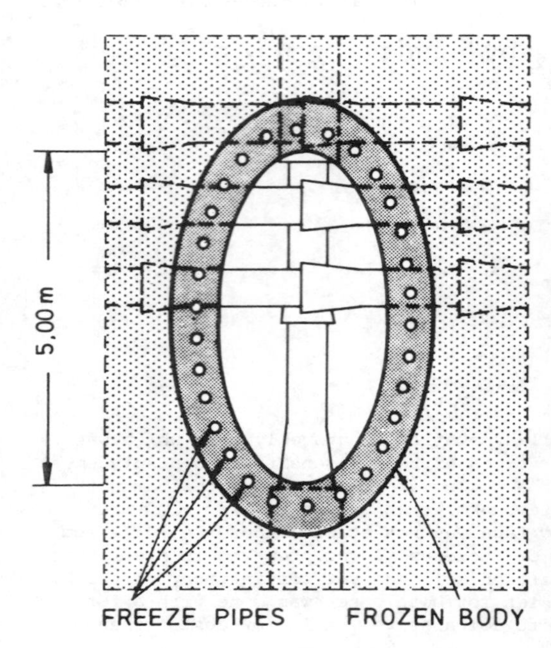

Fig. 2a Frozen pit, ground plan

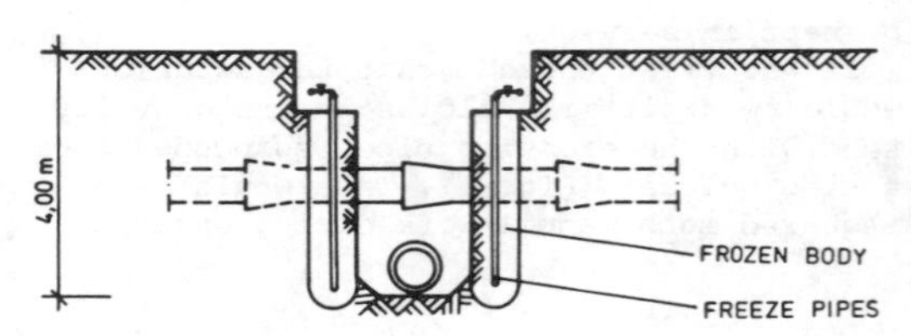

Fig. 2b Frozen pit, cross section

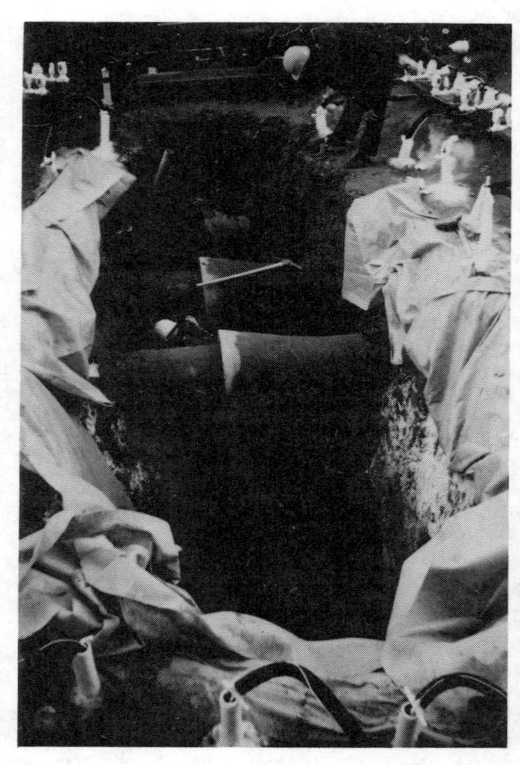

Fig. 4 Excavated pit

Fig. 3 Freezing phase

- negligible settlement of the three crossing pipes as subsoil was disturbed only within about 1.7 m of their length
- a smaller pit so the road could be kept in service
- a smaller volume to be compacted, especially below the pipes
- less construction time
- no permanent stabilisation of subsoil that could trouble future earth work.

3 UNDERPASS OF RAILWAY EMBANKMENT

A new road was to be constructed underpassing the railway line from Frankfurt to Basel/Switzerland (Ground plan see Fig.5). This railway is crossed by 230 trains per day and has to be kept in service, because there is no diversion. At the building site, the train speed was reduced from 160 km/h to 70 km/h. The maximum interruption time was 8 hours on the weekends and 20 and 30 minutes, respectively, on weekdays and at night.

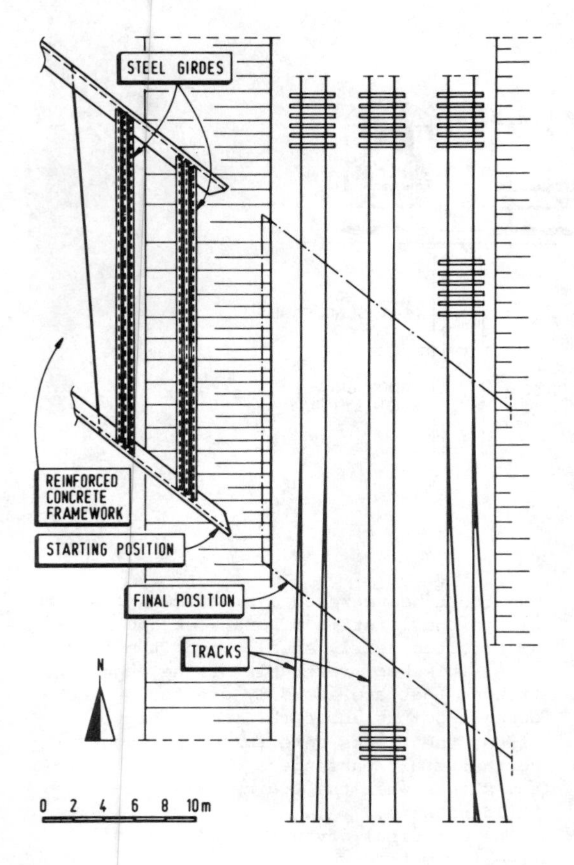

Fig. 5 Ground plan

As the road level needed to be as high as possible because of ground water, a 2600 t heavy concrete framework was constructed beside the embankment and then jacked through it by a hydraulic system with 4000 t maximum force (Fig. 6). The tracks were kept up by steel girders which were supported at one end by the concrete framework and at the other end by the earth embankment. The earth could then be removed in sections as the concrete framework was jacked forward. However, as the railway embankment consisted of 3 m of loose fill overlying sandy gravel with some silt, the slope stability of the working face was not secured under the load of the passing trains.

To improve stability, it was planned to drive vertical wooden or steel piles from the track level to transmit the train loads to deeper strata (Fig. 7). This is a very time-consuming method, and working periods between trains are very short (see above). Further, if difficulties arise while driving a pile, it might be necessary to cut it when a train is due and restart with a new pile after the train has passed.

The main disadvantage of the piling system, however, is that the slope has to be cut vertically to remove the piles. This period is very critical, and earlier experience has shown that some neighbouring piles are likely to fall out immediately after removing the first pile of a row, which is followed by a local slope collapse. It would then be necessary to stop all trains until the concrete framework was jacked close enough to the new slope crown to support the steel girder adequately.

To avoid these disadvantages, a horizontal frozen slab was prepared directly below the tracks and the steel girders (Fig. 8). This slab transferred the track load by bending force away from the slope edge and distributed the load in the track direction, thus increasing the slope stability at the working section. Further, the frozen slap damped the dynamic load from the trains.

As the concrete framework moved forward, the frozen slab had to be cut off. This was done by an hydraulic percussion excavator standing inside the framework (Fig. 9). The freeze pipes were cut off together with the frozen soil. As LN_2 was fed from one side (Fig. 10) and the gas outlet was on the other side at the working face (Fig. 8), the comparatively soft copper pipes were cut together with the

Fig. 6 Concrete framework in front of the embankment

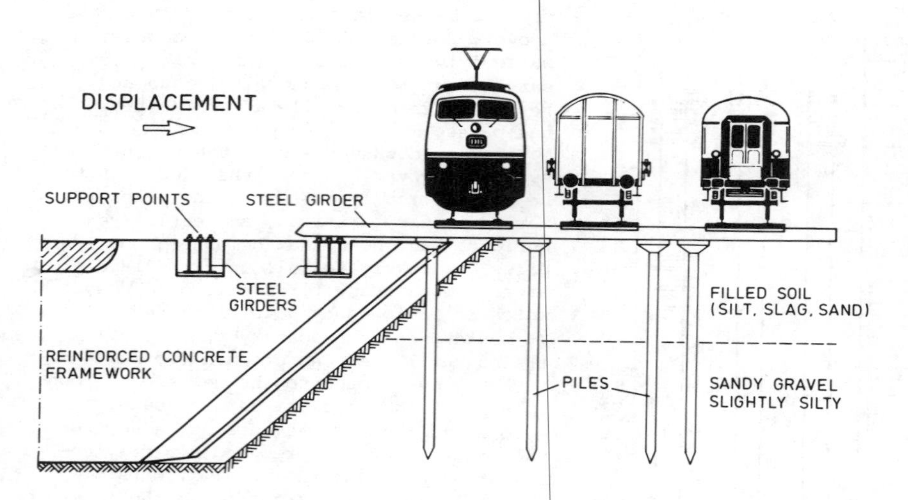

Fig. 7 Solution with piles, cross section

soil. The freeze pipes were spaced at 0.9 m and were approximately 24 m long in the beginning.

To control LN_2 intake, a thermo-couple was placed near the outlet of each freeze pipe. As the working face moved forward, these thermo-couples were pulled back with cables towards the intake side before cutting the next section of the pipe. To control further the frozen body's thickness and temperature, six vertical measurement pipes, each fitted with three thermocouples, had been driven from the embankment surface into the area to be frozen.

The frost heave reached a maximum of 2.2 cm within the lifetime of one week and was distributed nearly equally over the frozen area. This was preferable to the peaks and troughs that are usually created by the support points above the piles. At both sides, the tracks were ramped by wedges so that train passengers could not feel any shocks when the trains passed the building site.

The principal advantages of the soil freezing method were:

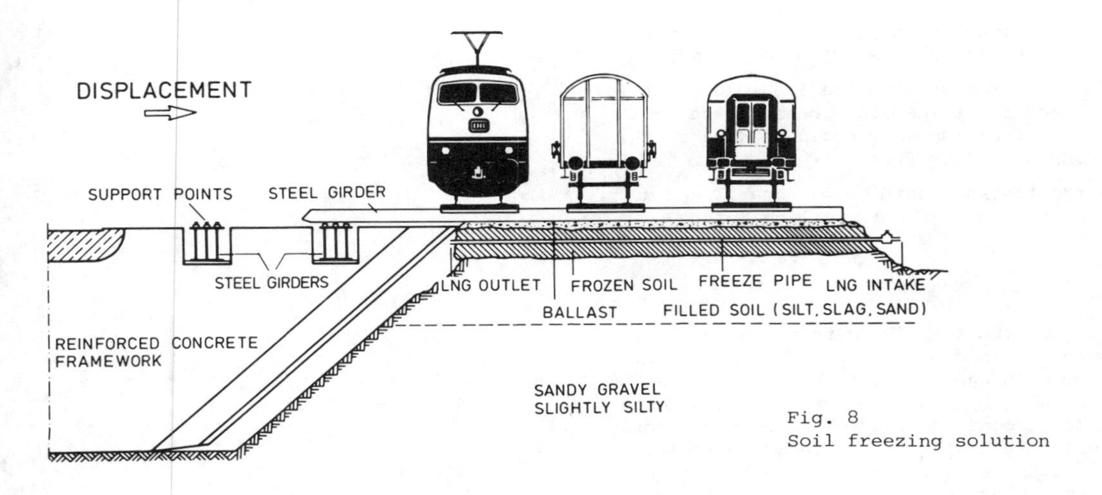

Fig. 8
Soil freezing solution

Fig. 9
Cutting off the frozen plate

Fig. 10
LNG supply

- no critical state (when stability could not be proven to exist at any time)
- except for driving the 1.5 m long vertical pipes for temperature measurement, all soil freezing work could be done in the absence of traffic
- the tracks' position was stabilized much better than with the piling method.

4 DESIGN FUNDAMENTALS

Both of the soil freezing cases described were designed using a creep formula (presented in another contribution to this Symposium by the same author) which takes into account the viscoplasticity of frozen soil. Especially in the second example, the comparatively thin frozen body resulted from taking into account the very short loading periods of the passing trains. Use of static design formulas would have led to uneconomical frozen bodies and possibly to unacceptable frost heave.

5th International Symposium on Ground Freezing, Jones & Holden (eds)
© 1988 Balkema, Rotterdam. ISBN 90 6191 824 3

Ground freezing solves a tunnelling problem at Agri Sauro, Potenza, Italy

A.Balossi Restelli
Geotechnical Consultant, Milano, Italy

G.Tonoli & A.Volpe
Rodio S.p.A., Casalmaiocco, Milano, Italy

ABSTRACT : In the paper the authors describe the consolidation works carried out by grouting and freezing with liquid nitrogen during the excavation of the Agri Sauro tunnel near Potenza, Italy. During construction, at chainage 2630 m, a large cave-in occurred with an inflow of about 6000 m^3 of silty-sandy soil which completely buried the shield; at that chainage, the tunnel lies at 150 m below ground surface and 70 m below the water table.

Remedial works consisting of filling all the voids and cavities created by the flow of unstable soil into the tunnel, grouting to recompress and consolidate the soil around it and freezing treatment carried out from within the tunnel, were undertaken. These works made it possible both to reach the shield and to restore the safety conditions of the tunnel so that excavation could be resumed.

INTRODUCTION

The Agri Sauro Tunnel belongs to a hydraulic system of intake and water conveyance structures for diverting the waters of the rivers Agri and Sauro into Sinni dam reservoir.

The 6700 m long tunnel is driven with a fully automatized shield. It consists of a 4 m diameter 6.5 m long steel cylindrical shell that incorporates all the motors, hydraulic jacks, conveyor belts, erector arms for placing the concrete segments, etc.

A rotating cutter head, fitted to the cylindrical shell, is equipped with four triangular openings allowing the entrance into the cylinder of the excavated soil.

The total thrust provided by the hydraulic jacks is 1500 tons.

The forward motion of the shield is given by the jacks pushing against the completed part of the lining. As the shield advances, the tunnel is lined at the rear of the shield by placing pre-fabricated concrete segments that are forced against the soil when a trapezoidal key segment is inserted.

The assembled circular rings have a 1.20 m length. The shaped joints between the rings are protected by special rubber packing.

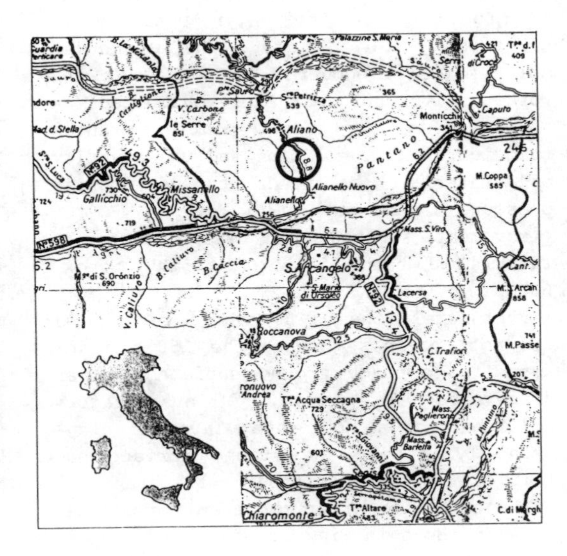

Fig. 1 - Work site location

As the tunnel lining proceeds, fluid mortar is injected through 4 holes into the annular space left around the periphery of the lining.

Tunnelling progressed regularly on a length of approx. 2000 m through a geological formation called "Aliano". This consists of layers of silty clay inclined 25°-30° interbedded with fine sands of moderate thickness (from a meter to a few meters). Along the tunnelled stretch the formation was found to be reasonably dry and no serious problems were caused by the water table.

Further on the hydrogeological conditions deteriorated and the advance of the shield began to be seriously hampered. Various expedients were employed to cope with the ever increasing difficulties arising from conditions much more severe than those revealed by the preliminary investigation.

The rate of advance was purposely slowed down and great care was exerted during construction. In spite of this, a number of segments (9 as was later discovered) situated immediately behind the shield could not withstand the strong water and soil pressure and collapsed with an inrush of fine silty clayey material into the tunnel.

The event started with a small leakage between two segments, which was impossible to stop due to high water pressure in coincidence with a fault. The inrush of solid material occurred so soon that the working site could hardly be evacuated.

Filling of the tunnel continued until friction along the lining and the quantity of material balanced the hydraulic pressure (70 m head).

Movement stopped only when 600 m of tunnel had been filled up with approx. 6000 m^3 of soil.

Fortunately the cavity above the tunnel did not progress through the 150 m thick deposits to reach ground surface. It was discovered by means of investigatory drilling that a 40 m thick layer of claystone lying near the ground surface had bridged the cavity without collapsing.

The situation after the event is schematically shown in Fig. 2.

An attempt was made to remove the material from the tunnel and reach the heading, while the water table was lowered by pumping from a drainage well drilled with great difficulty from the ground surface. The attempt not only failed but reactivated the movement of soil, which gave further evidence of the seriousness and complexity of the situation.

Thereafter radical remedial works entailing the application of special subsoil engineering techniques were investigated, designed and ultimately carried out by Rodio-Milan. The works, lasting 15 months among difficulties of any kind, were finally successful.

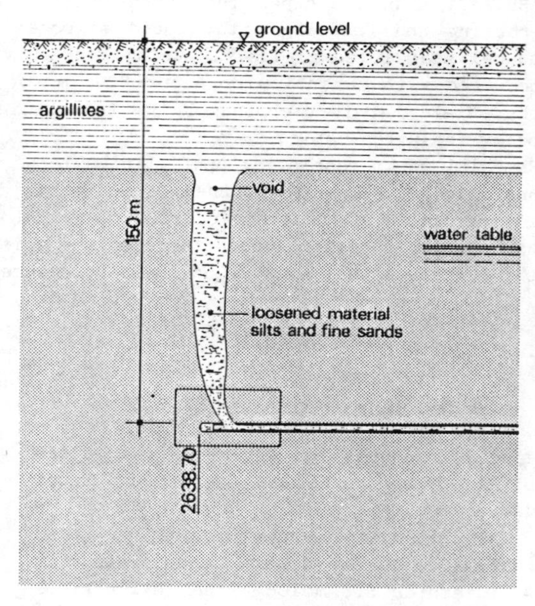

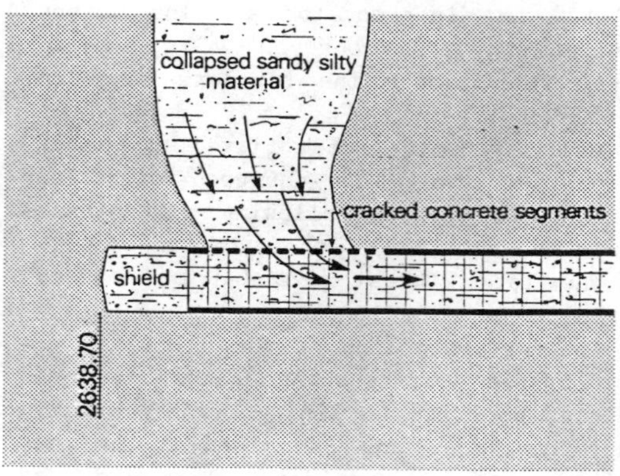

Fig. 2 - Schematic of tunnel after cave-in occurred at chainage 2630

The material filling the tunnel could be removed up to the collapsed lining segments, which were replaced, and the shield freed.

EMERGENCY WORKS

Exploratory boreholes drilled from ground surface revealed the extent of the cavity and the grain size of the dislocated soil mass.

Seven piezometers were also installed to monitor the behaviour of the water table.

The choice of emergency treatments was difficult due to :

- the complexity of the situation and uncertainties as to the real extent and configuration of the cavity
- the large hydraulic pressure (7 bar) exerted by the water table on the tunnel
- the Main Contractor's urgent need to get the shield rescued
- the need to limit the costs.

Consequently the operations were to be so phased as to avoid useless works and to resort to the smallest possible work quantities.

The various and diversified works are hereinafter briefly described in the sequence they have been carried out.

FILLING OF THE CAVITY

Filling was made by means of holes drilled from ground surface.

It was not an easy work because the drill rods frequently broke being unguided after entering the cavity. Besides it was necessary to inject small quantities of cement grout at a time to avoid possible overloading of other rings along the tunnel, as well as wastage of grout.

RECOMPRESSION AND CONSOLIDATION GROUTING AROUND THE TUNNEL

Section and layout of fig. 3 illustrate the location of the grout holes drilled from the surface down to max. 150 m depth to recompress and consolidate the ground above and alongside a 26 m long stretch of tunnel. The holes were equipped with steel sleeved pipes.

The considerable depths of the holes and the checking of hole alignement caused serious problems. Additional holes were installed to ensure that the whole mass of dislocated ground was treated.

Laboratory testing on soil samples had indicated that the sandy pockets within the mass could be penetrated by chemical

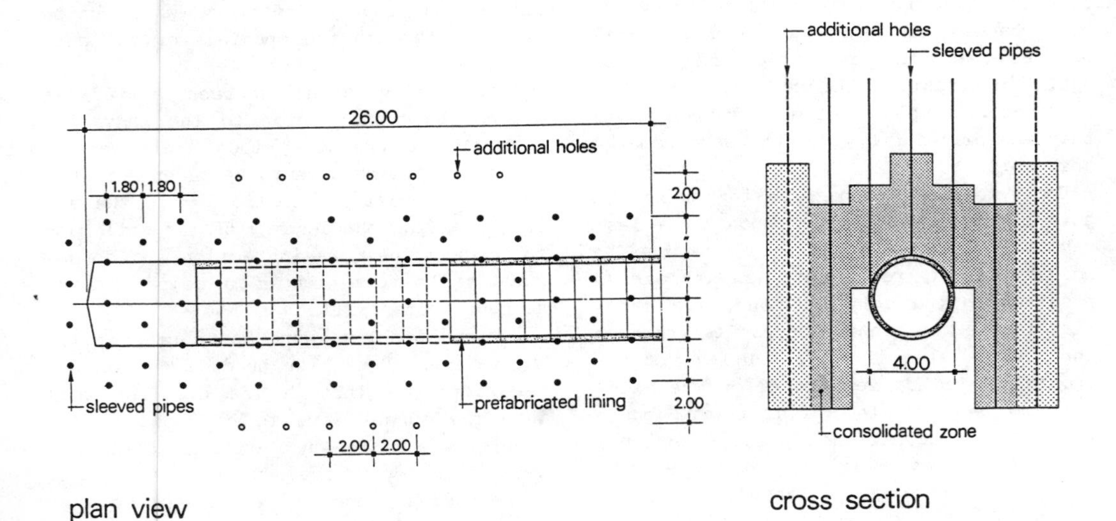

Fig. 3 - Recompression and consolidation grouting around the tunnel, carried out from the surface

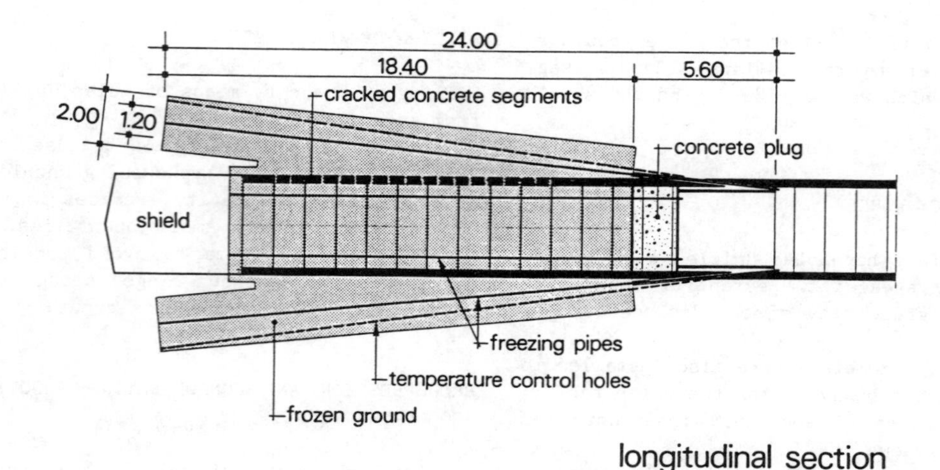

Fig. 4 - Scheme of freezing treatment by liquid nitrogen

grouts. Silica gels and acrylic resins (Siprogel) were therefore injected after a preliminary treatment with cement grouts. Water/cement ratio and bentonite content of these grouts were varied based on ground response to grouting.

It was concluded, at the end of these difficult and complex operations, that a satisfactory degree of consolidation had been achieved and therefore the removal of the material from inside the tunnel was resumed. When the first collapsed ring had been reached, an inflow of water and soil occurred and the excavation had to be again interrupted. The construction of a new concrete plug became necessary and this was placed fifty meters away from the accident.

Employment of ground freezing using liquid nitrogen was decided upon as a last resort. To enable freezing operations to be carried out from a position as near as possible to the critical zone, the material inside the tunnel was grouted through horizontal sleeved pipes using cement grouts and silica gels to allow the material to be excavated to a forward position. It was possible to place a concrete plug 10 m away from ring N. 2179 (the first of the rings collapsed, at a distance of 2,623 m from the entrance of the tunnel).

GROUND FREEZING

The scheme of the conical freezing treatment is shown in Fig. 4.

Thirty six, 24 m long, freezing pipes have been installed in a conical pattern so as to form a 1.2 m thick wall of frozen ground all around the tunnel. Twelve horizontal freezing pipes have been placed within the conical pattern to create a frozen ground plug at the rear of the shield and prevent inflow of water under pressure through the openings of the cutter head.

Additional grouting has been carried out around the tunnel prior to the installation of the freezing pipes. This was an indispensable measure to be taken for this specific freezing operation, with the aim of increasing the density of the soil mass to improve the freezing action and consequently the formation of a compact homogeneous frozen wall.

Serious problems were met during drilling under a 70 m water head. Flows of water and fines through the drilling rods into the tunnel were to be avoided. The problem was solved by drilling cased holes and employing preventers.

The casings were not withdrawn but left in the ground to accomodate the 2" freezing pipes.

Ground freezing was carried out regularly and according to programme.

Liquid nitrogen was pumped to the freezing pipes from ground surface through an insulated pipe placed into a vertical service hole. The nitrogen gases were conveyed to the surface through another service hole, as shown in Fig. 5.

Fig. 6 shows the layout of the freezing system installed in the tunnel, Fig. 7 provides a view of some heads of the freezing pipes.

Consumption of liquid nitrogen during ground freezing amounted to 1200 l per cubic meter of soil.

Maintenance operations were planned on the basis of the temperatures recorded by thermo-couples installed in the ground.

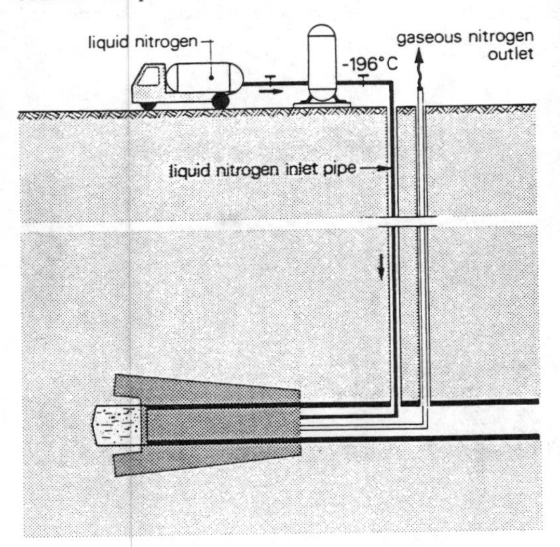

Fig. 5 - Freezing plant schematic

Fig. 6 - Freezing system installed in the tunnel

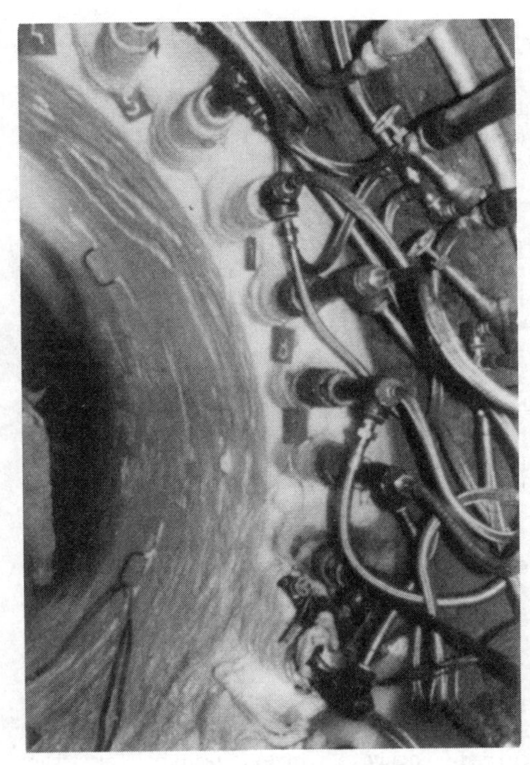

Fig. 7 - Detail of some freezing pipe heads and freezing connections

Maintenance was carried out on the average every two days and at night during interruption of the excavation.

Each maintenance required from 30,000 to 40,000 l of liquid nitrogen.

EXCAVATION AND PLACING OF STEEL REINFORCING LINING

Excavation into the critical zone started soon after formation of the frozen wall.

The excavation progressed safely and with remarkable regularity in spite of the difficulties caused by the collapsed concrete segments that had to be carried away and by the unfavourable local conditions (confined space and low temperatures).

Fig. 8, 9 and 10 show the working face. Fragments of a collapsed concrete lining ring can be seen, as well as grout veins in the soil mass previously recompressed and stabilized by grouting.

The face advanced by 1.00 m long stretches. The collapsed concrete rings were replaced by steel segments to obtain a new lining for the tunnel. The segments consisted of two articulated jointed semi-cylinders to enable transportation within the tunnel.

After placement against the frozen ground, each steel segment was welded to the preceding completed one.

Injections of mortar through special holes followed, with the purpose of filling all the spaces between the three circular steel ribs (T section) located behind the steel segments.

The procedure continued for 15 m until the steel lined tunnel reached the rear of the shield. It was found that 9 concrete rings had caved-in.

Lastly, the shield filled with collapsed soil was freed and its equipment removed and overhauled.

MEASUREMENTS AND CONTROLS

The complex and critical situation caused by the partial collapse of the tunnel required accurate investigations of the stresses and strains in the concrete and steel linings.

A series of measurements has been conducted by Geoanalysis of Turin to measure and monitor the :

- stresses in the lining by means of pressure cells and door-stoppers
- deformation of the rings by means of convergence measurements and measurements with mechanical strain-meters
- development of the deformation of the steel segments and counter-ribs (fig. 11) strengthening the lining in the proximity of the collapsed concrete rings.

Vibrating wire strain-meters were employed.

Considerable compression stresses have been measured at the lateral intrados by means of flat-jacks placed into the concrete lining.

In the meantime tensile stresses were found in the upper key measured by means of door-stoppers.

All these tests were carried out in the concrete rings adjacent to the collapsed ones.

These findings confirmed the trend of the rings to get out-of-round as evidenced by the measurements of the deformations.

Fig. 8 - General view of tunnel in the collapsed zone

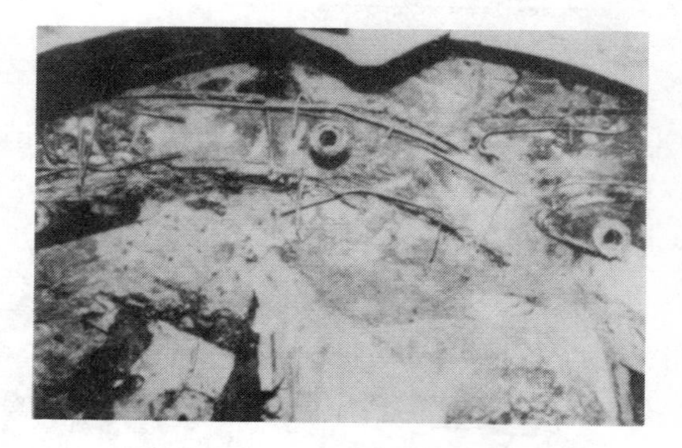

Fig. 9 - Excavation under protection of frozen soil. Some freezing pipes and collapsed concrete segments can be seen

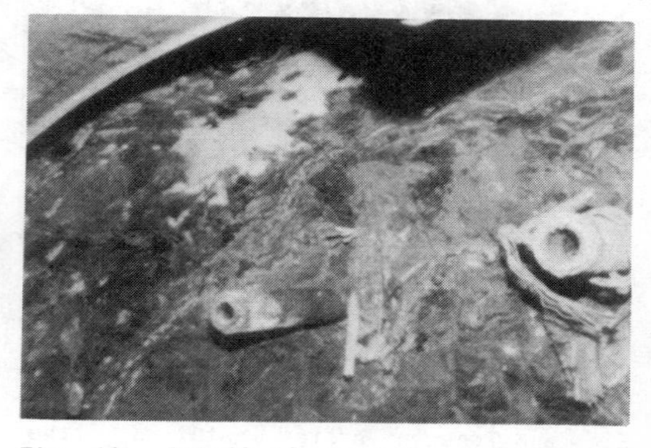

Fig. 10 - Detail of grouted and frozen soil, during excavation

Fig. 11 - Reinforcing steel structure placed in correspondence of concrete rings in critical conditions of stability

CONCLUSIONS

All the various data provided by the instrumentation have been extremely useful for :

- accurately planning the works and the sequence of the operations before construction and so ensure an ever increasing degree of safe execution of the works
- detecting, in the course of the works, the development of stresses and strains and check them with those planned
- long term monitoring of the effects on the tunnel structures of ground freezing and de-freezing. This makes it possible to judge now the long term stability of the stretch of tunnel where the emergency works have been carried out.

ACKNOWLEDGEMENTS

Owner : Cassa per il Mezzogiorno - Italian Government
Main Contractor : Agri Sauro Società Consortile a r.l., Caserta, Italy with Impresa DEL FAVERO S.p.A., Trento, Italy as Sponsor
Grouting and Freezing Works : RODIO S.p.A., Casalmaiocco (Milano), Italy
Monitoring and Measurements : GEOANALYSIS s.r.l., Torino, Italy.

5th International Symposium on Ground Freezing, Jones & Holden (eds)
© 1988 Balkema, Rotterdam. ISBN 90 6191 824 3

Author index

5th International Symposium on Ground Freezing, Jones & Holden (eds)
 ISBN 90 6191 824 3

Keyword index

Page numbers refer to the first page of each paper